Carbocationic
Polymerization

Carbocationic
□ Polymerization □

JOSEPH P. KENNEDY

Institute of Polymer Science
The University of Akron
Akron, Ohio

ERNEST MARÉCHAL

Laboratoire de Synthèse Macromoléculaire
Université Pierre et Marie Curie
Paris, France

A WILEY-INTERSCIENCE PUBLICATION

JOHN WILEY & SONS
New York · Chichester · Brisbane · Toronto · Singapore

Copyright © 1982 by John Wiley & Sons, Inc.

All rights reserved. Published simultaneously in Canada.

Reproduction or translation of any part of this work
beyond that permitted by Sections 107 or 108 of the
1976 United States Copyright Act without the permission
of the copyright owner is unlawful. Requests for
permission or further information should be addressed to
the Permissions Department, John Wiley & Sons, Inc.

Library of Congress Cataloging in Publication Data:

Kennedy, Joseph Paul, 1928–
    Carbocationic polymerization.

    "A Wiley-Interscience publication."
    Includes bibliographical references and index.
    1. Polymers and polymerization.   I. Maréchal,
Ernest, 1931–   joint author.   II. Title.
QD381. K38   547'.28   80-26366
ISBN 0-471-01787-6

Printed in the United States of America

10 9 8 7 6 5 4 3 2 1

To my Honeybee
and
à Josette, Luc et Sylvie
. . . pour tous les week-ends sacrifiés

□ Preface □

This is an interdisciplinary book written for the organic chemist who wants to relate knowledge of cationic reactions of small molecules to the science of large molecules, for the physical chemist who wishes to apply basic chemical–physical principles to polymerization mechanisms, for the polymer scientist who wants comprehensive up-to-date critical information about a large segment of vigorously growing polymer science and technology, for the research entrepreneur who is on the lookout for well-defined but unexploited leads, for the industrial researcher who wants to survey the technology of cationic polymerization processes leading to useful products, and for the student who is searching for working relations between abstract ideas, contemporary research in polymer science, and, ultimately, some of today's important technologies.

If a change is slow, not sudden or abrupt, it is difficult to perceive that it is in progress; once the change is recognized it becomes a "quiet revolution." The area of cationic polymerization is undergoing such a quiet revolution right now. It will be some time before it is generally perceived as such because outlooks, like habits, change slowly. Indeed, our resolve to write this book, a most painful exercise, stemmed in part from our conviction that this quiet revolution should be exposed to the polymer community.

During the past decade the elucidation of cationic polymerization has undergone such a rapid increase and the exploitation of new mechanistic information in terms of new products and processes in the research laboratory was so rapid that serious students of this field had to reassess completely their views with regard to the capabilities, particularly preparative capabilities, of this discipline.

Polymer scientists and technologists at large know carbocationic polymerization as a well-established but colorless segment of polymer science offering some commercially attractive possibilities (witness butyl rubber) and allowing the preparation of some quaint structures (such as $-CH_2CH_2C(CH_3)_2-$ by isomerization polymerization), but one that is better avoided because it is a bewildering, unmappable maze beset by some insurmountable experimental difficulties such as the necessity of working at very low (cryogenic) temperatures. It is one of the objectives of this book to dispel this completely false and distorted notion and to rejuvenate the field by showing by many examples the tremendous promise and unexploited possibilities offered by carbocationic polymerizations. In view of the very large number and variety of cationically responsive monomers, low cost and high efficiency of cationic initiating systems, usually rapid reactions, and modest investment required for cationic manipulations, the lack of entrepreneurship in exploiting cationic polymerization is rather surprising. One possible reason for lagging interest in cationic polyreactions may be two decades of colorless research mired in difficultly reproducible kinetic minutiae and fruitless basic investigations. Another factor is that quantum jumps such as Szwarc's discovery of "living" polymerizations or Ziegler and Natta's discovery of stereoregulating catalysts, discoveries that created unexpected new vistas in other areas of ionic polymerizations, eluded the field of carbocations.

Another objective of this book is to present a unified, interlocking, in many respects new view of carbocationic polymerization. Although select parts of this discipline have been reviewed in the past by several authors, the whole field as such has not yet been comprehensively and critically examined in a book written by one or two authors.

We start by asking, in Chapter 1, why carbocationic polymerizations? What is so special about this science? We find some unique answers in the fields of chemistry, structure–property relationships and technology. In the next chapter we define terms, describe basic concepts, and lay down foundations to be built on when we turn to the discussion of mechanisms. In Chapter 3 we proceed to phenomenology to acquaint the reader with what carbocationic polymerizations *are* by examining monomers, initiators, coinitiators, and solvents. The next, long chapter (4) concerns the chemistry and mechanisms of the important elementary events: initiation, propagation, chain transfer, and termination. In Chapter 5 on kinetics, an attempt is made to combine these mechanistic steps and kinetic expressions are examined. The following chapter on copolymerization and reactivity starts with a comprehensive compilation and evaluation of all monomer pairs copolymerized by carbocationic initiation and proceeds to a discussion of experimental and theoretical determinations of reactivity. A review of relative reactivity relations is given and the influence of experimental parameters on reactivity is examined.

In Chapter 7 carbocationic step-growth polymerizations are discussed. The following chapter examines in detail the chemistries leading to recently developed sequential (block and graft) copolymers. Chapter 9 is devoted to macromolecular engineering and a glance toward the future. We conclude that the time for tailoring physical–mechanical–chemical properties by carbocationic techniques (i.e., macromolecular engineering) has arrived and develop a framework for the synthesis of new sequential, functional, telechelic polymers. The book ends with a survey of industrial processes employing carbocationic polymerizations currently in use.

Our most sincere thanks to Dr. P. Borzel, Professor T. Higashimura, Dr. I. Puskas, and Dr. W. A. Vredenburgh, and co-workers for letting us have material prior to publication, and to Professor P. Sigwalt and Dr. T. Kelen for useful criticisms. Special thanks is due to Mrs. M. Israel for heroism during her most competent typing of our battle-scarred manuscript.

JOSEPH P. KENNEDY
ERNEST MARÉCHAL

Akron, Ohio
Paris, France
November 1981

□ Contents □

Carbocationic
Polymerization

□1□
Why Carbocationic Polymerization?

Why should one be interested in carbocationic polymerizations? Justification for the writing of this book lies in answering this question satisfactorily.

1.1 □ ADVANTAGES AND USES OF CARBOCATIONIC POLYMERIZATION

Cationic polymerization comprises an important body of techniques for the synthesis of a great variety of *useful* polymers; it provides a unique route to many high, medium, and low molecular weight materials with unique structures exhibiting a unique combination of properties; and last but not least it is a vigorously growing segment of polymer science which occupies many researchers, whose steady flow of discoveries bespeaks an intellectually challenging, indeed underexplored field and assures the long-range health of the discipline. A closer look at some of these points is rewarding.

Cationic polymerization is a useful branch of science that contributes greatly to the wealth, safety, and comfort of mankind. Many hundreds of people all over the world are gainfully employed by industries practicing cationic polymerizations, and if those who are involved in occupations only indirectly related to these industries, such as tire makers, who work with halobutyl rubber inner liners and compounders, who use pinene resins in pressure-sensitive tapes, are included, this number would probably be closer to many thousands.

Cationic polymerizations have two main roots in industry: technologies based on carbocationic polymerizations and those based on cationic heterocyclic ring-opening polymerizations. The latter fall outside the scope of this book, which focuses only on carbocationic or carbenium ion polymerization processes.

No doubt the largest carbocation-based polymerization industry by volume and monetary value to date is butyl rubber and halogenated (chlorinated and brominated) butyl rubber manufacture. Butyl rubber is a general-purpose elastomer obtained by copolymerizing isobutylene and a conjugated diene (isoprene); the halobutyls are specialty rubbers obtained by halogenating butyl rubber. These materials are used in a variety of applications in the tire, automotive, building, and construction industries.

Carbocation polymer industries of more modest scope include petroleum and indene–coumarone resin manufacture, polymerization of β-pinene, α-pinene, and mixed terpenes, and limited quantities of styrene and α-methylstyrene polymerization by acid initiators. In this class belong also "polybutenes" and "polyisobutenes" used, for example, as oils, viscosity index improvers, and additives, and low molecular weight (liquid) poly-butadienes used in specialty coatings. Similarly, certain vinyl ethers are cationically polymerized and employed in adhesive formulations, pressure-sensitive tapes, blending agents, and additives.

Cationic polymerization is an important, integral part of the fabric of synthetic polymer chemistry. Its relative importance contrasted with the other techniques within the panorama of polymer syntheses, as judged by the overall number of publications and patents by synthesis field (*Chemical Abstracts* 1975/76) or by estimating the number of practitioners active in this field, may be illustrated by the pie chart in Figure 1.1. Thus, somewhat arbitrarily and subjectively, we estimate that about 5 to 7% of all scientific publications and patents in the field of polymerizations concern cationic polymer and oligomer syntheses, preparations, modifications, derivatizations, and degradations, that is, a sizable chunk of polymer science.

In addition to the sheer number of reports in this field, one readily perceives a steady influx of novel ideas and original developments which should ensure the undiminished growth of this discipline at least in the near future. Indeed, an important mission of this book is to present and discuss some of these recently formalized subjects in an organized fashion.

Even a cursory examination of this field will convince the reader that cationic polymerization provides unique ways for the synthesis of unique structures. Although quite often two or more polymer synthesis techniques provide essentially the same structures (e.g., polystyrenes by free radical, anionic, or cationic methods; polybutadiene and polyacrylonitrile by free radical or anionic methods) and the ultimate decision as to which of the methods is to be preferred is made on the basis of nonchemical considerations, a large number of structures and materials exist that can be

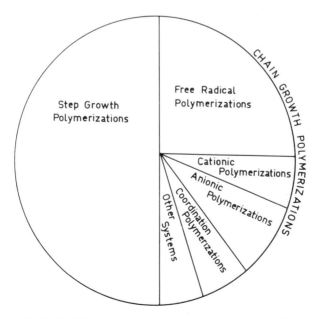

Figure 1.1 □ Relative importance of polymer synthesis techniques.

obtained *exclusively* by carbocationic methods. For example, the aforementioned polyisobutylene,

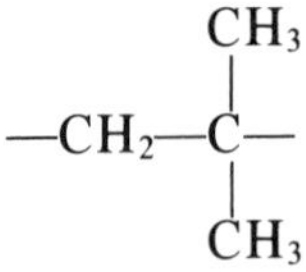

$$-CH_2-\overset{\displaystyle CH_3}{\underset{\displaystyle CH_3}{C}}-$$

or its telechelic derivative,

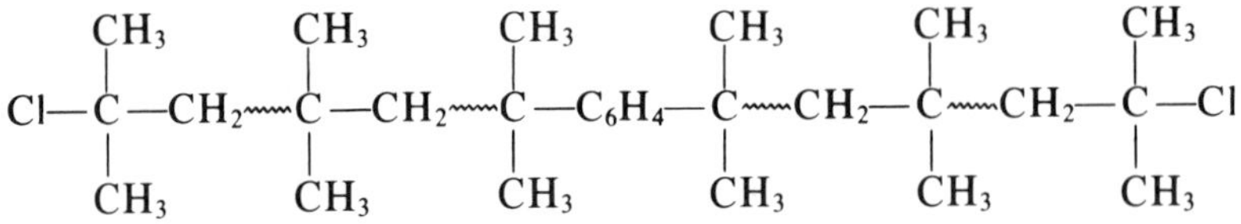

$$Cl-\overset{CH_3}{\underset{CH_3}{C}}-CH_2\text{---}\overset{CH_3}{\underset{CH_3}{C}}-CH_2\text{---}\overset{CH_3}{\underset{CH_3}{C}}-C_6H_4-\overset{CH_3}{\underset{CH_3}{C}}\text{---}CH_2-\overset{CH_3}{\underset{CH_3}{C}}\text{---}CH_2-\overset{CH_3}{\underset{CH_3}{C}}-Cl$$

poly(isobutylene-*co*-isoprene),

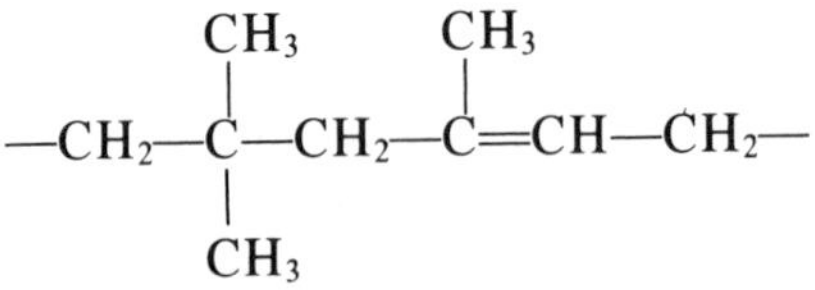

$$-CH_2-\overset{\displaystyle CH_3}{\underset{\displaystyle CH_3}{C}}-CH_2-\overset{\displaystyle CH_3}{C}=CH-CH_2-$$

poly(β-pinene),

$$-CH_2-C\underset{CH_2\text{------}CH_2}{\overset{CH\text{------}CH_2}{<}}CH-\overset{CH_3}{\underset{CH_3}{C}}-$$

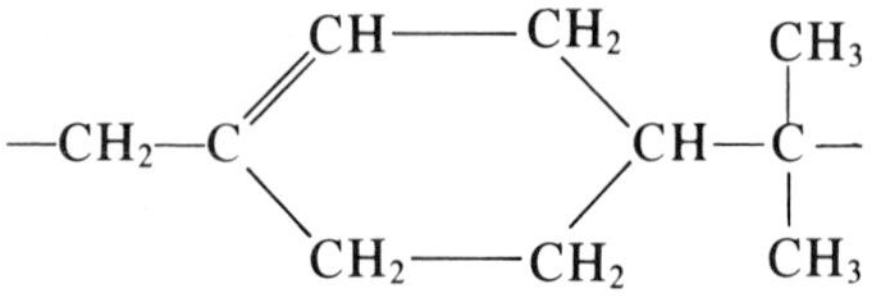

cationically obtained poly(4-methyl-1-pentene), which is formally a perfectly alternating copolymer of ethylene and isobutylene,

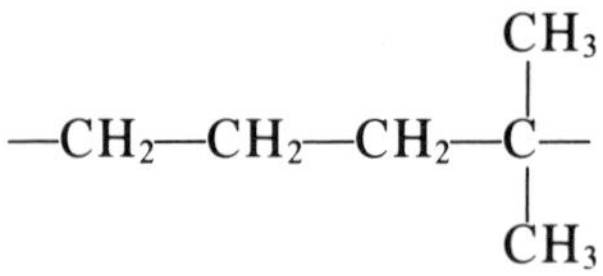

$$-CH_2-CH_2-CH_2-\overset{\displaystyle CH_3}{\underset{\displaystyle CH_3}{C}}-$$

cationically obtained poly(3,3-dimethyl-1-butene),

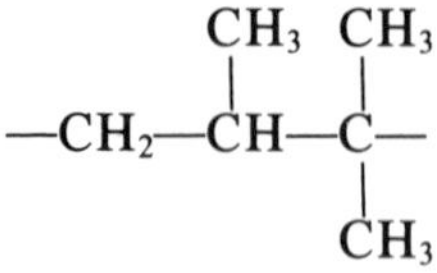

$$-CH_2-CH-\overset{\displaystyle CH_3\;\;\;CH_3}{\underset{\displaystyle CH_3}{C}}-$$

polynortricyclene,

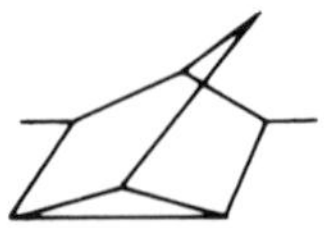

poly(vinyl isobutyl ether),

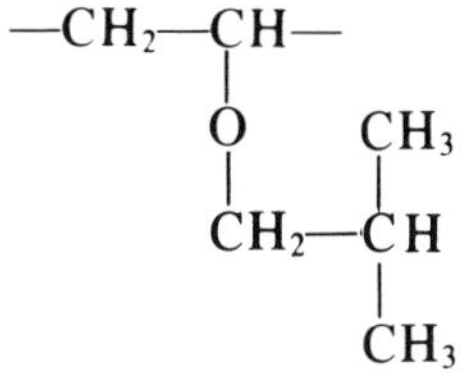

$$-CH_2-CH-$$

$$\begin{array}{cc} O & CH_3 \\ | & | \\ CH_2-CH \\ & | \\ & CH_3 \end{array}$$

and many other structures to be discussed below can be prepared only by cationic techniques. Obviously, these unique structures exhibit a unique combination of properties which render some of them uniquely useful for a variety of applications.

1.2 □ PROBLEMS, CHALLENGES, AND THE FUTURE

In spite of the tremendous potentialities offered by carbocationic polymerizations—consider the very large number of monomers demonstrated to give polymers by cationic means (Kennedy, 1975) and the variety of relatively inexpensive Brønsted and Lewis acids used to induce these reactions—the tangible results in terms of commercial products and scientific quantum jumps are rather meager.

Overviewing the entire panorama of cationic polymerization science from its earliest days (Bishop R. Watson, 1789), one perceives that although the field was well defined by the time butyl rubber came onstream some 40 years ago and it has shown remarkable vitality and uninterrupted progress since the early 1940s, it did not generate a scientific–technological revolution as did stereoregular or "living" polymerizations.

Although it is very difficult to analyze and pinpoint the reason(s) for this lack of exploitable breakthroughs, part of the answer must lie in the insufficient understanding of cationic mechanisms in general and carbocationic polymerizations in particular. Serious students of scientific–technical progress tend to accept the premise that barring discovery by lucky coincidence or serendipity, major technological advances usually come by hard, painstaking work at the drawing board. To create new, desirable materials or processes one must start by understanding key phenomena and proceed by integrating new information with existing knowledge. To translate this prescription for progress in our field of inquiry, the first step must be the elucidation of the detailed mechanism of cationic polymerizations. If a mechanistic step such as initiation or termination is understood, its control and ultimate exploitation become possible.

Although great advances have been made in our understanding of the mechanisms of elementary events, important fundamental questions remain and basic problems need solutions. A few remarks relative to some of the major gaps in our fundamental understanding of the field are in order.

The very high intrinsic reactivity of polymerization-active carbocations is a fundamental strength *and* weakness of cationic polymerizations. Although high reactivity is desirable for fast rates, the rates are sometimes unmeasurably fast and kinetic analysis becomes extremely difficult if not impossible. Extreme reactivity is disadvantageous on account of the many side reactions and by-products it may cause. Because of the high reactivity of growing carbocations the polymerization medium must be of highest purity and the presence of even traces of impurities may have detrimental consequences. Carbocations may be rapidly deactivated by a large number of oxygen-, nitrogen-, or sulfur-containing chemicals, such as water, alcohols, aldehydes, ketones, ethers, amines, sulfides, and organic and inorganic acids, which may reach the polymerization reactor with the monomer or solvent. These contaminants are severe rate poisons or molecular weight depressors even in minute quantities.

Chain transfer to monomer is also due to the very high reactivity of the propagating carbocation and is more important in cationic polymerizations than in anionic, free radical, or coordination polymerizations. Chain transfer to monomer is often undesirable because it reduces molecular weights and prevents terminal functionalization of macromolecules. Indeed, considerable research has been done on minimizing or eliminating chain transfer to monomer, particularly since the utility of telechelic polymers and sequential copolymers has been demonstrated. Carbocationic polymerizations that proceed in the absence of chain transfer have recently been developed and thus the synthesis of a variety of telechelic and sequential graft and block polymers became a reality (Kennedy, 1977).

Cationic polymerizations often stop before complete monomer consumption, indicating the presence of chain termination, and if higher conversions are desired in these systems initiator addition is continued. In spite of the considerable scientific–technological significance of the problem of termination, very few studies have been carried out to inquire why carbenium polymerizations stop and what determines the ultimate yield. Recent research in this area led to important advances, particularly for the synthesis of block copolymers.

The great ease with which carbocations undergo a wide spectrum of isomerizations is further illustration of the very high reactivity of these intermediates. Although controlled rearrangement of growing carbocations has been employed for the synthesis of a variety of well-defined, even crystalline, structures by "isomerization polymerization" (Kennedy, 1967), isomerization is often undesirable because it competes with head-to-tail chain growth and gives rise to ill-defined, undesirable structures.

High carbocation reactivity may also cause a number of other disturbing and largely uncharted side reactions and by-products, for example, hydride ion transfer resulting in branching, intra- and intermolecular cyclizations particularly with conjugated diolefins, and alkylation of aromatic rings giving rise to branching during polystyrene polymerizations.

In addition to the better understanding and control of these parasitic reactions, numerous fundamental problems remain with regard to the detailed mechanistic elucidation of elementary events. Thus the first and foremost task is to elucidate details of ion generation ("priming"), cationation, propagation, the various chain transfer steps, particularly that of chain transfer to monomer, and termination. Although the exploration of elementary events of ring-opening polymerizations proceeding by oxonium ions is far advanced and in many respects well understood, many details of vinyl or carbenium ion polymerizations still remain obscure. The mechanisms of carbocation formation and initiation by Brønsted acids and stable carbocation salts are satisfactorily understood; however, these initiators are of limited importance for monomers of medium or low activity. In contrast, details of initiation by Lewis acid-based initiator systems which produce high rates and high polymers are insufficiently understood. The efficiency of coinitiation by Lewis acids is usually much less than 100%; the reason(s) for this can only be surmised. The effect on the polymerization of the large amounts of Lewis acids that survive initiation is obscure. The interaction between the cationogen (cation source) initiator and Lewis acid coinitiator is still the subject of controversy and continued investigation. Light has been shown to affect initiation but the underlying mechanism is a matter of conjecture.

The knowledge of absolute rate constants of elementary steps is still the exception rather than the rule, and the effect of reaction conditions on these is still poorly understood. For example, the effect of temperature on the rate of polymerizations carried out in polar media is conceptually unavailable because the effects of temperature on polarity (dielectric constant) and rate cannot be separated.

Even in instances when k_p can be determined, the interpretation of the information is difficult, particularly the dependence of k_p on temperature, because of the negative activation energies found. Such phenomena cannot be explained without knowledge of the exact nature of the active species and their state of solvation, that is, their ionicity.

Mechanism studies are rendered very difficult because information concerning the nature and concentration of active sites is lacking. Spectroscopic investigations, for example, by NMR, UV, or IR, are of little use, particularly in vinyl polymerization systems, because of the low concentration of active sites. Experimentation with model compounds is promising in this area.

In contrast to anionic polymerizations where the understanding of the nature of growing anions is far advanced, the physical–chemical definition of polymerization-active carbocations is in a pitiful state. Very little is known about the solvation of growing species, the equilibria between the various solvated carbocations, or the aggregation of charged species. The interaction between the active cations and counteranion is also uncharted territory and, indeed, very little is known about counteranions in general. The "inorganic chemistry" of cationic polymerizations that is, the chemistry focusing on the counteranions, is yet to be developed.

Inevitably, progress in cationic polymerizations would be accelerated by a close dialogue between knowledge-oriented academic researchers and mission-oriented pragmatic industrial researchers. Unfortunately this well appreciated but seldom practiced symbiotic dialogue does not seem to work (witness the pitifully small number of papers delivered by industrial researchers during the Fourth International Symposium on Cationic Polymerization; Kennedy, 1976). In the absence of an increase in the rate of communication between these two camps, it can only be hoped that conventional communication channels will provide sufficient flow of easily digestible information in both directions to bring about cross-fertilization and ultimately scientific–technological breakthroughs.

□ REFERENCES

Kennedy, J. P. (1967), *Encyclopedia of Polymer Science and Technology*, Wiley, New York, Vol. 7, p. 754.

Kennedy, J. P. (1975), *Cationic Polymerization of Olefins: A Critical Inventory*, Wiley-Interscience, New York.

Kennedy, J. P. (1976), Fourth International Symposium on Cationic Polymerization, *J. Polym. Sci. Polym. Symp.* **56**.

Kennedy, J. P. (1977), "Cationic graft copolymerizations," *J. Appl. Polym. Sci., Polym. Symp.* **30**.

Watson, Bishop R. (1789), *Chemical Essays*, 5th ed., Vol 3. London.

□ 2 □
Definitions, Terminology, and Nomenclature

2.1 ☐ CARBOCATIONS, COUNTERANIONS, AND CARBOCATIONIC POLYMERIZATIONS

As judged by studying the recent literature on carbocationic polymerizations, authors in this field have almost without exception adopted Olah's proposal (1972) and use the term "carbenium ion" for the trivalent trigonal sp^2-hybridized carbocation, the parent of which is $CH_3^\oplus$. In contrast, the term "carbonium ion" is reserved for pentavalent carbocations, the parent of which is $CH_5^\oplus$. According to this nomenclature the suffix *onium* increases the valence state or coordination number of the central atom by one unit; for example, oxonium ion designates a three-coordinated oxygen-centered cation and ammonium ion indicates a four-coordinated nitrogen-centered cation. Carbenium ions and carbonium ions, as well as the species $CH_4^\oplus$, which has been postulated to exist in the gas phase by mass spectroscopy, are carbocations:

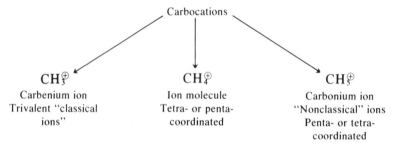

Carbocations

$CH_3^\oplus$	$CH_4^\oplus$	$CH_5^\oplus$
Carbenium ion	Ion molecule	Carbonium ion
Trivalent "classical	Tetra- or penta-	"Nonclassical" ions
ions"	coordinated	Penta- or tetra-
		coordinated

     Carbocationic polymerizations are polyaddition chain reactions in which the growing species is a carbenium ion. These polymerizations belong to the class of cationic polymerizations, that is, polyadditions in which the chain carrier is an electrophile. In addition to carbocationic polymerizations, the class of cationic polymerizations also includes oxonium ion, sulfonium ion, and ammonium ion polymerizations, for example; however, a discussion of these systems is outside the scope of this volume.

     Electroneutrality in cationic polymerization is maintained by negatively charged conjugate bases or *counteranions*. These species may be anions, for example, $Cl^\ominus$, $Br^\ominus$, and $CCl_3COO^\ominus$, obtained in polymerizations initiated by Brønsted acids or large coordinatively satisfied stable conjugate bases of Lewis acid–Lewis base reactions: $BF_3 + H_2O \rightleftharpoons H^\oplus BF_3OH^\ominus$. In some cases, for example, in high-energy induced carbocationic polymerizations, counteranions may be solvated free electrons, or in heterophase cationic processes they may be ill-defined entities embedded in solids.

2.2 ☐ INITIATORS, COINITIATORS, AND INITIATING SYSTEMS

Entities that induce carbocationic polymerizations are usually consumed during the polymerization reaction and usually become part of the polymer molecule. Thus, in line with the classical concept of catalysis, these materials are not catalysts and the usage of the terms "catalysis" or "catalysts" in this context should be discontinued. It is urged that the term "initiator" be adopted for chemicals that initiate cationic polymerizations in general and carbocationic polymerizations in particular.

More than 30 years ago Evans, Plesch, Polanyi, and co-workers (Evans et al., 1946a, b; Evans and Polanyi, 1947; Plesch, 1947) found that BF_3 requires the presence of traces of water to initiate the polymerization of isobutylene and coined the term "cocatalyst" to describe this phenomenon. Since this discovery evidence has been rapidly accumulating to indicate that the purest (dryest) Friedel–Crafts acids more often than not are unable to induce carbocationic polymerizations and that they usually (although not always) require for initiation the presence of protic acids (protogens) or cation sources (cationogens). Also, many protic acid/Lewis acid combinations have been recognized to be among the strongest acids (e.g., HF/BF_3, $HBr/AlBr_3$, H_2O/BF_3) and most efficient carbocationic initiator systems. In this historical context it is then understandable that many practitioners of carbocationic polymerizations refer to Friedel–Crafts acids as "catalyst" or "initiator" and proton sources as "cocatalyst" or "coinitiator." This misleading terminology unfortunately is widely used in spite of evidence to the contrary; that is, the true initiating species is provided by the cationogen and the role of the Friedel–Crafts acid is to help generate the first cation.

The earliest investigators (Evans et al., 1946; Evans and Polanyi, 1947; Plesch et al., 1947) formulated the currently accepted view that the Friedel–Crafts acid assists ion generation by removing and coordinatively stabilizing the halide or hydroxy anion:

$$H_2O + BF_3 \rightleftharpoons H^\oplus BF_3OH^\ominus$$

and that initiation could not be accomplished in the absence of cationogen (H_2O):

$$H^\oplus + CH_2{=}C(CH_3)_2 \rightarrow CH_3{-}C^\oplus(CH_3)_2$$

That certain Friedel–Crafts acids under carefully controlled conditions may initiate carbocationic polymerizations in the absence of purposefully added protogens is a much later development (see Chapter 4).

From the scientific and technological point of view, the most important carbocationic initiating systems are combinations of cationogens and Friedel–Crafts acids. Logically, then, cation sources, be these proton sources such as Brønsted acids or carbenium ion sources such as organic halides, are termed *initiators*, and the Friedel–Crafts acids or Lewis acids that assist in generating the initiating entity are termed *coinitiators*.

The term *"initiating system"* is reversed to describe two-component systems consisting of the cationogen that provides the initiating species and the Lewis acid that determines the nature of the counteranion.

In many instances putative water in conjunction with Friedel–Crafts acids has been proved to be the initiator, for example, H_2O/BF_3 or $H_2O/TiCl_4$. Initiation by water is demonstrated by "stopping experiments" (see Section 4.1), a very demanding and time-consuming undertaking. In many well-known systems (e.g., $AlCl_3$/isobutylene) justifiable suspicion exists, although without experimental proof, for initiation by ubiquitous water. Initiating systems in which water is suspected to be the initiator, although this has not been demonstrated by experiment, are denoted by quotation marks around the symbol for water, as in "H_2O"/$AlCl_3$.

In a few instances "autoionization" of Friedel–Crafts acids has been postulated to be involved in initiation. In these systems both the cation and anion arise from the same Friedel–Crafts acid, for example, $2AlBr_3 \rightleftharpoons AlBr_2^{\oplus} AlBr_4^{\ominus}$, and the initiator and coinitiator are identical. In others, combinations of Friedel–Crafts acids may function as initiating systems, for example, $AlBr_3/TiCl_4 \rightleftharpoons AlBr_2^{\oplus} TiCl_4Br^{\ominus}$.

If Brønsted acids, for example, $HClO_4$ and H_2SO_4, are used alone as initiating systems for carbocationic polymerizations the terminology is degenerate; that is, both the initiating system and initiator are identical. According to this terminology initiation of pseudocationic polymerizations by protic acids, that is, when propagation proceeds in the absence of carbocations, requires the presence of only an initiator but not a coinitiator.

In a few special instances this terminology should be used with circumspection. For example, in the case of direct $TiCl_4$-initiated polymerizations,

Table 2.1 ☐ Nomenclature of Cationic Initiating Systems

Initiating System	Initiator	Initiating Species	Coinitiator	Counteranion
H_2SO_4	H_2SO_4	$H^{\oplus}$	—	$HSO_4^{\ominus}$
$HClO_4$ (pseudo-cationic polymer-izations)	$HClO_4$	$HClO_4$	—	—
H_2O/BF_3	H_2O	$H^{\oplus}$	BF_3	$BF_3OH^{\ominus}$
"H_2O"/$AlCl_3$	"H_2O"	$H^{\oplus}$	$AlCl_3$	$AlCl_3OH^{\ominus}$
$(C_6H_5)_3C^{\oplus}SbCl_6^{\ominus}$	$(C_6H_5)_3CCl$	$(C_6H_5)_3C^{\oplus}$	$SbCl_5$	$SbCl_6^{\ominus}$
t-$BuCl/Et_2AlCl$	t-$BuCl$	t-$Bu^{\oplus}$	Et_2AlCl	$Et_2AlCl_2^{\ominus}$
Cl_2/BCl_3	Cl_2	$Cl^{\oplus}$	BCl_3	$BCl_4^{\ominus}$
$AlBr_3$	$AlBr_3$	$AlBr_2^{\oplus}$	$AlBr_3$	$AlBr_4^{\ominus}$
$AlBr_3 \cdot TiCl_4$	$AlBr_3$	$AlBr_2^{\oplus}$	$TiCl_4$	$TiCl_4Br^{\ominus}$
$TiCl_4$	Olefin	$H^{\oplus}$ or $R^{\oplus}$	$TiCl_4$	$TiCl_5^{\ominus}$
I_2	Olefin, vinyl ether	$H^{\oplus}(?)$	I_2	$I_3^{\ominus}$
$MgCl_2$, clays various acidic solids		?	?	?

γ-ray irradiation, or electroinitiated polymerization of olefins, the identification of the first cation-producing species may not be obvious and would demand careful analysis of the system.

Examples in Table 2.1 serve to illustrate the nomenclature.

The use of *ad hoc* terminologies, for example, dual acids, syncatalysts, to denote initiator/coinitiator systems should be discontinued.

2.3 □ ABBREVIATION OF MULTICOMPONENT SYSTEMS

Often it is necessary to denote concisely a polymerization system comprising a variety of components operating at a temperature or temperature range. These situations may be conveniently expressed by listing system components in the sequence initiator/coinitiator/monomer/solvent/temperature, for example, $H_2O/BF_3/i-C_4H_8/CH_2Cl_2/ - 78°C$.

2.4 □ A NOTE ON THE DEFINITION OF FRIEDEL–CRAFTS ACIDS

Although the terms "Friedel–Crafts halides" or "Friedel–Crafts reactions" are commonly used by chemists all around the world, it is very difficult, if not impossible, to define these expressions comprehensively and precisely. This difficulty persists even in spite of the efforts of such eminent scientists as C. K. Ingold* and G. A. Olah[†]. Evidently chemists know what a Friedel–Crafts halide or acid or reaction is, but cannot define it.

The definition of Friedel–Crafts acids became even more vague when Natta et al. (1959) coined the term "modified Friedel–Crafts catalysts" to denote halides of multivalent metals in their highest valence state in which the halogens are partly substituted by organic groups. This useless terminology should be abandoned.

Further remarks on the nature of Friedel–Crafts acids appear in Sections 3.3 and 4.1.

---

*"Friedel–Crafts reactions embrace all electrophilic reactions catalyzed by electron-deficient compounds—Lewis acids—whether these are molecules or cations, and include such reactions as are likewise catalyzed by those proton acids which are strong enough to act somewhat like Lewis acids, perhaps like the proton they donate, if it were free when it would be a Lewis acid (as proton acids themselves are not)" (Ingold, 1963).

[†]"Friedel–Crafts type reactions [are] those processes which proceed under the general conditions laid down by the pioneering investigators, and which can also be carried out by the later realization of the general acid-catalyzed nature of the reactions" (Olah, 1963a) and ". . . . we consider Friedel–Crafts type reactions to be any substitution, isomerization, elimination, cracking, polymerization, or addition reactions taking place under the catalytic effect of Lewis acid type acidic halides (with or without cocatalyst) or proton acids" (Olah, 1963b).

☐   **REFERENCES**

Evans, A. G., Holden, D., Plesch, P. H., Polanyi, M., Skinner, H. A., and Weinberger, M. A. (1946a), *Nature* **157**, 102.

Evans, A. G., Meadows, G. W., and Polanyi, M. (1946b), *Nature* **158**, 94.

Evans, A. G. and Polanyi, M. (1947), *J. Chem. Soc.*, 252.

Ingold, C. K. (1963), in *Friedel–Crafts and Related Reactions*, G. A. Olah, Ed., Interscience, New York, Vol. 1, p. vii.

Natta, G., Dall'Asta, G., Mazzanti, G., Giannini, U. and Cesca, S. (1959), *Angew. Chem.* **71**, 205.

Olah, G. A. (1963a), in *Friedel–Crafts and Related Reactions*, G. A. Olah, Ed., Interscience, New York, Vol. 1, p. xi.

Olah, G. A. (1963b), in *Friedel–Crafts and Related Reactions*, G. A. Olah, Ed., Interscience, New York, Vol. 1, p. 29.

Olag, G. A. (1972), *J. Am. Chem. Soc.* **94**, 808.

Plesch, P. H., Polanyi, M., and Skinner, H. A. (1947), *J. Chem. Soc.*, 257.

□3□

Phenomenology of Carbocationic Polymerization

3.1 □ THE ACTIVE SPECIES

The Nature of Polymerization-Active Carbocations

Formation of Carbocations

Carbocations having the greatest interest to the polymer chemist are those of intermediate stability. Highly stable cations, for example, pyrenium and anthracenium, are of little interest for polymer-synthetic purposes since they are unable to initiate the polymerization of olefins. Even if such ions had some demonstrated utility for the polymerization of highly active monomers, for example, N-vinylcarbazole and 9-vinylanthracene, more reactive and more readily available cations could produce the same polymers with faster rates.

At the other end of the spectrum, highly reactive cations such as $CH_3^{\oplus}$ and $CH_3CH_2^{\oplus}$ are also only of theoretical interest since their generation by conventional or even moderately expensive techniques is inefficient and difficult. Simple alkyl ions would be desirable initiating species; however, their agressiveness during the polymerization reaction would lead to numerous unacceptable side reactions and by-products. This contention is borne out by reference to Olah's work on the polycondensation of methane (Olah et al., 1971, 1974), which showed that extremely reactive unstabilized alkyl cations indiscriminately attack σ bonds and, consequently, give rise to polymer degradation and isomerization. Similar results have also been obtained by Kennedy (Kennedy et al., 1974a), who attempted to produce "living" carbenium ions by utilizing "super acid" techniques.

Carbenium ions useful for polymerizations may be generated by three principal techniques:

1 □ *Heterolytic bond scission of σ bonds*

$$R—X \rightarrow R^{\oplus} + X^{\ominus}$$

Since carbenium ions are of inherently low thermodynamic stability, this process is favorable only if relatively stable counteranions arise. The important cationogen/Friedel–Crafts acid initiating systems, for example, $t\text{-BuCl}/Et_2AlCl \rightarrow t\text{-Bu}^{\oplus}Et_2AlCl_2^{\ominus}$, produce cations by this technique. Similarly, less investigated systems that also produce cations by heterolytic fission include the decomposition of larger onium ions, for example, diazonium, $R\text{-}\overset{\oplus}{N}{\equiv}N \rightarrow N_2 + R^{\oplus}$, and mercuronium, $R\text{-}Hg^{\oplus} \rightarrow Hg + R^{\oplus}$, which are driven by the formation of highly stable neutral species, the ionization of hydrocarbons by hydride transfer $R^{\oplus} + R'H \rightarrow RH + R'^{\oplus}$, and fragmentation or cracking.

2 □ *Electrophilic addition of a cation to a neutral molecule*

$$H^{\oplus} \text{ or } R^{\oplus} + C{=}C \rightarrow H—C—C^{\oplus} \text{ or } R—C—C^{\oplus}$$

where $R^{\oplus}$ can be a small carbenium ion or propagating chain end. These reactions are of greatest interest for polymer synthesis because they in fact constitute initiation (when $R^{\oplus}$ is a small carbenium ion) or propagation (when $R^{\oplus}$ is an active chain end) of a polymerization chain.

3 □ Oxidation of a radical

$$R^{\cdot} \rightarrow R^{\oplus} + e$$

This process can occur in solution or in the gas phase by electron bombardment, for example. Carbenium ions generated in this manner may be visualized during high-energy irradiation induced carbocationic polymerization.

The concentration of carbocations in solution is in general very low ($\sim 10^{-3}$ to $10^{-6}\,M$), which explains the dearth of data relative to thermodynamic parameters. In contrast, numerous organic compounds when exposed to electron bombardment produce sufficient concentration of carbocations for fundamental investigations, and gas phase data have often been used to approximate thermodynamic properties for solution systems.

Relative Stability of Carbocations

Carbocations are inherently unstable, reactive species. In spite of sustained efforts by eminent chemists, the definition of carbocation stability (reactivity) in solution in thermodynamic or kinetic terms is elusive, and polymer scientists have to contend with relative stability scales derived by comparisons of large numbers of various cationic reactions. The fundamental difficulty in comparing reactivities of carbocations obtained in different systems is largely due to medium and counteranion effects on charge dispersion. Quantization of medium effects on carbocation stability is impossible at the present time.

Several methods proposed for the estimation of relative carbocation stabilities are briefly outlined:

1 □ Determination of free energy of carbocation formation □ The free energy of carbocation formation may be estimated from the dissociation of alkyl halides in the gas phase. Owing to the thermodynamics of the system only the entropy term (whose order of magnitude can be obtained from the absolute entropies of the precursor molecules and ions formed) is relatively unimportant and can be neglected in the first approximation.

The enthalpy required to dissociate an alkyl halide RX into $R^{\oplus}$ and $X^{\ominus}$ can be resolved into three terms:

$$RX \xrightarrow{\Delta H_1} R^{\cdot} + X^{\cdot} \xrightarrow{\Delta H_2} R^{\oplus} + e + X^{\cdot} \xrightarrow{\Delta H_3} R^{\oplus} + X^{\ominus}$$
$$\underset{\Delta H_g}{\underline{\hspace{8cm}}}$$

Table 3.1 ☐   **Enthalpies of Carbocation Formation from Alkyl and Aryl Bromides[a]**

Carbocation	ΔH_g (kcal/mole)
Methyl	214
Allyl	152
Benzyl	147
p-Cl-Benzyl	151
o-Me-Benzyl	142
m-Me-Benzyl	145
p-NO$_2$-Benzyl	139
m-NO$_2$-Benzyl	164
m-C≡N-Benzyl	164
p-C≡N-Benzyl	161

[a] After Bethell and Gold (1967).

where ΔH_1, ΔH_2, and ΔH_3 are the enthalpy changes due to homolytic bond cleavage, ionization potential, and electron affinity, respectively, and $\Delta H_g = \Delta H_1 + \Delta H_2 + \Delta H_3$. Table 3.1 shows representative enthalpies of carbocation formation calculated in this manner from some alkyl bromides. The very high enthalpy values are due to the cleavage of neutral molecules in the gas phase. Values for solution data would be much smaller owing to the gain in solvation energies. Relative enthalpy data are valuable in comparing relative stabilities of carbocations in the same solvent.

2 ☐   *Ionization potentials of radicals* ☐   A relative carbocation stability scale may be obtained by comparing ionization potentials of parent radicals. Representative values are shown in Table 3.2.

3 ☐   *Solvolysis Rates of S_N1 Reactions* ☐   Transition states of first-order nucleophilic substitutions (S_N1) are usually visualized to be carbocations:

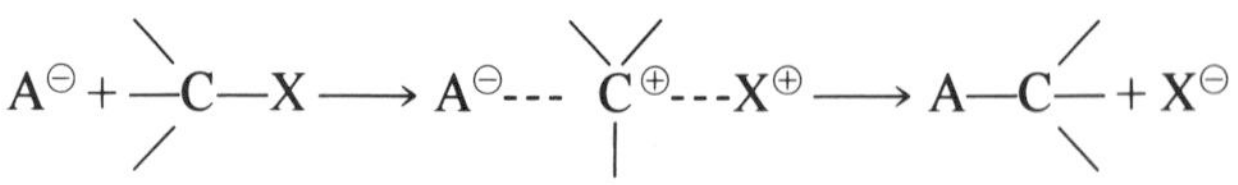

$$A^{\ominus} + {-}\overset{\diagdown}{\underset{\diagup}{C}}{-}X \longrightarrow A^{\ominus}{-}{-}{-}\overset{\diagdown\diagup}{\underset{|}{C^{\oplus}}}{-}{-}{-}X^{\oplus} \longrightarrow A{-}\overset{\diagup}{\underset{\diagdown}{C}}{-} + X^{\ominus}$$

The free energy difference between the starting molecule and the transition state is proportional to the relative reaction rates of similar compounds. In this sense, solvolysis rates of S_N1 reactions have often been used to compare relative stabilities of carbocations (Baker, 1974).

4 ☐   *Comparison of carbocation formation enthalpies from the corresponding alcohols* ☐   Enthalpies of carbocation formation have been determined using different solvents, particularly the HSO$_3$F–SbF$_6$ system (Arnett, 1973; Arnett et al., 1980). A good correlation seems to exist between relative

Table 3.2 □ **Ionization Potentials of Free Radicals**

Radical	Ionization Potential (kcal/mole)
$CH_3^{\cdot}$	229.4 ± 0.7
$CH_3CH_2^{\cdot}$	202.5 ± 1.2
$(CH_3)_2CH^{\cdot}$	182.2 ± 1.2
$(CH_3)_3C^{\cdot}$	171.8 ± 1.2
$C_6H_5CH_2^{\cdot}$	178.9 ± 1.8
$p\text{-}NCC_6H_5CH_2^{\cdot}$	197.9 ± 2.3
$p\text{-}MeOC_6H_5CH_2^{\cdot}$	157.7 ± 2.3
$(C_6H_5)_2CH^{\cdot}$	168.8 ± 2.3
$CH_2{=}CH^{\cdot}$	217.9 ± 1.2
$CH_2{=}CHCH_2^{\cdot}$	188.2 ± 0.7

^{a}After Barker (1974).

carbenium ion stabilities obtained by this method and those by solvolysis rates of corresponding alkyl halides.

5 □ **Quantum chemical methods** □ Quantum chemical methods have been employed to estimate carbocation stability. The calculated values must be carefully analyzed because they are obtained by the use of highly simplified models, neglecting many electronic interactions and totally disregarding solvation effects. The methods more sophisticated than Hückel's (such as CNDO) are also limited because necessary information relative to bond lengths and angles in carbocations is lacking. Recent theoretical chemical investigations (STO-3G method) gave good values for these parameters (± 0.02 Å, $\pm 2°$) for some carbocations (Hehre, 1975).

Structure Effects Influencing Carbocation Stability

Carbocation stability is determined by microarchitecture and by the nature of atoms in the vicinity of the carbenium center. Electronic and steric effects are of particular importance.

With regard to electronic effects, factors that affect charge delocalization by definition also affect carbocation stability. Increasing the size of the carbenium ion increases charge delocalization and consequently augments stability. Electron-donating groups stabilize cations by charge dispersion; this effect is apparent by examining the data in Table 3.2.

Conjugation, particularly by electron-donating groups in the para position of aromatic rings, tend to stabilize carbenium ion centers. For example, cation stability would increase along the following sequence:

$$p\text{-}CH_3O{-}C_6H_4{-}\overset{\oplus}{C}H_2 > (C_6H_5)_2\overset{\oplus}{C}H > C_6H_5\,\overset{\oplus}{C}H_2 \gg CH_3{-}\overset{\oplus}{C}H_2 \gg \overset{\oplus}{C}H_3$$

Heteroatoms with unshared electrons, for example, O, S, and N, adjacent to the carbenium ion center are strongly stabilizing by charge delocalization:

$$R—O—\overset{\oplus}{C}H_2 \leftrightarrow R—\overset{\oplus}{O}{=}CH_2, \quad R_2{=}N—\overset{\oplus}{C}H_2 \leftrightarrow R_2{=}\overset{\oplus}{N}{=}CH_2$$

Steric effects may also significantly affect carbocation stability. For example, the solvolysis rate of $C_6H_5–CHClCH_2R$, where R=methyl, ethyl, isopropyl and *tert.*-butyl, decreases as 540, 125, 27, and 1, respectively, most likely because nonbonded compression between the ortho hydrogens of the ring and the increasingly large substituents in the transition state inhibits coplanarity, that is, orbital overlap:

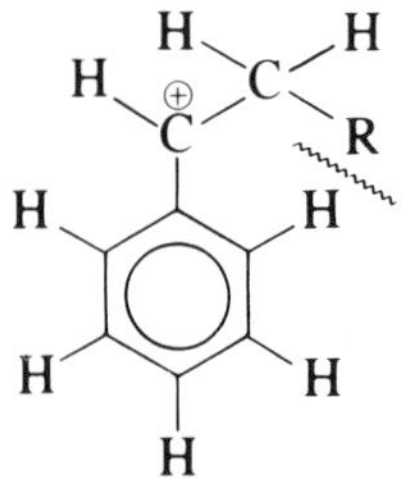

A similar situation is found in cationic polymerization of alkyl vinyl ethers (Yuki et al., 1969). An increase of the reactivity is observed with increasing size of the alkyl group. This is because when the size of R in $CH_2{=}CH–O–R$ increases, the overlap between the π system and unshared electron on the oxygen atom becomes more difficult and consequently resonance stabilization decreases.

Since resonance leads to a reduction in olefinic character of the vinyl group, a decrease of resonance increases reactivity. The oxygen atom is assumed to be sp^3-hybridized, leading to the following conformations:

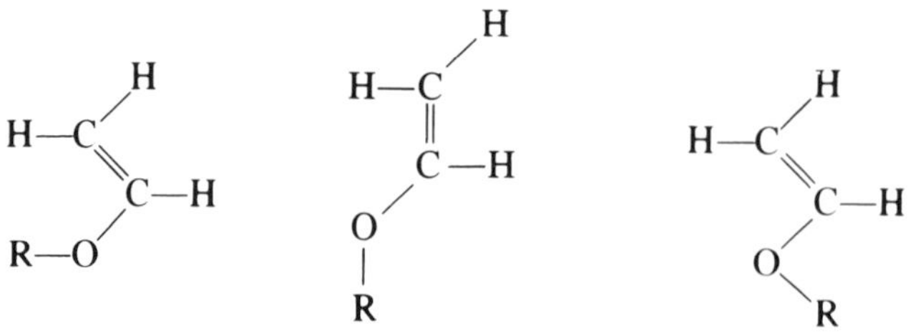

Examination of H–NMR spectra of vinyl alkyl ethers (Hatada et al., 1968) shows that the contribution of gauche conformation increases with increasing size of R and is the major conformer for *tert*-butyl vinyl ether.

On the other hand, the presence of bulky substituents on a cationogen may increase the rate of ionization and consequently increase cation stability. For example, increasing the size of substituents in RR'R''–C–X facilitates ionization due to strain relief in the transition state:

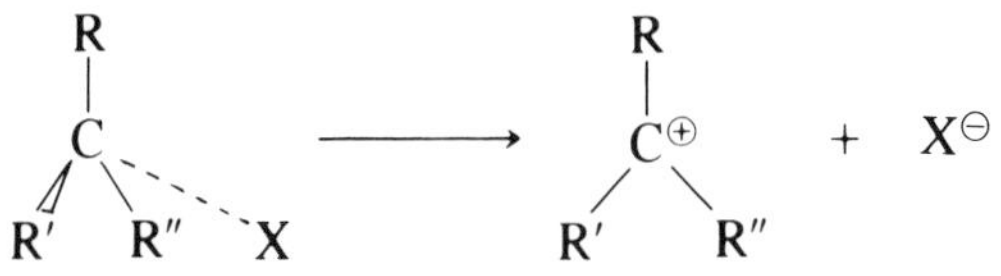

Carbocation Stability in Solution

Relative to the gas phase, solvents stabilize carbocations by charge dispersion.

Although many solvent effects are well recognized and have been investigated in great detail, although indeed in some instances the presence of solvents or their nature was found to be critical for the success of carbocationic polymerizations, no quantitative theory has become available to guide researchers in this area. The importance of understanding solvent effects on ionic polymerizations cannot be overemphasized. In spite of the tremendous influence of solvents on ionic polymerizations in general and cationic polymerizations in particular, only qualitative or at best semiquantitative theories are available.

The fountainhead of all theories remains the Born equation:

$$\Delta G = -\frac{q^2}{r}\left(1 - \frac{1}{\varepsilon}\right)$$

according to which the change in free energy ΔG is determined by the radius r and charge q of the ion transferred from the gas phase into a solvent with dielectric constant ε. Thus the energy of solvation of a cation increases with increasing ε and is inversely proportional to r. A variety of modifications of this equation have been proposed to overcome some of the basic shortcomings of the theory, that is, the estimation of r and ε. It has been recognized that ε is insufficient for the characterization of solvent power of a liquid; however, nothing better has been advanced. It is still extremely difficult to define what is meant by a "polar" solvent because of the multiplicity of special effects that may operate and thus complicate or even completely obviate predictions derived from the Born equation, for example, distortion of the dielectric continuum by placing a charged particle in it, mixing solvent and nonsolvent (monomer), solvent shell saturation, solvent shell restructuration due to charged entities, solvent polarizability, geometry effects, aggregation effects, nonuniform charge distribution, and preferential solvation.

Interaction between and ion and the solvent is influenced not only by the size of the ion but also by the charge distribution in the carbocation/solvent system; inversely, solvation affects charge distribution of the ion. For example, isobutylene polymerization proceeds rapidly in CH_3Cl but not in CH_3I diluent. Evidently the charge remains on the cation in CH_3Cl while it is transferred to iodine in CH_3I: $\sim CH_2-C(CH_3)_2---^{\oplus}I-CH_3$ (Kennedy and Trivedi, 1978).

The macroscopic dielectric constant is of only very limited significance in characterizing a liquid for its use in carbocationic polymerizations. The reaction profile is determined by the microscopic electrostatic field at the locus of the collision of the cation with the monomer, as well as the polarizabilities and steric effects of the ingredients, including the solvent.

In addition to polarity, solvent nucleophilicity is of paramount importance in determining the outcome of polymerization reactions. Again, in spite of its great significance, "nucleophilicity" is a vague term related to the coordinating ability of organic compounds to cations. Highly nucleophilic solvents may be inhibitors because they may coordinate with Lewis acids, be these Friedel–Crafts acids or carbenium ions. Thus olefins cannot be polymerized in diethyl ether because of oxonium ion formation; however, vinyl ethers readily polymerize in this medium.

Although solvation can be very useful in many instances, strong cation solvation may retard or inhibit polymerization. Thus polymerization of vinyl ethers by $BF_3 \cdot OEt_2$ proceeds faster in hexane than in Et_2O (Eley and Johnson, 1964), or that of α-methylstyrene is inhibited by even small quantities of Et_2O (Worsfold and Bywater, 1957).

Complicated self-solvation effects may arise in nonpolar media. In these instances monomer may replace solvent in the solvate shell of the ion and may give rise to peculiar reaction orders. In polar media the nature of the solvent may determine the initiator. For example, water was found to be initiator for $SnCl_4$/styrene/CCl_4; however, alkyl halides such as t-BuCl were ineffective. In contrast, alkyl halides were active initiators in the $SnCl_4$/styrene/1,2-dichlorethane system (Colclough and Dainton, 1958).

In first approximation, solvents with low polarity should give rise to faster propagation reactions because of charge dispersion in the transition state and favorable desolvation energetics:

$$\sim C^{\oplus} + C{=}C \rightarrow [\sim \overset{\delta\oplus}{C} --- C{\vcentcolon=}\overset{\delta\oplus}{C}]^{\ddagger} \rightarrow \sim C{-}C{-}C^{\oplus}$$

If propagation occurs by ion pairs, the energetics that control dissociation of the active ion pair may require polar solvent participation with unpredictable consequences that have not yet been examined for vinyl cationic polymerizations. It appears that the overall rate of carbocationic polymerizations is affected in a complex manner by solvation: whereas initiation rate is enhanced by increasing solvent polarity on account of increasing rate of ion generation, propagation rate may decrease (see also Section 4.2).

The Active Species in Carbocationic Polymerizations

Ions and Ion Pairs

Electroneutrality demands that carbocations always be accompanied by counteranions. The distance between the carbocation and counteranion is important because it largely determines species reactivity. Depending on the distance between the charged particles, which in turn is determined by the intrinsic properties of the ions and experimental conditions, a continuous spectrum of ionicities exists ("Winstein spectrum"):

$$R\text{--}X \rightleftharpoons R^{\oplus}X^{\ominus} \rightleftharpoons R^{\oplus}/X^{\ominus} \rightleftharpoons R^{\oplus}\|X^{\ominus} \rightleftharpoons R^{\oplus} + X^{\ominus}$$

| Contact (tight) ion pairs | Solvent separated ion pair | Solvated (lose) ion pair | Free ions |

Free carbocations can exist only under the purest conditions in the absence of even traces of water (Williams et al., 1966). In the absence of moisture high polymerization rates can be obtained ($\sim 10^8$ mole/sec) and the energy of activation is close to zero. Free ions exist in γ-ray radiation-induced polymerizations of certain olefins, for example, α-methylstyrene, β-pinene, cyclopentadiene. Irradiation of styrene produced 10^{-10} and 10^{-11} M ions per 10^6 and 10^4 rad/hr, respectively (Metz, 1969). Obviously at such low ion concentrations impurities must be most carefully removed.

In the majority of cationic polymerization systems reported in the literature, the propagating species are probably associated ion pairs. The accurate definition of system ionicity in terms of actual species in solution and their concentration is almost impossible. When active species are free ions, an order of magnitude of their concentration can be obtained (Plesch, 1971) provided numerous assumptions are made.

The possibility for ion aggregation also exists:

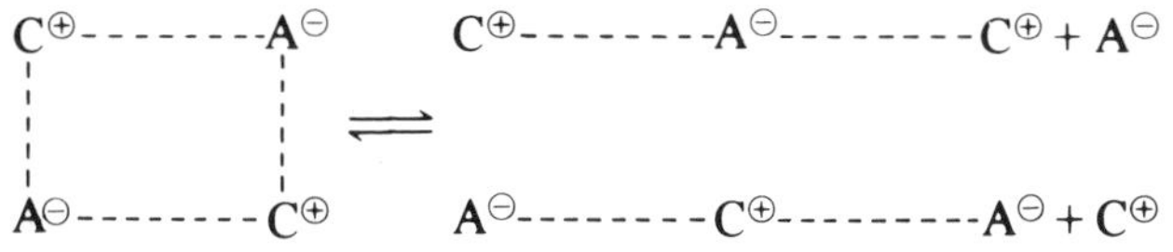

$$
\begin{array}{ccc}
C^{\oplus}\text{-------}A^{\ominus} & C^{\oplus}\text{--------}A^{\ominus}\text{--------}C^{\oplus} + A^{\ominus} \\
& \rightleftharpoons \\
A^{\ominus}\text{-------}C^{\oplus} & A^{\ominus}\text{--------}C^{\oplus}\text{--------}A^{\ominus} + C^{\oplus}
\end{array}
$$

Multiple ion clusters have been postulated to occur in anionic polymerizations, particularly in hydrocarbon solvents (Margerison and Newport, 1963); however, the cationic polymerization literature is devoid of studies along these lines.

The thermodynamics of the equilibrium which exists between associated ions and free ions,

$$C^{\oplus}\!\cdot\!A^{\ominus} \overset{K_d}{\rightleftharpoons} C^{\oplus} + A^{\ominus}$$

has been studied (Denison and Ramsey, 1955) and K_d has been determined (Fuoss, 1934, 1958; Fuoss and Accasina, 1959; Gilkerson, 1956) by conductivity measurements.

It would be of great interest to expand fundamental understanding of polymerizing carbocation ionicities and to formulate these systems in terms of relationships such as the following:

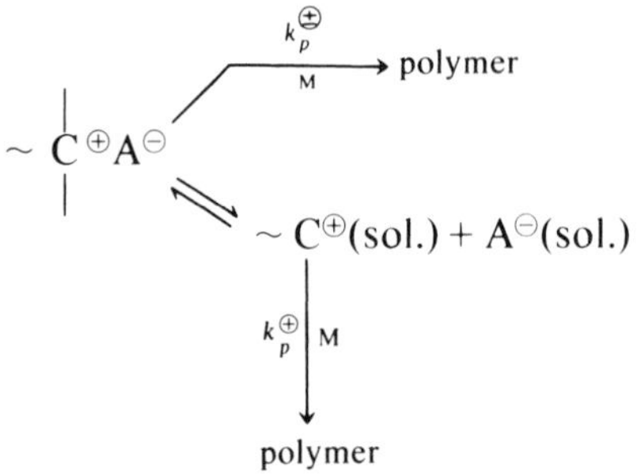

In the field of anionic polymerizations significant advances have been made in the definition of k_p because the active species are stable, or "living." Unfortunately this has not yet been possible with cationic systems because of the relatively high reactivity of these species.

Carbocations and Active Species in Propogation

Hunter and Yohe (1933) suggested that the propagating species in carbocationic polymerizations are either free carbenium ions or associated ion pairs. This formalism immediately became successful because it explained most experimental observations available not only in the field of polymerization chemistry but also in related areas of organic chemistry. Certain olefins in the presence of acidic Friedel–Crafts acids give rise to ions, but this does not necessarily prove that ions must be responsible for propagation (Plesch, 1968).

Several studies have concerned the examination of evidence for the presence of carbenium ions in olefin polymerizations (Evans et al., 1951, 1955, 1957a, b, 1959; Metz, 1961; Pepper, 1949; Pepper and Reilly, 1962; Gandini and Plesch, 1965a–c; Jordan and Treloar, 1961; Upadyhyay et al., 1965; Pac and Plesch, 1967). The presence of carbocations during propagation of aromatic monomers has been satisfactorily established (Plesch and Gandini, 1966). Table 3.3 is a compilation of systems in which propagating carbocations have been proved to exist.

The situation is much less satisfactory with aliphatic monomers and *direct* proof for the existence of carbocations during propagation of these olefins has not been obtained. Spectroscopic studies are of little use because of the very low concentration of the active species; electrical conductivity, kinetic measurements, and structural investigations give only

Table 3.3 ☐ Aromatic Polymerization Systems in which the Presence of Ionic Chain Carriers has been Established[a]

Monomer	Solvent	Diagnosis
Styrene	CH_2Cl_2, $(CH_2Cl)_2$	Kinetics, spectroscopy
	CH_2Cl_2	Kinetics, spectroscopy
	CH_2Cl_2, $(CH_2Cl)_2$	Kinetics, spectroscopy
p-Methoxystyrene	CH_2Cl_2	Rate, color, conductivity
	CH_2Cl_2	Rate, color, conductivity
	CH_2Cl_2	Rate, color, conductivity
1,1-Diphenylethlene	C_6H_6	Kinetics, color
	$C_2H_5NO_2$	Kinetics, color
	C_6H_6	Kinetics, spectroscopy
	C_6H_6	Kinetics, spectroscopy
1-Phenyl-1(p-methoxy-phenyl)ethylene	C_6H_6	Kinetics, color
1,1-Di-p-tolylethylene	C_6H_6	Kinetics, color

The monomer column corresponds to these reagent entries (second implicit column):

$HClO_4$
$R^{\oplus}ClO_4^{\ominus}$
H_2SO_4
$HClO_4$
$AgClO_4 + CH_3C_6H_4Br$
$Ph_3C^{\oplus} + ClO_4^{\ominus}$
CCl_3CO_2H
CCl_3CO_2H
$SnCl_4 \cdot H_2O$
$SnCl_4 \cdot HCl$
CCl_3CO_2H
CCl_3CO_2H

[a] After Plesch and Gandini (1966).

circumstantial evidence. Thus Zlamal et al. (1957) studied isobutylene polymerization by the "H_2O"/$AlCl_3$ initiating system using ethyl chloride diluent at $-78°C$ in the presence of various polar components. That the $\overline{DP}$ of the polymers increased with decreasing conductivity of initiator solution (prior to monomer addition) may be construed as strong evidence for propagation by ionic species (Plesch, 1964; Plesch and Gandini 1966). Kinetic studies can readily be interpreted by assuming the presence of free and paired carbocations (Plesch, 1964; Biddulph, et al., 1965; Norrish and Russell, 1952). The structure of polymers is an excellent repository of mechanistic information; thus Lewis acid-initiated isomerization polymerizations, for example, 3-methyl-1-butene, can be explained only by assuming carbocationic intermediates (Kennedy and Thomas, 1962a).

Pseudocationic Polymerizations

According to Plesch and Gandini (1966) and Gandini and Plesch (1964, 1965a–c), certain Brønsted acid-initiated polymerizations may propagate in the absence of carbocations. In addition to their data, those published by Pepper et al. (Pepper and Reilly, 1961, 1962, 1966; Burton and Pepper, 1961; Hayes and Pepper, 1961; Albert and Pepper, 1961) concerning the polymerization of styrene by sulfuric and perchloric acid using polar solvents, for example, methylene chloride, seem to be in concord with this interpretation.

Originally Pepper et al. (Burton and Pepper, 1961; Hayes and Pepper, 1961; Albert and Pepper, 1961) interpreted their observations in terms of conventional carbocations and proposed the following scheme:

1 □   *Initiation*

$$CH_2{=}CH{-}C_6H_5 + H_2SO_4 \xrightarrow{k_1} \underbrace{CH_3{-}\overset{\oplus}{C}H{-}C_6H_5}_{P_1^{\oplus}} + H_2SO_4^{\ominus}$$

2 □   *Propagation*

$$P_x^{\oplus}HSO_4^{\ominus} + M \xrightarrow{k_2} P_{x+1}^{\oplus}HSO_4^{\ominus}$$

3 □   *Transfer*
Spontaneous

$${\sim}CH_2{-}\overset{\oplus}{C}H{-}C_6H_5HSO_4^{\ominus} \xrightarrow{k_3} {\sim}CH{=}CH{-}C_6H_5 \text{ or cyclized}$$
$$\text{product} + H_2SO_4$$

To monomer

$${\sim}CH_2{-}\overset{\oplus}{C}H{-}C_6H_5\,HSO_4 + M \xrightarrow{k_4} {\sim}CH{=}CH{-}C_6H_5 + P_1^{\oplus} + HSO_4^{\ominus}$$

4 □ Termination

$$\sim CH_2\overset{\oplus}{-CH}-C_6H_5HSO_4^{\ominus} \xrightarrow{k_5} \sim CH_2-CH-OSO_3H$$
$$\underset{C_6H_5}{|}$$

Although none of the products were chemically identified, the scheme explained in kinetic terms the experimental observations.

Steady state kinetics ($k_1 \ll k_2$) did not agree with the data and the authors assumed rapid initiation ($k_1 \gg k_2$) and equally rapid termination. These assumptions led to the proposal of rapid formation of active species at a concentration practically equal to that of the added initiator followed by first-order decrease by k_5. The existence of termination was suggested by incomplete consumption. Although most observations could be explained with these assumptions, difficulties were encountered in explaining the unusually high activation energy of propagation ($E = 8.5$ kcal/mole), that high acid concentrations (tenfold on acid) did not affect k_p, and that k_1 was found to be the same for free ions and ion pairs over a broad concentration range.

Gandini and Plesch (1966) and Gandini et al. (1964, 1965a, b) reinvestigated this system and by spectrophotometry and conductivity measurements coupled with kinetic studies showed that ions were absent during propagation and that they formed only when the styrene concentration was less than four times that of the acid. Styrene addition resulted in propagation and in instantaneous disappearance of carbocations, which, however, reappeared when the styrene concentration decreased to ~4 [St]/[HClO$_4$]. These authors confirmed that large amounts of water had no effect on the kinetics; however, if [H$_2$O] > [HClO$_4$] ions remained absent at the end of the polymerization even at low styrene concentrations. The carbocation formed below 4[HClO$_4$]/[St] could be characterized.

To explain these phenomena Plesch (1966) proposed the concept of "pseudocationic polymerizations," according to which the propagating species are esters. Initiation occurs by

$$1/4[HClO_4]_4 + 5CH_2{=}CH \rightarrow CH_3{-}CH{-}OClO_3 \cdot 4CH_2{=}CH \rightarrow$$
$$\underset{C_6H_5}{|} \qquad\qquad \underset{C_6H_5}{|} \qquad\qquad\qquad \underset{C_6H_5}{|}$$

$$\xrightarrow{St} CH_3{-}CH\text{wwww}CH_2{-}CH{-}OClO_3 \cdot 4CH_2{=}CH$$
$$\underset{C_6H_5}{|} \qquad\qquad \underset{C_6H_5}{|} \qquad\qquad \underset{C_6H_5}{|}$$

Ester formation is very rapid and the rate determining step is propagation.

Table 3.4 is a compilation of systems for which pseudocationic polymerization have been proposed.

Table 3.4 □ Systems that Proceed by Pseudocationic Polymerization

Monomer	Initiator	Solvent	References
Styrene	$HClO_4$	CH_2Cl_2	Gandini and Plesch, (1964)
	$HClO_4$	$(CH_2Cl)_2$	Gandini and Plesch, (1965a)
	$HClO_4$	$EtNO_2$	Gandini and Plesch, (1965b)
	$HClO_4$	$EtNO_2 + (CH_2Cl)_2$	Gandini and Plesch, (1965b)
	H_2OSnCl_4	CH_2Cl_2	Gandini and Plesch, (1965a)
	CF_3COOH	CH_2Cl_2	Gandini and Plesch, (1965a)
	$C_6H_5CH(CH_3)ClO_4$	CH_2Cl_2	Gandini and Plesch, (1965a)
p-Methoxystyrene	CF_3CO_2H	CH_2Cl_2	Gandini and Plesch, 1965b)
Acenaphthylene	$HClO_4$	CH_2Cl_2	Gandini and Plesch, (1965b)
	H_2SO_4	CH_2Cl_2	Gandini and Plesch, (1965b)
N-Vinylcarbazole	$HClO_4$	CH_2Cl_2	Gandini and Plesch, (1965b)
	H_2SO_4	CH_2Cl_2	Gandini and Plesch, (1965b)
1,1-Di-p-methoxyphenyl-ethylene	CCl_3COOH	C_6H_6	Evans et al. (1955, 1957b)
1,1-Diphenylethylene	$HClSbCl_3$	C_6H_6	Evans et al. (1961)

Certain systems can be polymerized by either pseudocationic or conventional carbocationic route; the nature of the mechanism depends on reaction conditions. For example, at high acid concentration a pseudocationic system may become conventional and the rate of polymerizations may increase a hundredfold. Or, at very low temperatures ($-90°C$) the rate of conventional carbocationic polymerization may remain high, whereas that of pseudocationic mechanism becomes very slow (Gandini and Plesch, 1965a).

The proposition of pseudocationic polymerization is not a fundamental departure from cationic polymerization mechanisms; rather it should be viewed as a polymerization in which the active species are covalent instead of being ionic. These reactions are at the other end of the ionicity spectrum, which starts with polymerizations in which the propagating entities are free ions.

Types of Electrophilic Reactions in Carbocationic Polymerizations

According to Olah (1974), carbocations may react with π donors (alkenes, arenes), σ donors (C–C or C–H bonds), or n donors (unshared electron pairs on oxygen, sulfur, etc. atoms). This classification of electrophilic reactions can readily be extended to carbocationic polymerizations as follows:

• Reaction with π donors

   Initiation: $C^{\oplus} + C{=}C \rightarrow C{-}C{-}C^{\oplus}$

   Propagation: $\sim C^{\oplus} + C{=}C \rightarrow \sim C{-}C{-}C^{\oplus}$

• Reaction with σ donors

   Chain transfer to polymer:

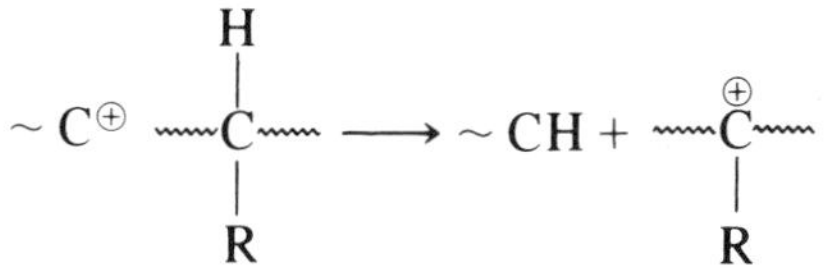

   Isomerization (internal σ donor):

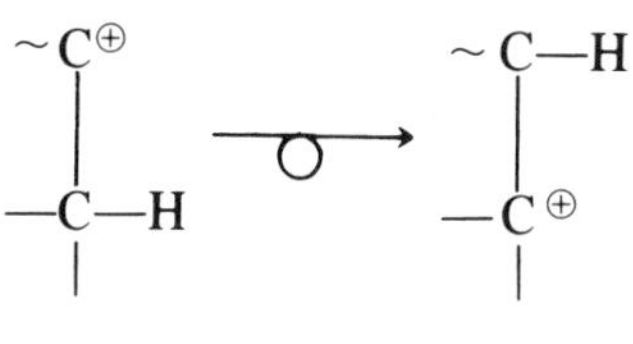

Termination with alkylaluminum counteranions:

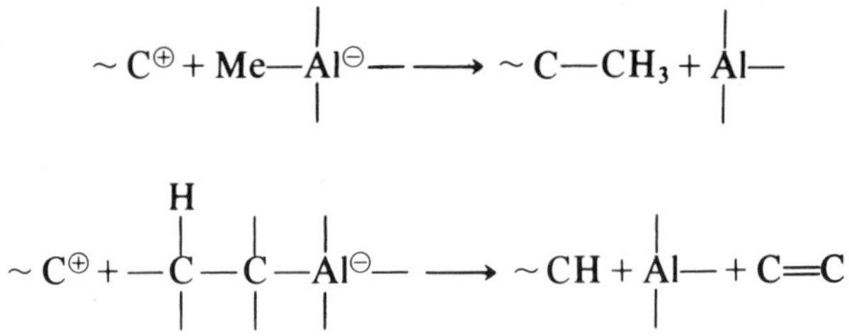

$$\sim C^{\oplus} + Me{-}\underset{|}{\overset{|}{Al}}{}^{\ominus}{-} \longrightarrow \sim \overset{|}{C}{-}CH_3 + \overset{|}{\underset{|}{Al}}{-}$$

$$\sim C^{\oplus} + {-}\overset{\overset{\displaystyle H}{|}}{\underset{|}{C}}{-}\overset{|}{\underset{|}{C}}{-}\overset{|}{Al}{}^{\ominus}{-} \longrightarrow \sim \overset{|}{CH} + \overset{|}{\underset{|}{Al}}{-} + C{=}C$$

Degradation (chain scission via carbonium ions):

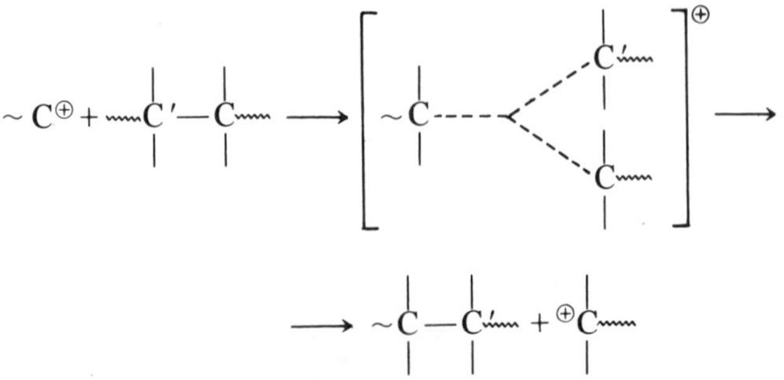

$$\longrightarrow \sim \overset{|}{\underset{|}{C}}{-}\overset{|}{\underset{|}{C'}}\!\!\sim + {}^{\oplus}\overset{|}{\underset{|}{C}}\!\!\sim$$

Chain transfer to monomer represents an interesting combination of internal σ donor-external π donor participation ($G^{\ominus}$ is counteranion):

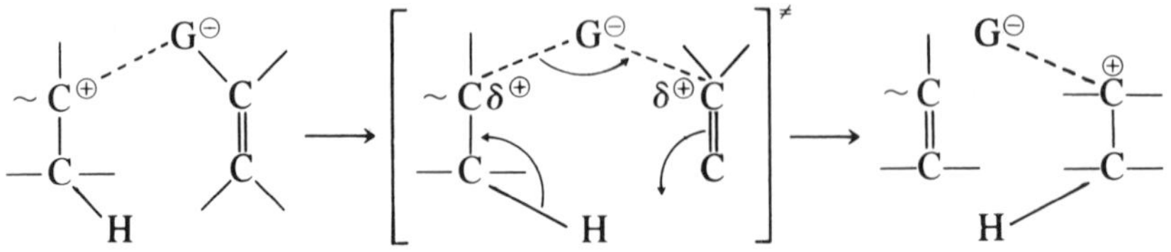

• Reaction with *n* donors

Chain transfer to solvent (X = halogen, e.g., CH_2Cl_2):

$$\text{\textbackslash\textbackslash}C^{\oplus} + R{-}X \rightarrow \text{\textbackslash\textbackslash}C{-}X + R^{\oplus}$$

Termination with $BCl_4^{\ominus}$ counteranion:

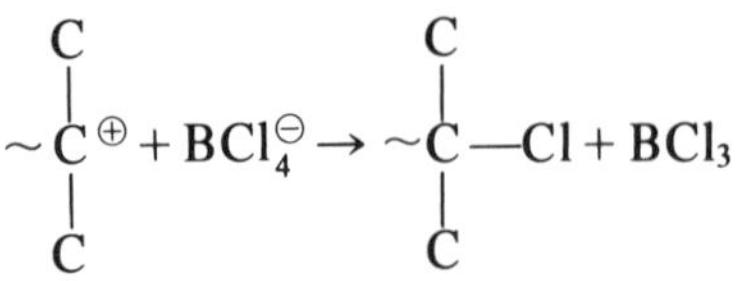

$$\sim \overset{\overset{\displaystyle C}{|}}{\underset{\underset{\displaystyle C}{|}}{C}}{}^{\oplus} + BCl_4^{\ominus} \rightarrow \sim \overset{\overset{\displaystyle C}{|}}{\underset{\underset{\displaystyle C}{|}}{C}}{-}Cl + BCl_3$$

Termination with impurity (e.g., H_2O):

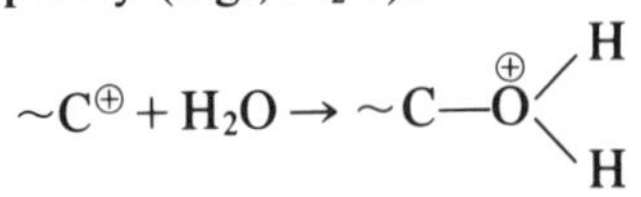

$$\sim C^{\oplus} + H_2O \rightarrow \sim C{-}\overset{\oplus}{O}\!\!\diagup^{\displaystyle H}_{\diagdown H}$$

3.2 □ MONOMERS

Any chemical that can be cationated and in turn is able to cationate another molecule of this chemical is a cationically polymerizable monomer or, briefly, "cationic monomer." To obtain high polymers from cationic monomers the proton (cation) affinity of the monomer must be sufficiently high and the carbocation must be sufficiently stable to sustain propagation. In addition to cationic reactivity and relative stability, other factors must also be conducive for high polymer formation; that is, transfer and termination should be relatively slow in comparison to propagation, and steric hindrance should not be prohibitive.

The structure of the cationic monomer determines the success or failure of polymerization; this section concerns a brief discussion of this matter.

Electronic Characteristics of Cationic Monomers

Cationic olefin polymerization occurs by repetitive attack by an electrophile (carbocation, proton) on the π system of the vinyl monomer. The vinyl monomer must be sufficiently nucleophilic to support reaction.

A thermodynamic evaluation of nucleophilicity required for polymerization may be carried out by examining the enthalpies of protonation or carbocationation. Enthalpies of protonation and carbocationation of ethylene, propylene, and isobutylene in the gas phase are shown in Table 3.5. According to these data, all three polymers are cationically polymerizable, and ethylene polymerization is thermodynamically favored over that of isobutylene. In contrast, from the kinetic point of view, the reactivity order is ethylene < propylene < isobutylene, which corresponds to the relative stability of carbocations:

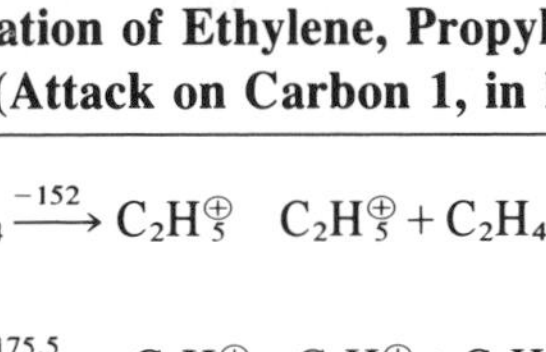

$$
\sim \overset{\oplus}{C}H_2 <\sim \underset{CH_3}{\overset{\oplus}{C}H} <\sim \underset{CH_3}{\overset{CH_3}{\underset{|}{\overset{|}{C}}{}^{\oplus}}}
$$

Table 3.5 □ **Enthalpies of Protonation and Carbocationation of Ethylene, Propylene and Isobutylene (Attack on Carbon 1, in kcal/mole)**[a]

$$H^{\oplus} + C_2H_4 \xrightarrow{-152} C_2H_5^{\oplus} \quad C_2H_5^{\oplus} + C_2H_4 \xrightarrow{-35} C_4H_9^{\oplus}$$

$$H^{\oplus} + C_3H_6 \xrightarrow{-175.5} n\text{-}C_3H_7^{\oplus} \quad C_3H_7^{\oplus} + C_3H_6 \xrightarrow{-30.5} C_6H_{13}^{\oplus}$$

$$H^{\oplus} + i\text{-}C_4H_8 \xrightarrow{-189} t\text{-}C_4H_9^{\oplus} \quad t\text{-}C_4H_9^{\oplus} + C_4H_8 \xrightarrow{-24} t\text{-}C_8H_{17}^{\oplus}$$

[a] After Sawada (1972).

Judging by the favorable protonation and cationation enthalpies of ethylene and propylene, these monomers could be expected to produce high polymer readily. However, these olefins in fact yield only oligomers under forcing cationic conditions because a variety of poorly controllable secondary reactions favorably compete with propagation, such as isomerization, proton transfer and elimination, and chain transfer. For example, protonation and ethylation of ethylene are rapidly followed by energetically favorable isomerization:

$$H^{\oplus} + CH_2{=}CH_2 \rightarrow CH_3\overset{\oplus}{C}H_2 \xrightarrow{\ CH_2{=}CH_2\ }$$
$$CH_3{-}CH_2{-}CH_2{-}\overset{\oplus}{C}H_2 \rightarrow CH_3{-}CH_2{-}\overset{\oplus}{C}H{-}CH_3 \rightarrow etc.$$

These transformations follow the favorable enthalpy differences between primary to secondary (-22 kcal/mole), and secondary to tertiary ($\simeq -33$ kcal/mole) carbocations. Also the reactions between ethylene and substituted carbocations are much less favorable than that between ethylene and the ethyl cation:

$$R_1^{\oplus} + CH_2{=}CH_2 \rightarrow R_2^{\oplus} \qquad \Delta H_{298}^{\circ}\,(kcal/mole)$$

where $\Delta H_{298}^{\circ} = -69.5$ for $R_1^{\oplus} = Me^{\oplus}$; -35.0 for $Et^{\oplus}$; -22.5 for $n\text{-}Pr^{\oplus}$; -8.5 for $i\text{-}Pr^{\oplus} - 2.5$ for $n\text{-}Bu^{\oplus}$; 0.0 for $s\text{-}Bu^{\oplus}$; and $+7.5$ for $t\text{-}Bu^{\oplus}$.

Propylene can be polymerized by $HBr/AlBr_3$ to a very broad molecular weight distribution product (Fontana, 1959). According to IR evidence numerous secondary reactions (Ketley and Harvey, 1961; Immergut et al., 1961) dominated by hydrogen transfer compete with propagation.

Conjugation with the carbenium ion center helps to disperse the positive charge and tends to increase monomer reactivity. This effect is particularly strong when conjugation and electron-donating groups cooperate. Thus the reactivity sequence of styrene derivatives is as follows:

styrene $< \alpha$-methylstyrene $< p$-methoxystyrene $< p$-dimethylaminostyrene

It is a truism, nonetheless often overlooked, that the double bond system must be the most nucleophilic site in the monomer. If more than one nucleophilic site exists in a monomer and the π electron system of the double bond does not represent the most nucleophilic site, the other site(s) may complex with the electrophile (proton, cation, Lewis acid). This may explain why, for example, acrylates or vinyl acetate are not cationically polymerizable:

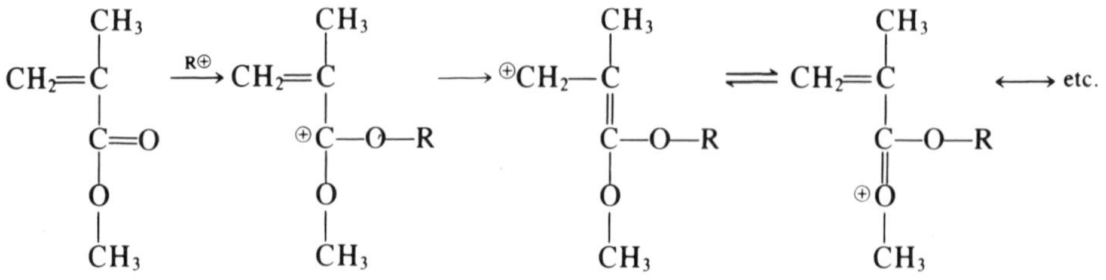

In addition, with these monomers the substitutent not only preferentially complexes the electrophile but may even reduce the nucleophilicity of the double bond by electron attraction.

Nucleophilicity may also be affected by reaction conditions. Maréchal et al. (Tortai and Maréchal, 1971a; Tortai et al., 1971b; Maréchal, 1973) found that 5-methoxyindene did not polymerize below −50°C; however, polymerization was instantaneous above this threshold temperature. Also, ^{1}H-NMR showed the existence of complexes between the methoxy group and Lewis acid at low temperatures. According to calculation the reactivities of indene, which under the same conditions rapidly produces high polymer, and 5-methoxyindene are about equal. Evidently below −50°C preferential complexation occurs with a nucleophilic site other than the double bond system; however, at elevated temperatures the activation energy difference favors polymerization which thus becomes competitive with complexation.

The presence of a second nucleophilic site does not prevent polymerization if it augments electron density of the double bond by donation or by stabilization of the ion formed. This is the case, for example, with vinyl ethers, the polymerizability of which is very high:

$$\overset{\underset{-}{|}}{R-O-}\overset{|}{C}=\overset{|}{C} \xrightarrow{+H\oplus} \overset{\underset{-}{|}}{R-O-}\overset{\oplus}{\underset{|}{C}}-\overset{|}{\underset{|}{C}}-H \longleftrightarrow \overset{\underset{-}{|}}{R-O}=\overset{\oplus}{\underset{|}{C}}-\overset{|}{\underset{|}{C}}-H$$

Although vinyl ethers readily copolymerize with each other, they are reluctant to yield statistical copolymers with vinyl monomers of lesser nucleophilicities.

A second electron-donating center may prevent polymerization if it renders the cation excessively stable. For example, vinylamine is highly active cationically; however, the cation obtained is overstabilized by the formation of an ammonium salt:

$$\underset{NH_2}{\overset{|}{CH_2}=CH} \xrightarrow{H\oplus} \underset{NH_2}{\overset{\oplus}{\overset{|}{CH_3}-CH}} \longleftrightarrow \underset{\underset{\oplus}{NH_2}}{\overset{\|}{CH_3}-CH} \longleftrightarrow \underset{\underset{\oplus}{NH_3}}{\overset{|}{CH_2}=CH}$$

Steric Prohibition of Vinyl Cationic Polymerization

A monomer may be highly reactive toward cationic attack but nonetheless unable to produce high polymer owing to steric compression in the transition state. This is the case, for example, with 1,1-diphenylethylene:

$$\underset{C_6H_5}{\overset{C_6H_5}{\overset{|}{CH_2}=C}} + H^\oplus \longrightarrow \underset{C_6H_5}{\overset{C_6H_5}{\overset{|}{CH_3}-C^\oplus}} \longrightarrow \text{dimerization}$$

Although the proton affinity of 1,1-diphenylethylene is quite high, ~200 kcal/mole, only the dimer stage can be reached because of the sterically "buried" carbenium ion.

The reactivity of 3-methylindene is significantly higher than that of indene; nonetheless, cationic polymerization does not proceed beyond the trimer because of interference between the monomer and the penultimate unit in the growing chain (Sigwalt and Maréchal, 1966). In contrast, the monomer is readily copolymerizable with indene, which relieves the steric inhibition to 3-methylindene incorporation:

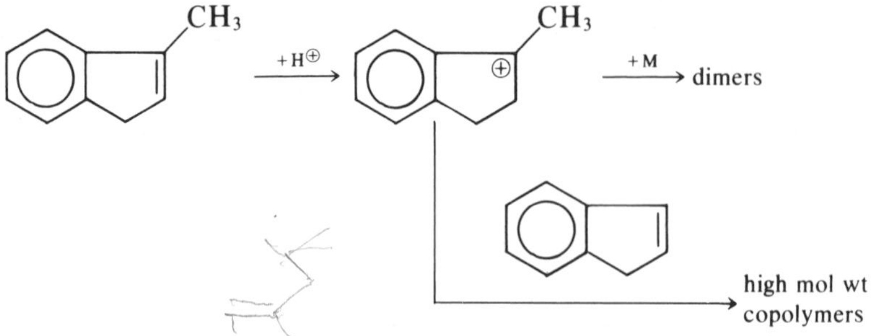

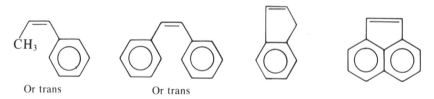

Steric compression also inhibits the polymerization of cationically highly active β-substituted styrenes.

Steric inhibition to propagation can be advantageous for a variety of purposes. For example, 2,4,4-trimethyl-1-pentene, a structural relative to isobutylene, has been used as a model compound to study mechanistic details of isobutylene polymerization. This hindered olefin mimics the polymerization of isobutylene; however, owing to steric compression, it can undergo only one propagation step (Kennedy and Gillham, 1971). This material is also useful in molecular weight control of polyisobutylene production.

1,2-Disubstituted olefins are reluctant to polymerize because of steric compression in the transition state; however, if the substituents are part of a cyclic structure polymerization may occur. Thus whereas *cis*- or *trans*-1,2-diphenylethylene (stilbene) does not polymerize, indene and acenaphthylene readily produce high polymer:

Maréchal et al. (Belliard et al., 1972a, b; Cohen et al., 1973) have shown that the reactivity of acenaphthylene is partly due to strain relief (acenaphthylene → acenaphthene) during polymerization.

Steric hindrance strongly affects the polymerization behavior of norbornane derivatives. Thus norbornene rapidly produces high polymers under mildest conditions (EtAlCl$_2$, EtCl, −78°C) whereas methylenenorbornane, which is more reactive to cationation than the former, yields only oligomers under the same conditions (Kennedy and Makowski, 1967, 1968):

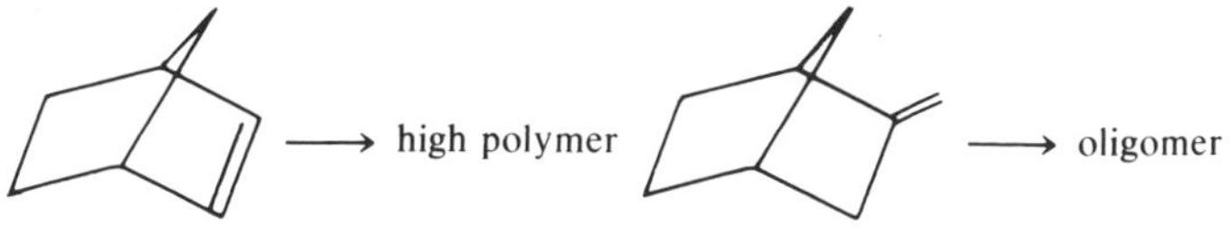

The great propensity of carbocations for isomerizations is well known. If a mechanism exists by which a sterically encumbered carbenium ion can rearrange to a sterically more favorable one, propagation may proceed and high polymer may form. For example, 3,3-dimethyl-1-butene or 2,3-dimethyl-1-butene, monomers that cannot be polymerized by any other mechanism except cationic because this route provides a favorable escape by rearrangement of the sterically compressed transition stage, give rise to high polymer upon initiation by "H$_2$O"/AlCl$_3$ (Van Lohuizen and De Vries, 1968; Kennedy et al., 1964b):

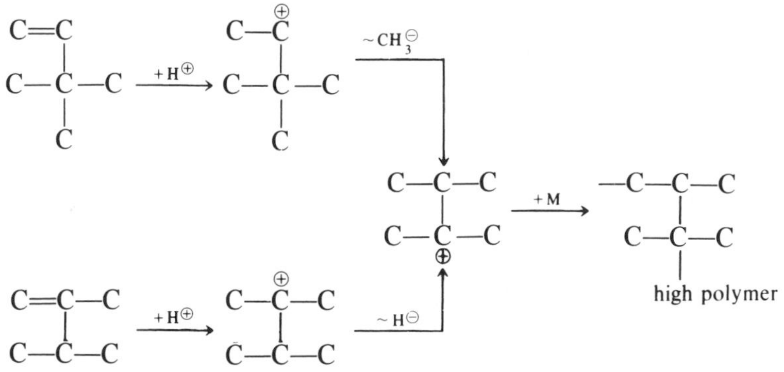

Isomerizations also allow an "escape of steric trap" in the polymerization of β-pinene or methylenenorbornene (Kennedy, 1964a; Kennedy and Makowski, 1967, 1968; Sartori et al., 1963). In the latter instance the steric effect leads from a thermodynamically relatively more stable structure to a kinetically more active transition state:

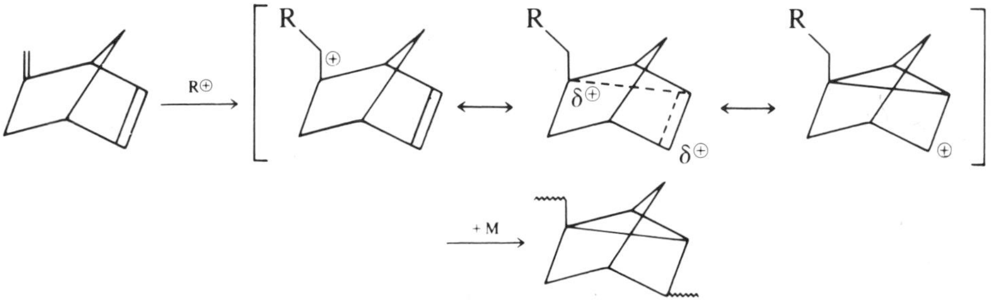

Monomers Containing More than One Nucleophilic Site Capable of Polymerization

The cationic polymerization of monomers that contain more than one site of comparable nucleophilicities capable of polymerization (for example, conjugated or nonconjugated dienes, trienes) more often than not produce ill-defined mixtures of cyclized, branched, crosslinked polymers. Only in special instances can well-defined uniform structures be obtained from conjugated dienes, for example, 2,5-dimethyl-2,4-hexadiene (Moody, 1961). Certain nonconjugated dienes, for example, 2,6-dimethyl(or diphenyl)-1,6-heptadiene may undergo inter-intramolecular polymerization and lead to fairly homogeneous structures (Field, 1960; Marvel and Gall, 1960).

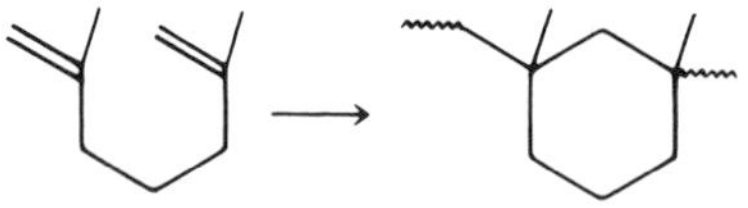

Similarly, reasonably well-defined polymerization may be obtained with dienes containing one highly and one less reactive double bond, for example, 2-methyl-1,5-heptadiene. In this molecule the C_1 double bond is far more reactive than the C_5 unsaturation and preferential propagation involving the C_1 double bond may produce chains with pendant unsaturation owing to the less reactive C_5 double bond system.

Cationic Monomers

Table 3.6 is a list of cationic monomers, the polymerization of which has been described, together with some pertinent information such as polymerization behavior and molecular weight. This table is an updated condensed version of a comprehensive compilation of cationic monomers published some time ago (Kennedy, 1975). The organization of the table follows the previous list; that is, starting with ethylene the monomers are ordered by increasing branching, first aliphatic, then aromatic compounds. For further information the reader should consult the comprehensive list (Kennedy, 1975) or original literature sources.

3.3 □   INITIATORS, COINITIATORS, AND INITIATING SYSTEMS

Carbocationic polymerizations may be initiated by chemical or physical methods. Chemical methods of initiation make use of protic or Brønsted acids, stable cation salts, Friedel–Crafts acid-based initiating systems, some miscellaneous cationogens, and charge transfer complexes. Physical methods of initiation employ external sources of energy, for example, radiation and electrical, to generate initiating cations or cation radicals in

Table 3.6 ☐ Cationic Monomers

Monomer	Experimental Conditions	Mol wt	Comments
Ethylene	AlCl$_3$, HF/BF$_3$/SO$_2$, etc.	Low (oils)	Numerous isomerizations
Propylene	BF$_3$, AlCl$_3$, AlBr$_3$: promotors (H$_2$O, HF, HBr); H$_3$PO$_4$	Low (oils)	Numerous isomerizations; broad mol wt distribution
1-Butene	Various Lewis acids (mainly HBr/AlBr$_3$ and H$_3$PO$_4$/BF$_3$)	Low (oils)	Hydride transfers during polymerization (side groups n-propyl and n-butyl)
2-Butene	H$_3$PO$_4$ and Lewis acids	Low (oils)	Rearrangements (neopentyl groups present in the polymer)
1-Pentene, 1-hexene, 1-dodecene, 1-cetene	HBr/AlBr$_3$	Low	Ill-defined structures
3-Methyl-1-butene	Strong protic acids; AlCl$_3$, -80°C AlCl$_3$, -130°C	Low (oils) High	Ill-defined structures Crystalline structure:
3-Methyl-1-pentene	AlCl$_3$/EtBr/-75 to -55°C	[η] = 0.20 − 0.34	Rearranged structure:

$$\sim CH_2CH_2-\underset{\displaystyle CH_3}{\overset{\displaystyle CH_3}{C}}-CH_2CH_2-\underset{\displaystyle CH_3}{\overset{\displaystyle CH_3}{C}}\sim$$

obtained by hydride transfer

$$\sim CH_2CH_2-\underset{\displaystyle CH_3}{\overset{\displaystyle CH_2CH_3}{C}}\sim$$

Table 3.6 □ *(Continued)*

Monomer	Experimental Conditions	Mol wt	Comments
3-Ethyl-1-pentene	AlCl$_3$/EtCl/ −78°C	≈ 10,000	Probable rearrangement: 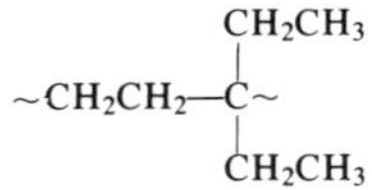 $\sim CH_2CH_2$—$\underset{\underset{CH_2CH_3}{\vert}}{\overset{\overset{CH_2CH_3}{\vert}}{C}}\sim$
3,3-Dimethyl-1-butene	BF$_3$ (gas phase) AlCl$_3$/EtCl AlCl$_3$/EtCl/ −100°C	Trimers Trimers DP ≈ 68	Partial isomerization to: 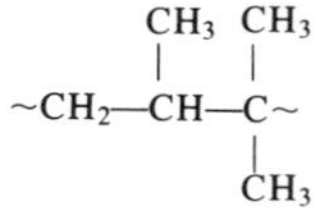 $\sim CH_2$—$\underset{\overset{\vert}{CH_3}}{\overset{\overset{CH_3\ \ CH_3}{\vert\ \ \ \vert}}{CH}}$—C$\sim$
4-Methyl-1-pentene	AlCl$_3$/EtCl/ −78°C	Maximum at −78°C ($\overline{M}_v \approx 2\cdot10^6$)	Contributing structures: 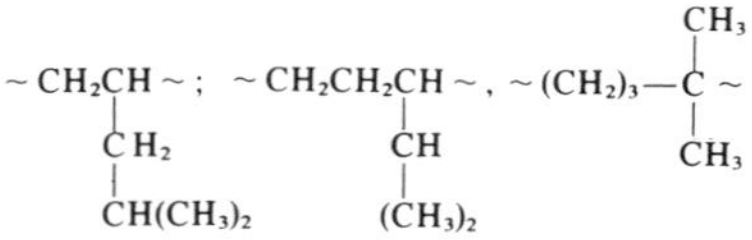 $\sim CH_2CH\sim$; $\sim CH_2CH_2CH\sim$, $\sim (CH_2)_n$—$\underset{\overset{\vert}{CH_3}}{\overset{\overset{CH_3}{\vert}}{C}}\sim$ $\underset{CH(CH_3)_2}{\overset{\vert}{CH_2}}$ $\underset{(CH_3)_2}{\overset{\vert}{CH}}$
4-Methyl-1-hexene	AlCl$_3$/EtCl/ +10 to −130°C	$\overline{M}_v = 273,000$ $\overline{M}_n > 40,000$	Amorphous rubber. Main structural unit: 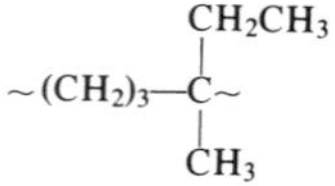 $\sim (CH_2)_3$—$\underset{\overset{\vert}{CH_3}}{\overset{\overset{CH_2CH_3}{\vert}}{C}}\sim$

38

Monomer	Conditions	Property	Structure / Notes
4,4-Dimethyl-1-pentene	AlCl₃/various temperatures in the range − 130 to 0°C AlCl₃ (solid)/in bulk 0°C	Low	Hydride and methide migrations: 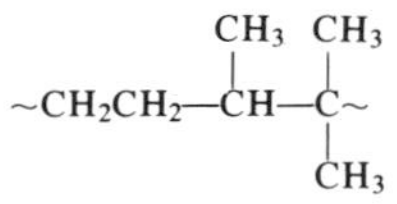 $\sim CH_2CH_2-CH-\overset{\displaystyle CH_3}{\underset{\displaystyle CH_3}{\overset{\displaystyle \mid}{\underset{\displaystyle \mid}{C}}}}-\overset{CH_3}{\underset{CH_3}{C}}\sim$
5-Methyl-1-hexene	AlCl₃/MeCl/− 73°C AlCl₃ (solid)/in bulk 0°C	$\overline{M}_n \simeq 6{,}000$	At − 73°C, 50% of structure: $\sim(CH_2)_4-\overset{CH_3}{\underset{CH_3}{C}}\sim$
5-Methyl-1-heptene	AlBr₃/EtCl/− 75°C	$[\eta] = 0.42$	Mixture of various monomer units
6-Methyl-1-heptene	AlBr₃/EtCl/− 78°C	$\overline{M}_n \simeq 22{,}000$	Partial rearrangements
Isobutylene	Strong protic acids or acidic clays	Oligomers	Composite structure, resulting from the various possible rearrangements
	AlCl₃ (see Kennedy, 1975)	10^3-10^4 $> 5 \cdot 10^4$	Dimers: a mixture of 2,4,4-trimethyl-2-pentene (4 parts) and of 2,4,4-trimethyl-1-pentene (1 part); trimers contain 5 isomers Polybutenes (viscous oils) Polyisobutylenes (see Kennedy, 1975)
2-Methyl-1-butene	AlCl₃ (solid)/bulk/− 80°C BF₃/propane/− 78°C AlCl₃ or AlBr₃/EtCl/− 175°C Al(O-i-Pr)₂(O-s-Bu) or BF₃/EtCl/− 45°C	Oligomers $\overline{M}_n = 320{,}000$ $\overline{M}_v = 300{,}000$	
2-Methyl-2-butene	TiCl₄/CH₂Cl₂/− 78°C	Oligomers	
2-Methyl-1-pentene	AlCl₃ or AlBr₃/EtCl/low temperatures	$\overline{M}_n \simeq 336{,}000$	

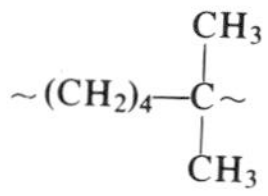

Table 3.6 □ (*Continued*)

Monomer	Experimental Conditions	Mol wt	Comments
2-Ethyl-1-butene	$AlCl_3$/EtCl-vinyl chloride/$-175°C$	Liquid oligomers	
2,3-Dimethyl-1-butene	$AlCl_3$/EtCl/-110 to $-175°C$	150,000 to 175,000	Main structure: 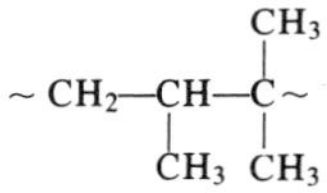 $\sim CH_2-CH-\underset{\underset{CH_3}{\overset{\displaystyle CH_3}{\vert}}}{C}\sim$ $\qquad\qquad CH_3$
2,4,4-Trimethyl-1-pentene	$AlCl_3$/mixtures of EtCl and vinyl chloride/$-175°C$	Oils	Dimers only; polymerization impossible owing to steric compression
2,5-Dimethyl-1,5-hexadiene	BF_3; $AlBr_3$ or $TiCl_4$/ bulk and in various solvents/20 to $-78°C$	$\overline{M}_n = 450–2{,}050$	Major monomer unit: 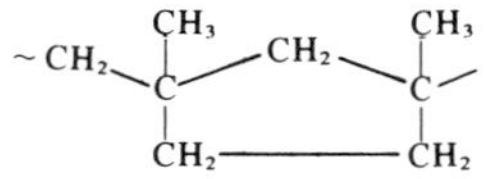 $\sim CH_2-\underset{\underset{CH_2}{\vert}}{\overset{\overset{CH_3}{\vert}}{C}}\quad CH_2-\underset{\underset{CH_2}{\vert}}{\overset{\overset{CH_3}{\vert}}{C}}\sim$
Methylenecyclohexane	BF_3/CH_2Cl_2/$-78°C$	Solid	Ill-defined oligomers
Vinylcyclopropane	Various Lewis acids/-50 to $-78°C$	2000–25,000	Possible units: 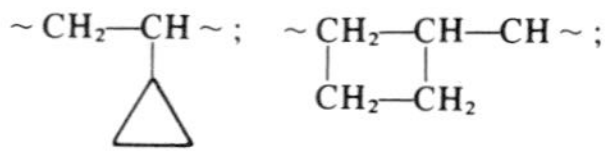 $\sim CH_2-CH\sim$; $\sim CH_2-CH-CH-CH\sim$; $\qquad\qquad\qquad\qquad\qquad CH_2-CH_2$ $\triangle$ $\sim CH_2-CH=CH-CH_2CH_2\sim$
Isopropenylcyclopropane	$SnCl_4$/atmospheric and high pressure $\simeq$ 14,000 kg/cm²	DP $\simeq 3$–4	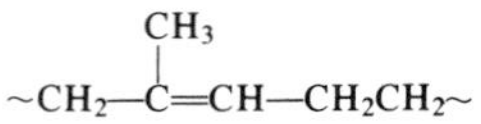 $\qquad\quad CH_3$ $\qquad\qquad \vert$ $\sim CH_2-C=CH-CH_2CH_2\sim$
1,1-Dicyclo-propylethylene	$AlBr_3$/EtCl/$-78°C$ $AlEtCl_2$/MeCl/-35 to 100°C	Solid	Rearranged units (25%): 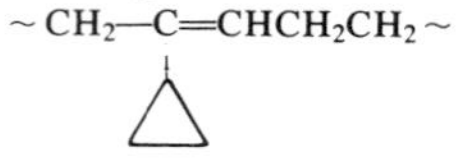 $\sim CH_2-C=CHCH_2CH_2\sim$ $\triangle$

Monomer	Catalyst	$\overline{M}$	Products
Methylenecyclobutane	AlCl$_3$; TiCl$_4$; SnCl$_4$; BF$_3$	$\overline{M}_w \simeq 1200$ to 1500	Depending on the nature of the Lewis acid, the following monomer units are obtained: 
Cyclopentene and cyclohexene	BF$_3$ + HF	Dimers, trimers, tetramers, resins	No polymerization
3-Methylcyclopentene	AlCl$_3$/EtCl/− 20°C AlCl$_3$/EtCl/− 20°C		Slowly gives a polymer: 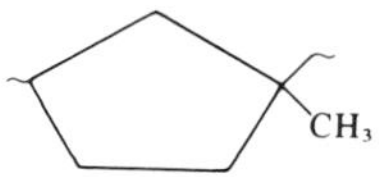
1-Methylcyclopentene	BF$_3$ or BF$_3$·camphor in hexane or CHCl$_3$/0°C	Viscous oil	
Methylenecyclopentane	AlCl$_3$/EtCl/− 135°C	Trimer	
Vinylcyclopentane	AlBr$_3$/EtCl/− 78 to − 100°C	$\overline{M}_n \simeq 950$	No rearrangement
3-Vinylcyclopentene	Alkylaluminums	1900–3600	Partially soluble polymers
Allylcyclopentane	AlBr$_3$/EtCl/− 78°C	$\simeq 1300$	Partial rearrangement to: 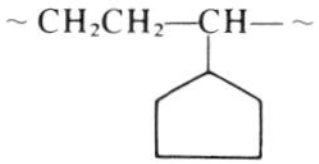
3-Allylcyclopentene	EtAlCl$_2$; EtAlCl$_2$/PhCH$_2$Cl	1370–1420	Bicyclic polymer structures
1-Methylcyclohexene	AlCl$_3$/C$_6$H$_6$/40 to 45°C	Oligomers (mainly dimers)	

Table 3.6 □ (Continued)

Monomer	Experimental Conditions	Mol wt	Comments
3-Methylcyclohexene	AlCl₃/EtCl/−20°C AlCl₃/EtCl/−78°C	Dimers + trimers Polymer	Semicrystalline solid: 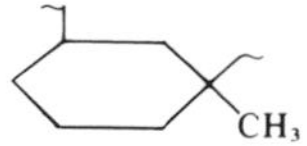
Methylenecyclo-hexane	AlCl₃ or BF₃/EtCl−vinyl chloride/−150 to −180°C	$[\eta] = 0.1-0.4$	
2 or 3 Methyl Methylene-cyclohexanes	AlCl₃/EtCl−vinyl chloride/−175°C	$[\eta] = 0.05-0.17$	Poly(4-Methylmethylenecyclohexane); crystallizes readily
Vinylcyclohexane	AlCl₃/EtCl/ +7 to −100°C AlBr₃/EtCl/−50°C	2000–6500 10,500	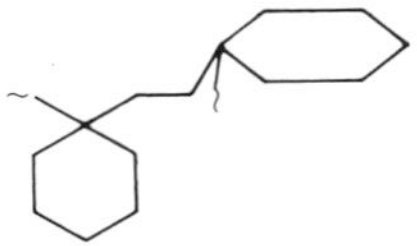
4-Vinylcyclohexene	Various Lewis acids/chlorinated solvents	$[\eta] = 0.03-0.11$	Polymer units: 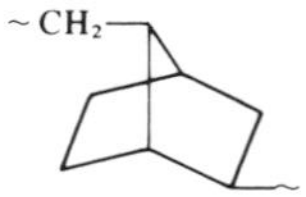 ~CH₂~
Allylcyclohexane	AlBr₃/EtCl/−78°C	$\overline{M}_n \simeq 13,000$	Mainly 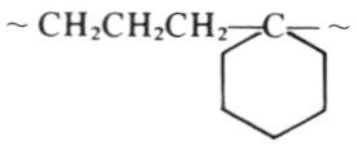 ~CH₂CH₂CH₂~C~
d-Limonene 1,4-Dimethylene-cyclohexane	Lewis acids BF₃/CH₂Cl₂/−80°C to room temperature	$[\eta] = 0.041$ at −78°C 0.045 at −40°C	Rearrangements; structure not established Complex repeat units; for latest information see Sebenik and Harwood (1981).

Monomer	Conditions		Structure / Remarks
1-Methylene-4-vinylcyclohexane	$BF_3/CH_2Cl_2/-70°C$	$[\eta] = 0.08$	Copolymer of 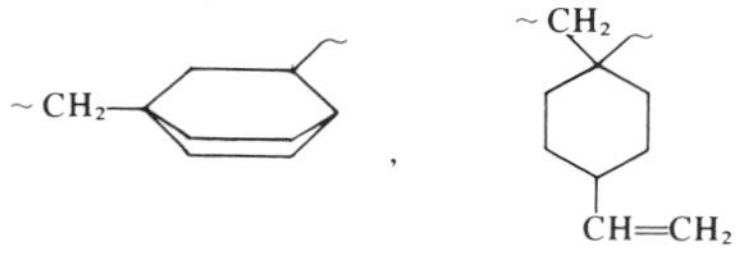
cis-1,2-Divinyl-cyclohexane	$AlCl_3$ or $Et_2AlCl-t-BuCl$ or $Et_2AlCl-PhCH_2Cl$ or $Et_2AlCl-CH_3OCH_2Cl$	1100–3000	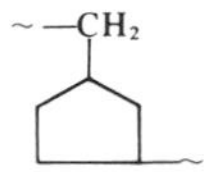
2-Allyl-1-methylene-cyclohexane	$BF_3/CH_2Cl_2/-70°C/46$ h	—	Soluble (95%) 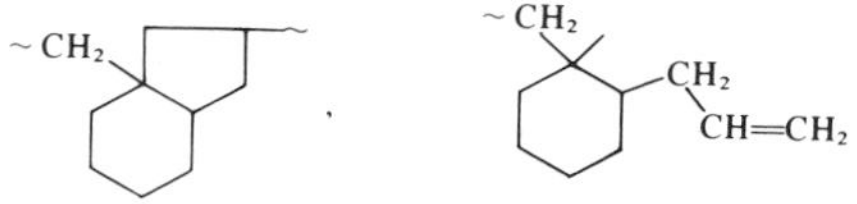
(CH$_2$)$_n$ $\quad$ C=CH$_2$ 	Various Lewis acids	DP $\simeq 6$	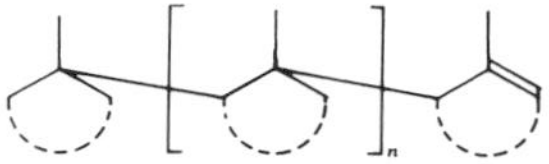
1,3-Butadiene	Various Lewis acids		Sometimes linear (with Al sesquichloride in C_6H_6); more often crosslinked and cyclized
	$AlEt_2Cl/EtCl/-80°C$	—	Amorphous powder; degree of unsaturation 67%
	$C_6H_5MgBr-TiCl_4(2:1)$	—	"Cyclopolybutadienes" (ladder structure) and trans-1,4 units
	$AlEt_2Cl-TiCl_4/n$-heptane	> 3,000,000	Insoluble (cycles and ladder structures)
	$AlEt_2Cl-TiCl_4/benzene$		Soluble
	$AlEt_3-AlCl_3$–cobalt octoate		98% 1,4 structure
	$AlEt_2Cl + AlEtCl_2$–cobalt octoate		98% 1,4 structure
Isoprene	Various Lewis acids		50–80% of theoretical unsaturation: 90% trans-1,4; the balance, 1,2 and 3,4
	$CCl_3COOH/TiCl_4/CH_2Cl_2/20°C$ $EtAlCl_2-TiCl_4$ or $AlBr_3/CH_2Cl_2/61$ and 20°C	$\simeq 10,000–3,000,000$	Microgel

Table 3.6 □ (Continued)

Monomer	Experimental Conditions	Mol wt	Comments
2,3-Dimethyl-1,3-butadiene	AlEtCl₂	8000	White powdery polymer. Structure largely unknown
1,3-Pentadiene	BF₃/MeCl/−78°C	—	Elastomeric materials by copolymerization with butadiene
1,4-Dimethyl-1,3-butadienes (cis, cis; trans, trans; cis, trans)	BF₃·OEt₂	Oligomers	No isomerization
	SnCl₄, TiCl₄, SbCl₅, WCl₆		Partial isomerization
trans-2-Methyl-1,3-pentadiene	Lewis acids/benzene, heptane, or CH₂Cl₂/−80 to 25°C	67,000	Mainly trans-1,4 enchainment
cis- and trans-3-Methyl-1,3-pentadiene	SnCl₄/CHCl₃/room temperature	Oligomers	
4-Methyl-1,3-pentadiene	AlEt₂Cl-t-BuCl/n-heptane/−80°C	Oligomers	Mainly 1,2 (81% 1,2 and 19% trans-1,4 enchainment
2,4-Dimethyl-1,3-pentadiene	H₂SO₄/0°C	Oligomers	
2,5-Dimethyl-2,4-hexadiene	BF₃/petroleum ether/−78 to −50°C	High	Crystalline powder, probable structure:
2-Methyl-6-methylene-2,7-octadiene (Myrcene)	BF₃·OEt₂, or TiCl₄/n-heptane/−78 to 0°C	DP 4-6	Possible unit.

Monomer	Conditions	Property	Remarks
1-Phenyl-1,3-butadiene	TiCl₄/EtCl/−70°C	—	Predominantly 1,4 enchainment
2-Phenyl-1,3-butadiene	SnCl₄/CHCl₃/−70°C	—	Predominantly 1,4 enchainment
1-Phenyl-4-methyl-1,3-butadiene	SnCl₄/petroleum ether/−24°C; Lewis acids		Trans-1,4 enchainment
1,2,3,4-Tetraphenyl-1,3-butadiene	SbCl₅	Trimer	
2-Chloro-1,3-butadiene			Studied only as a comonomer. Crosslinking
trans-1,3,5-Hexatriene	BF₃ or AlCl₃; BF₃·OEt₂ or SnCl₄	η_{inh} (0.25 g in 100 ml benzene) = 0.29–0.81	Little if any cyclic product
2,6-Dimethyl-2,4,6-octatriene	BF₃/CH₂Cl₂/−70 to −101°C; TiCl₄/n-heptane/−70 to −15°C	3000–3500	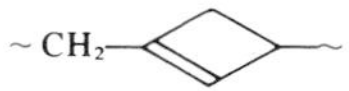 ~CH₂— (structure)
Methylenecyclobutene	BF₃; BF₃·OEt₂; AlEt₂Cl		CH₃ (structure)
1-Methyl-3-methylene-cyclobutene	BF₃ (in bulk); BF₃·OEt₂/n-hexane; Et₂AlCl–HBr/toluene/−78°C	3800–9200	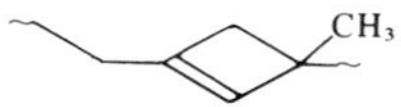
Cyclopentadiene	TiCl₃On-Bu;	$[\eta] = 1.4$	Polymer rapidly oxidizes and crosslinks in air. Mainly 1,2 and 1,4 enchainments
1 or 2-Methycyclopentadienes	CCl₃COOH or H₂O/TiCl₄ or SnCl₄; BF₃·OEt₂; SnCl₄ and TiCl₄	$[\eta] = 2.76$; $[\eta] \approx 0.8$	Polymerizations of isomer mixtures. Structure mainly 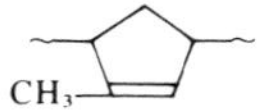 (structure, CH₃)
1,3-Dimethyl-cyclopentadiene	BF₃·OEt₂; SnCl₄; TiCl₄/toluene or CH₂Cl₂/−70°C and 0°C	$[\eta] = 0.1–1.0$	Contributing structures: 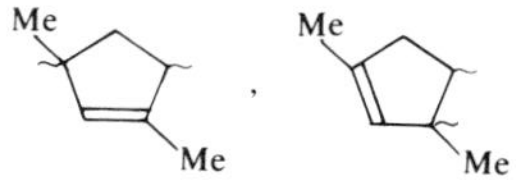 (Me structures)

45

Table 3.6 □ (Continued)

Monomer	Experimental Conditions	Mol wt	Comments
2,3-Dimethyl-cyclopentadiene	BF$_3$·OEt$_2$/toluene/−78°C; SnCl$_4$/C$_2$H$_4$Cl$_2$/−78°C	[η] = 0.76 [η] = 0.14	Main structure (≃ 100%): 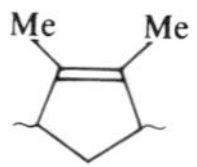
1,2-Dimethyl-cyclopentadiene	BF$_3$·OEt$_2$/toluene/−78°C SnCl$_4$/C$_2$H$_4$Cl$_2$/−78°C	—	Main structure (≃ 100%) 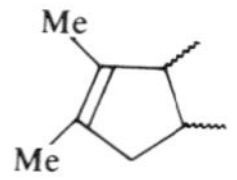
Allylcyclopentadiene	BF$_3$·OEt$_2$; SnCl$_4$; TiCl$_4$; AlBr$_3$/toluene or CH$_2$Cl$_2$/−78 to 0°C	[η] = 0.1–0.3	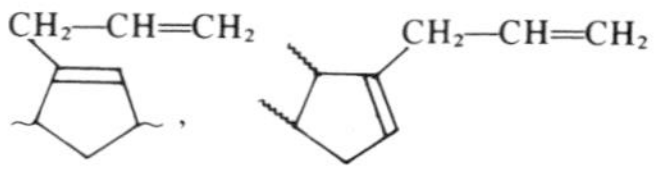
Methallylcyclopentadiene 6,6-Dimethylfulvene	BF$_3$·OEt$_2$/CHCl$_3$/room temperature SnCl$_4$·5H$_2$O; AlCl$_3$; SbCl$_5$, FeCl$_3$/CHCl$_3$/room temperature	[η] = 0.16 1000–5000	65% 1,4 and 35% 1,2 Soluble powder crosslinks in air
6-Methyl-6-ethyl-fulvene	Same as for the preceding monomer	5500	
1,3-Cyclohexadiene	BF$_3$, PF$_5$, or TiCl$_4$; CCl$_3$COOH/SnCl$_4$ or BF$_3$·OEt$_2$/CH$_2$Cl$_2$/0°C	[η] = 0.04–0.12	1,2 and 1,4 enchainments. Branching

Monomer	Conditions	Property	Structure
1-Methylenecyclohexane	BF₃/bulk/−20°C; BF₃·OEt₂, TiCl₄, Et₂AlCl, AlCl₃, or VCl₄/n-hexane or toluene/−78°C	$[\eta] = 0.35–0.94$	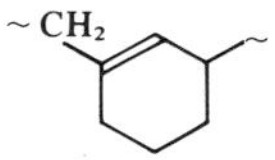 ~CH₂⌒ plus minor 1,2 enchainment
1-Vinylcyclohexene	BF₃·OEt₂ or CCl₃COOH/SnCl₄/toluene or CH₂Cl₂/0°C	$\simeq 1200$	(with CCl₃COOH/SnCl₄ some 1,2 structure present) ~CH₂—CH
1,2-Dimethylenecyclohexane	BF₃·OEt₂/in bulk/dry ice temperature	$\geqslant 100{,}000$	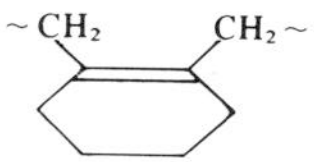 ~CH₂⌒ / CH₂~
cis,cis-1,3-Cyclooctadiene	CCl₃COOH/TiCl₄/CH₂Cl₂/−78°C	1500–10,000	
p-Xylylene	BF₃; SbCl₅; AlCl₃; TiCl₄; H₂SO₄; CH₃COOH/hydrocarbon/−78°C	—	(branched structure in toluene) ~CH₂— CH₂CH₂— CH₂— ~CH₂~
1-Methyl-4-isopropyl-1,3-cyclohexadiene and 1-methyl-4-isopropyl-1,3-cyclohexadiene	CCl₃COOH	DP 9–10	
α-Pinene; β-Pinene	Various Lewis acids/toluene 40–45°C; Various Lewis acids; BF₃/EtCl/CHClF₂/−75°C	$\leqslant 1000$; $[\eta] = 0.13$	α-pinene → limonene which polymerizes Repeat unit 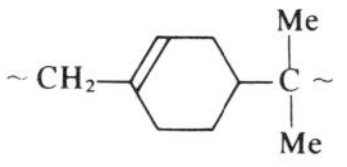 ~CH₂— (ring with C(Me)Me)

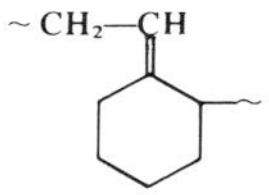

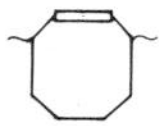

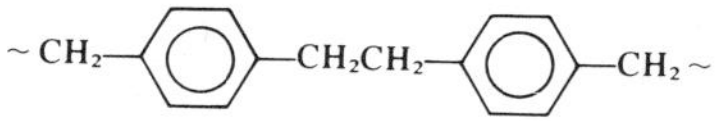

Table 3.6 □ (Continued)

Monomer	Experimental Conditions	Mol wt	Comments
Norbornene 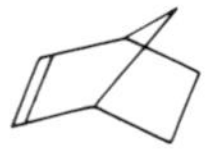	AlEt$_2$Cl/EtCl/−78 to −100°C	$M_n \simeq 1740$ and 1940	Repeat unit 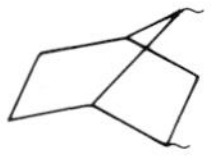
5-Methylnorbornene 2-Methylenenorbornane	Lewis acids AlEtCl$_2$/EtCl/−30 to −100°C	Oligomers Oligomers	Very low yield in solid polymer Very reactive but does not propagate because there is a severe steric compression in the transition state of propagation
Camphene	Lewis acids	Oligomers	
2-vinylnorbornane		$M_n = 885$	Low yield 1,2 enchainment
Isopropenyl-2-norborane	EtAlCl$_2$	Oligomers	No polymerization
Methylenenorbornene 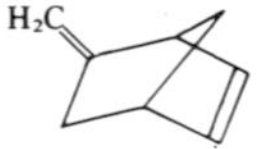	EtAlCl$_2$ or AlBr$_3$/n-heptane/−78°C; VCl$_4$/n-heptane/−20°C	$[\eta] \simeq 0.3$	 by IR and NMR
Norbornadiene 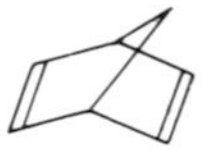	AlCl$_3$/EtCl/−123 to +40°C	$\overline{M}_n$ of soluble fraction obtained at −123°C = 5520	At −123°C completely soluble, main repeat unit 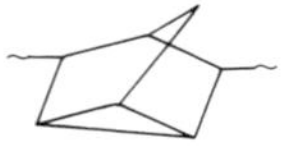

48

Monomer	Conditions	$\overline{M}_n$	Notes
2-Vinylnorbornene 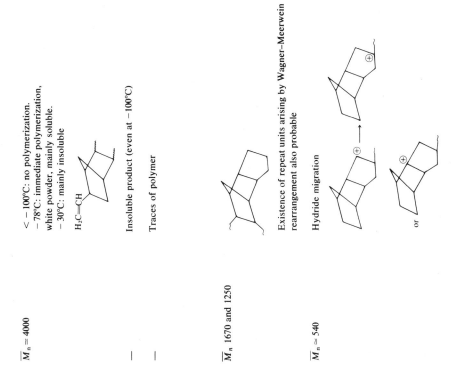 H₂C=CH	EtAlCl₂/EtCl/ −135 to −30°C	$\overline{M}_n \simeq 4000$	< −100°C: no polymerization. −78°C: immediate polymerization, white powder, mainly soluble. −30°C: mainly insoluble H₂C=CH
2-Isopropenyl-norbornene	EtAlCl₂/EtCl/ −100 to −30°C	—	Insoluble product (even at −100°C)
5-Methylbicyclooct-2-ene 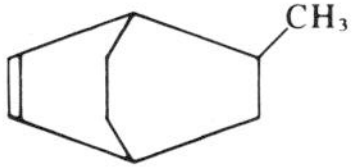 CH₃	BF₃·OEt₂/ −78 to 0°C	—	Traces of polymer
1,2 Dihydroendodicyclopentadiene	AlEtCl₂-t-BuCl (1 : 1)/CH₂Cl₂/ −78 to −20°C	$\overline{M}_n$ 1670 and 1250	
9,10-Dihydroendodicyclopentadiene	EtAlCl₂-t-BuCl/CH₂Cl₂/ −78°C	$\overline{M}_n \simeq 540$	Existence of repeat units arising by **Wagner–Meerwein** rearrangement also probable. Hydride migration

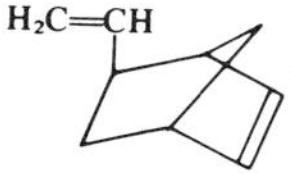

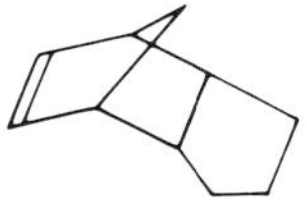

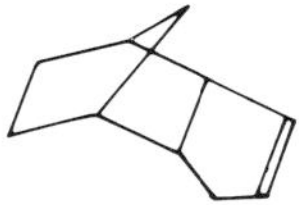

or

49

Table 3.6 □ *(Continued)*

Monomer	Experimental Conditions	Mol wt	Comments
Dicyclopentadiene	BF₃·OEt₂/room temperature	$\overline{M}_n \simeq 1450$ (exo) $\overline{M}_n \simeq 820$ (endo)	
(endo or exo)			
			by Wagner–Meerwein rearrangement
Styrene	Varied	Various	See Kennedy (1975)
o-, m-, or p-Methylstyrenes	Various Lewis acids/I₂/EtCl/Protic acids	M_n in the range 1000–3000	Copolymerizes only; similar to p-methylstyrene
p-Ethylstyrene	H₂O/SnCl₄/CH₂Cl₂/0°C	—	More reactive than p-methylstyrene
p-Dodecylstyrene	Various Lewis acids	$\overline{M}_n \simeq 4700$	Colorless amorphous solid; no isomerization
o- or p-Isopropylstyrene	H₂O/SnCl₄/CH₂Cl₂/0°C	—	Rate of polymerization about 5 times that of styrene
p-tert-Butylstyrene	Various Lewis acid/CH₂Cl₂ or EtCl/ −78 to 0°C	$\overline{M}_n \sim 7600$ 48,000–74,000	
2,4-Dimethylstryene			
2,3,4-Trimethylstyrene	Various Lewis acids	$[\eta] = 0.06$	
2,4,6-Trimethylstyrene	Various Lewis acids	$[\eta] = 2.55$ $(M_n > 10^6)$	
2,4,5-Trimethylstyrene	Various Lewis acids	$[\eta] = 0.60$	
3,4,5-Trimethylstyrene	Various Lewis acids	$[\eta] = 0.45$	
2-Methyl-5-iso-propylstyrene	AlCl₃/n-hexane/20°C	1000–5000	
2-Isopropyl-5-methyl-styrene	AlCl₃/n-hexane/20°C	1000–5000	
2,6-Dimethyl-4-butyl-	BF₃/CCl₄/ −15°C/or		
2,6-Dimethyl-4-butylstyrene	AlCl₃/nitrobenzene/ −15°C	29,000 and 13,500	

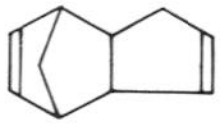

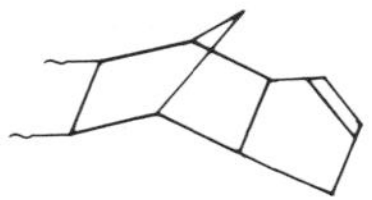

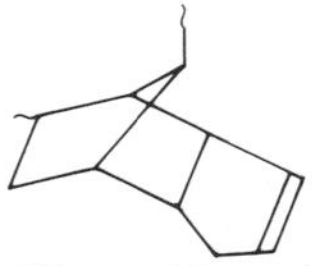

Monomer	Conditions	Molecular weight / $[\eta]$	Remarks
o-, m- or p-Fluorostyrenes	Various Lewis acids, chlorinated solvents		Mainly copolymerization
o-, m- or p-Chlorostyrenes	Various Lewis acids, chlorinated solvents	10,000–20,000	
3,4- or 2,5-Dichlorostyrenes	Various Lewis acids	1400–14,700 for p-chlorostyrene	Copolymerization
p-Bromostyrene	$H_2O/SnCl_4/CH_2Cl_2/0°C$	15,000	
p-Cyanostyrene	$H_2O/SnCl_4/CH_2Cl_2/0°C$	19,100	
o-, m- or p-Hydroxystyrenes	$BF_3 \cdot OEt_2/CH_2Cl_2$/various temperatures	$[\eta] = 0.04\text{–}0.5$	
p-Methoxystyrene	Various Lewis acids: $BF_3 \cdot OEt_2/CH_2Cl_2/-78°C$	$[\eta] \approx 3$	
	Triphenylmethylhexachloroantimonate/$CH_2Cl_2/-70°C$	$[\eta] = 2.3$, $\overline{DP} = 2720$	
o-Methoxystyrene	$BF_3 \cdot OEt_2$/chlorinated solvents	$\overline{DP} = 3300$	
m-Methoxystyrene	$SnCl_4/CCl_4/0°C$	—	
4-Methoxy-2,5-dimethylstyrene	Various Lewis acids	$[\eta] = 2.9$	Copolymerizations only
4-Methoxy-2,6-dimethylstyrene		$[\eta] = 2$	
4-Methoxy-3,5-dimethylstyrene		$[\eta] = 4$	
4-Methoxy-2,3,5,6-tetramethylstyrene	Various Lewis acids	$[\eta] = 0.1$	With $TiCl_4$: no polymerization; with $SnCl_4$: insoluble product
3,4-Dimethoxystyrene		$[\eta] = 0.1\text{–}1.1$	With BF_3 partial demethylation; insoluble product Low molecular weights with all isomers
2,3,4-Trimethoxystyrene	Various Lewis acids	$[\eta] = 0.25$	
4-Methoxy-2(5)-methyl-5(2)-isopropylstyrene	$AlCl_3/n$-hexane/20°C	1000–5000	
p-Benzyloxystyrene	$H_2O/SnCl_4/CH_2Cl_2/0°C$	$\overline{M}$ = twice $\overline{M}$ of poly-(2-methoxystyrene)	
m-Nitrostyrene	Various Lewis acids	<1000	No polymerization
p-Nitrostyrene	Various Lewis acids		Very reactive but there is a complex formation between $SnCl_4$ and N of amino group
p-Dimethylaminostyrene	$H_2O/SnCl_4/CH_2Cl_2/0°C$	$\overline{M}_n = 4000$	

Table 3.6 ☐ (Continued)

Monomer	Experimental Conditions	Mol wt	Comments
α-Methylstyrene	Various	Various	See Kennedy (1975)
α-Methyl-p-methyl-styrene			Copolymerizations
α-Methyl-p-methoxy-styrene	H₂SO₄ (43%)	Dimer	
α-Ethylstyrene		Dimers and trimers η(at 0.2 g/l) 0.3–0.8	Dimers and trimers due to steric hindrance Soluble
p-Diisopropenylbenzene	BuLi-TiCl₄-HCl		
Diphenylethylene	Various Lewis and protic acids	Dimers 1000–3000	Polymerization without rearrangement
Propenylbenzenes	Various Lewis acids		Copolymerizations only
cis- and trans-p-Methyl-β-Methylstyrene	BF₃·OEt₂/CH₂Cl₂/30°C		
cis- and trans-β-Ethylstyrene	SnCl₄	—	Copolymerizations only
cis- and trans-β-Propyl-styrene	SnCl₄	Dimers	Copolymerizations only
cis- and trans-α,β-Dimethylstyrene	SnCl₄/0°C	—	cis → dimers; trans → no polymerization
cis- and trans-4-Methoxy-propenylbenzene	Various Lewis acids BF₃·Et₂O	$[\eta] = 1.4$	
cis- and trans-4-Ethoxy-propenylbenzene	Various Lewis acids BF₃·Et₂O	$[\eta] = 0.15$	
4-Isopropoxypropenyl-benzene	Various Lewis acids BF₃·Et₂O	$[\eta] = 0.15$	

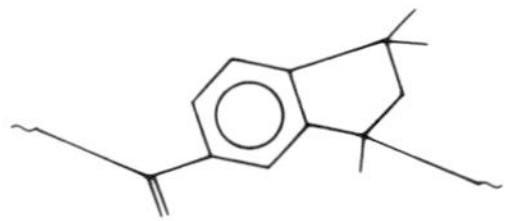

Monomer	Catalyst	Molecular weight / $[\eta]$	Remarks
4-tert-Butoxypropenyl-benzene	Various Lewis acids	—	Polymerizes with BF_3.
	$BF_3 \cdot Et_2O$	$[\eta] = 0.1$	Polymer is insoluble
2-Methyl-4-methoxy-propenylbenzene	Various Lewis acids	—	
	$BF_3 \cdot Et_2O$		
Indene	Lewis and protic acids	$[\eta] \approx 0.2$	
Monomethylindenes			
1	Various Lewis and protic acids	$\overline{M}_v \approx 30,000$	
2	Various Lewis and protic acids	$[\eta] = 0.04$	
3	Various Lewis and protic acids	Dimers	
5	Various Lewis and protic acids	$[\eta] \approx 0.5\text{–}0.8$	
6	Various Lewis and protic acids	$1\text{–}4.5$	
7	Various Lewis and protic acids	$0.5\text{–}0.8$	
Dimethylindenes			
1,1	Various Lewis and protic acids	$\overline{M}_v \approx 800$	
2,3	Various Lewis and protic acids	$\overline{M}_v \approx 800$	
4,6	Various Lewis and protic acids	$[\eta] = 1$	
4,7	Various Lewis and protic acids	$[\eta] = 3.5$	
5,6	Various Lewis and protic acids	$[\eta] = 3.3$	
5,7	Various Lewis and protic acids	$[\eta] \approx 0.07\text{–}0.2$	
6,7	Various Lewis and protic acids	$[\eta] \approx 0.4\text{–}0.5$	
4,6,7-Trimethylindene	Various Lewis and protic acids	$[\eta] = 2$	Insoluble polymers
4,5,6,7,-Tetra methylindene	Various Lewis and protic acids	$[\eta] = 0.2\text{–}0.4$	Insoluble and nonmelting polymers
5-Vinylindene	Various Lewis and protic acids	—	According to the conditions: soluble or insoluble
1-Phenylindene	Various Lewis and protic acids		
Bis(1-indenyl)	Various Lewis and protic acids	$[\eta] = 1.2$	Very low yields
Bis(2-indenyl)	Various Lewis and protic acids	Very low	Fluorescent polymer containing probably stilbene units
Bis(3-indenyl)	Various Lewis and protic acids	$[\eta] = 0.08$	
1,2-Di(1-indenyl)-ethane	Various Lewis and protic acids	$\overline{M}_n \approx 4600$	
1,2-Di(3-indenyl)-ethane	Various Lewis and protic acids	$\overline{M}_n \approx 1800$	Polymer crosslinks on heating

Table 3.6 □ (Continued)

Monomer	Experimental Conditions	Mol wt	Comments
1,4-Di(3-indenyl)-ethane	Various Lewis and protic acids	$\overline{M}_n \sim 5000-14,500$	Polymer crosslinks on heating
1,4-Di(3-indenyl)-butane	Various Lewis and protic acids	$\overline{M}_n = 1700$	One unsaturation per repeat unit
Benzofulvene	TiCl$_4$/CH$_2$Cl$_2$/ −72°C	$[\eta] = 0.2$	One unsaturation per repeat unit
Methyl and ethyl-benzofulvenes	TiCl$_4$/CH$_2$Cl$_2$/ −72°C	$[\eta] = 0.15$	
Benzylidenindene	SbCl$_5$/CHCl$_3$	DP = 6	
Cinnamylindene	SbCl$_5$ and SnCl$_4$/CHCl$_3$	DP = 4	Two fractions (DP = 12, DP = 4)
Cinnamylidenefluorene	SbCl$_5$/CHCl$_3$/room temperature	DP = 8	
1-Isopropylidene-3a,4,7,7a-tetrahydroindene	Various Lewis acids and AlEt$_2$Cl–BuCl(1:3)	$\overline{M}_n = 15,700$	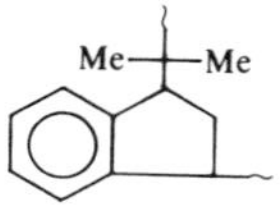
Fluoroindenes	Various Lewis acids		
5		$[\eta] = 0.4$	
6		$[\eta] = 0.6$	
7		$[\eta] = 0.25$	
Chloroindenes			
5		$[\eta] = 0.2$	
6	Various Lewis acids	$[\eta] = 0.2$	
7		$[\eta] = 0.5$	
4,7		$[\eta] = 0.2$	
Bromoindenes			
6	Various Lewis acids	$[\eta] = 0.4$	Crosslinks on heating
7	Various Lewis acids	$[\eta] = 0.1$	Insoluble polymers except with TiCl$_4$/EtCl
3- and 4-Bromopropylindenes	Various Lewis acids	$\overline{M}_n = 5860$ and 5400	
Methoxyindenes	Various Lewis acids		
4		$[\eta] = 1$	Polymerizes above −60°C
5		$[\eta] = 0.1$	Below −60°C a stable "initiator–monomer" complex is formed
6		$[\eta] = 0.35-0.45$	

Monomer	Catalyst / conditions	Properties	Remarks
Allylbenzene	AlCl₃/MeCl/ and AlBr₃/EtCl	$\overline{M}_n = 4700$	Rearrangement to —CH₂—CH—~ with CH₃ (benzene ring) Polyalkylated products, ill-defined structures
p-Allyltoluene	AlBr₃/CS₂		
3-Phenyl-1-butene	AlCl₃/EtCl/ − 78°C	$\overline{M}_n = 570$	Complex structures
o- or p-Methoxyallylbenzenes	H₂SO₄ (43%)	Dimers	Various dimers
1,2-Dihydronaphthalenes	CCl₃COOH/SnCl₄/(CH₂Cl₂)₂/30°C		Low yield. Partially soluble. Mainly studied as a comonomer
Acenaphthylenes	Various Lewis acids BF₃·OEt₂	125,000–200,000	
Methylacenaphthylenes	Various Lewis acids	Trimers	
1		$\overline{M}_n \simeq 56,000–90,000$	
3		$[\eta] = 0.55$	
5		$[\eta] = 0.1$	
		$[\eta] = 3$	
1-Vinylnaphthalene	Various Lewis acids		
4-Methoxy-1-vinyl-naphthalene	Various Lewis acids		
4-Fluoro-1-vinyl-naphthalene	Various Lewis acids	$[\eta] = 0.6$	
	CH₂Cl₂/dry ice-acetone temperature		
4-Chloro-1-vinyl naphthalene	Various Lewis acids	$[\eta] = 0.2$	
	CH₂Cl₂/Dry Ice-acetone temperature		
4-Bromo-1-vinyl-naphthalene	Various Lewis acids		
	CH₂Cl₂/Dry Ice-acetone temperature		
2-Vinylnaphthalene	Various Lewis acids	$[\eta] = 0.1$	
	CH₂Cl₂/Dry Ice-acetone temperature		
1-Vinylanthracene	Various Lewis acids	$[\eta] = 0.04$	
	CH₂Cl₂/Dry Ice-acetone temperature	$\bar{x}_n = 8$	
2-Vinylanthracene	Various Lewis acids	$[\eta] = 0.05$	
	CH₂Cl₂/Dry Ice-acetone temperature	$\bar{x}_n = 11$	
9-Vinylanthracene	Various Lewis acids	$[\eta] = 0.12$	—CH₂CH= oxidizable polymer ~CH₂—CH= (anthracene structure)
	CH₂Cl₂/Dry Ice-acetone temperature	$\bar{x}_n = 34$	
4-Vinylbiphenyl	Various Lewis acids	$M_n = 63,000$	
	CH₂Cl₂/Dry Ice-acetone temperature		
2-Vinylfluorene	Various Lewis acids	$\overline{M}_n = 290,000$	
	CH₂Cl₂/Dry Ice-acetone temperature		

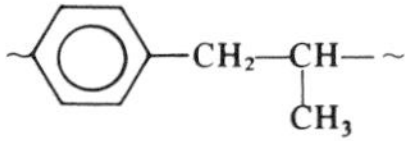

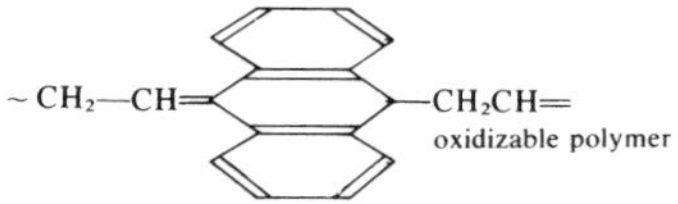

the charge. This section concerns chemicals used for initiation of carbo-cationic polymerizations. Examination of the various energy sources such as cobalt-60 sources, UV lamps, and electrode systems used in physical methods fall outside the scope of this examination. The chemistry of initiation is discussed in Chapter 4.

Protic or Brønsted Acids

Protic acids are potential sources of proton and are by definition initiators of carbocationic polymerizations.

The demarcation line between protic acids and Lewis acids, of which Friedel–Crafts acids are a subclass is sometimes blurred (Gillespie, 1963). Thus BF_3 may be regarded to increase the acidity of liquid HF either because of the formation and ionization of a hypothetical protic acid HBF_4,

$$BF_3 + HF \rightleftharpoons HBF_4 + HF \rightleftharpoons H_2F^{\oplus}BF_4^{\ominus}$$

or because the BF_3 complexes with $F^{\ominus}$ and leaves behind $H^{\oplus}$:

$$BF_3 + 2HF \rightleftharpoons H_2F^{\oplus} + BF_4^{\ominus}$$

An important distinction between protic and Lewis acids is the behavior of these acids with the same base B. Although all protic acids lead to the same species $BH^{\oplus}$, Lewis acids plus this base give rise to different complexes exhibiting different physicochemical properties. Olah views protic acids as Friedel–Crafts catalysts of the Brønsted–Lowry type, for example, H_2SO_4 and HF (Olah, 1963, pp. 315, 624).

For this presentation, it is useful to differentiate between protic and Lewis acids. Even strong protic acids that rapidly protonate olefins usually produce only low molecular weight products. In practice, very few protic acids produce high molecular weight polymer and then only of highly reactive monomers, for example, N-vinylcarbazole and α-methylstyrene. In contrast, Friedel–Crafts acid-based initiating systems often lead to rapid polymerization and high molecular weight material.

With regard to initiating ability, the most important property of protic acids is acid strength, characterized by the equilibrium

$$HA \rightleftharpoons H^{\oplus} + A^{\ominus}$$

or preferably by

$$HA + S \rightleftharpoons HS^{\oplus} + A^{\ominus}$$

where S = solvent, since free protons do not exist in the condensed phase. Because [S] is constant, acidity is defined by the equilibrium constant:

$$K = \frac{[HS^{\oplus}][A^{\ominus}]}{[HA]}$$

Since only undissociated acid HA and solvated proton $HS^{\oplus}$ exist in solution, protonation of olefin is visualized to occur by a process akin to nucleophilic displacement involving proton transfer within acid–olefin coordination complexes (Gillespie, 1963):

$$HA + \overset{|}{\underset{|}{C}} = \overset{|}{\underset{|}{C}} \rightarrow [A{-}H \cdots \overset{|}{\underset{|}{C}} {\cdots} \overset{|}{\underset{|}{C}}] \rightarrow H{-}\overset{|}{\underset{|}{C}}{-}\overset{|}{\underset{|}{C}}{}^{\oplus} + A^{\ominus}$$

and not by direct attack of the π double bond system on the proton:

$$H^{\oplus} \overset{\frown}{C}{=}C \rightarrow H{-}C{-}C^{\oplus}$$

The latter symbolism occasionally used in this book is merely an abbreviation and should not be misconstrued to indicate protonation by free proton in the liquid phase. For a detailed discussion of Brønsted acidities the reader is referred to the many specialized treatises on this subject.

There seems to be only one quantitative study in the literature concerning a correlation of acid strength with overall polymerization rate and oligomer molecular weight. Wichterle et al. (1959) found a linear correlation between the acid strength H_0 (Hammett acidity function) of H_2SO_4 and H_2O/BF_3, and the rate constant k for isobutylene polymerization and $\overline{DP}$.

Table 3.7 shows the protic acids most frequently used for initiation together with some secondary or parasitic reactions these acids are known to induce under polymerization conditions. A comprehensive compilation of protic acids and their properties and uses is available (Olah, 1963, p. 315).

Table 3.7 □ Frequently Used Brønsted Acids and Types of Reactions They Induce

Brønsted acid	Reactions Induced
HBr	Polymerization, addition
HCl	Polymerization, addition, alkylation
HF	Polymerization, addition, alkylation, acylation, isomerization
$HClO_4$	Polymerization, alkylation, acylation
$HClSO_3$	Polymerization
$HFSO_3$	Polymerization
H_2SO_4	Polymerization, alkylation, acylation, isomerization
H_3PO_3	Polymerization, acylation, alkylation
$CH_2ClCOOH$	Polymerization, acylation
$CHCl_2COOH$	Polymerization
CCl_3COOH	Polymerization
CF_3COOH	Polymerization, acylation, isomerization
Alkanesulfonic acids	Polymerization, alkylation
CF_3SO_3H	Polymerization

Hydrogen halides HA (A=F, Cl, Br, I) as a group are rather undesirable initiators for the synthesis of high molecular weight polyolefins or poly(alkyl vinyl ethers). In a few instances HCl and HBr have been used to initiate polymerization of styrene using polar solvents (Pepper, 1959); however, the molecular weights obtained were rather low.

Protic acids with less nucleophilic conjugate bases, such as H_2SO_4 and H_3PO_4, have been used in olefin oligomerizations (Güterbock, 1959). The polymerization of styrene with H_2SO_4 and $HClO_4$ proceeds by pseudocationic mechanism (see Section 3.1), which also yields low molecular weight product (Gandini and Plesch, 1968).

Chlorosulfonic acid has been used to polymerize styrene (Pepper, 1959; Okamura et al., 1959) and butadiene (Marvel et al., 1951). Fluorosulfonic acid has been used for styrene (Pepper, 1959; Okamura et al., 1959). Trichloracetic acid has been studied by Brown and Mathieson, (1958) for the polymerization of styrene and α-methylstyrene, by Evans et al. (1955) for various diphenylethylenes, and by Eisler et al. (1952) for cyclopentadiene. Dichloroacetic acid has been used as an initiator for styrene (Brown and Holmes, 1956; Brown and Mathieson, 1957a, b, 1958) and α-methylstyrene (Brown and Mathieson, 1958). Throssel et al. (1956) studied the polymerization of styrene by trifluoroacetic acid. Alkanesulfonic acids are initiators for isobutylene (Proell et al., 1948). Recently Sawamoto et al. (1976, 1977) used CF_3SO_3H to polymerize styrene.

Nitroform, $HC(NO_2)_3$, possesses a rather acidic hydrogen and has been used to initiate the polymerization of N-vinylcarbazole (Penczek et al., 1968).

Stable Cation Salts

Certain stable cations, or more accurately carbocation salts, have been shown to be initiators of cationic polymerizations. The rationale for using these entities was the desire to simplify initiation by avoiding the sometimes complicated ion generation step. Various stable, well-defined carbenium ion salts have been used to initiate the polymerization of styrene, α-methylstyrene, N-vinylcarbazole, and alkyl vinyl ethers. Table 3.8 lists representative stable carbocation salts together with the monomers whose polymerizations they induce.

Advantages offered by these salts to the fundamental investigator are their ease of preparation and characterization and their utility in kinetic studies. For example, Higashimura et al. (1967) helped to elucidate the kinetics of initiation of styrene polymerization by adding trityl salt $[(C_6H_5)_3CSnCl_5]$ to the monomer in polar solvent and determining the rate of disappearance of the trityl cation absorption by UV spectroscopy.

A severe limitation of initiation by stable cation salts is that this method can be used to advantage only with relatively highly reactive monomers (see above). In view of the obviously unfavorable energetics, initiation by

Table 3.8 □ Some Stable Carbocation Salts Used for Polymerizations

Stable Carbocation Salt	Monomer	References
Cycloheptatrienyl$^{\oplus}$ $ClO_4^{\ominus}$	N-Vinylcarbazole	Bowyer et al. (1971)
$SbCl_6^{\ominus}$	N-Vinylcarbazole	Bowyer et al. (1971)
$SbCl_6^{\ominus}$	N-Ethyl-3-vinylcarbazole	Jackson (1975)
$ClO_4^{\ominus}$	p-Methoxystyrene	Goka and Sherrington (1975)
$SbCl_6^{\ominus}$	Isobutyl vinyl ether	Subira et al. (1976)
$SbCl_6^{\ominus}$	Chloroethyl vinyl ether	Ledwith et al. (1975)
$SbCl_6^{\ominus}$	Cyclohexyl vinyl ether	Ledwith et al. (1975)
$BF_4^{\ominus}$	Isobutyl vinyl ether	Bawn et al. (1971)
$BF_4^{\ominus}$	Ethyl vinyl ether	Ledwith et al. (1975)
$BF_4^{\ominus}$	$tert$-Butyl vinyl ether	Ledwith et al. (1975)
$BF_4^{\ominus}$	Cyclohexyl vinyl ether	Ledwith et al. (1975)
$BF_4^{\ominus}$	Methyl vinyl ether	Ledwith et al. (1975)
$BF_4^{\ominus}$	Chloroethyl vinyl ether	Ledwith et al. (1975)
$(C_6H_5)_3C^{\oplus}$		
$SbCl_6^{\ominus}$	N-Vinylcarbazole	Cotrel et al. (1973)
$SbCl_6^{\ominus}$	p-Methoxystyrene	Cotrel et al. (1976)
$SbCl_6^{\ominus}$	Styrene	Sauvet et al. (1967)
$SbF_6^{\ominus}$	Styrene	Johnson and Pearce (1976)
$SnCl_5^{\ominus}$	Styrene	Higashimura and Okamura (1960)
$HgCl_3^{\ominus}$	Styrene	Sambhi and Treloar (1965)
$BF_4^{\ominus}$	o-Divinylbenzene	Aso et al. (1968)
$SbCl_6^{\ominus}$	Cyclopentadiene	Sauvet et al. (1974)
$BF_4^{\ominus}$	Chloroethyl vinyl ether	Bawn et al. (1971), Chung et al. (1975)
$BF_4^{\ominus}$	Ethyl vinyl ether	Chung et al. (1975)
$SbCl_6^{\ominus}$	Isopropyl vinyl ether	Ma et al. (1979)
$[SbCl_6^{\ominus}\overset{\oplus}{C}(C_6H_5)_2C_6H_4CH_2]_2$	Styrene	Villesange (1980)

stable cations, for example, trityl cation, or relatively less reactive monomers such as 3-methyl-1-butene and 4-methyl-1-pentene appears to be a hopeless undertaking.

Friedel–Crafts Acid-Based Initiating Systems

The Problem of Defining Friedel–Crafts Acids

Although Friedel–Crafts acids are constituents of the most important cationic initiating systems and are used in innumerable other industrially useful processes, for example, Ziegler–Natta coordination polymerization, butyl rubber manufacture, alkylations, acylations, and isomerizations, and are the subject of countless scientific investigations, a succinct scientifically valid definition or classification based on some fundamental property of these materials has not yet been developed.

Friedel–Crafts acids are by definition Lewis acids, that is, acceptors that possess at least one vacant orbital capable of accepting a pair of electrons of a donor Lewis base, or are able to produce such an orbital under the influence of a Lewis base (Luder and Zuffanti, 1961). The vacant orbital may be partially occupied by electron back-donation from atoms covalently bound to the central metal atom, for example,

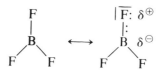

Besides Friedel–Crafts acids, Lewis acids also embrace charge-carrying electrophiles such as protons, inorganic cations, carbocations, oxonium ions, and sulfonium ions. Evidently, Friedel–Crafts acids represent a subclass of Lewis acids.

Various definitions valiantly offered, quoted by Olah in his magnificent compendium on *Friedel–Crafts and Related Reactions* (Olah, 1963), are cumbersome and out of focus, and are circumlocutions rather than definitions. Indeed, it appears that it will not be possible to formulate a satisfactory all-encompassing definition of Friedel–Crafts acids mainly because of the enormous variety and breadth of the field. This situation is quite reminiscent of the well-known dilemma of the judge who, when confronted with the task of defining pornography, opined, "I don't know how to define pornography, but I sure know it when I see it."

A functional definition of Friedel–Crafts acids useful for students of cationic polymerization may be as follows: Friedel–Crafts acids are metal halides, organometal halides, or organometals of Lewis acidic nature that in conjunction with cationogens coinitiate carbocationic polymerizations. Another, broader but still not all-inclusive definition useful for polymer-oriented investigators is based on the ability of Friedel–Crafts acids MX_n to coordinate with various conjugate bases $B^\ominus$, such as anions, conjugate bases of Brønsted acids, halogens, and alkyl halides ($B^\ominus + MX_n \rightleftharpoons BMX_n^\ominus$); that is, Friedel–Crafts acids are metal halides, organometal halides, and organometals capable of coordinatively stabilizing conjugate bases by complexation and thus increasing the lifetime of associated electrophiles.

Table 3.9 lists some frequently used Friedel–Crafts acids in cationic polymerizations together with their electronic structures and characteristics. Serious students of this field are again referred to Olah's extensive tomes (1963), in which all conceivable aspects of Friedel–Crafts acids and their complexes have been tabulated and discussed. The following discussion complements this source and has been organized with polymer-oriented scientists in mind.

Table 3.9 □ Frequently Used Friedel–Crafts Acids in Cationic Polymerization

Group of Central Atom	Electronic Structure of Central Atom	Physical-Chemical Characteristics	Remarks
Group IIA BeCl$_2$	$[1s^2, 2s^2] \rightarrow$ $[1s^2, 2s^1, 2p^1]$ Coordination 4 sp^3 hybridization	Linear molecule, partially dimeric Soluble in many organic solvents	Used in the same reactions as AlCl$_3$ but not as efficient
Group IIB ZnCl$_2$	$3d^{10}, 4s^2$ Coordination 4 sp^3 hybridization	Predominantly covalent in solution	Some alkylation and isomerization. Chloromethylation. Fairly selective and mild catalyst
CdCl$_2$	$4d^{10}, 5s^2$ Coordination 4 sp^3 hybridization	Weaker covalent character than zinc halides. Soluble in some organic solvents (i.e., acetonitrile)	Halogenation; rather weak FC acid
HgCl$_2$	$5d^{10}, 6s^2$	Essentially covalent	Alkylation, acylation, isomerization. Very mild FC acid
Group IIIA BF$_3$	$2s^1, 2p^1$; structure sp^2, tendency to sp^3 (tetrahedral structure)	Covalent and monomeric. Forms stable complexes with H$_2$O	Alkylation, rearrangement, isomerization (paraffinic and unsaturated hydrocarbons, alkylbenzenes, stereospecific isomerization of olefins), acylation of aromatics
BCl$_3$		Monomeric—does not form stable complexes with H$_2$O or HX	Alkylation and acylation particularly with organic fluorides
AlCl$_3$	$3s^2p^1$	Ionic structure; dimer insoluble or sparingly soluble in alkyl halides CS$_2$ and SO$_2$	Used in almost all FC reactions. Very sensitive to impurities which decrease or promote its efficiency

Table 3.9 □ (*Continued*)

Group of Central Atom	Electronic Structure of Central Atom	Physical-Chemical Characteristics	Remarks
$AlBr_3$		Dimer—more soluble than $AlCl_3$. Complexes formed easily	More efficient than $AlCl_3$, partly due to its higher solubility
R_3Al, R_2AlX, and $RAlX_2$		Pyrophoric compounds. $EtAlCl_2$ (crystalline solid); Et_2AlCl (liquid); Et_3Al (liquid)	Cocatalysts in Ziegler–Natta polymerizations. Agents for the dimerization and oligomerization of olefins
Group IIIB $GaX_3(Cl, Br)$	$4s^2$, $4p^1$	Ionic solid—covalent in the liquid state. Highly insoluble in organic solvents	Very efficient for alkylation and acylation
$InX_3(Cl, Br)$	$5s^2$, $5p^1$	Far less ionic than gallium halides	Alkylation, acylation, halogenation. Considerably weaker than the correspondant gallium hallides.
Group IVA $SnCl_4$	$5s^2$, $5p^2$ Coordination 6 sp^3 hybridization	Liquid	Acylation and alkylation, chloromethylation, mild catalyst
$SnBr_4$		Solid	As above
Group IVB $TiCl_4$	$3d^2$, $4s^2$	Liquid. Monomeric (checked in CH_2Cl_2). Stable complexes formed in the presence of oxygen donors and halides	Alkylation, acylation. Generally less active than $AlCl_3$
$TiBr_4$ $ZrCl_4$	$4d^2$, $5s^2$	Forms complexes as $TiCl_4$ but they are far less stable than the titanium analogues	Less efficient than $TiCl_4$. Alkylation, acylation, isomerization,

Group VA			
VCl_4	$3d^3, 4s^2$	Viscous liquid. Monomeric	Very active in alkylation and acylation
NbF_5	$4d^4, 5s^1$	Solid	Active in alkylation and acylation
TaF_5	$6d^3, 6s^2$	Solid	Active in alkylation and acylation
Group VB			
AsF_5	$4s^2, 4p^3$	Gas ($bp - 53°C$). Covalent	Alkylation, acylation
AsF_3	$4s^2, 4p^3$	Liquid	Alkylation, acylation
SbF_5	$5s^2, 5p^3$	Viscous liquid. Covalent	Effective FC acid in alkylation, acylation. Powerful halogenating agent
	Coordination 6 $(SbF_6^\ominus)$	Associated molecules. One of the strongest Lewis acids	
$SbCl_5$	$5s^2, 5p^3$	Covalent; liquid	Effective in alkylation, acylation. Powerful halogenating agent
	Coordination 6 $(SbCl_2^\ominus)$		
$SbCl_3$		Covalent; solid	Low activity
$BiCl_3$	$6s^2, 6p^3$	Soluble in numerous solvents	Alkylation, acylation. Fairly effective
Group VIA			
WCl_5	$3d^4, 6s^2$	Crystalline solid soluble in many organic compounds but often reacts with the solvent. Covalent	Considerable activity in alkylation, acylation, ring chlorination
Group VIIIB			
$FeCl_3$	$3d^6, 4s^2$	Dimeric in the solid state. Soluble in many organic solvents. Monomeric in donating solvents (e.g., OEt_2)	Alkylation, acylation. Milder than $AlCl_3$

Acidity of Friedel–Crafts Acids and Nucleophilicity of Counteranions

It would be highly desirable to possess a reliable, preferably absolute, scale of Friedel–Crafts acidities and with this information to predict the behavior of Friedel–Crafts acids or cationogen/Friedel–Crafts acids combinations as cationic initiating systems. In principle, the utility of Friedel–Crafts acids as coinitiators in carbocationic polymerizations could be estimated by an acidity scale based on ΔH_{AB} and K_a values of the equilibrium leading to ion-generation:

$$AB + MX_n \xrightleftharpoons{K_a} A^{\oplus} + MX_n B^{\ominus} \qquad -\Delta H_{AB}$$

where MX_n = Friedel–Crafts acid coinitiator and AB = cationogen, for example, protic acid, alkyl halide, or halogen (see Chapter 4). Unfortunately, such an absolute scale of Friedel–Crafts acidities does not exist and in spite of the sustained efforts of many eminent investigators has not been developed.

The principal difficulty preventing an objective scaling of Friedel–Crafts acidities resides in the large number and variety of factors determining and/or influencing this property, among which are the electronic nature of the central metal, the nature and electronegativity of ligands on the metal, size and geometry of the assembly, reorganization and back-donation energetics, such as with boron halides, and dissociation energetics with aluminum-containing Friedel–Crafts acids. Beyond these variables Friedel–Crafts acidities are also influenced by the nucleophilicity, size, and geometry of the donor base as well as the strength of donor–acceptor complexes and their solvation energies.

Among these various factors the electronic nature of the central metal strongly influences electron deficiency that is, acid strength of Friedel–Crafts acids (Burton and Prail, 1954). Covalent compounds of Groups II and III elements, that is, Be, Zn, B, Al, and Ga, are good acceptors. The vacant $2p$ orbital of B in boron halides is partially occupied by back-donation by nonbonding electrons leading to contributions by double bonded structures:

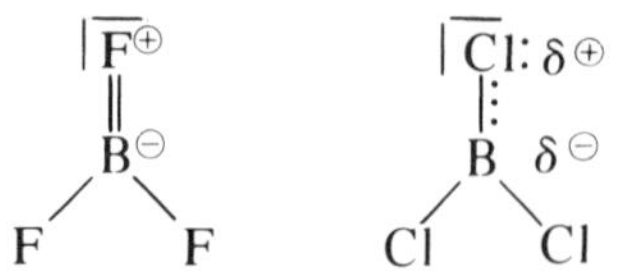

which help to rationalize the unexpected experimental observation (Brown and Holmes, 1956) that the Lewis acidity sequence in the boron series is $BF_3 < BCl_3 < BBr_3 < BI_3$. Most aluminum-containing Friedel–Crafts acids do not have vacant $3p$ orbitals on the central aluminum owing to association, for example,

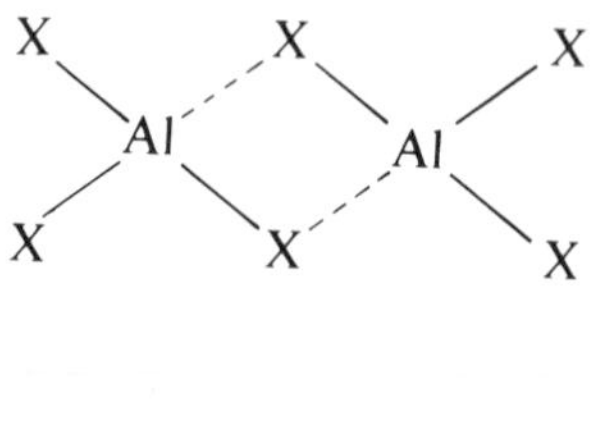

Transition metal compounds have vacant d orbitals, and depending on their number may accept one or more donors:

$$FeBr_3 + Br^{\ominus} \rightarrow FeBr_4^{\ominus}$$
$$TiF_4 + 2\ Me_2SO \rightarrow TiF_4(Me_2\ SO)_2$$
$$ZrF_4 + 3\ F^{\ominus} \rightarrow ZrF_7^{3\ominus}$$

Electronegativity of ligands covalently bound to the metal significantly affects overall Friedel–Crafts acidity. Increasing the number of electronegative groups around the central metal increases Friedel–Crafts acidity. For example, Friedel–Crafts acidity increases by replacing the less electronegative methyl groups around boron with highly electronegative fluorines: (Burn and Green, 1943),

$$Me_3B < Me_2BF < MeBF_2 < BF_3$$

It would be expected that Friedel–Crafts acidity would increase by replacing less electronegative iodines with more electronegative fluorines around a central metal M:

$$MF_n > MCl_n > MBr_n > MI_n$$

Interestingly, however, this is the exception rather than the rule. For example, in the boron halide series heats of complexation with pyridine and nitrobenzene (Brown and Holmes, 1956), heats of formation of crystalline complexes (Greenwood and Perlins, 1960a, b), NMR spectroscopy (Deters et al., 1968), and gas phase calorimetry (Mente et al., 1975) show that in fact the acceptor strength of these acids is just the opposite of the expected one:

$$BF_3 < BCl_3 < BBr_3 < BI_3$$

An explanation for this anomaly may be increased electron back-bonding or back-donation and associated reorganization energy along the F to I sequence (Brown and Holmes; 1956; Cotton and Letto, 1959; Kakubari et al., 1974).

Similar observations have also been made with aluminum-containing Friedel–Crafts acids for which it has been found that acidity increases as $AlCl_3 < AlBr_3 < AlI_3$ (Plotnikov and Yakubson, 1938).

Size and geometry, that is, steric factors, exert important influence on Friedel–Crafts acidity. Large ligands or increasing size of halogen atoms surrounding the central metal can effectively block the electron-deficient site and reduce acidity. For example, the high intrinsic Friedel–Crafts acidity of BI_3 is difficult to exploit on account of steric inaccessibility of the buried small boron acceptor site blocked by surrounding large iodine ligands.

Crystal lattice energy may also affect Friedel–Crafts acidity. For example, among the four aluminum halides, AlF_3 has no discernible activity as a Friedel–Crafts acid on account of its very high lattice energy, which prevents its dissolution in organic solvents or other liquids (the ionic character of the AlF_3 network is ~78%). The utility and/or efficiency of $AlCl_3$ is also limited by relatively high lattice energy (ionic character of $AlCl_3$ network is ~43%). This energy is considerably reduced by replacing a chlorine by an organic group; indeed monoalkylaluminum dihalides are liquids, and these compounds are useful, active coinitiators of cationic reactions. $AlBr_3$, only 34% ionic, is the most active coinitiator among the aluminum halides. In contrast, electrophilic properties of AlI_3 are reduced owing to the metallic character of and steric hindrance by iodine.

Association tendency of Friedel–Crafts acids should also be considered. Thus, in the solid, liquid, and even in the gaseous state most aluminum-containing Friedel–Crafts acids remain associated as dimers or trimers. The dissociation energies of the aluminum halides are (in kcal/mole); AlF_3, 48; $AlCl_3$, 24; $AlBr_3$, 26.5; and AlI_3, 22.5. Obviously, dissociation energetics determine activity of these Friedel–Crafts acids.

It has been shown that only the monomer forms of aluminum-containing Friedel–Crafts acids are active. According to Walker (1960, 1961) the rate of dissociation of the dimer to monomer $Al_2X_6 \rightleftharpoons 2\,AlX_3$ controls the kinetics of subsequent reactions. In the presence of agents capable of complexing with AlX_3 the dissociation energy is considerably reduced. For example, the dissociation energy of Al_2Br_6 is 26.5 kcal/mole; however, the energy of association between $AlBr_3$ and MeBr or xylene is 20 kcal/mole, and that with 2-pentene is 18 to 19 kcal/mole. If the dissociation of Al_2Br_6 is rate determining, the rate will be multiplied by the factor $\exp(26{,}500 - 20{,}000/2T)$; that is, it will be about 5,000 times faster than in the absence of these complexing agents. Walker (1960, 1961) considered these complexing agents as "cocatalysts" and subdivided various donors (bases) in relation to $AlBr_3$ into three classes. Class 1 donors, for which the donor–$AlBr_3$ bond is < 15 kcal/mole, for example, paraffins, benzene, and naphthenes, do not exhibit cocatalytic activity, evidently because the donor–$AlBr_3$ bond is too weak to promote significant dissociation of Al_2Br_6. Class 2 comprises donors, for example, alkyl halides, olefins, aromatic compounds, and ethers, that produce 15 to 27 kcal/mole complexation energy. These bases give rise to significant concentrations of donor–$AlBr_3$ complexes the dissociation energies of which are lower than that of Al_2Br_6. Class 3 includes oxygen-containing compounds, for example, water, alcohols, and ethers, that readily dissociate Al_2Br_6 but form complexes that are more stable than the Al_2Br_6 dimer itself; that is, the complexation energy exceeds 27 kcal/mole. Since $AlBr_3$ monomer is absent in these systems cocatalytic effect is also absent.

Careful NMR and model studies have indicated (Kennedy and Milliman, 1969) that only the Me_3Al monomer (or "half-opened dimer") is active in the t-BuCl/Me_3Al/isobutylene/MeCl system.

Combinations of Brønsted acids with Friedel–Crafts acids are among the strongest acids known. For example, mixtures of HF and SbF_6 or BF_4 in nonnucleophilic solvents such as SO_2 are many orders stronger than concentrated sulfuric acid (Olah, 1963). Evidently, the nucleophilicity of counteranions $SbF_6^\ominus$ or $BF_4^\ominus$ in these systems is so low that they are reluctant to accept protons (or other electrophiles) which therefore are forced to attack even such weakly electrophilic sites as C–C or C–H bonds, for example,

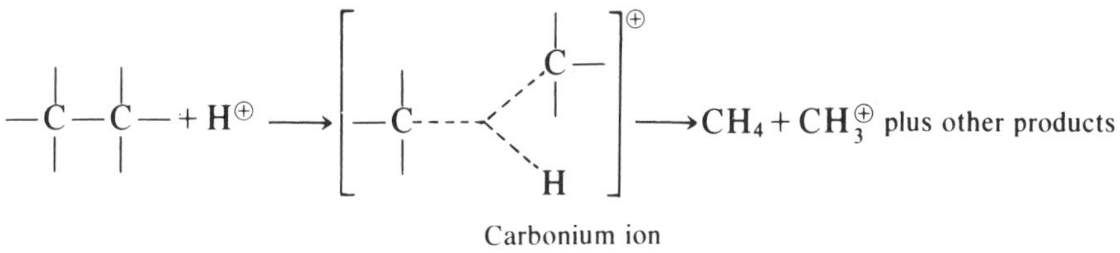

$$-\overset{|}{\underset{|}{C}}-\overset{|}{\underset{|}{C}}- + H^\oplus \longrightarrow \left[-\overset{|}{\underset{|}{C}}---\overset{\overset{|}{C}-}{\underset{H}{\diagdown}} \right]^\oplus \longrightarrow CH_4 + CH_3^\oplus \text{ plus other products}$$

Carbonium ion

The carbonium ion contains a three-centered two-electron bond and may decay to give methane and methyl carbenium ion. The latter, the most reactive carbenium ion, may in turn attack C–C or C–H bonds,

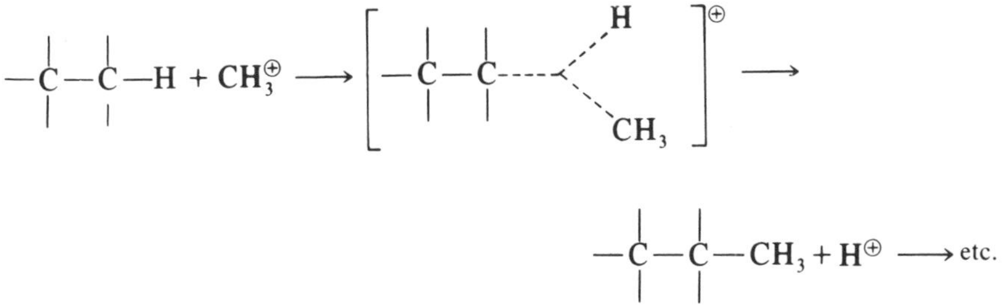

$$-\overset{|}{\underset{|}{C}}-\overset{|}{\underset{|}{C}}-H + CH_3^\oplus \longrightarrow \left[-\overset{|}{\underset{|}{C}}-\overset{|}{\underset{|}{C}}---\overset{\overset{H}{\diagup}}{\underset{CH_3}{\diagdown}} \right]^\oplus \longrightarrow$$

$$-\overset{|}{\underset{|}{C}}-\overset{|}{\underset{|}{C}}-CH_3 + H^\oplus \longrightarrow \text{etc.}$$

and lead to higher molecular weight products (Olah, 1974). The condensation of paraffins under the influence of HSO_3F/SbF_5 combinations has been shown to lead to higher branched paraffins of complicated structures (Roberts and Calihan, 1973a, b).

The extremely high acidity of these so-called super acid systems (HSO_3F/SbF_5 combinations in low nucleophilic solvents) is undesirable for the synthesis of high molecular weight materials since polymer growth and degradation would proceed simultaneously. Further, these acids would induce a great variety of cationic reactions, for example, isomerization, elimination, and cationation, and the resulting product would be a hopelessly complicated mixture of polyhydrocarbons (Kennedy et al., 1974a).

Although extremely high acidity may be counterproductive in high polymer synthesis, low acidities may be equally undesirable. Obviously, cationation will not proceed with very weak acids. In the presence of weak acids, proton concentration is low and the rate of protonation may be impractically slow. Increasing acidity would produce increasingly acceptable initiating systems; however, in the presence of super acids protonation of sigma bonds may commence, which may lead to unacceptable degradation. Thus it appears that there exists an intermediate acidity strength for useful cationic polymerization initiation.

In view of these complexities efforts have been made to construct relative Friedel–Crafts acidity scales based on various characteristics of Friedel–Crafts acids; such empirical acidity scales have occasionally been used in specific instances. Prominent among such semiquantitative acidity scales are those based on various acceptor properties of Friedel–Crafts acids, for example, heats of formation with pyridine bases (Greenwood and Perkins, 1960a, b), coordination and ionization with trityl chloride (Baaz and Gutman, 1963; Burg and Green, 1943), chloride ion affinities (Baaz and Gutman, 1963), color change of Hammett indicators (Hawke and Steigman, 1954; Saegusa, 1969; Benesi, 1956), and IR stretching frequency of CO complexes, notably benzophenone, acetophenone (Szusz and Chalandon, 1958), ethyl acetate (Lappert, 1962), and more recently xanthone (Cook, 1961, 1963a,b; Kagiya et al., 1968). Furukawa et al. (1974) used the latter acidity scale to compare $\Delta H^{\neq}$ and $\Delta S^{\neq}$ values obtained from cationic copolymerization of styrene and p-substituted styrene; however, the correlation seemed to depend on the reference carbocation. Several activity scales constructed on the basis of polyisobutylene yield (Seymour, 1943) or molecular weight (Chalmers, 1932) or polystyrene molecular weight (Okamura et al., 1958) were obtained under nonoptimal, not strictly comparative conditions and are of little predictive significance. For example, whereas Chalmers' (Chalmers, 1932) oft-quoted scale ($BF_3 > AlBr_3 > TiCl_4 > TiBr_4 > BCl_3 > SnCl_4 > BBr_4$) lists BF_3 to be the most efficient Friedel–Crafts acid, way ahead of BCl_3, according to recent experiments (Kennedy et al., 1977) carried out under strictly comparative conditions, BCl_3 is a far more efficient coinitiator than BF_3 in terms of both polyisobutylene yield and molecular weight.

The concept of hard and soft acids and bases (HASAB) may become valuable for predicting initiating cationogen/Friedel–Crafts acid combinations. According to the HASAB principle (Pearson, 1973), hard acids prefer to bind to hard bases and soft acids prefer to bind to soft bases.

In general, hardness is characteristic of ionic bonds whereas softness is characteristic of covalent bonds (Symposium, 1965). Materials with small atomic radius, high nuclear charge, low polarizability, and slight tendency to oxidize tend to be hard; for example, among the hard acids are protons, cations of small radius ($Li^{\oplus}$), small carbenium ions, and molecules in which substituents on a central atom induce a high positive charge at the central atom (BF_3, $AlCl_3$, $TiCl_4$). Unfortunately, only few Friedel–Crafts acids used as coinitiators for carbocationic polymerization have been studied in this regard. Among the hard acids of interest are, in order of decreasing hardness,

$$H^{\oplus} \gg BF_3 > AlCl_3$$

and soft acids (by decreasing hardness)

$$I^{\oplus} \simeq Br^{\oplus} > I_2 \simeq Br_2 \simeq HCN \simeq \text{chloranil} \simeq (CN)_2C=C(CN)_2$$

Similarly, hard bases of interest are

$$H_2O, OH^\ominus, F^\ominus > CH_3COO^\ominus, PO_4^{3\ominus}, SO_4^{2\ominus} > Cl^\ominus, ClO_4^\ominus > ROH, RO^\ominus, R_2O$$

and soft bases

$$R_2S, RSH, RS^\ominus > C_2H_4, C_6H_6 > H^\ominus, R^\ominus$$

Thus according to the HASAB principle interaction between H_2O and BF_3 or $AlCl_3$ should be quite favorable and, in fact, H_2O/BF_3 and "H_2O"/$AlCl_3$ are among the most frequently used initiating systems. Regrettably, our store of knowledge is limited with regard to application of the HASAB principle to Friedel–Crafts acid-based initiating systems. It is hoped that investigations along these lines will be carried out and HASAB characterization of cationogens and Friedel–Crafts acids will lead to new insight into the design of useful initiating systems.

A semiquantitative method to estimate directly the enthalpy ΔH of the reaction $A + B \rightarrow AB$ developed by Drago et al. (Mc Millin and Drago, 1972; Drago et al., 1971, 1972, 1974; Marks and Drago, 1975) may be of use for quantizing Friedel–Crafts acidities. Drago's equation

$$-\Delta H = E_A E_B + C_A C_B$$

contains four empirical constants, E_A, C_A, E_B, and C_B, characteristics of acids and bases (subscripts A and B, respectively). Table 3.10 shows E_A and C_A values reported for various acids, and E_B and C_B data for different bases.

From the data in Table 3.10 enthalpies of reactions arising during

Table 3.10 □ Drago's Empirical Constants for the Calculation of ΔH of Friedel–Crafts Acid–Base Reactions[a]

Constant	Acid									
	$SbCl_5$	BF_3[b]	Iodine	ICl	IBr	$I^\oplus$	$CH_3^\oplus$	$Et^\oplus$	n-$Pr^\oplus$	$Ph^\oplus$
E_A	7.38	7.96, 9.88	1	5.10	2.41	68.72	62.1	51.49	48.93	57.08
C_A	5.13	3.08, 1.62	1	0.830	1.56	4.57	7.30	7.05	7.08	7.77

	$F^\ominus$	$Cl^\ominus$	$Br^\ominus$	$I^\ominus$	$Me^\ominus$	$Et^\ominus$	$OH^\ominus$	$Ph^\ominus$
E_B	2.94	2.47	2.47	2.46	3.24	3.29	3.10	2.80
C_B	10.38	9.66	8.59	7.50	12.07	11.66	11.32	12.86

[a] After Drago et al. (1971).

[b] The two E_A and C_A values have been obtained by applying two different assumptions in the calculation.

initiation with Friedel–Crafts acid-based systems may be estimated (ΔH in kcal/mole):

$$BF_3 + F^{\ominus} \rightarrow BF_4^{\ominus}$$ $-\Delta H = 55.47$ using 7.96 and 3.08
         45.86 using 9.88 and 1.62

$$BF_3 + OH^{\ominus} \rightarrow BF_3OH^{\ominus}$$ $-\Delta H = 59.54$ using 7.96 and 3.08
         48.97 using 9.88 and 1.62

$$SbCl_5 + Cl^{\ominus} \rightarrow SbCl_6^{\ominus}$$ $-\Delta H = 67.78$

Thus according to Drago's method, the relative stability sequence for the three counteranions for which data are available is

$$SbCl_6^{\ominus} > BF_3OH^{\ominus} > BF_4^{\ominus}$$

These data should be corrected by solvation energy contributions (see Section 3.4). Also, it should be kept in mind that by considering the enthalpy of the reaction

$$B^{\ominus} + C_o \rightleftharpoons [BC_o]^{\ominus}$$

we assumed that iniation involved cationation by

$$A^{\oplus}[BC_o]^{\ominus} + \overset{|}{\underset{|}{C}}=\overset{|}{\underset{|}{C}} \rightarrow A-\overset{|}{\underset{|}{C}}-\overset{|}{\underset{|}{C}}{}^{\oplus} [BC_o]^{\ominus}$$

and not by an alternative route, for example,

$$C_o + \overset{|}{\underset{|}{C}}=\overset{|}{\underset{|}{C}} \longrightarrow [\, C_o \leftarrow \overset{-C-}{\underset{-C-}{\|}} \,] \overset{AB}{\longrightarrow} A-\overset{|}{\underset{|}{C}}-\overset{|}{\underset{|}{C}}{}^{\oplus} [BC_o]^{\ominus}$$

where C_o is the coinitiator and AB the initiator.

Nucleophilicity of counteranions arising from Friedel–Crafts acids would be another measure of the electron-accepting tendency (acidity) of these materials. Although efforts have beem made to quantize nucleophilicities of various anions for aqueous media (Swain and Scott, 1953), similar nucleophilicity scales have not been devised for counteranions derived from Friedel–Crafts acids or anions in nonaqueous media. It would be most desirable to conduct fundamental studies in this direction. Efforts have been made by Heublein et al. (1979) to quantize counteranion stabilities by the use of conductivity and spectroscopic characteristics of stable trityl ion salts.

Though an absolute scale of nucleophilicities of counteranions needed by polymer scientists does not exist, workers in this field generally agree that nucleophilicities of counteranions derived from Friedel–Crafts acids are low. A gradation among these species would be valuable in helping to evaluate outcome of cationic polymerizations in terms of yields, conversions, and product molecular weight. Extended molecular weight studies yielded valuable information relative to counteranion nucleophilicities of Friedel–Crafts acids (Kennedy and Rengachary, 1974b; Kennedy and Trivedi; 1978; Deters et al., 1968). Thus molecular weight of poly-isobutylenes obtained under carefully comparable conditions decreased as $Me_2AlCl_2^{\ominus}$ > $Me_2AlClBr^{\ominus}$ > $Et_2AlCl_2^{\ominus}$ > $Et_2AlClBr^{\ominus}$ > $MeAlCl_2OH^{\ominus}$ > $Et_3AlBr^{\ominus}$ > $Et_3AlCl^{\ominus}$ (Kennedy and Trivedi, 1978) and in another series of experiments (Kennedy and Trivedi, 1978) as $Et_2AlCl_2^{\ominus}$ > $Et_2AlClBr^{\ominus}$ > $Et_2AlBr_2^{\ominus}$ > $Et_2AlClI^{\ominus} \approx Et_2AlBrI^{\ominus}$ > $Et_2AlI_2^{\ominus}$. In the latter series of experiments the number and nature of alkyl groups was constant and only the halogen was changed. These and other data were interpreted to mean that the molecular weight of polyisobutylene is determined by the nucleo-philicity of counteranion, which in turn is determined by charge delo-calization over electronegative groups. For counteranions with an equal number of similar alkyl groups, nucleophilicity is determined only by the nature of halogens; that is, nucleophilicity decreases as $Me_2AlCl_2^{\ominus}$ > $Me_2AlClBr^{\ominus}$, $BF_3OH^{\ominus}$ > $AlCl_3OH^{\ominus}$, $Et_2AlCl_2^{\ominus}$ > $Et_2AlClBr^{\ominus}$ > $Et_2AlBr_2^{\ominus}$ > $Et_2AlClI^{\ominus} \sim Et_2AlBrI^{\ominus}$ > $Et_2AlI_2^{\ominus}$. Evidently counteranion nucleophilicity is determined not by its actual size but "effective" size, that is, the area over which charge delocalization occurs (Kennedy and Trivedi, 1978).

At present no absolute fundamentally meaningful Friedel–Crafts acidity scale exists; indeed it cannot exist in view of our incomplete quantitative understanding of factors influencing electron deficiency. Even empirical Friedel–Crafts acidity scales that seek to compare compounds of different central metals have only very limited significance and should not be construed to be more than what they are, rough guideposts of some nebulous quantity akin to Friedel–Crafts acidity. Acidity scales comparing compounds of the same central metal may have lasting significance, for example, $BF_3 < BCl_3 < BBr_3 < BI_3$, $AlCl_3 < AlBr_3 < AlI_3$, or $Me_3Al < Me_2AlCl < MeAlCl_2 < AlCl_3$; but even in these cases great circumspection is required to examine acidities under strictly comparable optimal con-ditions.

The ultimate goal, the prediction of the behavior of Friedel–Crafts acid-based initiating systems from thermodynamic (enthalpy) data or first principles, is still far in the furure, but relative, empirical tests and examination modes are available that may give at least some qualitative quideposts of Friedel–Crafts acids and on careful inspection of various structural parameters may explain their behavior.

Reactivity of Friedel–Crafts Acid-Based Initiating Systems

The literature of cationic polymerizations induced by Friedel–Crafts acid-based initiating systems is replete with references to "catalyst" reactivities, activities, or efficiencies without providing suitable definitions for these terms. Sometimes these terms are meant to convey relative amounts of polymer produced; sometimes they refer to molecular weights or relative rates of initiation or overall rates, or simply to purely subjective visual observations made during experiments. It would be desirable to develop a uniform mode of expression to describe reactivities or efficiencies of Friedel–Crafts acid-based initiating systems. A useful quantity may be the ratio initial rate/initiator concentration or, in the absence of this information, amount of polymer formed/amount of initiator used, which is easy to obtain by experiment.

Admittedly, obtaining meaningful comparable reactivities of Friedel–Crafts acid-based initiating systems is quite cumbersome, but once available, it provides valuable insight into the nature of these materials. For example, the reactivity of various aluminum-containing Friedel–Crafts coinitiators have been examined in detail (Kennedy and Trivedi, 1978). With the t-BuX/Me$_3$Al/isobutylene/MeI (X=Cl, Br, I) system polymer yields were affected by reagent introduction sequence and an examination of the results indicated the presence of complexation between Me$_3$Al and i-C$_4$H$_8$. According to a combination of results on initiator efficiencies (monomer conversion in M/initiator concentration in M), polymerization rates and yields, and floor temperatures, initiator reactivities decrease as t-BuCl > t-BuBr $\gg$ t-BuI $\approx$ 0, and are affected by solvent as MeCl > MeBr $\gg$ MeI = 0.

Similar detailed studies with the t-BuX/Et$_2$AlX/isobutylene/MeX (X = Cl, Br, I) system at various temperatures and concentrations have revealed the following reactivity sequences.

With Et$_2$AlCl:
t-BuCl > t-BuBr $\gg$ t-BuI = 0, and depending on the solvent MeCl > MeBr > MeI = 0.

With Et$_2$AlBr:
t-BuCl > t-BuBr > t-BuI, and depending on solvent MeCl > MeBr $\gg$ MeI = 0.

With Et$_2$AlI:
t-BuI > t-BuBr > t-BuCl, and depending on solvent MeCl > MeBr $\gg$ MeI = 0.

Coinitiator reactivity decreases as EtAlCl$_2$ > Et$_2$AlX > Me$_3$Al and for Et$_2$AlX as Et$_2$AlI > Et$_2$AlBr > Et$_2$AlCl.

3.4 □ SOLVENTS

Solvents are seldom "extras" on the stage of cationic polymerizations. Usually they are cast in two important supporting roles: they are heat sinks and polar media necessary for efficient polymerizations.

Several specific requirements must be satisfied by a useful cationic polymerization solvent:

1 □ The solvent should not complex or react with active species. Although isobutylene polymerization initiated by t-BuCl/Et$_2$AlCl readily

occurs in CH_3Cl, the rate is much reduced in CH_3Br and stops altogether in CH_3I (Kennedy and Trivedi, 1978). Evidently the growing carbocation strongly complexes with methyl iodide $\sim \overset{\oplus}{C}(CH_3)_2 + MeI \rightarrow \sim C(CH_3)_2-I^{\oplus}-Me$, which prevents propagation. Similarly, isobutyl-lene cannot be polymerized using ether solvents, presumably because of oxonium ion formation.

2 □ Solvents should be preferentially polar to promote ion generation. Great care must be exercised in selecting and interpreting the effect of polar solvents; misinterpretations due to oversimplification abound.

The overall rate of cationic polymerizations is usually determined by rate of ion generation and propagation, and the effect of solvent polarity on these rates is opposite. Ion generation $R–X \rightarrow R^{\oplus} X^{\ominus}$ must proceed through a transition state that is more polar than the starting material:

$$R\!-\!X \rightarrow [R^{\delta\oplus}\text{----}X^{\delta\ominus}]^{\ddagger} \rightarrow R^{\oplus} + X^{\ominus}$$

Polar solvents preferentially stabilize the transition state relative to the initial state, reduce the activation energy of ion generation, and con-sequently increase the rate. Conversely, cationation and propagation in-volve a spreading of the charge in the transition state:

$$R^{\oplus} + C\!=\!C \rightarrow [R^{\delta\oplus}\text{---}C\!\cdots\!C^{\delta\oplus}]^{\ddagger} \rightarrow R\!-\!C\!-\!C^{\oplus}$$

where $R^{\oplus}$ = initiating or propagating cation; in this case polar solvents preferentially stabilize the initial state, that is, increase the activation barrier (desolvation energy required), and thus reduce the rate of these reactions.

3 □ The solvents should be liquids of rather low viscosity at the poly-merization temperature, which in these polymerizations may be quite low. Properly chosen solvent mixtures allow liquid phase polymerizations as low as $-180°C$, for example, isobutylene polymerization using "H_2O"/$AlCl_3$ in a ternary mixture of ethyl chloride–vinyl chloride–propane (Kennedy and Thomas, 1962b).

4 □ To maintain homogeneous experimental conditions the solvent should preferentially dissolve all the ingredients of the charge and the polymer that is formed. Under special circumstances it may be advantageous to carry out slurry polymerizations, for example, butyl rubber manufacture, where the high heat capacity medium (MeCl) efficiently removes the heat of polymerization and allows high conversions to be reached at low system viscosities.

5 □ The solvent should be readily removable after polymerization is complete.

Table 3.11 lists frequently used solvents, together with some important physical parameters.

Nonpolar solvents, *n*-hexane, for instance, should be used with extreme

Table 3.11 ☐ **Physical Parameters of Frequently Used Solvents in Cationic Polymerization**[a]

Solvent	θ_{mp}	θ_{bp}	D_4^{20}	$\varepsilon(°C)$	$a \times 10^2$ or (α)	t, t'
Ethylene	−181	−103.7	—	—	—	—
Ethane	−183.3	−88.63	—	—	—	—
Propane	−189.9	−42.07	0.585(−45)	1.61 (0)	0.200	−90, +15
n-Butane	−138.9	−0.5	—	—	—	—
n-Hexane	−95	69	0.660	1.890(20)	0.155	−10, 50
Cyclohexane	6.6	80.7	0.779	2.023(20)	0.160	10, 60
Benzene	5.5	80.1	0.879	2.284(20)	0.200	10, 60
Toluene	−95	110.6	0.867	2.379(25)	0.0455(α)	−90, 60
Methyl chloride	−97.7	−24.2	0.916	12.6 (−20)	$\varepsilon_t = 12.6 - 0.061(t + 20) + 0.0005(t + 20)^2$	−70, −20
Ethyl chloride	−136.4	12.27	0.898	16.50 (−72)	$\varepsilon_t = -12.95 - 5.93 \cdot 10^{-8}t$	−72, −2
Methylene chloride	−95.1	40	1.327	9.08 (20)	$\varepsilon = (3320/t) - 2.24$	−2.24
Chloroform	−63.5	61.7	1.483	4.806(20)	0.160(α)	0, 50
Carbon tetrachloride	−23	76.5	1.594	2.238(20)	0.200	−10, 60
1,2-Dichloroethane	−35.4	83.5	1.235	—	—	—
Chlorobenzene	−45.6	132	1.106	5.708(20)	0.130(α)	0, 80
o-Dichlorobenzene	−17	180.5	1.305	9.93 (25)	0.194(α)	0, 50
m-Dichlorobenzene	−24.7	173	1.288	5.04 (25)	0.120(α)	0, 50
Nitromethane	−17	100.8	1.137	35.9 (20)	—	—
Nitroethane	−50	115	1.045	28.06 (30)	11.4	30, 35
Nitrobenzene	5.7	210.8	1.204	34.82 (25)	0.225	10, 80
Carbon dioxide	−56.6 (at 5.2 atm)	−78.5(sub.)	—	1.60(50 atm, 20°C)	0.268	−90, 130
Carbon disulfide	−110.8	46.3	1.263	2.641(20)	—	—
Sulfur dioxide	−72.7	−10	—	17.6(−20)	0.287(α)	−65, −15

[a] θ_{mp} = melting temperature; θ_{bp} = boiling point (760 torr); D_4^{20} = density at 20°C relative to water at 4°C; ε = dielectric constant at the temperature given in parentheses. The dependence of ε on temperature is $a = d\varepsilon/dt$ or $\alpha = d \log_{10} \varepsilon/dt$; a or $\log(\alpha)$ values are shown, $\log(\alpha)$ values indicated by (α); t, t' is the range of temperatures between which a or α is considered applicable.

circumspection, particularly in kinetic investigations. Since all monomers are more polar than n-alkanes autosolvation by monomer will occur, which may unduly complicate interpretation of kinetic data. For similar reasons polymerization of alkyl vinyl ethers in the less polar CH_2Cl_2 should be avoided.

Carbon disulfide is an interesting solvent. Judging by its dielectric constant (2.641 at 20°C) it is a nonpolar liquid; however, it is highly polarizable and may function as a polar medium. It is difficult to purify this solvent and it forms a dangerous, easily ignitable mixture with air.

Many researchers prefer to work with CH_2Cl_2. This liquid is a good solvent for many momomer/polymer systems, is fairly polar ($\varepsilon \sim 9$), and has a conveniently low freezing point ($\sim -95°C$) and low boiling point (40°C), facilitating rapid removal after polymerization. It is nonflammable and relatively nontoxic, and appears to be noncarcinogenic.

Satisfactory formulation of the concept of polarity is a perennial, unsolved problem. It has often been pointed out that the dielectric constant is only a crude indicator of polarity. The ion-solvating power of a liquid is determined by a variety of interlocking parameters for example, unshared electron pairs in the molecule and their accessibility, the shape and geometry of the structure, and steric hindrance of autoassociation. The dielectric constant in the vicinity of the active species is undoubtedly strongly reduced owing to the orientation of solvent molecules by the electric field of the cation or ion pair.

Certain workers have employed mixtures of solvents to study the effect of ε on rate constants. The overall ε was changed by changing the volume ratio of pure components such as CCl_4 and $C_6H_5-NO_2$. This technique is fraught with danger since it implies that the ε of a mixture is the same in the bulk of the system and in the vicinity of an active ionic species. This assumption is valid only if the components of the mixture are very similar (which is of little experimental interest). However, if the solvents are quite dissimilar (i.e., one polar, one nonpolar) preferential solvation by the more polar component may occur, orientation of solvent in the vicinity of active species may arise, or destructuration of one of the solvents by the other may take place. It is impossible to evaluate these effects even qualitatively, let alone quantitatively.

In a series of publications which followed his work on calculating enthalpies of donor–acceptor systems (see Section 3.3), Drago (Nozari et al., 1973) derived equations to correct for solvation effects, in particular to estimate enthalpy variation when changing from CCl_4 to CH_2Cl_2.

□ REFERENCES

Albert, A., and Pepper, D. C. (1961), *Proc. Chem. Soc.* A **263**, 75.

Arnett, E. M. [1973], *Acc. Chem. Res.*, **6**, 404.

Arnett, E. M., Pientq, N., and Petro, C. (1980), *J. Am. Chem. Soc.* **102**, 398.

Aso, C., Kunitake, T., Matsuguma, Y., and Imaizumi, Y. (1968), *J. Polym. Sci. A1.* **6**, 3049

Baaz, M., and Gutman, V. in Olah (1963).

Baker, R. (1974), in *Reactivity, Mechanism and Structure in Polymer Chemistry*, A. D. Jenkins and A. Ledwith, Eds., Wiley, New York.

Bawn, C. E. H., Fitzsimmons, C., Ledwith, A., Penfold, J., Sherrington, D. C., and Weightman, J. A. (1971), *Polymer* **12**, 119.

Belliard, P., Cohen, S. and Maréchal, E. (1972a), *Int. Symp. Macromol., Helsinki—Prep.* 1, 155.

Belliard, P., and Maréchal, E. (1972b), *Bull. Soc. Chim. Fr.*, 4255.

Benesi, H. A. (1956), *J. Am. Chem. Soc.* **78**, 5490.

Bethell, D., and Gold, V. (1967), *Carbonium Ions*, Academic, New York, p. 6.

Biddulph, R. H., Plesch, P. H., and Rutherford, P. (1965), *J. Chem. Soc.*, 275.

Boyer, P. M., Ledwith, A., and Sherrington, D. C. (1971), *Polymer* **12**, 509.

Brown, H. C., and Holmes, R. R. (1956), *J. Am. Chem. Soc.* **78**, 2173.

Brown, C. P., and Mathieson, A. R. (1957a), *J. Chem. Soc.*, 3612, 3620, 3625, 3631.

Brown, C. P., and Mathieson, A. R. (1957b), *Trans Faraday Soc.* **53**, 1033.

Brown, C. P., and Mathieson, A. R. (1958), *J. Chem. Soc.*, 3445 and earlier papers.

Burg, B. A., and Green, A. A. (1943), *J. Am. Chem. Soc.*, **65**, 1838.

Burton, H., and Praill, P. F. G. (1954), *Chem. Ind.*, 90.

Burton, R. E., and Pepper, D. C. (1961), *Proc. Chem. Soc.* **A263**, 58.

Chalmers, W. (1932), *Can. J. Res.* **7**, 464 and 472.

Chung, Y. J., Rooney, J. M., Squire, R. D., and Stannett, V. (1975), *Polymer* **16**, 527.

Cohen, S., Belliard, P., and Maréchal, E. (1973), *Polymer* **14**, 352.

Colclough, R. O., and Dainton, F. S. (1958), *Trans. Faraday Soc.* **54**, 886.

Cook, D. (1961), *Can. J. Chem.* **39**, 1184.

Cook, D. (1963a), *Can. J. Chem.* **41**, 522.

Cook, D. (1963b), *Can. J. Chem.* **41**, 515.

Cotrel, R., Sauvet, G., Vairon, J. P., and Sigwalt, P. (1973), *Int. Symp. Cationic Polym.*, *Rouen.*

Cotrel, R., Sauvet, G., Vairon, J. P., and Sigwalt, P. (1976), *Macromolecules* **9**, 931.

Cotter, L. J., and Evans, A. G. (1959), *J. Chem. Soc.*, 2988.

Cotton, F. A., and Letto, J. R. (1959), *J. Chem. Phys.* **30**, 993.

Denison, J. T., and Ramsey, J. B. (1955), *J. Am. Chem. Soc.* **77**, 2615.

Deters, J. F., Mc Cusker, P. A., and Pilger, R. C., Jr. (1968), *J. Am. Chem. Soc.* **90**, 4583.

Drago, R. S., Vogel, G. C., and Needham, T. E. (1971), *J. Am. Chem. Soc.* **93**, 6014.

Drago, R. S., Nozari, M. S., and Vogel, G. C. (1972), *J. Am. Chem. Soc.* **94**, 90.

Drago, R. S., Nusz, J. A., and Courtright, R. C. (1974), *J. Am. Chem. Soc.* **96**, 2082.

Eisler, B., Farnsworth, S. D., Kendrick, E., Schnurmann, P., and Wassermann, A. (1952), *J. Polym. Sci.* **8**, 157.

Eley, D. D., and Johnson, A. F. (1964), *J. Chem. Soc.*, 2238.

Evans, A. G., Holden, D., Plesch, P. H., Polanyi, M., Skinner, H. A., and Weinberger, N. A., (1946) *Nature* **157**, 102.

Evans, A. G. (1951), *J. Appl. Chem.* **1**, 240.

Evans, A. G., Jones, N., and Thomas, J. H. (1955), *J. Chem. Soc.*, 1824.

Evans, A. G., Jones, N., Jones, P. M. S., and Thomas, J. H. (1956), *J. Chem. Soc.*, 2757.

Evans, A. G., and Lewis, J. (1957a), *J. Chem. Soc.*, 2975.

Evans, A. G., Jones, P. M. S., and Thomas, J. H. (1957b), *J. Chem. Soc.*, 104 and 2095.

Evans, A. G., and Lewis, J. (1959), *J. Chem. Soc.*, 1946.

Evans, A. G., James, E. A., and Owen, E. D. (1961), *J. Chem. Soc.*, 3532.

Field, N. D. (1960), *J. Org. Chem.* **25**, 1006.

Fontana, C. M. (1959), *J. Phys. Chem.* **63**, 1167.

Fuoss, R. M. (1934), *Trans. Faraday Soc.* **30**, 967.

Fuoss, R. M. (1958), *J. Am. Chem. Soc.* **80**, 5059.

Fuoss, R. M. and Accasina, F. (1959), *Electrolytic Conductance*, Interscience, New York.

Furukawa, J., Kobayashi, E., and Taniguchi, S. (1974), *Bull. Inst. Chem. Res. Kyoto Univ.* **52**, 472.

Gandini, A., and Plesch, P. H. (1964), *Proc. Chem. Soc.*, 246.

Gandini, A., and Plesch, P. H. (1965a), *J. Polym. Sci. B Polym. Lett.* **3**, 127.

Gandini, A., and Plesch, P. H. (1965b), *J. Chem. Soc.*, 4765.

Gandini, A., and Plesch, P. H. (1965c), *J. Chem. Soc.*, 4826.

Gandini, A., Giusti, P., Plesch, P. H., and Westermann, P. H. (1965d), *Chem. Ind.*, 122.

Gandini, A., and Plesch, P. H. (1968), *Eur. Polym. J.* **4**, 55.

Gilkerson, W. R. (1956), *J. Chem. Phys.* **25**, 1199.

Gillespie, R. J. (1963), in *Friedel–Crafts and Related Reactions*, G. Olah, Ed., Interscience, New York, Vol. 1, p. 169.

Goka, A. M., and Sherrington, D. C. (1975) *Polymer* **16**, 819.

Greenwood, N. N., and Perkins, B. G. (1960a), *J. Chem. Soc.*, 1141.

Greenwood, N. N., and Perkins, P. G. (1960b), *J. Chem. Soc.*, 1145.

Güterbock, H. (1959), *Polyisobutylen*, Singer, Berlin.

Hatada, K., Takeshita, M., and Yuki, H. (1968) *Tetrahedron Lett.* 4621.

Hawke, D. L., and Steigman, J. (1954), *Anal. Chem.* **26**, 1989.

Hayes, M. J., and Pepper, D. C. (1961), *Proc. Chem. Soc.* **A263,** 63.

Hehre, W. J. (1975), *Acc. Chem. Res.* **8**, 369.

Heublein, G., Spange, S., and Hallpap, P. (1979), *Makromol. Chem.*, **180**, 1935.

Higashimura, T., and Okamura, S. (1960), *Kobunshi Kagaku* **13**, 57.

Higashimura, T., Fukushima, T., and Okamura, S. (1967), *J. Macromol. Chem.* **A1**, 683.

Hunter, W. H., and Yohe, R. V. (1933), *J. Am. Chem. Soc.* **55**, 1248.

Immergut, E. H., Kollmanan, G., and Malatesta, E. (1961), *J. Polym. Sci.* **51**, 557.

Jackson, R. (1975), Ph.D. Thesis, Liverpool.

Johnson, A. F., and Pearce, D. A. (1976), *J. Polym. Sci.* **56**, 157.

Jordan, D. O. and Treloar, F. E. (1961), *J. Chem. Soc.*, 734, 737.

Kagiya, T., Sumida, Y., and Inoue, T. (1968), *Bull. Chem. Soc. Jap.* **41**, 767.

Kakubari, H., Konaka, S., and Kimura, M. (1974), *Bull. Chem. Soc. Jap.* **47**, 2337.

Ketley, A. D., and Harvey, M. C. (1961), *J. Org. Chem.* **26**, 4649.

Kennedy, J. P., and Thomas, R. M. (1962a), *Macromol. Chem.* **53**, 28.

Kennedy, J. P., and Thomas, R. M. (1962b), *Adv. Chem. Ser.* **34**, 111.

Kennedy, J. P. (1964a), *Encyclopedia of Polymer Science and Technology*, Wiley-Interscience, New York, Vol. 7, p. 754.

Kennedy, J. P., Elliot, J. J., and Hudson, B. E., Jr. (1964b), *Macromol. Chem.* **79**, 109.

Kennedy, J. P., and Makowski, H. S. (1967), *J. Macromol. Sci. (Chem)* **A1**, 345.

Kennedy, J. P., and Makowski, H. S. (1968), *J. Polym. Sci.* **C22**, 247.

Kennedy, J. P., and Milliman, G. E. (1969), *Adv. Chem. Ser.* **91**, 287.

Kennedy, J. P., and Gillham, J. K. (1971), *Polym. Prepr.*, **12**,(2), 463.

Kennedy, J. P., Melby, E., and Johnston, J. E. (1974a), *J. Macromol. Sci. Chem.* **A8**, 463.

Kennedy, J. P., and Rengachary, S. (1974b), *Adv. Polym. Sci.*, **14**, 1.

Kennedy, J. P. (1975), *Cationic Polymerization of Olefins: A Critical Inventory*, Wiley-Interscience, New York.

Kennedy, J. P., Huang, S. Y., and Feinberg, S. C. (1977), *J. Polym. Sci., Chem. Ed.* **15**, 2801.

Kennedy, J. P., and Trivedi, P. D. (1978), *Adv. Polym. Sci.* **28**, 83.

Lappert, M. F. (1962), *J. Chem. Soc.*, 542.

Ledwith, A., Lockett, E., and Sherrington, D. C. (1975) *Polymer* **16**, 31.

Luder, W. F., and Zuffanti, S. (1961), *The Electronic Theory of Acids and Bases*, Wiley, New York.

Ma, C. C., Kubota, H., Rooney, J. M., Squire, D. R., and Stannett, V. (1979), *Polymer* **20**, 317.

Mc Millin, D. R., and Drago, R. S. (1972), *Inorg. Chem.* **11**, 872.

Maréchal, E., Basselier, J. J., and Sigwalt, P. (1964), *Bull. Soc. Chim. Fr.* 1740.

Maréchal, E. (1973), *J. Macromol. Sci. Chem.* **A7**, 433.

Margerison, D., and Newport, J. P. (1963), *Trans. Faraday Soc.* **59**, 2058.

Marks, A. P., and Drago, R. S. (1975), *J. Am. Chem. Soc.*, **97**, 3324.

Marvel, C. S., Gilkey, R., Morgan, C. R., North, J. F., Rands, R. D. and Young, C. H. (1951), *J. Pol. Sci.* **6**, 483.

Marvel, C. S. and Gall, E. J. (1960), *J. Org. Chem.* **25**, 1784.

Mente, D. C., Mills, J. L., and Mitchell, R. E. (1975), *Inorg. Chem.* **14**, 123.

Metz, D. J. (1961), *J. Pol. Sci.* **50**, 497.

Metz, D. J. (1969), "Radiation induced ionic polymerization: Addition and condensation polymerization processes," *Adv. Chem. Ser.* **91**, 202.

Moody, F. B. (1961), *Polymer Prepr.* **2**, 285.

Norrish, K. G., and Russell, K. E. (1952), *Trans. Faraday Soc.* **48**, 91.

Nozari, M. S., Jensen, C. D., and Drago, R. S. (1973), *J. Am. Chem. Soc.* **95**, 3162.

Okamura, S., Higashimura, T., and Sakurada, H. (1958), *Kogyo Kagaku Zasshi* (*J. Chem. Soc. Japan–Ind. Chem.*) **61**, 1640.

Okamura, S., Higashimura, T. and Sakurada, Y. (1959), *Chem. High Polymers* (Japan) **16**, 49.

Olah, G. A. (1963), in *Friedel–Crafts and Related Reactions*, Vol. 1, G. A. Olah, Ed., Wiley-Interscience, New York.

Olah, G. A., Halpen, Y., Shen, J., Mo, Y. K. (1971), *J. Am. Chem. Soc.* **93**, 1251.

Olah, G. A. (1974), *Makromol. Chem.* **175**, 1039.

Olah, G. A. (1974), *Carbocations and Electrophilic Reactions*, Verlag Chemie, Wiley, New York.

Pac, J., and Plesch, P. H. (1966), *Symp. Macromol.* (Tokyo).

Pac, J., and Plesch, P. H. (1967), *Polymer*, **8**, 237.

Pearson, R. G. (1973), in *Hard and Soft Acids and Bases*, R. G. Pearson, Ed., Dowden Hutchinson, Stroudsburg, Pa., p. 72.

Penczek, S., Jagur-Grodzinski, J., and Szwarc, M. (1968), *J. Am. Chem. Soc.* **90**, 2174.

Pepper, D. C. (1949), *Trans. Faraday Soc.* **45**, 397.

Pepper, D. C. (1954), *Quart. Rev.* **8**, 88.

Pepper, D. C. (1959), *Int. Symp. Macromol. Chem., Wiesbaden*, paper III, A9.

Pepper, D. C., and Reilly, P. J. (1961), *Proc. Chem. Soc.*, 200 and 460.

Pepper, D. C., and Reilly, P. J. (1962), *J. Polym. Sci.* **58**, 639.

Pepper, D. C., and Reilly, P. J. (1966), *Proc. Roy. Soc. A* **291**, 41.

Plesch, P. H., Polanyi, M., and Skinner, H. A. (1947), *J. Chem. Soc.*, 257.

Plesch, P. H. (1964), *J. Chem. Soc.*, 104.

Plesch, P. H. (1966), *Pure Appl. Chem.* **12**, 117.

Plesch, P. H. and Gandini, A. (1966), *The Chemistry of Polymerization Processes*, Monograph 20, Society of Chemical Industry, London.

Plesch, P. H. (1968), *Prog. High Polym.* **2**, 137.

Plesch, P. H. (1971), *Adv. Polym. Sci.* **8**, 137.

Plotnikov, V. A. and Yakubson, S. I. (1938), *J. Phys. Chem., U.S.S.R.* **12**, 120.

Proell, W. A., Adams, C. E., and Schoemaker, B. H. (1948), *Ind. Eng. Chem.* **40**, 1129.

Roberts, D. T., Jr., and Calihan, L. E. (1973a), *J. Macromol. Sci. Chem.* **A7**, 1629.

Roberts, D. T., Jr., and Calihan, L. E. (1973b), *J. Macromol. Sci. Chem.* **A7**, 1641.

Saegusa, T., Imai, H., Ueshima, T., and Furukawa, J. (1965), *Shokubai (Tokyo)* 7, 43.

Saegusa, T. (1968), Imai, H., and Matsumoto, S., *J. Polym. Sci.* A6, 459.

Saegusa, T. (1969), in *Structure and Mechanism in Vinyl Polymerization*, pp. 283, 287.

Sambhi, M. S., and Treloar, F. E. (1965), *J. Polym. Sci. B* 3, 445.

Sartori, G., Valvarossi, A., Turba, V., and Lachi, M. P. (1963), *Chim. Ind. (Milan)*, 45, 1529.

Sauvet, G., Varion, J. P., and Sigwalt, P. (1967), *C. R. Acad. Sci.* 265c, 1090.

Sauvet, G., Vairon, J. P., and Sigwalt, P. (1974), *Eur. Polym. J.* 10, 501.

Sawada, H. (1972), *Thermodynamics of Polymerization*, Dekker, New York, p. 161.

Sawamoto, M., Masuda, T., and Higashimura, T. (1976), *Makromol. Chem.* 177, 2995.

Sawamoto, M., Masuda, T., and Higashimura, T. (1977), *Makromol. Chem.* 178, 1497.

Sebenik, A., and Harwood, H. J. (1981), *Polymer Prepr.* 22, 15.

Seymour, E. L. (1943), Thesis, Manchester University.

Sigwalt, P., and Maréchal, E. (1966), *Eur. Polym. J.* 2, 15.

Skinner, H. A. (1953), in *Cationic Polymerization and Related Complexes*, P. H. Plesch, Ed., Academic Press, New York, p. 28.

Subira, F., Sauvet, G., Vairon, J. P., and Sigwalt, P. (1976), *J. Pol. Sci. Symp.* 56, 221.

Swain, C. G., and Scott, C. B. (1953), *J. Am. Chem. Soc.* 75, 141.

Symposium on Hard and Soft Acids and Bases (1965), summaries of papers in *Chem. Eng. News* 43, 90.

Szusz, B. P., and Chalandon, P. (1958) *Helv. Chim. Acta* 41, 1332.

Throssel, J. J., Sood, S. P., Szwarc, M., and Stannett, V. (1956), *J. Am. Chem. Soc.*, 78, 1122.

Tortai, J. P., and Maréchal, E. (1971a), *Bull Soc. Chim.*, 2673.

Tortai, J. P., Mayen, M., and Maréchal, E. (1971b), *IUPAC Boston—Macromol. Prep.* 1, 172.

Upadhyay, J., Gaston, P., Levy, A. A., and Wassermann, A. (1965), *J. Chem. Soc.*, 3252; *Polym. Lett.* 3, 369.

Van Lohuizen, O. E. and De Vries, K. S. (1968), *J. Polym. Sci.* C16, 3943.

M. Villesange, G. Sauvet, J. P. Vairon and P. Sigwalt (1980), *Polym. Bull.* 2, 131.

Walker, D. G. (1960), *J. Phys. Chem.* 64, 939.

Walker, D. G. (1961), *J. Phys. Chem.* 65, 1367.

Wichterle, O., Kolinski, M. and Marek, M. (1959), *Coll. Czech. Chem. Commun.* 24, 2473.

Williams, F. K., Hayashi, K., and Okamura, S. (1966), *Polym. Prepr., ACS Div. Polym. Chem.* 7, 479.

Worsfold, W. J. and Bywater, S. (1957), *J. Am. Chem. Soc.* 79, 4917.

Yuki, H., Hatada, K., and Takeshita, M. (1969), *J. Polym. Sci.* A1 7, 667.

Zlamal, Z., Ambroz, L., and Vesely, K. (1957), *J. Polym. Sci.* 24, 285.

□ 4 □
The Chemistry of Carbocationic Polymerization

4.1 □ THE CHEMISTRY OF INITIATION

Definitions and Scope

Initiation of carbocationic polymerization involves, in principle, (1) ion generation or "priming" and (2) cationation. Ion generation is the production of a cation of sufficient activity for cationation of an olefin. Cationation is the addition of the first cation formed by priming to the π double bond system of an olefin. Ion generation as such is obviously unnecessary with stable cation salts that are already ionized. Initiation is complete on the formation of the first propagating carbenium ion. Subsequent to initiation polymerization proceeds by further repetitive cationation of olefin molecules.

Carbocationic polymerizations may be induced by a variety of methods:

  I. Chemical methods
    A. Two-electron (heterolytic) transpositions
      1. Brønsted (protic) acids
      2. Carbenium ion salts
      3. Friedel–Crafts acids
      4. Miscellaneous systems
    B. One-electron (homolytic) transpositions
      1. Direct radical oxidation
      2. Charge transfer polymerization
  II. Physical methods
      1. High-energy irradiation
      2. UV radiation
      3. High electric field: field emission and field ionization
      4. Electroinitiation

Chemical Methods

Chemical initiation of carbocationic polymerization may involve two-electron and one-electron transpositions. Chemical initiation by two-electron transposition occurs in systems in which the initiating entity is a proton or carbenium ion and in which ion generation proceeds by heterolytic bond breakage or dissociation of cation precursors. This group includes conventional initiating systems, for example, Brønsted acids, Brønsted acid/Friedel–Crafts acid combinations, and stable cation salts. In contrast, chemical initiation by one-electron transposition proceeds in polymerization induced by direct radical oxidation or charge transfer interactions involving electron donor monomers. Charge transfer polymerizations may be initiated by thermal or photochemical processes, for example, the polymerization of N-vinylcarbazole in the presence of chloranil (thermal) or fumaronitrile (photochemical).

Two-Electron (Heterolytic) Transportations

In somewhat of a departure from previous literature, chemical processes that induce vinyl carbocationic polymerizations are subdivided into two separate events, *ion generation*, or "priming," and *cationation*. This classification arises from consideration of fundamentally different chemistries, kinetics, and thermodynamics of these processes. Ion generation may occur via various mechanisms, for example, ionization of simple protic acids to proton and counteranion in polar nonnucleophilic media, ionization and dissociation of Brønsted acid/Friedel–Crafts acid combinations or halogen/Friedel–Crafts acid systems, or the more speculative self-ionization of certain Friedel–Craft acids. This can be represented schematically as follows:

- Ion generation or "priming"

$$HA \rightleftharpoons H^{\oplus} + A^{\ominus}$$

or

$$RX/MeX_n \rightleftharpoons R^{\oplus} + MeX_{n+1}^{\ominus}$$

or

$$X_2/MeX_n \rightleftharpoons X^{\oplus} + MeX_{n+1}^{\ominus}$$

or

$$2MeX_n \rightleftharpoons MeX_{n-1}^{\oplus} + MeX_{n+1}^{\ominus}$$

Subsequent to priming, that is, to the birth of an ion of *sufficient* activity for olefin cationation, cationation completes initiation. Depending on the nature of the initiator used, cationation may involve proton, carbenium ion, halonium ion, or electrophilic Friedel–Crafts acid fragments.

- Cationation

$$H^{\oplus} + C = C \rightarrow H - C - C^{\oplus}$$

or

$$R^{\oplus} + C = C \rightarrow R - C - C^{\oplus}$$

or

$$X^{\oplus} + C = C \rightarrow X - C - C^{\oplus}$$

or

$$MeX_{n-1}^{\oplus} + C = C \rightarrow MeX_{n-1} - C - C^{\oplus}$$

In the overwhelming majority of scientifically and technologically important chemically induced carbocationic polymerizations distinction between priming and cationation is obvious. Instances in which separation of these events is somewhat obscured by a plethora of other possibilities are discussed below. In a few polymerizations induced by physical methods, that is, ionizing radiation, a formal separation into ion generation and cationation processes is less discernible and they are not considered separately.

The subdivision of the overall initiation process into ion generation and cationation is, of course, still a gross simplification, and for a detailed discussion of polymerization mechanisms these events should be subdivided into series of complex processes culminating in the formation of the first propagating entity. Events or processes that have to be considered for a comprehensive understanding of ion generation may include, for example, dissociation of dimeric Friedel–Crafts acids (AlCl$_3$; AlBr$_3$), complexation between Friedel–Crafts acid and monomer, solvation, desolvation, displacement of solvent by initiator in Friedel–Crafts acid–solvent complexes, ionization of Friedel–Crafts acid–initiator complexes, and dissociation of such complexes leading to polymerization active ions. Similarly, cationation may also embrace a series of intermediates or transition states up to the formation of the first propagating species. Also, these processes may be chronologically separate or concerted. In select instances references to these detailed processes are made.

And another caveat: the chemistry of carbocationic polymerization initiation can be discussed only in terms of a purely experimental, indeed empirical, science devoid of fundamental calculations or quantitative mathematical equations. Organic chemical principles and intuition are used to account for a large body of seemingly disparate observations. One of the perennial problems that faces the theoretician is that practically all polymerizations worth investigating have been carried out in the liquid phase, and fundamental information available from gas-phase or theoretical work is of limited validity for carbocationic polymerization occurring in the condensed state. We are particularly far from quantizing the effect of the nature of solvent (polar/nonpolar) and counteranion on carbocationic reactions in general and polymerization in particular. In the absence of quantitative understanding of these influences there is little hope for a detailed quantitative description of any of the elementary events leading to polymer formation including initiation. Our best efforts are made throughout this treatment to apply as much as possible the fundamental principles of physical-organic chemistry and to use these as the basis for discussing details of cationic polymerization mechanism.

Brønsted (Protic) Acids

Surprisingly little fundamental (gas phase, quantitative) experimentation has been carried out relative to the mechanism of protonation of olefins by Brønsted (protic) acids. The first careful study of gas phase addition of HCl to propylene has only recently been published (Haugh and Dalton, 1975). The authors found the expected 2-chloropropane as the only product; however, they concluded that trace quantities of polar molecules such as water (!) serve as homogeneous catalysts, or under the dryest conditions surface catalysis takes over. Both the homogeneous and heterogeneous processes are first order in propylene and third order in hydrogen chloride.

These results are quite similar to much earlier ones (Mayo and Katz, 1947) obtained by using isobutylene and hydrogen chloride in heptane solution.

Brønsted or protic acids HA are proton sources and as such may function as cationic polymerization initiators. The overall reactivity of a Brønsted acid in regard to a particular olefin is mainly governed by the proton affinity of the olefin. The utility of a Brønsted acid for initiation of carbocationic polymerization is determined by the nucleophilicity of its conjugate base $A^{\ominus}$:

$$HA \rightleftharpoons H^{\oplus} + A^{\ominus}$$

If the nucleophilicity of the conjugate base (counteranion) is high, protonation of the olefin may rapidly be followed by ion pair collapse, a process equivalent to termination of a propagating chain:

$$H^{\oplus} + A^{\ominus} + C{=}C \rightarrow H{-}C{-}C^{\oplus} + A^{\ominus} \rightarrow H{-}C{-}C{-}A$$

In this sense the addition of a Brønsted acid across a double bond may be viewed as initiation followed by immediate termination or "polymerization without propagation." If the nucleophilicity of the conjugate base $A^{\ominus}$ is low, the life-span of the cation may increase and in the presence of a suitable olefin propagation may occur. For example, mixing HCl and isobutylene rapidly produces *tert*-butyl chloride. Evidently the nucleophilicity of $Cl^{\ominus}$ is high and the rapid collapse of the ion pair prevents polymerization. In contrast, with H_2SO_4 or H_3PO_4, that is, with Brønsted acids having less nucleophilic conjugate bases $HSO_3^{\ominus}$ or $H_2PO_4^{\ominus}$, isobutylene dimers or trimers are obtained (Butlerov, 1877; Güterbock, 1959; Eidus and Nefredov, 1960).

Brønsted acids alone can lead to rapid polymerization and high molecular weight product only with the most reactive cationic monomers. N-vinylcarbazole, for example, rapidly polymerizes with HCl in toluene solution (Jones, 1963). Although high molecular weight polyisobutylenes cannot be obtained by simple Brønsted acids alone, Brønsted acid/Friedel–Crafts acid combinations produce high polymers because the Friedel–Crafts acid is able to stabilize by complexation the highly nucleophilic conjugate bases of protic acids (see below).

It is of interest to consider thermodynamic aspects of propylene oligomerization (Sawada, 1972). The enthalpy change ΔH accompanying the addition of a halogen acid HA to olefin

$$HA + C{=}C \rightarrow H{-}C{-}C^{\oplus} + A^{\ominus}$$

is composed of

$$HA \rightarrow H^{\oplus} + A^{\ominus} \qquad\qquad \Delta H_1$$

$$H^{\oplus} + C{=}C \rightarrow H{-}C{-}C^{\oplus} \qquad\qquad \Delta H_2$$

$$H-\overset{|}{\underset{|}{C}}-\overset{|}{\underset{|}{C}}{}^{\oplus} + A^{\ominus} \rightarrow H-\overset{|}{\underset{|}{C}}-\overset{|}{\underset{|}{C}}{}^{\oplus}A^{\ominus} \qquad\qquad \Delta H_3$$

where $\Delta H_1 =$ enthalpy change for heterolytic dissociation of HA, $\Delta H_2 =$ proton affinity of olefin, and $\Delta H_3 = e^2/\varepsilon r$, the potential energy of two ions at a distance r in a medium of dielectric constant ε.

$$\Delta H = \Delta H_1 + \Delta H_2 + \Delta H_3$$

$$\Delta G = \Delta H - T\Delta S$$

Since ΔS is negative, ΔH must be negative for reaction to occur. For the propylene–HCl system:

$$\Delta H_1 = 330 \text{ kcal/mole}$$

$$\Delta H_2 = -175 \text{ kcal/mole}$$

$$\Delta H_3 = -130 \text{ kcal/mole}$$

According to these enthalpies ΔG is positive and polymerization cannot occur. However, it is well-known that HCl and HBr initiate oligomerization of olefins, including propylene, in suitable solvents. Evidently, ΔH_1 is strongly reduced by solvation (ΔH_1 for aqueous HCl is ~ 25 kcal/mole). Thus in conjunction with HCl the polar solvent may be regarded as the coinitiator because it reduces the energy of ion generation and renders initiation feasible.

Solvation by polar solvents may help to stabilize the charged intermediates and delay the collapse of the ion pair. Thus HCl or HBr in polar solvents, for example, ethylene dichloride, have been mentioned to polymerize styrene, but the yields and molecular weights are still very low (Pepper, 1959). In contrast, H_2SO_4 (Albert and Pepper, 1951), CCl_3COOH (Brown and Mathieson, 1957, 1958), CF_3COOH (Pepper, 1959), and $HClO_4$ (Pepper and Reilly, 1966), that is, Brønsted acids with relatively low nucleophilic conjugate bases, rapidly produce relatively high molecular weight polystyrenes.

The effect of solvent polarity on cationic initiation was dramatically illustrated by Throssel et al. (1956). These authors showed that the order of reagent addition in the CF_3COOH/styrene system is of critical importance for the success of polymerization: when CF_3COOH was added to styrene practically no polymerization occured; however, when styrene was added to CF_3COOH instantaneous polymerization to high molecular weight product ensued. Evidently in the former case ion generation could not occur because of the low polarity (dielectric constant) of the styrene medium, whereas in the latter instance, owing to the high polarity of CF_3COOH and stabilization by complexation, polymerization could proceed readily.

These simple examples forcefully illustrate that efficient cationic polymerizations to high molecular weight products can occur only in the presence of relatively stable, weakly nucleophilic conjugate bases (counteranions) so that collapse of the propagating ion pair is prevented or at least delayed so that propagation to high polymers can proceed.

Although Brønsted acids carry their own conjugate base, that is, have "built-in" counteranions, a large amount of work is being devoted to find ways to control (delay) externally the rate of ion pair collapse.

The rate of reaction between a particular cation and conjugate base to neutral species can be controlled to a modest degree by reaction conditions, that is, medium polarity, temperature, and concentrations. Under the least favorable conditions for polymerization (relatively low stability carbenium ion, highly nucleophilic conjugate base, nonpolar medium, high temperature) protonation may be rapidly followed by collapse of the ion pair. Under more favorable conditions (relatively stable carbenium ion, weakly nucleophilic conjugate base, polar medium, low temperature), rapid cationic polymerization to high molecular weight product may occur.

The collapse of the cation–conjugate base pair can also be delayed by nonspecific complexation with metals. Thus Aoki et al. (1968) studied the polymerization of isobutyl vinyl ether by HCl in the presence of metals, for example, Ni, Co, Fe and Ca and their oxides, V_2O_5, PbO_2, SiO_2, active Al_2O_3, and powdered glass. Polymerization did not occur and the vinyl ether was hydrochlorinated in the absence of these additives. Separate experiments showed that the metals and/or metal compounds M in the absence of HCl also did not induce polymerization. According to these authors such additives aid in the dissociation of HCl and in the stabilization of $Cl^\ominus$ by complexation:

$$HCl + M \rightarrow H^\oplus[Cl \rightarrow M]^\ominus \xrightarrow{\text{i-BuVE}} CH_3{-}\overset{\oplus}{\underset{\underset{OC_4H_9}{|}}{C}}H[Cl \rightarrow M]^\ominus \rightarrow polymer$$

which in turn tips the kinetics in favor of propagation.

The concentration of the Brønsted acid may also be important. For example, with α-methylstyrene the concentration of H_2SO_4 largely determines oligomerization *and* the composition of product mix (Hersberger et al., 1947). Similarly, the oligomerization of propylene gives quite different results with dilute or concentrated (92 to 109%) H_3PO_4 (Monroe and Gilliland, 1938; Langlois, 1953; Bethea and Karchmer, 1956). The minimum temperature for propylene oligomerization is 100 to 130°C lower with concentrated H_3PO_4 than that with dilute acid (Bethea and Karchmer, 1956). The reaction order in respect to monomer is 2 with dilute acid and with concentrated H_3PO_4, which may be due to different aggregates having different initiating activities in these systems (Fontana, 1963a; Sawamoto et al., 1975).

Temperature has been demonstrated to control the path of these oligomerizations. Lowering the temperature helps to stabilize active species and reduces the rate of proton elimination. For example, the interaction between styrene and CF_3SO_3H leads to high molecular weight product ($\bar{M}_n \sim 10^3$) at 0°C but to dimer at 50°C (Sawamoto et al., 1975).

The path of proton elimination from styryl cation can be controlled by the nature of the acid employed:

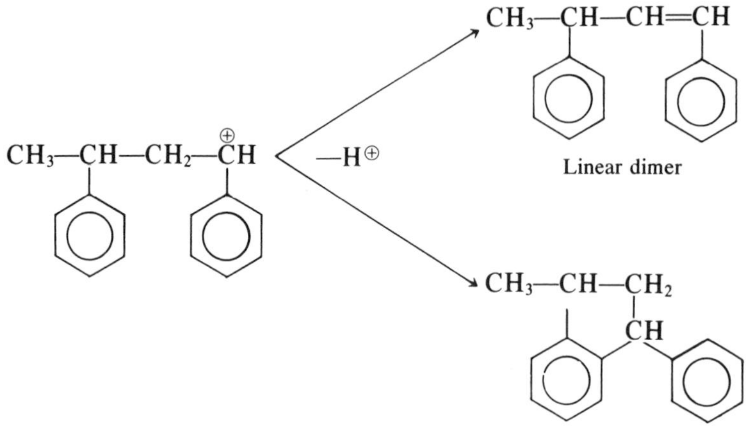

CH₃—CH—CH=CH

Linear dimer

CH₃—CH—CH₂
          CH

Cyclic dimer

CH₃—CH—CH₂—CH (⊕) —H⊕

The linear dimer is formed with high selectivity using CF_3SO_3H or CH_3COCl_4 in benzene solvent at 50°C (Sawamoto et al., 1975), whereas the cyclic dimer by product arises during polymerization with $HClO_4$ (Gandini and Plesch, 1968; Bertoli and Plesch, 1968; Bywater and Worsfold, 1966). Acetyl perchlorate also produces selectively linear dimer in various solvents (Higashimura and Nishii, 1977c). According to Higashimura and Nishii (1977c) multidentate anions are essential for the selective formation of the linear dimer:

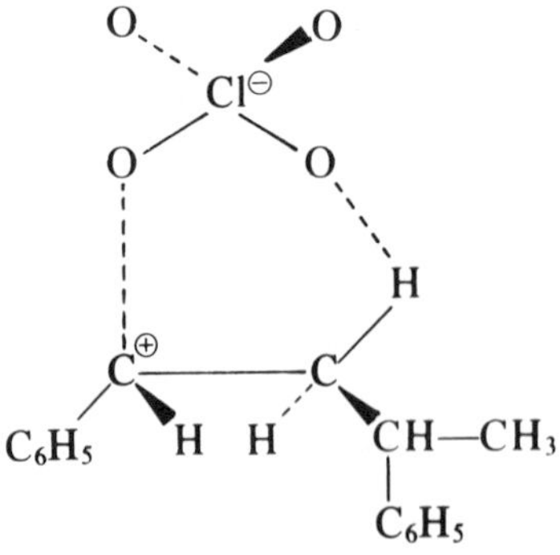

Since Friedel–Crafts acids may also produce multidentate anions ($AlCl_4^{\ominus}$ or $BF_4^{\ominus}$ could readily be substituted in this intermediate for $ClO_4^{\ominus}$), the essential feature for linear dimer formation may rather be the presence of oxygen-containing counteranions and the possibility for hydrogen bridge formation.

Dimerization of α-methylstyrene by Brønsted acids has also been studied; however, it was found to be less controllable than that of styrene. The dimerization of this monomer produces a liquid (linear dimer, L) and/or a solid (cyclized dimer, S); the former is formed by deprotonation whereas the latter arises by intramolecular alkylation:

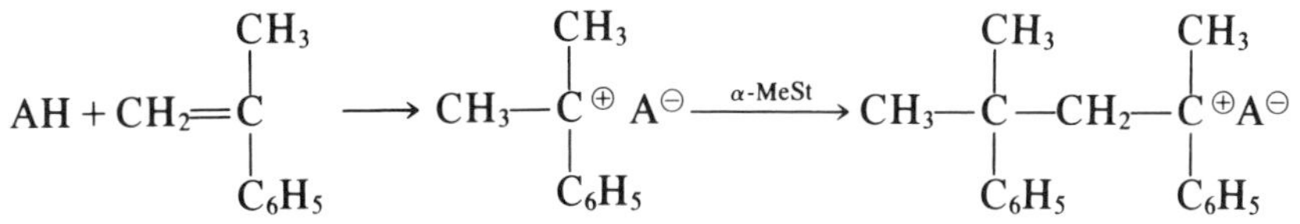

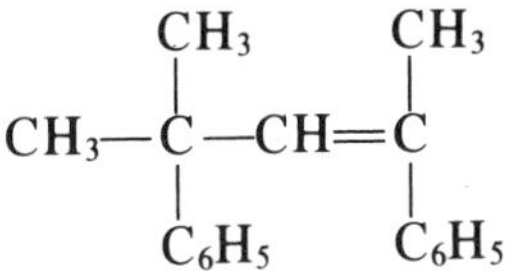

Linear dimer, L 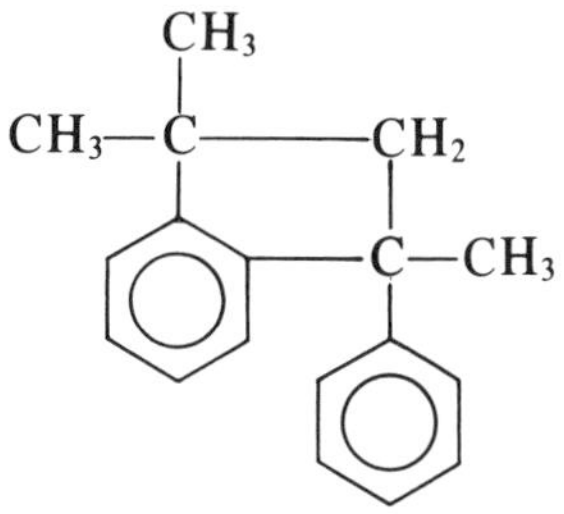   Cyclic dimer, S

In the presence of sulfuric acid, both acid concentration and temperature strongly affect the reaction path (Hersberger and Heiligmann, 1947). Table 4.1 shows some pertinent results. Neither the original patent source nor Bywater (1963), who analyzed these data, attempted to interpret the observations.

The patent literature is particularly rich in disclosures concerning protic acid systems, such as free or supported protic acids or protic acid mixtures in solution or in bulk, for the preparation of oligomers from olefins or olefin mixtures for adhesive, lubricant, viscosity improvers, oil additive, and similar applications. In spite of the obvious commercial significance of these products, systematic sustained research on oligomerization has not yet

Table 4.1 □ Influence of Reaction Conditions on the Nature of α-Methylstyrene Dimer[a]

Acid Concn. (%)	Temp. (°C)	Nature of dimer
< 60	20–80	Only L
70	< 40	Largely L
70	85	Largely S
80	> room	Only S

[a] After Hersberger and Heiligmann (1977).

reached the scientific literature, most likely because of the difficulties involved in the analysis of heterogeneous ill-defined blends of oligomers.

Stable Carbenium Ion Salts

Conceptually the simplest initiation of a carbocationic polymerization would be the cationation of a monomer by a stable, well-characterized free cation the structure of which is (almost) identical to that of the propagating cation derived from this monomer. This statement is illustrated by the *tert*-butyl cation–isobutylene or cumyl cation–α-methylstyrene systems:

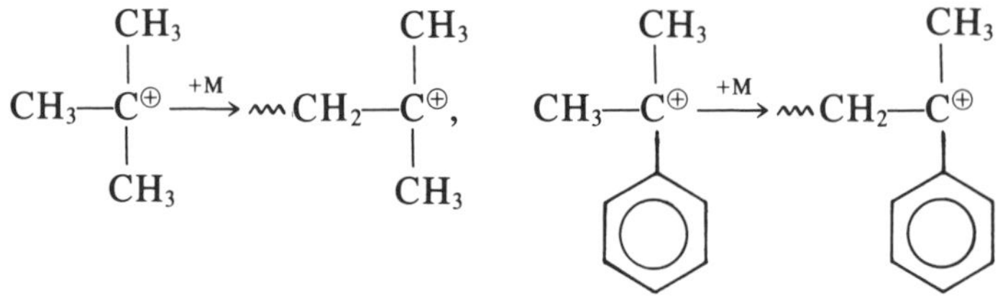

By the use of stable carbenium ion–counteranion salts in polar nonnucleophilic solvents in which these salts would exist as dissociated free ions, complicated ion generation could be eliminated and, provided side reactions were absent, valuable kinetic studies could be carried out. Experimentation along these lines has been carried out and insight into kinetic details of initiation has been obtained.

Elegant techniques for studying the kinetics of initiation of carbocationic polymerizations using stable carbenium ion salts have been developed by Ledwith (1969a); Ledwith and Sambhi (1965) and Ledwith and Sherrington (1975). These authors employed stable salts of trityl cation with anions such as $PF_6^{\ominus}$, $SbCl_6^{\ominus}$, and $AsF_6^{\ominus}$ for the polymerization of alkyl vinyl ethers, alkoxystyrenes, and N-vinylcarbazole (Ledwith, 1969b, Ledwith and Sherrington, 1975; Bawn et al., 1971; Cowell et al., 1970).

A variety of carbenium ion salts, for example, trityl, tropylium, trimethylpyrilium, triphenylpyrilium, and xanthylium, have been found to be indefinitely stable in crystalline form under conventional experimental conditions (Ledwith and Sherrington, 1975). Table 4.2 is a compilation of dissociation constants of various stable cation salts whose polymerization behavior has been investigated (Ledwith and Sherrington, 1975; Sherrington, 1969).

Several of these stable salts have been demonstrated to initiate carbocationic polymerization by well-defined direct addition to olefinic unsaturation. Polymerization initiated by stable cation salts readily leads to highly reproducible rates and has often been used in kinetic studies (cf. Chapter 5). For example, Sauvet et al. (1969, 1974) have shown that

Table 4.2 □ Ion Pair Dissociation Constants K_d for Various Stable Cation Salts[a]

Cation	Anion	Solvent	$10^4 K_d (M)$ −45°C	0°C	25°C	References
$Ph_3C^{\oplus}$	$SbCl_6^{\ominus}$	CH_2Cl_2			1.4	Kalfoglou and Szwarc (1968)
	$SbCl_6^{\ominus}$	CH_2Cl_2	5.3	3.1	1.9	Bowyer et al. (1971)
	$SbCl_6^{\ominus}$	CH_2Cl_2			7.0	Kubisa and Penczek (1971)
	$ClO_4^{\ominus}$	$(CH_2Cl)_2$			13(21°5)	Longworth and Mason (1966)
	$ClO_4^{\ominus}$	$(CH_2Cl)_2$			2.5	Lee and Treloar (1969)
	$ClO_4^{\ominus}$	Liq. SO_2			45	Lichin and Pappas (1957)
	$SbCl_4^{\ominus}$	CH_2Cl_2			0.7	Penczek and Kubisa (1973)
	$AsF_6^{\ominus}$	CH_2Cl_2			2.1	Pepper (1972)
	$SbF_6^{\ominus}$	CH_2Cl_2			1.7	Pepper (1972)
	$SnCl_5^{\ominus}$	CH_2Cl_2			1.5(19°C)	Subira et al. (1973)
$C_7H_7^{\oplus}$	$SbCl_6^{\ominus}$	CH_2Cl_2	0.73	0.3		Bowyer et al. (1971)
	$ClO_6^{\ominus}$	CH_2Cl_2		0.3		Bowyer (1972)
	$BF_4^{\ominus}$	CH_2Cl_2		0.7		Bowyer (1972)
	$Br^{\ominus}$	Liq. SO_2		13		Lichin and Pappas (1957)
$Me_2(PhCH_2)(Ph)N^{\oplus}$	$SbCl_6^{\ominus}$	CH_2Cl_2	1.2	0.59		Bowyer et al. (1971)
	$SbCl_6^{\ominus}$	CH_2Cl_2	0.51	0.49		Goka and Sherrington (1976)
$Et_4N^{\oplus}$	$SbCl_6^{\ominus}$	CH_2Cl_2	1.2	0.72		Bowyer et al. (1971)
	$SbCl_6^{\ominus}$	CH_2Cl_2	1.02	0.84		Goka and Sherrington (1976)
$(n\text{-}Bu)Et_3N^{\oplus}$	$SbCl_6^{\ominus}$	CH_2Cl_2	0.86	0.69		Goka and Sherrington (1976)

Table 4.2 □ (*Continued*)

Cation	Anion	Solvent	$10^4 K_d (M)$			References
			−45°C	0°C	25°C	
$(n\text{-Bu})(i\text{-}C_5H_{11})_3N^{\oplus}$	$SbCl_6^{\ominus}$	CH_2Cl_2	1.4	1.0		Bowyer et al. (1971)
	$BPh_4^{\ominus}$	CH_2Cl_2	2.8	2.5		Bowyer et al. (1971)
$(n\text{-Octyl})Et_3N^{\oplus}$	$SbCl_6^{\ominus}$	CH_2Cl_2	0.79	0.60		Goka and Sherrington (1976)
$(n\text{-Dodecyl})Et_3N^{\oplus}$	$SbCl_6^{\ominus}$	CH_2Cl_2	0.76	0.60		Goka and Sherrington (1976)
$(Cetyl)Et_3N^{\oplus}$	$SbCl_6^{\ominus}$	CH_2Cl_2	0.88	0.72		Goka and Sherrington (1976)
$Me_2(cetyl)(PhCH_2)N^{\oplus}$	$SbCl_6^{\ominus}$	CH_2Cl_2	0.71	0.54		Goka and Sherrington (1976)
$(p\text{-}BrC_6H_4)_3N^{\oplus}$	$SbCl_6^{\ominus}$	CH_2Cl_2	2.0	1.4		Goka and Sherrington (1976)
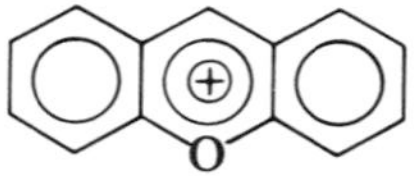	$SbCl_6^{\ominus}$	CH_2Cl_2	—	0.19		Bowyer et al. (1971)
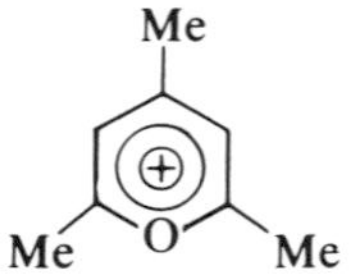	$SbCl_6^{\ominus}$	CH_2Cl_2	0.88	0.58		Bowyer et al. (1971)
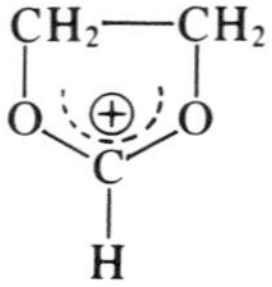	$AsF_6^{\ominus}$ or $SbF_6^{\ominus}$	CH_3NO_2			65	Penczek (1974)

$Et_3O^{\oplus}$	$SbCl_6^{\ominus}$	CH_2Cl_2	0.53(20°C)	Goethals (1974)
$Et_3S^{\oplus}$	$BF_4^{\ominus}$	CH_2Cl_2	0.36(20°C)	Drijvers and Goethals (1971)
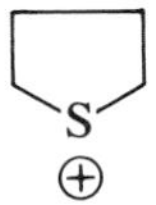 (ring)$S^{\oplus}$	$BF_4^{\ominus}$	CH_2Cl_2	0.53(20°C)	Drijvers and Goethals (1971)
	$BF_4^{\ominus}$	$C_6H_5NO_2$	135(20°C)	Drijvers and Goethals (1971)
	$BF_4^{\ominus}$	CH_2Cl_2	0.56(20°C)	Drijvers and Goethals (1971)
	$BF_4^{\ominus}$	$C_6H_5NO_2$	165(20°C)	Drijvers and Goethals (1971)

[a] After Ledwith et al. (1975).

initiation of cyclopentadiene polymerization by trityl hexachloroantimonate occurs by direct cationation:

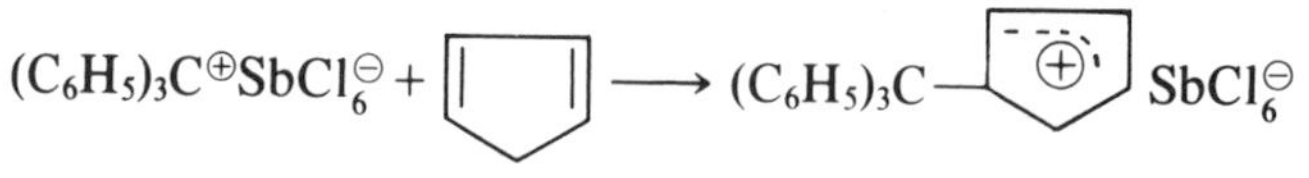

$$(C_6H_5)_3C^{\oplus}SbCl_6^{\ominus} + \boxed{} \longrightarrow (C_6H_5)_3C \boxed{\oplus} \; SbCl_6^{\ominus}$$

Very recently these workers (Villesange et al., 1980) have also found initiation of cyclopentadiene polymerization by cationation with a double-headed stable carbenium salt:

$$
\begin{array}{ccc}
C_6H_5 & & C_6H_5 \\
| & & | \\
SbCl_6^{\ominus\oplus}C-C_6H_4\!\!-\!\!(CH_2\!)_2\!C_6H_4\!-\!C^{\oplus}SbCl_6^{\ominus} \\
| & & | \\
C_6H_5 & & C_6H_5
\end{array}
$$

Similarly, initiation of styrene polymerization by the use of $(C_6H_5)_3CSbCl_6$ (Sauvet et al., 1967), $(C_6H_5)_3CSbF_6$ (Johnson and Pearce, 1976b), $(C_6H_5)_3CSnCl_5$ (Higashimura et al., 1967a) and $(C_6H_5)_3CHgCl_3$ (Sambhi and Treloar, 1967) have been concluded to occur by direct tritylation. Various groups of investigators have demonstrated initiation by direct trityl cation addition to p-methoxystyrene and vinyl alkyl ethers (Bawn et al., 1965b; Eley and Richards, 1949).

The possibility that trityl cation addition to olefin occurs at a para position on one of the phenyl rings and not at the sterically crowded central cation has been considered by Ledwith (1969a, b) and Kampmeier et al. (1966):

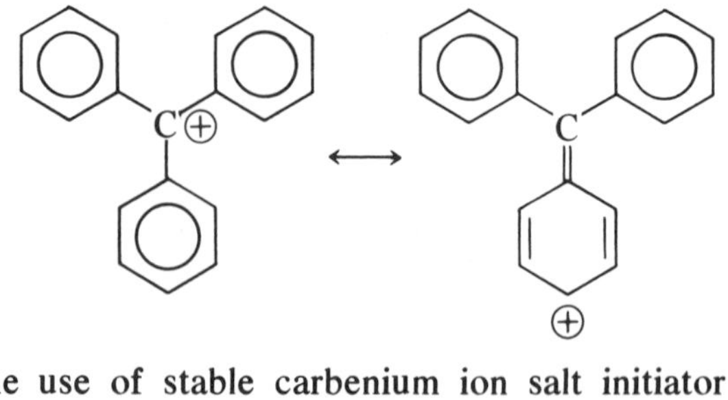

Although the use of stable carbenium ion salt initiator systems gave valuable insight into the initiation kinetics of some highly reactive, mostly aromatic and vinyl ether monomers, this technique could not be employed for the study of aliphatic olefin or diolefin polymerizations, most likely because the stability of these carbenium ion salts is too high to induce these reactions. It would therefore be of great interest to carry out similar investigations with more reactive carbenium salts, for example, $t\text{-Bu}^{\oplus}X^{\ominus}$. tert-Butyl cations have been proved to be efficient initiators for the polymerization of many olefins. Though research in this area is rendered much more difficult owing to the inherent instability of these ion pairs even at moderately low temperatures, experimentation at cryogenic temperatures appears feasible.

That *tert*-butyl carbocations exist has been firmly established by direct NMR observation in superacid medium, that is, HSO_3F diluent and $SbF_6^{\ominus}$ counteranion (Olah et al., 1964). However, polymerization would be difficult to carry out under these conditions because the super acids would rapidly attack and degrade high polymers during their formation in the reactor. In contrast, ion pairs such as $t\text{-}Bu^{\oplus} Me_3AlCl^{\ominus}$, $t\text{-}Bu^{\oplus}Et_2AlCl_2^{\ominus}$ almost certainly exist (Kennedy and Sivaram, 1973a; Kennedy et al. 1973b) and could be produced *in situ* by reacting *tert*-butyl halides with alkylaluminums at low temperatures in nonnucleophilic solvents. These ion pairs combine high reactivity toward many monomers with sufficient stability for rapid, efficient initiation, and would thus be amenable for kinetic studies.

Friedel–Crafts Acids

Introduction

Carbocationic polymerizations induced by Friedel–Crafts acids are no doubt the most important such processes from the technological–industrial point of view. Although a variety of chemical and physical initiation techniques have become available, industrial polymerization processes designed to yield oligomers or high polymers invariably employ Friedel–Crafts systems.

The single great advantage of Friedel–Crafts systems over those of Brønsted acids is their ability to prolong the lifetime of the kinetic chain and thus render propagation to high molecular weights possible. Owing to highly nucleophilic conjugate bases of Brønsted acids, carbocationic polymerization chains initiated by these acids are usually short and consequently lead to lowest molecular weight products. Friedel–Crafts acids are capable of coordinatively complexing conjugate bases of Brønsted acids and thus lead to remarkably stable counteranions. Mixtures of Brønsted acids with Friedel–Crafts acids, therefore, are among the strongest acids known and among a myriad of other applications found use as most effective initiators of carbocationic polymerizations.

This section concerns a concise review of fundamental information in the field of Friedel–Crafts-induced carbocationic polymerizations, and in particular, the organization of a great variety of experimental observations into a conceptually simple mechanistic system. First, information relative to cationogen/Friedel–Crafts acid systems is discussed. Subsequently, Friedel–Crafts acid systems that do not require the purposeful addition of cationogen will be examined. Finally, attempt is made to combine these seemingly disparate fields into a unified seamless web of initiation theory.

Cationogen/Friedel–Crafts Acid Systems

Cationogen = Brønsted Acids

STOPPING EXPERIMENTS. Systematic scientific investigation in the field of carbocationic polymerizations induced by Friedel–Crafts acid-based systems commenced with the discovery of the concept of coinitiation [Evans et al., 1946a, b, 1947a, b; Plesch et al., 1947]. Thus Evans et al. (1946b), in the course of their studies on the fundamentals of Lewis acid-induced polymerizations, demonstrated that purest BF_3 does not initiate the polymerization of isobutylene; however, the addition of a protogen (H_2O) to a quiescent BF_3/isobutylene system immediately produces high polymer. Experimentally, these authors increasingly purified (dried) the ingredients employed and noted that the rate of polymerization decreased with the degree of drying until in an essentially waterfree system polymerization did not occur at all ("stopping experiment"). In the wake of these now classical investigations, many authors interested in the elucidation of initiation mechanism of cationic polymerizations developed highly sophisticated techniques for the drying of chemicals and carried out series of stopping experiments. Table 4.3 summarizes some systems investigated in this manner.

A plethora of publications concern the problem of reproducibility of rates and molecular weights; the discrepancy between results obtained by various authors is usually attributed to variations in (unidentified) impurity levels. In spite of the improvements in techniques and heroic efforts to dry chemicals, irreproducibility still kept plaguing basic researchers. For example, as late as in 1965, some two decades after the introduction of stopping experiments, Plesch still obtained such severely scattered rate and molecular weight data in superpurified H_2O/$TiCl_4$/isobutylene/CH_2Cl_2/ + 18 to −91°C systems that he questioned the validity of even his own results (Biddulph et al., 1965). These matters have been discussed in detail (Kennedy, 1975b).

The following simple calculation illustrates the futility of "chasing the last trace of water" in a cationic polymerization system. The equation describing ion generation in a H_2O/MX_n system in the presence of a very small quantity of water is displaced toward the right:

$$H_2O + MX_n \overset{k_i}{\rightleftharpoons} H^\oplus + MX_nOH^\ominus$$

so that

$$[H_2O] = [H^\oplus]$$

Consider a polymerization where $[M] = 0.1$ mole/1, $k_i = 3$ mole·sec/1, and the DP_n of the polymer formed is 10^3 (these values are reasonably close to a styrene polymerization) and assume that a rate of 1% conversion per 24 hr can be determined with sufficient accuracy (a most conservative assumption). Under these conditions 10^{-8} mole of monomer is polymerized

Table 4.3 □ Stopping Experiments

Olefin	Medium	Temperature (°C)	Friedel–Crafts Acid Coinitiator	Added Compounds	References
Isobutylene	Gaseous phase	20	BF_3	H_2O; t-BuOH, acetic acid[a]	Evans and Meadows (1949)
	Gaseous phase	26	BF_3	D_2O[a]	Dainton and Sutherland (1949)
	Gaseous phase	−80	BF_3	H_2O[a]	Evans and Meadows (1958)
	Gaseous	20; −80	BF_3	H_2O[a]	Evans (1951)
	n-Heptane	−12.5	BF_3	H_2O[a]	Chmelir et al. (1965)
	n-Heptane	−8; −12	$TiCl_4$	H_2O[a]	Chmelir et al. (1965)
	CH_2Cl_2 or in bulk	−72 to −78	$TiCl_4$	[b]	Cheradame and Sigwalt (1964, 1970)
	EtCl	−93.5 to −63.5	$SnCl_4$	H_2O[a]	Norrish and Russel (1952)
	n-Hexane	−70; −103	$TiCl_4$	CCl_3COOH, CF_3COOH[a]	Biddulph and Plesch (1960)
	n-Heptane	−8 to −12	$TiCl_4$	H_2O[a]	Biddulph and Plesch (1960)
	EtCl	−93.5 to −63.5	$SnCl_4$	H_2O, D_2O, phenols[a]	Norrish and Russel (1952)
	EtCl	−63.5; −78.5		Measureable polymetrization rate even after exhaustive drying[b]	Bauer et al. (1970)
	EtCl	−78.5	$SnCl_4$	H_2O is retarder[b]	Norrish and Russel (1952)
	n-Heptane	20 to −60	$AlBr_3$	H_2O is retarder[b]	Chmelir and Marek (1967a)
	n-Heptane	21 to −55	$AlEtCl_2$	H_2O is inhibitor[b]	Solich et al. (1969)
	CH_2Cl_2	0 to −60	$AlCl_3$	H_2O is not an initiator	Beard et al. (1964)

Table 4.3 ☐ (Continued)

Olefin	Medium	Temperature (°C)	Friedel–Crafts Acid Coinitiator	Added Compounds	References
Isobutylene	n-Heptane	−13	AlBr₃, TiCl₄, etc	b	Chmelir and Marek (1968)
	n-Heptane	−14	AlBr₃, TiCl₄, etc.	b	Lopour and Marek (1970)
	n-Heptane	−14	AlI₃	b	Chmelir et al. (1967b)
	n-Heptane	−14	GaCl₃	b	Lopour and Marek (1970)
	n-Heptane	−14	GaBr₃	b	Lopour and Marek (1970)
Isoprene	n-Heptane	20 to −18	AlEtCl₂	b	Kössler et al. (1963)
	Benzene	20	AlEtCl₂	b	Kössler et al. (1963)
	Benzene, toluene, and n-heptane	21	AlBr₃	b	Matyska et al. (1966)
	n-Heptane	21	AlEtCl₂	b	Gaylord et al. (1966)
	n-Heptane, toluene, or benzene	21	AlBr₃	b	Gaylord et al. (1966)
	n-Heptane	21	AlCl₃	b	Gaylord et al. (1966)
	n-Heptane or benzene	21	AlCl₃	b	Matyska et al. (1965)

Monomer	Solvent	Temperature	Catalyst		Reference
Propylene	CH_2Cl_2	−35	BF_3	[b]	Szell and Eastham (1966)
Butene-1	CH_2Cl_2	−70 to +20	$TiCl_4$	H_2O	Sigwalt et al. (1973)
Butene-2	(CH_2Cl_2)	25	BF_3	[b]	Clayton and Eastham (1957, 1961)
1-Methylcyclopentene	n-Hexane	0	BF_3	[b]	Woolhouse and Eastham (1966)
					Schmitt and Schuerch (1961)
trans-2-Butene	Ethylene dichloride	25	BF_3	H_2O[a]	Eastham (1956)
Ethylene			$AlCl_3$	H_2O, HCl[a]	Ipatieff and Grosse (1936)
Styrene	CCl_4	0–25	BF_3	H_2O[a]	Clark (1953)
Styrene	Bulk		$AlCl_3$	H_2O[a]	Jordan and Treloar (1961)
	Hexane, toluene	−64 to 25.5	$TiCl_4$	H_2O, CCl_3COOH[a]	Plesch (1953b, c)
	CH_2Cl_2	−14; −29	$TiCl_4$	H_2O[a]	Longworth et al. (1959c)
	Benzene	25	$SnCl_4$	H_2O[a]	Pepper (1963)
	Nitrobenzene + CCl_4	25	$SnCl_4$	H_2O[a]	Overberger et al. (1958)
Isobutylene–styrene	CH_2Cl_2 Ethylene dichloride				
	Isopropyl dichloride	−95 to 20	$TiCl_4$	H_2O[a]	Longworth et al. (1960)
	Ethylidene dichloride				
Styrene–α-methylstyrene	EtCl	20	$SnCl_4$	H_2O[a]	Lyudvig et al. (1968)

Table 4.3 □ (Continued)

Olefin	Medium	Temperature (°C)	Friedel–Crafts Acid Coinitiator	Added Compounds	References
α-Methylstyrene	CH₂Cl₂	-72 to 10	TiCl₄	TiCl₄ is active alone, rate increases in presence of H₂O, HCl, or added TiCl₄[b]	Branchu et al. (1969)
	EtCl	55	SnCl₄	Questionable dryness	Dainton et al. (1953)
	MeCl	-50	AlEt₂Cl	Controls: isobutylene and styrene need initiator under the same conditions[b]	Kennedy (unpublished)
	(CH₂Cl)₂	25	BF₃		Armstrong et al. (1971)
3,4-Dimethoxystyrene	CH₂Cl₂	-72	BF₃·OEt₂	HCl, HF, CCl₃COOH[a]	Mayen and Maréchal (1972)
Cyclopentadiene	CH₂Cl₂	-43 to -70	TiCl₃OBu	TiCl₃OBu is active alone, rate increases in the presence of H₂O and strongly increases with HCl; second TiCl₃OBu addition is also positive[b]	Vairon and Sigwalt (1971)
	MeCl	-50	AlEt₂Cl	Controls: isobutylene and styrene need initiator under the same conditions	Kennedy (unpublished results)
Indene	CH₂Cl₂	-70	TiCl₄	TiCl₄ is active alone; rate increases in presence of H₂O or HCl	Cheradame et al. (1965, 1969)
	CH₂Cl₂	-30	SnCl₄	SnCl₄ is active alone, rate increases in presence of H₂O	Polton and Sigwalt (1969, 1970)
2-Methylindene	CH₂Cl₂	-72	TiCl₄	H₂O[a]	Maréchal et al. (1964)
1,1-Diphenylethylene	Benzene	30, 40, 55	SnCl₄	H₂O[a]	Evans and Lewis (1957)
	Benzene	40	Sb₂Cl₆	HCl[a]	Evans et al. (1961)

[a] The monomer–Lewis acid system does not polymerize alone, the addition of an initiator is necessary (Zlamal, 1959).

[b] Polymerization may occur through self-initiation in a two-component system (Kennedy, 1972d).

per second and the number of moles of initiated chains is 10^{-11}. Given this quantity and considering the equation for protonation (cationation),

$$H^{\oplus} + \overset{|}{\underset{|}{C}}=\overset{|}{\underset{|}{C}} \rightarrow H-\overset{|}{\underset{|}{C}}-\overset{|}{\underset{|}{C}}{}^{\oplus}$$

the water concentration in this polymerization system can be calculated:

$$[H_2O][M] \times k_i = [H_2O][0.1] \times 3 = 10^{-11}$$

Thus

$$[H_2O] \approx 4.10^{-11}$$

or

$$10^{-10} < [H_2O] < 10^{-12}$$

This water concentration is certainly far below any analytical detection limit and may be present in the equipment, despite most careful drying and baking, via diffusion of moisture from glass or metal walls. It therefore appears that barring extraordinary measures laboratory equipment on this earth cannot be sufficiently dry for a perfect "stopping" experiment.

In view of the large number of systems that could be demonstrated to require the presence of protogen for efficient polymerization, authors tended to believe that Friedel–Crafts acids alone would be unable to induce cationic polymerizations of olefins (Plesch, 1963). In light of recent work (see below) this view became obsolete. Nonetheless, the era of stopping experiments was useful in that it forcefully focused the attention of researchers on the importance of impurities in determining the outcome of polymerization. In retrospect, however, the cost of establishing superpure and superdry systems was so high in terms of both labor and materials that many researchers became discouraged and turned their attention to more cost-effective investigations. (For an analysis of these results, see "Basic" and "Industrial" Researches, pp. 93 and 111 in Kennedy, 1975b.)

A GENERAL SCHEME OF INITIATION WITH BRØNSTED ACID/FRIEDEL–CRAFTS ACID SYSTEMS. Stopping experiments demonstrated that many cationic olefin polymerizations require the presence of protogenic materials for initiation and thus rendered untenable mechanistic schemes that neglected to consider this requirement. For a brief history of these matters see Zlamal (1959).

According to the pioneering investigators (Plesch et al., 1947; Evans et al., 1946b; Evans and Polanyi, 1947a), the role of water is to generate the initiating proton, which in turn completes initiation by protonation of olefin:

$$BF_3 + HY \rightarrow F_3BHY$$

$$F_3BHY + CH_2{=}C(CH_3)_2 \rightarrow F_3BY^{\ominus} + CH_3{-}\overset{\oplus}{C}(CH_3)_2$$

This proposition has immediately been accepted and since its introduction in 1946–1947 was corroborated by numerous chemical and kinetic investigations.

Indeed, this scheme remains a cornerstone of mechanistic interpretation of protogens in carbocationic polymerization.

A careful review of pertinent publications in this field indicates that workers without exception have adopted the view that the role of protogen is to provide the *de facto* initiating entity and the role of Friedel–Crafts acids is merely to facilitate the initiation process. The path to the first initiating ion pair may be tortuous and many details remain to be elucidated.

In principle there exist at least four paths the initiator/co-initiator/monomer ternary system may take to produce the initiating entity:

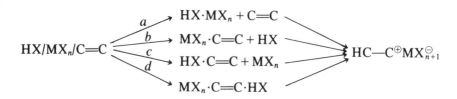

Since termolecular reactions are extremely unlikely, it must be assumed that the three reaction partners form binary complexes (paths *a*, *b*, *c*), which in turn interact with the third component to produce the initiating species. Path *d* exemplifies a hypothetical ternary complex which may rearrange to the initiating species. The original authors (Kennedy and Langer, 1964i; Zlamal, 1969) have analyzed many specific initiating systems in terms of this scheme. An extended discussion of these paths would be at best highly speculative on account of our ignorance of details, and therefore is beyond the scope of this chapter; however, some cautious generalizations should be made.

Among the possibilities presented by the above scheme, path *a* represents the "conventional" route by which Brønsted acids and Friedel–Crafts acids are visualized to interact. These species may form a binary complex $H^{\oplus} \cdot MeX_{n+1}^{\ominus}$ (ion generation or priming), which in turn would protonate the olefin and thus produce the true initiating species. The exploration of path *a* is considerably complicated because the existence of entities such as HBF_4 and $HAlB_4$ has never been demonstrated.

Path *a* is probably operational in most "open" polymerizations, that is, reactions carried out in equipment not rigorously protected from adventitious moisture. Since most Friedel–Crafts acids are readily hydrolyzable and thus generate Brønsted acids *in situ*, open systems must contain substantial amounts of strong Brønsted acids.

It should be emphasized, however, that our understanding of the detailed reaction mechanism of this path, or of the other paths for that matter, is extremely fragmentary. Regrettably, we have very little knowledge as to the rates of the elementary events, the architecture or geometry of the complex anion $MX_{n+1}^{\ominus}$, the ionicity of the species involved, the effect of solvent, temperature, concentration on equilibria, and so on. The physical chemistry of Friedel–Crafts acids in nonaqueous solvents is extremely complicated and even in the presence of solid characterization data the question remains how far are these data meaningful for polymerizations. For example, most aluminum-containing Friedel–Crafts acids exist as dimers in the solid phase and in solution, and their dimerization tendency is known to affect polymerization details. However, the exact nature of the effect is still a matter of speculation (see Section 3.3).

Path *b* must in principle be involved in stopping experiments that is, when the Brønsted acid is added last to quiescent Friedel–Crafts/monomer systems. The existence of rather weak Friedel–Crafts acid/olefin complexes has been demonstrated (Kennedy and Langer, 1964i; Longworth and Plesch, 1959a; Clayton and Eastham, 1961; Taft et al., 1955; Fontana, 1963; Brackman and P. H. Plesch, 1955; Bonner et al., 1958; Fairbrother, 1941; Brown and Mathieson, 1957, 1958; Nakane et al., 1962). The collision between these complexes and the added protogen must produce the initiating ion pair; however, again, the detailed mechanism of this path is completely obscure.

With these matters in mind it is proposed that paths *a* or *b* are operational in most systems in which participation of Brønsted acids for initiation has been demonstrated. Paths *c* and *d* are possible in principle; however, their operational significance is limited to special circumstances (Kennedy and Langer, 1964i; Mass et al., 1923; Mass, 1918, 1921, 1925, 1930; Mayo and Katz, 1947; Kilpatrick and Luborsky, 1953; Brown and Brady, 1953).

SCOPE AND LIMITATION OF BRØNSTED ACID/FRIEDEL–CRAFTS ACID INITIATING SYSTEMS. Brønsted acid/Friedel–Crafts acid combinations are valuable, efficient initiators for carbocationic polymerizations. Although the necessity of Brønsted acids can be demonstrated only by laborious stopping experiments, their influence is all important in regard to rates, molecular weights, and molecular weight distributions in polymerizations carried out under conventional open conditions. Conceivably, carbocationic polymerizations may occur even in the absence of added protic acid (see below); nonetheless, the rates of these reactions would be orders of magnitude slower than those in the presence of such acids.

The nature of Brønsted acids in various Brønsted acid/Friedel–Crafts acid systems has been carefully examined. Polton and Sigwalt (1970) compared polymerization rates of HCl and $H_2O/SnCl_4/indene/CH_2Cl_2/-30°C$ systems and found HCl a more efficient initiator than H_2O. Similarly, Kennedy and Gillham (1972e) investigated a series of Brønsted acid/Et_2AlCl/isobutylene/$CH_3Cl/-50°C$ systems and obtained the following sequence of initiator efficiencies: $HCl > H_2O > CCl_3COOH > CH_3OH > CH_3COCH_3$. Acetone was considered to react through its enolic form. This series follows approximately the acidity of the initiators used.

Di- and triorganoaluminum compounds are of particular interest for the polymerization of olefins and diolefins not only because they produce remarkably high molecular weight polymer at relatively high temperatures but also because they permit the carrying out of stopping experiments in open equipment. Thus Et_3Al or Et_2AlCl/isobutylene/CH_3Cl charges can be stirred for extended periods in conventional equipment under a blanket of nitrogen without polymerization occuring; however, the addition of Brønsted acids to such quiescent systems immediately initiates vigorous polymerization and high molecular weight product is formed. The reason(s) for this exceptional position of certain organoaluminum compounds is not immediately apparent; however, this property proved extremely valuable in the synthesis of sequential copolymers (Chapter 8).

Mixtures of water with trialkylaluminums exhibit some unusual properties valuable for initiating cationic olefin polymerizations. Saegusa et al. (1964) obtained remarkably high molecular weight polystyrenes ($\bar{M}_v \sim 200,000$) at $-78°C$ by the use of premixed Et_3Al–H_2O systems ($Et_3Al : H_2O = 1.0$ to 0.66). According to Hammett indicator studies (Lee and Treloar, 1969) the $1:1$ Et_3Al/H_2O mixture is a stronger acid than $AlCl_3$. The chemistry of the interaction between water and Et_3Al is quite complex and the role of water as an initiator in these systems is far from understood.

In line with the initiation mechanism (see above), Brønsted acid/Friedel–Crafts acid combinations give rise to polymers with proton head groups. Since hydrogen head groups are of limited interest for preparative purposes, these initiating systems are of little importance for polymer derivatization and functionalization. In contrast, carbenium ion precursor/Friedel–Crafts acid systems are of considerable interest in postpolymerization preparative chemistry.

Cationogen = Carbenium Ion Source

INITIATION DETAILS WITH RX/MX$_n$ SYSTEMS. Initiation of carbocationic polymerization consists of the addition of the first carbenium ion to olefin monomer (cationation by carbenium ion). In this sense protonation of an olefin and the formation of the first carbenium ion (first step) may be regarded as a prelude to cationation by a carbenium ion (second step):

$$H^{\oplus} + C{=}C \rightarrow H{-}C{-}C^{\oplus}$$
$$H{-}C{-}C^{\oplus} + C{=}C \rightarrow H{-}C{-}C{-}C{-}C^{\oplus}$$

As analyzed above, in order to shorten and simplify the path of initiation it appeared desirable to use preformed carbenium ion salts. Thus *in situ* prepared transitory carbenium ions capable of rapidly and efficiently initiating vinyl polymerization promised to be of even higher utility since these cations are far more reactive, can be generated more easily and cheaply, and represent a much larger group of species than stable carbenium ion salts.

With great insight Pepper (1949) some 30 years ago suggested that transitory carbenium ions arising from alkyl halides in the presence of Friedel–Crafts acids could initiate carbocationic polymerizations by $RX + MX_n \rightarrow R^{\oplus}MX_{n+1}^{\ominus}$.

The thermochemistry of this process was the subject of a largely theoretical study (Plesch, 1968). Occasionally authors have invoked initiation by alkyl halides in conjunction with some Friedel–Crafts acids, for example, $tert$-BuCl/SnCl$_4$/styrene/(CH$_2$Cl)$_2$/25°C (Colclough and Dainton, 1958a, b) RF (R = i-Pr, t-Bu, cyclohexyl)/BF$_3$/propylene, 1-butene, isobutylene/ − 80°C (Olah et al., 1960). Beard et al. (1964) investigated the AlCl$_3$/isobutylene/CH$_2$Cl$_2$/0 to − 60°C system and found that water was not needed for polymerization. Initiation by solvent was suggested to occur:

$$CH_2 = C(CH_3)_2 + AlCl_3 \cdot CH_2Cl_2 \rightarrow ClCH_2{-}CH_2{-}\overset{\oplus}{C}(CH_3)_2AlCl_4^{\ominus}$$

To demonstrate initiation by alkyl halides, the systems had to be laboriously superdried (stopping experiments) to eliminate adventitious rapid polymerization by protogens. However, interest in these systems and/or concept of initiation remained low because alkyl halide/Friedel–Crafts acid pairs did not offer any apparent advantage or incentive over simple Brønsted acid/Friedel–Crafts acid combinations. This picture completely changed by the discovery of controlled initiation with alkyl halides in conjunction with di- and triorganoaluminum compounds (Kennedy and Gillham, 1972e; Kennedy and Tornquist, 1968c; Saegusa, 1969).

In a series of patents and publications Kennedy et al. (1959, 1966a, b, 1967b, 1969a, b, 1970, 1972a, 1973a, 1974d) described that olefins, diolefins, or mixtures thereof can be stirred with certain organoaluminum compounds, for example, Et$_3$Al, Me$_3$Al, or Et$_2$AlCl, for extended periods in bulk or in solution without visible change; however, the addition of various alkyl halides to these quiescent charges immediately induces rapid and sometimes even explosive polymerizations. Importantly, these experiments lead to high molecular weight products and were carried out under conventional laboratory conditions, that is, under a blanket of dry inert gas.

Most likely, spontaneous initiation with the relatively strong Friedel–Crafts acids AlCl$_3$ and EtAlCl$_2$ occurs in less than superdry (routinely purified) systems because the concentration of moisture or other cationogenic impurity suffices to provide the necessary H$^{\oplus}$ or R$^{\oplus}$ concentration for initiation; that is, the equilibrium constant K is large:

$$K = \frac{[H^{\oplus}][AlCl_3OH^{\ominus}]}{[H_2O][AlCl_3]}$$

With the relatively weaker acids Et$_2$AlCl or Et$_3$Al the cationogen level in routinely dried systems is insufficient to produce measurable initiation rates—that is, the equilibrium constant for these Friedel–Crafts acids is small—and the addition of relatively large amounts of cationogen for initiation becomes necessary.

In the t-BuCl/Et$_2$AlCl/isobutylene system initiation is visualized to occur by

Ion generation: t-BuCl + Et$_2$AlCl → t-Bu$^\oplus$Et$_2$AlCl$_2^\ominus$

Cationation: t-Bu$^\oplus$ + CH$_2$=C(CH$_3$) → t-Bu—CH$_2$—C$^\oplus$(CH$_3$)$_2$

A semiquantitative sequence of alkyl halide initiation efficiencies has been obtained for the RCl/Et$_2$AlCl/isobutylene/CH$_3$Cl/ − 50°C system, where RCl was a series of alkyl chlorides (Kennedy and Tornqvist, 1968c). Figure 4.1 shows comparative initiator efficiencies expressed in grams of polyisobutylene per mole RCl. Initiator efficiency reflects the relative rates of initiation, propagation, and termination, all of which are affected by reaction conditions. The abscissa in Figure 4.1 is dimensionless and was constructed by listing a qualitative sequence of relative cation stabilities. The maximum in the initiator efficiency curve suggests the presence of at least two effects: on the left side of the maximum, initiator efficiency is mainly determined by carbenium ion availability, on the right side by carbenium ion stability. The cations on the left side of the curve would be efficient initiators; however, their concentration is too low. On the right side, carbenium ion concentration is sufficient; however, their stability is too high, limiting yield.

An essentially similar set of results has also been obtained with the RCl/Et$_2$AlCl/styrene/CH$_3$Cl/ − 50°C system (Kennedy, 1959, 1969b).

These experiments seem to indicate that most efficient initiation occurs with that transitory carbenium ion the structure of which is closest to the

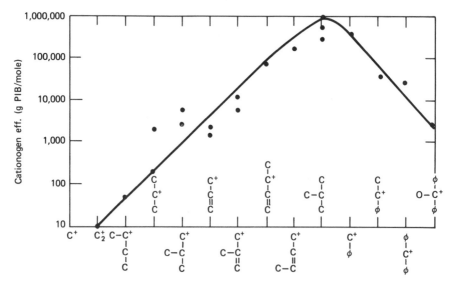

Figure 4.1 ☐ Polymerization of isobutylene with various initiating ions in conjunction with Et$_2$AlCl.

propagating cation. Thus the most efficient initiation was obtained with
t-BuCl for isobutylene and with 1-chloroethylbenzene for styrene.

Alkyl, benzyl, and tertiary halides, particularly chlorides, have been found
to be most useful in conjunction with alkylaluminums for polymerization
initiation of olefins, for example, styrene and isobutylene. However, unexpected exceptions to this generalization have been reported; for example,
alkyl bromides were found to be much less active initiators than alkyl
chlorides (Kennedy, 1970). The original literature should be consulted for
the interpretation of this observation (Kennedy, 1970). Evidently, the nature
of the individual components must be carefully considered for the design of
efficient initiating systems.

An interesting development is initiation by difunctional initiators. For
example,

$$Cl \!\!\succ\!\!\!\diagup\!\!\!\diagdown\!\!\!\diagup\!\!\!\prec\!\! Cl$$

has been used to great advantage in conjunction with $(C_6H_5)_3Al$ and
isobutylene in the preparation of α, ω-diphenylpolyisobutylenes (Kennedy
and Chung, 1980). The initiating species is probably akin to
$(C_6H_5)_3AlCl^{\ominus}(CH_3)_2C^{\oplus}-CH_2CH_2CH_2-C^{\oplus}(CH_3)_2(C_6H_5)_3AlCl^{\ominus}$.

Saegusa et al. (1964) introduced CH_3CO- and CH_3OCH_2-head groups into
polystyrene by initiating the polymerization of this monomer by CH_3COCl
or CH_3OCH_2Cl/Et_3Al mixtures.

PREPARATIVE SIGNIFICANCE OF RX/MX$_n$ SYSTEMS. The discovery that certain
organic halides in conjunction with aluminum-containing Lewis acids rapidly produce high polymer under conventional conditions led to new dimen-
sions in synthetic polymer chemistry. Among the new products that became
readily available by this technique are randomly and terminally functional
(telechelic) polymers, block copolymers, and graft and bigraft copolymers.
This section concerns an outline of direct synthesis of terminally functional
polymers; preparative possibilities by postpolymerization functionalization
and sequential copolymers are discussed separately (Chapter 8).

Stated in simplest terms, tertiary, allyl, and benzyl halides in conjunction
with certain Lewis acids, for example, Et_2AlCl, Et_3Al, and BCl_3, generate
transitory carbenium ions which initiate olefin polymerization and thus
become head groups of polymers. These polymerizations can be carried out
under conventional laboratory conditions and initiation by $H^{\oplus}$ (arising from
adventitious H_2O) is avoided or minimized. The equations above, for
example, show the incorporation of the t-Bu head group into polyisobutylene. Figure 4.1 shows carbenium ions obtained from the corresponding
organic chlorides plus Et_2AlCl used to initiate isobutylene polymerizations
and to introduce various head groups into polyisobutylene (Kennedy et al.,
1968a). Similar work has also been carried out with styrene (Kennedy, 1959).
Details of initiation kinetics by the $CH_3C_6H_4CH_2Cl/Et_2AlCl$ and Et_3Al

systems are also available (Reibel et al., 1979). Initiation by certain allyl chlorides, that is, 1-chloro-3-methyl-2-butene and 3-chloro-2-methyl-1-butene, are of interest because they may give rise to polymers with unsaturated head groups.

Triethylaluminum was found to be an efficient coinitiator for the polymerization of styrene, α-methylstyrene, and isobutyl vinyl ether in the presence of CH_3OCH_2Cl and CH_3COCl (in CH_2Cl_2 at $-78°C$) initiators, and evidence for CH_3OCH_2- and CH_3CO- head groups was found (Saegusa et al., 1964). Ion generation evidently involves, for example,

$$CH_3OCH_2Cl + Et_3Al \rightleftharpoons CH_3\overset{\oplus}{O}CH_2 + Et_3AlCl^{\ominus}$$

A similar system CH_3OCH_2Cl/Et_2Zn efficiently polymerized isobutyl vinyl ether (27.1% conversion, in CH_2Cl_2 at $-78°C$). This is one of the rare reports for coinitiation by Et_2Zn, a very weak Friedel–Crafts acid.

Similarly to alkylaluminums, BCl_3 has also been found to initiate olefin polymerization only in the presence of initiators, for example, Brønsted acids or certain organic halides (Kennedy et al., 1977a, c, 1978a, c). However, the nature of the halide initiators found to be effective in conjunction with BCl_3 and with alkylaluminums is quite different. For example, allyl, benzyl, or *tert*-butyl chloride are inactive with BCl_3/isobutylene, whereas substituted allyl and benzyl chlorides are quite effective initiators (Kennedy et al., 1977). RCl/BCl_3 initiating systems are valuable for purposes of molecular engineering since, in addition to controlled initiation, they also lead to controlled termination by chlorination in the absence of chain termination (see Section 4.4). For example, the $(CH_3)_2C=CH–CH_2Cl/BCl_3/$ isobutylene/$CH_2Cl_2/-78°C$ system gives rise to $(CH_3)_2C=CH–CH_2–$wPIB $w CH_2–C(CH_3)_2Cl$, a bifunctional telechelic polymer (Kennedy et al., 1977c, 1979f).

Although not Friedel–Crafts acids, certain silver salt-induced polymerizations should be mentioned here because of the great similarity between the chemistries of initiation involved. Thus the polymerization of olefins and vinyl ethers in polar solvents (e.g., methylene chloride), can be induced by mixing the monomers with suitable silver salts, such as $AgBF_4$, $AgClO_4$ (coinitiators) and adding alkyl halides, such as *t*-butyl, acetyl chloride (initiators) to the mixture. Ion generation most likely involves:

$$RX + AgG \rightarrow R^{\oplus}G^{\ominus} + AgX \downarrow$$

where R = alkyl, aralkyl, acyl group; X = F, Cl, Br, I; and G = BF_4, ClO_4, SbF_6, PF_6, NO_3, CF_3COO. The reaction is driven by the precipitation of silver halide. Olah et al. (1960) were the first who polymerized olefins in this manner. These authors mixed propylene, isobutylene, and other olefins with anhydrous $AgBF_4$ at $-80°$ and added alkyl (isopropyl, *n*-butyl, *t*-butyl) or acyl (acetyl, propionyl) chloride or bromide to the mixture. Heat evolution and the formation of white silver halide precipitate indicated rapid polymerization.

Franta et al. (1977) polymerized isobutylene and styrene by benzylic halides in the presence of $AgSbF_6$ and Kennedy and Plamthottam (1980) obtained poly(isobutyl vinyl ethers) with allylic chlorines and $AgPF_6$. Fundamentally similar experiments and results have been reported by Gandini and Plesch (1965) and Kagiya et al. (1969), who mainly employed $AgClO_4$ in conjunction with a large variety of organic halides for the polymerization of styrene. The latter investigators dissolved $AgClO_4$ in the monomer and added stoichiometric amounts of organic halides at 0°C. Rapid polymerization occurred with t-BuBr and t-BuI and linear conversion–time plots were obtained which extrapolated back to the origin, suggesting rapid initiation. In contrast, with $RX = t$-BuCl, $(C_6H_5)_3CCl$, $C_6H_5CH_2Br$, or i-PrI, for example, the conversion–time plots were more or less convex to the abscissa, indicating rate acceleration. These and other data were analyzed to mean that, in the first approximation, initiating activity of organic halides in conjunction with $AgClO_4$ is inversely proportional to their heterolytic dissociation energy. The same order of efficiencies has been observed by Burgess et al. (1978) for the cationic initiation of THF polymerization by $AgClO_4$ in conjunction with various organic halides.

Similar investigations have been carried out with the $Ph_3CCl/$ $HgCl_2/styrene/1, 2$-dichloroethane/30°C system (Sambhi, 1970). According to spectrophotometry the equilibrium

$$Ph_3CCl + HgCl_2 \rightleftharpoons Ph_3C^\oplus HgCl_3^\ominus$$

is immediately established and $[Ph_3C^\oplus HgCl_3^\ominus]$ decreases with time as a consequence of slow tritylation of styrene. Evidently free ions are absent in the medium and the counteranion remains in the vicinity of the growing cation. The presence of Cl in the polymer indicates termination involving chlorine abstraction by the styryl cation from $HgCl_3^\ominus$.

This method seems to be a reliable although expensive route for the introduction of various head groups into polyolefins. Although the chemistry appears to be straightforward, direct chemical proof for initiator-fragment incorporation is needed to accept unreservedly these propositions.

Cationogen = Halogen

Halonium ions ($X^\oplus$, where X = chlorine, bromine, iodine) have often been postulated to be intermediates involved in halogenation assisted by Friedel–Crafts acids, for example,

This and similar formalisms prompted investigations focusing on the use of halogens as initiators in cationic polymerization; Kennedy et al. (1968d,

1972c, 1973c) first used Cl_2 and Br_2/Me_3Al and Et_2AlCl combinations to initiate the polymerization of isobutylene and isobutylene–isoprene charges in polar solvent, and on the basis of results of model experiments proposed the following initiation mechanism:

$$Cl_2 + Me_3Al \rightleftharpoons Cl^{\oplus}Me_3AlCl^{\ominus}$$

$$Cl^{\oplus} + CH_2{=}C(CH_3)_2 \rightarrow Cl{-}CH_2{-}C^{\oplus}(CH_3)_2$$

Experimentally immediate polymerization occurred when solutions of Cl_2 in CH_2Cl_2 were added to quiescent Me_3Al/isobutylene/CH_2Cl_2 systems at low temperatures, and high polymer was obtained. In a series of publications (Baccaredda et al., 1973; Priola et al., 1975a, b; Giusti et al., 1975; Cesca et al., 1975a, b; Di Maina et al., 1977; Magagnini et al., 1977), which ostensibly resulted from an effort to develop a commercial isobutylene-isoprene copolymerization (butyl rubber) process, a team of industrial–academic researchers confirmed and greatly extended these findings. In addition to generating a large amount of fundamental information concerning interaction between alkylaluminums and halogens and mechanistic details of isobutylene polymerization with such combinations, these authors described the synthesis of high-quality butyl rubbers by Cl_2/Et_2AlI combinations.

Initiation by chloronium ion has also been found to occur with Cl_2/BCl_3 combinations. Recently Kennedy and Chen (1979d) obtained α, ω-dichloropolyisobutylenes by the use of this system and used both terminal chlorines in subsequent derivatizations:

$$Cl_2/BCl_3/i\text{-}C_4H_8/CH_2Cl_2/ - 78°C \rightarrow Cl^{\oplus}BCl_4^{\ominus} + i\text{-}C_4H_8 \rightarrow$$

$$Cl{-}CH_2^{\oplus}(CH_3)_2 \xrightarrow{+M} Cl{-}CH_2{-}C(CH_3){-}\text{\small www}PIB\text{\small www}CH_2(CH_3)_2Cl$$

$$\longrightarrow \text{block copolymers}$$

These authors also used ICl to initiate the polymerization; however, the identity of the initiating halonium ion, $I^{\oplus}$ or $Cl^{\oplus}$, could not be ascertained. In view of the difficulty in purifying ICl from traces of Cl_2 (contamination by $2\,ICl \rightleftharpoons I_2 + Cl_2$), claims that the initiating ion is $I^{\oplus}$ (Baccaredda et al., 1973) must be carefully scrutinized.

Obviously, an important consequence of initiation by halonium ions is the possibility of convenient synthesis of polymers with halogen head groups.

A special case of ion formation has been proposed to occur with $SbCl_5$. According to Brake et al. (1969), when a solution (nature of the solvent unidentified) of $(p\text{-}CH_3O{-}C_6H_4)_2C{=}CH_2$ is rapidly mixed with excess $SbCl_5$ the products are $(p\text{-}CH_3OC_6H_4)_2C^{\oplus}{-}CH_2Cl$ and $SbCl_6^{\ominus}$. The identity of the cation was demonstrated by UV and NMR spectroscopy, and that of the counteranion by preparation of salts such as $NH_4^{\oplus}SbCl_6^{\ominus}$. The participation of Cl_2 was ruled out. Although the authors prefer the above formalism and regard $SbCl_5$ as a $Cl^\oplus$ donor, they concede that chlorination

$$(p\text{-}CH_3O\text{—}C_6H_4)_2C\text{==}CH_2 + SbCl_5 \rightarrow \underset{I}{(p\text{-}CH_3O\text{—}C_6H_4)_2\overset{\overset{\displaystyle Cl}{\displaystyle |}}{C}\text{—}CH_2Cl} + SbCl_3$$

followed by ionization with excess $SbCl_5$

$$I + SbCl_5 \rightarrow (p\text{-}CH_3O\text{—}C_6H_4)_2\overset{\oplus}{C}\text{—}CH_2Cl$$

cannot be ruled out. It would be of interest to investigate the $Cl^\oplus$-donating ability of $SbCl_5$, in combination with other $Cl^\ominus$-accepting Friedel–Crafts acids, for polymerizations.

Cationogen = Miscellaneous Compounds; BF₃·OR₂ Complexes

Boron trifluoride etherate complexes often used for the polymerization of olefins, diolefins, and vinyl ethers belong to the class of Brønsted acid/Friedel–Crafts acid initiating systems. Most investigations have been carried out with the reasonably stable $BF_3 \cdot OEt_2$ complex; however, other etherates, such as O-n-Bu$_2$, have also been used occasionally.

In spite of the large number of polymerizations induced by these complexes, only two reports concern the chemistry of these reactions. Worsfold and Bywater (1957) investigated the polymerization of α-methylstyrene and Giusti et al. (1970) that of styrene using $BF_3 \cdot OEt_2$ in the presence and absence of H_2O. The data obtained by both groups indicate that although $BF_3 \cdot OEt_2$ may be active, $BF_3 \cdot H_2O$ is probably a much more active initiating system. The hydrate may form by the following displacement:

$$BF_3 \cdot OEt_2 + H_2O \rightleftharpoons BF_3 \cdot H_2O + Et_2O$$

and the role of the ether is to reduce the concentration of free BF_3 ($BF_3 + OEt_2 \rightleftharpoons BF_3 \cdot OEt_2$) and/or proton:

$$BF_3 \cdot H_2O \rightleftharpoons H^\oplus BF_3OH^\ominus + OEt_2 \rightleftharpoons HO^\oplus Et_2BF_3OH^\ominus$$

Evidently many practitioners have found the use of the relatively less reactive but conveniently applicable $BF_3 \cdot OEt_2$ complex more advantageous than the highly reactive "H_2O"/BF_3 system.

Direct Initiation by Friedel–Crafts Acids (Purposeful Addition of Cationogen Unnecessary)

Not long ago eminent investigators believed that "isobutylene is not polymerized by metal acids unless an ionogenic substance, the cocatalyts, is present" (Plesch, 1963). For the last ten years or so evidence has increasingly been accumulating that in many systems Friedel–Crafts acids alone are able to induce carbocationic polymerizations without the purposeful addition of cationogen. These systems are said to operate by "direct initiation." In view of many recently described direct initiation systems the above statement has become obsolete.

A close scrutiny of pertinent publications allows discernment of three fundamentally different theories concerning the interpretation of direct initiation by Friedel–Crafts acids. An analysis of these systems reveals that these phenomena and corresponding interpretations at first glance unrelated, are in fact fundamentally similar and can be explained by a common characteristic of Friedel–Crafts acids.

Halometalation: The Sigwalt-Olah Theory

At the dawn of the science of cationic olefin polymerization Hunter and Yohe (1933) proposed that polymerization of isobutylene in the presence of AlCl$_3$ commences by metalation:

$$AlCl_3 + CH_2{=}C(CH_3)_2 \rightarrow Cl_3\overset{\ominus}{Al}{-}CH_2{-}\overset{\oplus}{C}(CH_3)_2 \xrightarrow{+M} Cl_3\overset{\ominus}{Al}{\sim\sim}CH_2\overset{\oplus}{C}(CH_3)_2$$

This theory was soon abandoned because it could not explain the function of protic acid initiators, for example. Recently, however, elements of this theory were resurrected and modified to explain a series of observations made with superpurified monomer/Friedel–Crafts acid systems.

Cheradame and Sigwalt (1964) observed that most carefully purified TiCl$_4$/isobutylene/CH$_2$Cl$_2$/ $-70°$C systems can be stirred in one arm of an H-shaped glass vessel without any reaction occurring; however, polymerization ensues on distillation of the mixture into the second arm of the apparatus. Since polymerization occurred only upon condensation from the gas phase, the authors described this phenomenon as "polymerization by condensation." Subsequently, similar observations have also been made with the TiCl$_4$/indene/CH$_2$Cl$_2$/ $-70°$C (Chéradame et al., 1973), Cl$_3$Ti n-Bu/cyclopentadiene/CH$_2$Cl$_2$ (Vairon and Sigwalt, 1971), and TiCl$_4$/1,1-diphenylethylene/CH$_2$Cl$_2$/ $-70°$C (Sauvet et al., 1978) systems. It has been suggested (Diem and Kennedy, 1978) that photolysis (or thermolysis) of TiCl$_4$ to TiCl$_3$ + Cl⁻ followed by HCl initiator formation (by $(Cl^. + \overset{|}{\underset{|}{C}}H \rightarrow HCl + \overset{|}{\underset{|}{C}}^.$ may at least partly account for polymerization by condensation. Recently this hypothesis has been further discussed by Gandini et al. (1980).

Sigwalt and co-workers continued to be interested in details of direct initiation of olefin polymerization by Friedel–Crafts acids and in a lecture delivered at the Third International Symposium of Cationic Polymerization proposed a theory to explain available experimental data (Sigwalt, 1974). According to this theory direct initiation occurs by halometalation as follows:

$$MX_n + CH_2\!\!=\!\!\underset{|}{\overset{|}{C}} \rightarrow X_{n-1}M\!\!-\!\!CH_2\!\!-\!\!\underset{|}{\overset{|}{C}}\!\!-\!\!X \quad (M = \text{metal})$$

Depending on the substituents, the metallo-organic compound may ionize in the presence of a second MX_n or may eliminate HCl; in the latter case HCl plus a second MX_n (or olefin·MX_n complex) produce the conventional initiating species:

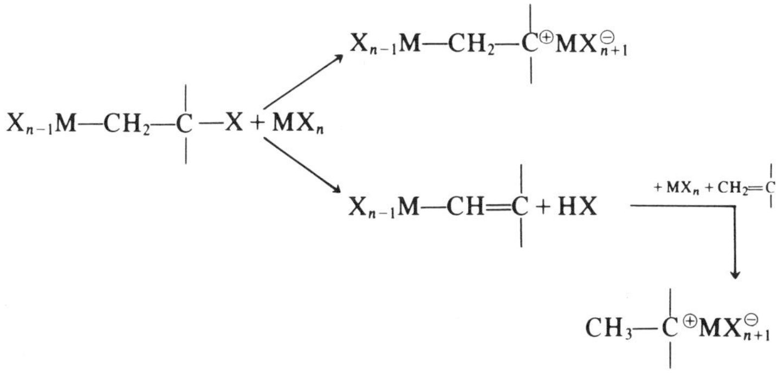

Recent work by Sauvet et al. (1978), on the dimerization of 1, 1-diphenylethylene by $TiCl_4$ in CH_2Cl_2 seems to conform to this scheme. Satisfactorily reproducible kinetics in purest $TiCl_4$/olefin systems achieved by these investigators is strong indication for the controversial existence of direct initiation.

Coincidentally, at the same meeting Sigwalt presented his theory, Olah, independently and purely by intuitive reasoning, proposed a conceptually identical mechanism to account for the polymerization of isobutylene by SbF_5 and $AlBr_3$ (Olah, 1974). Although the absence of protogen in a SbF_5/isobutylene system has never been demonstrated, purest (dryest) $AlBr_3$/isobutylene combinations have been claimed to produce polymer (Korshak and Lebedev, 1948; Chmelir M., Marek M. and Wichterle, O.(1967b).

Sigwalt (1974), Sauvet et al. (1978), and Olah (1974) point out that according to their mechanism the halometalated intermediate may lose HX so that even if the system was originally free of protogen, it can still rapidly revert to typical proton initiation.

Autoionization: The Korshak–Plesch–Marek Theory

According to the theory of direct initiation by autoionization first proposed by Korshak and Lebedev (1948) and resurrected by Longworth and Plesch (1959a), Plesch (1973, 1974), Ghanem and Marek (1972), Chmelir and Marek (1968), and Lopour and Marek (1970), ion generation involves autoionization of Friedel–Crafts acid aggregates (usually dimers) followed by initiation by addition of the electrophillic moiety to the olefin. For example, initiation in the $(AlBr_3)_2$/isobutylene/n-heptane/-30 to $-78°C$ system may occur by

$$Al_2Br_2 \rightleftharpoons AlBr_2^{\oplus}AlBr_4^{\ominus}$$

$$AlBr_2^{\oplus}AlBr_4^{\ominus} + CH_2{=}C(CH_3)_2 \rightarrow Br_2Al{-}CH_2{-}\overset{\oplus}{C}(CH_3)_2AlBr_4^{\ominus}$$

According to Chmelir and Marek (1967a), isobutylene is polymerized by $AlBr_3$ in n-heptane in the absence of added ionogen, whereas under the same (stopping experiments) conditions BF_3 and $TiCl_4$ are unable to induce polymerization, except after addition of H_2O.

The autoionization of Friedel–Crafts acids and the significance of these processes to cationic olefin polymerization has been investigated in detail by Plesch by a variety of techniques (Longworth and Plesch, 1958, 1959b; Plesch, 1973, 1974).

According to Plesch (1974), cationation of isobutylene polymerization in CH_3Cl or CH_2Cl_2 media involves direct addition of $AlCl_2^{\oplus} +$ $CH_2{=}C(CH_3)_2 \rightarrow Cl_2Al{-}CH_2{-}\overset{\oplus}{C}(CH_3)_2$ followed by rapid chain growth. A cornerstone of the theory is that un-ionized, excess $AlCl_3$ is immediately complexed with olefin to yield inactive complexes, which explains low $AlCl_3$ efficiencies (Kennedy and Thomas, 1960b; Di Maina et al., 1977). It is hard to accept this proposition without some direct (spectroscopic) evidence for the presence of necessarily large quantities of inactive, strong $AlCl_3 \cdot i\text{-}C_4H_8$ complexes. That these hypothetical complexes must be uncommonly strong and difficult to break is indicated by the nonresumption of polymerization in a quiescent but unquenched $AlCl_3/i\text{-}C_4H_8/CH_3Cl$ system by diluting with CH_3Cl or by the addition of H_2O or HCl in CH_3Cl solutions (Kennedy et al., 1967b). If Plesch's theory is correct, not only should these additions produce more polymer by releasing $AlCl_3$ from the $AlCl_3 \cdot i\text{-}C_4H_8$ complex,

$$AlCl_3 \cdot i\text{-}C_4H_8 + HX \rightleftharpoons AlCl_3 \cdot HX + i\text{-}C_4H_8 \rightarrow \text{initiation, polymerization}$$

but also the polymer head group $Cl_2AlCH_2C(CH_3)_2P$ (P = polymer) should act as a coinitiator in conjunction with HX and produce higher yields:

$$PAlCl_2 + HX \rightleftharpoons [H^{\oplus}PAlCl_2X^{\ominus}] \xrightarrow{+M} \text{initiation, polymerization}$$

Subsequently Chmelir and Marek (1968) and Lopour and Marek (1970) discovered that the rate of $AlBr_3$-initiated polymerization of isobutylene can be significantly enhanced by the addition of a second Friedel–Crafts acid, for example, $TiCl_4$, VCl_4, $SbCl_5$, $SnCl_4$, which on its own is unable to induce polymerization. In concord with their earlier results, the authors proposed ion generation by mixed Lewis acid:

$$AlBr_3 + TiCl_4 \rightleftharpoons AlBr_3 \cdot TiCl_4 \rightleftharpoons TiCl_3^{\oplus}AlBr_3Cl^{\ominus}$$

followed by rapid initiation

$$TiCl_3^{\oplus}AlBr_3Cl^{\ominus} + CH_2{=}C(CH_3)_2 \rightarrow Cl_3Ti{-}CH_2{-}\overset{\oplus}{C}(CH_3)_2AlBr_3Cl^{\ominus}$$

The idea of autoionization producing the initiating entity occurs in a somewhat modified form also in Gippin's work on the synthesis of high molecular weight rubbery poly-cis-1, 4-butadienes (Gippin, 1962, 1965). Analysis of his results led Gippin to the conclusion that the best initiating systems were obtained from mixtures of, for example, Et$_2$AlCl and EtAlCl$_2$ in conjunction with cobalt salts in polar solvents. According to this author initiation involves the addition of Et$_2$Al$^{\oplus}$ to butadiene complexed by cobalt in cisoid conformation:

Allylic Self-Initiation: The Kennedy Theory

Kennedy (1972b) comprehensively reviewed reports concerning stopping experiments and noted that olefins possessing an allylic hydrogen, for example, isobutylene, isoprene, α-methylstyrene, indene, and cyclopentadiene, cannot be "stopped" from polymerizing by exhaustive drying, whereas styrene and butadiene, that is, monomers that do not have allylic hydrogen atoms, can be prevented from polymerizing by drying. This author proposed the following possibility of allylic self-initiation in superdry systems:

A shortcoming of this proposition is that it cannot account for the dimerization of 1, 1-diphenylethylene in superdry systems (Sauvet et al., 1978).

Conclusions Relative to Direct Initiation

Direct initiation implies the existence of polymerization systems free even of the last trace of protogenic materials. To *prove* the complete absence of protons in any system on earth is an inherently exasperating task. Observing polymerization in superpure Friedel–Crafts acid/olefin systems, purists may argue that initiation may have been due to the presence of infinitesimal but sufficient amounts of proton provided by, say, chemi- or physisorbed water, surface OH groups, or slow diffusion of water from the glass entrapped during fluxing (see discussion above on stopping experiments). However, if this rather philosophical objection is ignored and the writings of authors who examined this problem in great detail are accepted, it appears that in many systems direct initiation by Friedel–Crafts acid is possible and that the statement quoted at the beginning of this section has become obsolete.

It also appears, however, that even in direct initiation the fundamental role of Friedel–Crafts acids remains the same: to assist the formation of the first electrophile capable of cationating olefin. Assistance may take the form of internal HX generation, or autoionization, that is, when the initiator and coinitiator species are identical, or self-initiation when the Friedel–Crafts acid is directly involved in the generation of an allylic cation. In this sense direct initiation systems are but special cases of Friedel–Crafts acid-coinitiated polymerizations, whose existence can be demonstrated only under ultrapure conditions.

Miscellaneous Methods

In addition to the above well-investigated chemical initiating systems involving protogens and carbenium ion sources, some well-known but less understood systems include iodine and various acidic solids such as clays and earths.

Inorganic Complexes

In a pioneering paper that remained forgotten somehow, Olah et al. (1960) first demonstrated that the crystalline complex NO_2BF_4, produced by reacting dinitrogen pentoxide with anhydrous HF in nitromethane medium and saturating the solution with BF_3, readily polymerized 1-butene at $-80°C$. Infrared spectroscopy indicated the presence of nitro groups in the liquid polymer.

The initiation mechanism is most likely identical to those discussed for Friedel–Crafts acids. It assumes the participation of nitronium ions $NO_2^\oplus$ and consequently leads to NO_2 head groups. Because NO_2 incorporation could also be due to chain transfer, direct chemical proof for the absence of adventitious protic impurities should be obtained. This interesting lead is worthy of further investigations.

The demonstration of NO_2BF_4-initiated 1-butene polymerization is of further significance because it points the way to a plethora of other possibilities for the initiation of olefin polymerization by well-characterized stable inorganic complexes.

Iodine

That iodine polymerizes vinyl ethers was mentioned by Wislicenus some 100 years ago (Wislicenus, 1878). The polymerization of styrene neat and in chlorinated solvents has been studied by various authors (Trifan and Bartlett, 1959; Okamura et al., 1961). Giusti and co-workers investigated the polymerization of styrene (Giusti and Andruzzi, 1966a) and acenaphthylene (Giusti and Andruzzi, 1966b) by iodine and on the basis of dilatometric and conductivity studies concluded that these processes are pseudocationic. This was confirmed by Plesch (1966). Initiation was preceded by an introduction period that could be reduced or eliminated by the addition of HI. These authors suggest that initiation probably involves the *in situ* generation of HI by the following reactions:

$$C_6H_5CH{=}CH_2 + I_2 \rightleftharpoons C_6H_5CHICH_2I \rightleftharpoons C_6H_5CH{=}CHI + HI$$

In contrast, the polymerization of vinyl ethers is visualized to proceed by a conventional cationic mechanism; however, details of initiation are controversial. Kinetic (Eley and Saunders, 1952), spectroscopic (Higashimura et al., 1961; Eley et al., 1964, 1968), and conductometric (Eley et al., 1968) studies have not been conclusive in this regard. Recently careful works by Ledwith and Sherrington (1971) and Johnson and Young (1976a) clarified aspects of the first stage in iodine-initiated polymerization of ethyl, *n*-butyl and isobutyl vinyl ethers in various solvents. According to these workers an equilibrium is established between $CH_2{=}CHOR$ and I_2, and equilibrium constant and enthalpy data indicate that the first product is a 1, 2-diiodide and not a charge transfer complex, as suggested by some investigators (Eley et al., 1968; Andrews and Keefer, 1964; Kanoh et al., 1965). Rate studies, spectrophotometry, and experiments with ethanol addition seemed to confirm these conclusions. The first step of initiation was proposed to involve the following reactions:

$$CH_2{=}CH{-}OR + I_2 \rightleftharpoons CH_2{=}CH{-}OR{\cdot}I_2 \qquad \text{(fast)}$$

$$CH_2{=}CH{-}OR{\cdot}I_2 + I_2 \rightarrow ICH_2{-}CH_2(OR)I + I_2 \qquad \text{(slow)}$$

According to Ledwith and Sherrington (1971), initiation is completed by the formation of the ion pair:

$$ICH_2{-}CH(OR)I + I_2 \rightleftharpoons ICH_2{-}CH{=}\overset{\oplus}{O}RI_3^{\ominus}$$

Styrene (Okamura et al., 1962b), and some styrene derivatives, for example, *p*-methoxystyrene and *p*-methylstyrene (Kanoh et al., 1965), have been polymerized by iodine and kinetic details have been studied; however, the chemistry of initiation remains obscure. The assembled kinetic information did not agree with the proposition that initiation involves $2\,I_2 + M \rightarrow [\text{complexes}] \rightarrow I{-}M^{\oplus}I_3^{\ominus}$.

It is of interest that among the halogens only iodine is able to induce polymerization of olefins, whereas with bromine, chlorine, or fluorine, various substitutions or addition occur. This phenomenon may be rationalized by assuming that the smaller anions $F^{\ominus}$, $Cl^{\ominus}$, and $Br^{\ominus}$ are relatively hard bases and as such rapidly interact with the relatively hard acid carbenium ion in the transition state. In contrast, the larger $I^{\ominus}$ or $I_3^{\ominus}$ anions are much softer bases and therefore add more slowly to the hard carbenium acid, allowing competition by the monomer for the electrophile.

In contrast with earlier propositions recent investigators do not postulate even the transitory existence of ionodium cations $I^{\oplus}$ (Johnson and Young, 1976a).

Miscellaneous Systems Including Acidic Solids

According to Biswas and Mishra (1973, 1975) and Biswas and Kabir (1978) oxychlorides such as $POCl_3$ produce low molecular weight poly(α-methylstyrene) and benzoyl chloride induces facile polymerization of isobutyl and n-butyl vinyl ethers by cationic mechanism.

Aoki et al. (1971) initiated cationic polymerization of isobutyl vinyl ether and to a lesser extent that of styrene by the use of various combinations of alkyl halides/reduced nickel, alkyl halides/nickel oxides, and even carbon tetrachloride/nickel carbonyl; however, these interesting leads have not been followed up.

A variety of complex solids have been described to yield high molecular weight polyisobutylenes. Also, many petrochemical processes have been developed for the oligomerization of light olefins to lubricating oils by numerous acidic solids, for example, oxides and clays.

High molecular weight polyisobutylenes have been obtained by Wichterle et al. (1959a, b, 1961) and Marek (1959) at relatively high temperatures with solids prepared by reacting in n-hexane solution an aluminum alcoholate (preferably a sec-butyl alcoholate) with BF_3 and activating the precipitate formed with $TiCl_4$. This work has been discussed by Plesch (1963) and Kennedy and Langer (1964i) and repeated and verified by others (Imanishi et al., 1967); it was shown that growth occurred on the surface of the catalyst and not in the homogeneous phase (Wichterle et al., 1959a). The high molecular weights were explained (Imanishi et al., 1967) by assuming the existence of a counteranion in the solid phase. This work has been continued by Japanese investigators (Yamamoto et al., 1967a–e) and further studied by researchers of the Nippon Oil Company (1965, 1967, 1971) and Miyoshi et al. (1958), who developed complex systems for example, $Fe(O-n-Bu)_3 + BF_3$, that produced very high molecular weight polyisobutylenes and polyisobutylene copolymers. Representative data are compiled in Table 4.4. Unusual mixtures of BF_3 with alkoxides and other chemicals, for example, $ZnEt_2$ have been described by investigators of the Sumitomo Chemicals Co. (Matsushima, 1972a; Matsushima and Ueno, 1972b). For example, binary systems of $(RO)_2Al–O–Zn–O–Al(OR)_2/BF_3$ at $-65°C$ give polyisobutylenes with $\bar{M}_v = 1,210,000$, whereas BF_3 alone under the same conditions yielded only $\bar{M}_v = 95,000$.

Ball-milled, anhydrous $MgCl_2$ also produces at a low rate high molecular weight polyisobutylenes (Table 4.4; Addecott et al., 1967).

The chemistries of these polymerization reactions are largely unknown.

Table 4.4 □ Molecular Weights of Polyisobutylenes Obtained by Various Complex Systems at −20°C

System	Diluent	$\bar{M}_v \times 10^{-3}$	Reference
Al(O-s-Bu)$_3$/BF$_3$/BiCl$_3$ (aged)	n-C$_6$H$_{14}$	180	Wichterle et al. (1959a)
Al(O-s-Bu)$_3$/BF$_3$/TiCl$_4$ (fresh)	n-C$_6$H$_{14}$	264	Wichterle et al. (1961)
Al(O-s-Bu)$_3$/BF$_3$ (1:3)	n-C$_6$H$_{14}$	258	Nippon Oil Company (1967)
Ti(O-n-Bu)$_3$/Br/BF$_3$ at −25°C	n-C$_6$H$_{14}$	200	Nippon Oil Company (1967)
Ti(O-i-Pr)$_2$(O-n-Bu)Cl/BF$_3$ (1:2) at −25°C	CH$_3$Cl	510	Nippon Oil Company (1967)
MgCl$_2$ (ball-milled, anhydrous)	n-C$_6$H$_{14}$	250	Addecott et al. (1967)
AlCl$_3$	EtCl	30	Wichterle (1959a)
TiCl$_4$	CH$_2$Cl$_2$	30	Kennedy (1979a)

Conceivably, initiation is due to adventitious protogens and usually high molecular weight polymers are obtained because the counteranion is embedded in the solid phase and as such is reluctant to participate in chain breaking events, be these by transfer or termination. Although chain breaking, particularly termination, would require counteranion participation by cumbersome surface reactions, propagation may occur in the homogeneous phase largely in the absence of counteranions. In this sense polymerizations initiated by solids may be regarded to involve "free" propagating carbocations, somewhat similar to those prevailing in polymerizations induced by high-energy irradiation. Thus it may not be a coincidence that very high molecular polyisobutylenes can be obtained both by γ-irradiation and by certain complex chemical initiating systems.

In addition to these complex solids giving rise to high molecular weight products, hosts of acidic solids, for example, Friedel–Crafts acids, clays, and earths, have been described to oligomerize light olefins. For example, the polymerization of propylene and butylenes by "solid phosphoric acid" (P_2O_5–SiO_2, 60:40) to liquids boiling in the gasoline range is one of the oldest processes in petroleum refining (Anonymous, 1964; Jones, 1956). Various synthetic lubricating oils are prepared from olefins including ethylene with solid $AlCl_3$ under moderate pressures from room temperature to ~ 150°C (Hall and Nash, 1938; Zorn, 1958). Lubricating oils have also been produced from light olefins by the use of clays, such as fuller's earth (Marschner, 1940), silica-alumina (Gayer, 1933), and activated clay (Houdry, 1940). Certain synthetic molecular sieves (zeolites) contain acidic sites and have been observed to induce olefin and vinyl ether oligomerization (Norton, 1964; Barrer and Oei, 1973; Borson et al., 1972).

Although there is a dearth of mechanistic information, conceptually these processes do not differ from those discussed in detail in this chapter. In line with the assumption that many acidic solids are *de facto* Brønsted acids (Emmet, 1958), initiation of these oligomerizations is most likely by protons. Low molecular weights are due to the usually high temperatures employed in these operations and the structure of the oligomers can be explained by conventional isomerization, alkylation, or hydride shift, for example.

One-Electron (Homolytic) Transpositions

Introduction

In contrast to more conventional chemical initiation involving two-electron or heterolytic transpositions (see preceding section), important recently developed initiating systems, that is, initiation via direct radical oxidation and charge transfer, operate by one-electron or homolytic transpositions. In direct radical oxidation relatively stable radicals are oxidized either *in situ* or in a separate step to initiating cationic species:

$$R\cdot \xrightarrow{-e} R^{\oplus} \xrightarrow{+M} \text{polymer}$$

Distinct from polymerizations induced by direct oxidation of radicals are polymerizations initiated by various charge transfer or electron transfer reactions, so-called charge transfer polymerizations. Polymerizations in this class are induced by charge transfer complexes D·A comprising an electron donor monomer D and an electron acceptor A. Suitable D·A charge transfer complexes under the influence of thermal or photo-chemical energy may be converted to reactive initiating cation radicals:

$$D\cdot A \xrightarrow{E} D^{\cdot\oplus}A^{\cdot\ominus} \xrightarrow{+M} \text{polymer}$$

A fundamental difference between initiation by conventional two-electron transpositions and one-electron transpositions is that whereas in the former systems ion generation involves the heterolytic conversion of neutral molecules to proton or carbenium ion, in the latter radicals or cation radicals are intermediates.

Direct Radical Oxidation

Although the possibility of cation formation by direct oxidation of radicals that is, $R\cdot -e \rightarrow R^{\oplus}$, has long been recognized, only very recently has this route been exploited for the initiation of carbocationic polymerizations. Thus in a series of pioneering recent papers Abdul-Rasoul et al. (1978a, b) and Ledwith (1978, 1979) described experiments that demonstrate initiation of the polymerization of several cationically polymerizable monomers, that is, *n*-butyl vinyl ether, in the presence of a radical source, oxidizing agent, and either thermal or light energy. Clearly, some radicals are more readily oxidized than others and carbon-centered radicals bearing electron-releasing substituents, for example, benzyl, allyl, alkoxy, nitrogen, and sulfur, are especially easily oxidized.

Suitable oxidants were proposed to be fragmentable salts, for example, $(C_6H_5)_2I^{\oplus}PF_6^{\ominus}$, which in addition to oxidizing the radical would supply the counteranion for cationic initiation and propagation to ensue. Thus this new initiation method involves thermal or photochemical radical generation, oxidation of the radical to cation with simultaneous generation of a suitable counteranion (ion generation), followed by cationation. Thermal process (Abdul-Rasoul et al., 1978b):

$$(C_6H_5)_2C(OH)-C(OH)(C_6H_5)_2 \xrightarrow{80°C} 2(C_6H_5)_2\dot{C}-OH$$

$$(C_6H_5)_2\dot{C}-OH + (p\text{-}CH_3C_6H_4)_2I^{\oplus}PF_6^{\ominus} \rightarrow (C_6H_5)_2CO + H^{\oplus}PF_6^{\ominus}$$

$$+ CH_3C_6H_4I + CH_3\dot{C}_6H_4$$

Photochemical process (Ledwith, 1978):

$$C_6H_5COCHC_6H_5 \xrightarrow{h\nu} C_6H_5\dot{C}O + \dot{C}HC_6H_5$$
$$\quad\quad\quad | \quad\quad\quad\quad\quad\quad\quad\quad\quad |$$
$$\quad\quad\quad OCH_3 \quad\quad\quad\quad\quad\quad\quad OCH_3$$

$$C_6H_5\dot{C}H + (C_6H_5)_2I^{\oplus}PF_6^{\ominus} \rightarrow C_6H_5CHPF_6^{\ominus} + C_6H_5I + \dot{C}_6H_5$$
$$\quad | \quad\quad\quad\quad\quad\quad\quad\quad\quad\quad\quad |$$
$$\quad OCH_3 \quad\quad\quad\quad\quad\quad\quad\quad OCH_3$$

The fate of the aromatic radicals is obscure. In a thermal bulk polymerization of n-butyl vinyl ether at 50°C, 75% conversion to polymer was obtained after 3 hr using azobisisobutyronitrile radical source in the presence of $(p\text{-}CH_3C_6H_5)_2I^{\oplus}PF_6^{\ominus}$. Similarly, in a photochemical polymerization of this monomer at 25°C after 10 min irradiation at 366 nm 98% conversion was obtained with 2, 2-dimethoxy-2-phenylacetophenone and the same salt as above.

Besides fragmenting aromatic salts, silver salts have also been found to be suitable radical oxidants. The polymerization of n-butyl vinyl ether readily occurred in the presence of $C_6H_5COC(OCH_3)_2C_6H_5$ and $AgPF_6$ plus light. The initiating cation is proposed to arise by photolysis of the radical source followed by oxidation of the radical:

$$
\begin{array}{ccc}
C_6H_5-C(OCH_3)_2 & & C_6H_5-\dot{C}(OCH_3)_2 \\
| & h\nu & \\
C=O & \xrightarrow{\quad} & \dot{C}=O \\
| & & | \\
C_6H_5 & & C_6H_5
\end{array}
$$

$$C_6H_5-\dot{C}(OCH_3)_2 + AgPF_6 \rightarrow C_6H_5-\overset{\oplus}{C}(OCH_3)_2PF_6^{\ominus} + Ag°$$

For example, 95% poly(n-butyl vinyl ether) was obtained after 10 mins at 25°C in the presence of $\sim 10^{-3} M$ radical source and $10^{-2} M$ $AgPF_6$ (Abdul-Rasoul et al., 1978b).

This lead may open new vistas for carbocationic polymerizations.

Charge Transfer Polymerizations

Many organic reactions proceed by charge transfer complex intermediates. According to the theory of electronic states of charge transfer complexes mainly developed by Mulliken and Peason (1969), combinations of D and A molecules may form D·A pairs in which an electron originally on the D may partially or completely be transferred to the A. The degree of electron transfer is determined by the nature of the D and A (e.g., ionization potential, electron affinity) and experimental conditions (e.g., solvent polarity, temperature, light). Complete electron transfer produces ion radicals which in the presence of polar solvent may completely dissociate:

$$D \cdot A \underset{solv}{\rightleftharpoons} D^{\cdot \oplus} + A^{\cdot \oplus}$$

The presence of ground state $D \cdot A$ complexes is characterized by charge transfer absorption bands usually in the UV or visible region. Chemical reactions may proceed by ground state $D \cdot A$ or excited $(D^{\oplus} A^{\ominus})^*$ complexes. Excitation can occur thermally or by photochemical means, and may involve selective excitation of D or A which in the presence of ground state partners may also lead to reactive excited state (E = energy; $*$ = excited state):

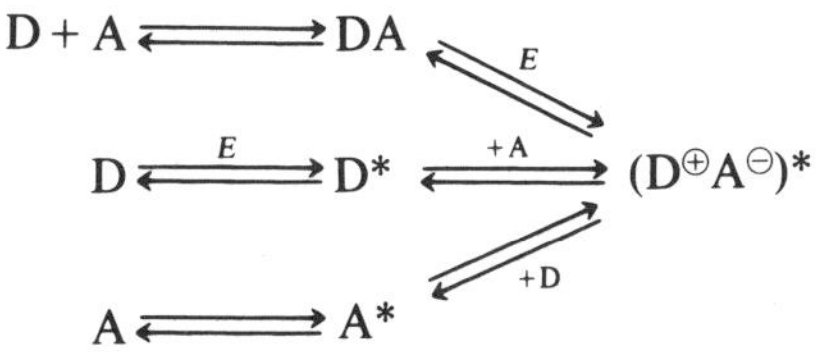

$$
\begin{array}{l}
D + A \rightleftharpoons DA \\
D \xrightleftharpoons{E} D^* \xrightleftharpoons{+A} (D^{\oplus} A^{\ominus})^* \\
A \rightleftharpoons A^*
\end{array}
$$

Discussion of the theory, detection, and characterization of the large variety of charge transfer complexes, that is, ground state, excited, exciplex, or zwitterionic, and their involvement in numerous kinds of chemical reactions is outside the scope of this book; the reader is referred to original literature (Mulliken and Peason, 1969; Hyde and Ledwith, 1974). A survey of pertinent charge transfer polymerizations follows. An excellent detailed discussion of this subject has recently been published (Shirota et al., 1978a, b).

Thermally Induced Charge Transfer Polymerization

Almost simultaneously and independently of each other two groups of investigators (Scott et al., 1963; Ellinger, 1963) discovered that N-vinylcarbazole rapidly polymerizes in the presence of traces of a large variety of organic electron acceptors, for example, p- and o-chloranil, tetracyanoethylene, 2, 3-dichloro-5, 6-dicyanobenzoquinone, and halogens, and with great insight stated their belief that initiation occurs by radical cations (CTC = change transfer complex):

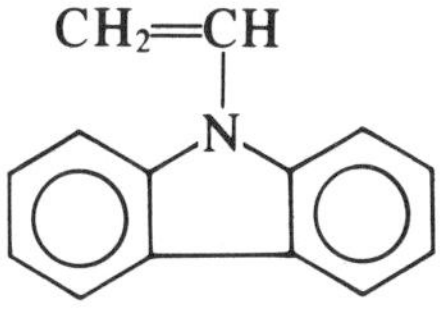

$$\text{(N-vinylcarbazole)} + A \rightleftharpoons [CTC] \rightarrow [R^{\cdot \oplus}] \xrightarrow{+M} \text{polymer}$$

These pioneering studies have been followed up and extended by a large number of detailed investigations by other workers and led to the development of a new field in polymer chemistry, namely, thermally and photoinduced charge transfer polymerizations.

In charge transfer polymerization electron transfer from the donor monomer to an acceptor molecule determines ion generation. Conceptually, ion generation by charge transfer may occur on combination of suitable pairs of monomers provided the donor D has a low ionization potential and the acceptor A a high electron affinity, and the available thermal energy is sufficient to promote electron transfer. If both partners D and A are polymerizable monomers, cationic and anionic polymerization may mutually be feasible merely by mixing the monomers (Hyde and Ledwith, 1974):

$$D + A \rightleftharpoons [D \cdot A] \rightleftharpoons \underset{\underset{\text{cationic}}{\downarrow +M}}{D^{\cdot \oplus}} + \underset{\underset{\text{anionic}}{\downarrow +M}}{A^{\cdot \ominus}}$$

cationic anionic
pzn. pzn.

The classical example for such a system is the alkyl vinyl ether (D)/1, 1-dicyanoethylene (vinylidene cyanide, A) pair (Gilbert et al., 1956), which was shown to polymerize spontaneously on mixing and give rise to mixtures of homopolymers. The mechanism of this unique polymerization system was studied in detail by Stille and Chung (1975a, b), who corroborated Gilbert et al.'s (1956) observations and concluded that the alkyl vinyl ethers and vinylidene cyanide polymerize separately by cationic and anionic process, respectively.

$$CH_2{=}CH{-}OR + CH_2{=}C(CN)_2 \rightleftharpoons [CTC] \longrightarrow$$

$$\underset{\underset{\text{cationic}}{\downarrow +M}}{[CH_2{=}CH{-}OR]^{\cdot \oplus}} + \underset{\underset{\text{anionic}}{\downarrow +M}}{[CH_2{=}C(CN)_2]^{\cdot \ominus}}$$

cationic anionic
homopzn. homopzn.

It is astonishing that these processes can occur in the same reactor.

Although besides alkyl vinyl ethers several other types of cationically polymerizable electron donor monomers have been described, for example, N-vinylcarbazole, N-vinylindole, vinylnaphthalene, and vinylanthracene, additional anionically polymerizable electron acceptors have not been described, and in this sense the alkyl vinyl ether/vinylidene cyanide pair (Gilbert et al., 1956) remains unique.

Besides this exceptional system in which cationic and anionic polymerization occur in the same reactor, the overwhelming majority of charge transfer polymerizations involve an electron-rich donor monomer and a nonpolymerizable electron acceptor. Many efficient electron acceptors are known to organic chemists, such as maleic anhydride, o- and p-chloranil, bromanil, tetracyanoethylene, large number of quinones, and tetranitromethane (Andrews and Keefer, 1964), and have been found to be

active in initiating cationic polymerizations via charge transfer intermediates. For example, polymerization of styrene in liquid sulfur dioxide in
the presence of electron acceptors such as maleic anhydride plus cumene
(Matsuda and Abe, 1968a, b) and aromatic hydrocarbons such as anthracene plus oxygen (Nagai et al., 1968) was proposed to involve charge
transfer complexes and to initiate by cation radicals; however, detailed
mechanisms remain obscure.

In spite of the large number of studies concerning cationic initiation by
charge transfer, considerable uncertainty exists as to the detailed
mechanism of these reactions. Notable exceptions are systems studied by
Stille et al. (1975a–c), Aoki et al. (1970), Oguni et al. (1974), Tarvin et al.
(1972), Ledwith and Sambhi (1965), Ledwith (1969a); Bawn et al. (1964,
1965a, b), Natsuume et al. (1970), Tada et al. (1973a, b), Fujimatsu et al.
(1970), and Shirota et al. (1978a, b).

Among cationic polymers investigated with regard to charge transfer
initiation, N-vinylcarbazole has probably received the most attention
(Bawn et al., 1964; Scott et al., 1963; Shirota et al., 1978a, b). Sustained
interest in this molecule,

is due to its extraordinarily high cationic reactivity and to the unusual
mechanical-electronic properties of poly(N-vinylcarbazole). From the
scientific–mechanistic point of view the polymerization of N-vinylcar-
bazole is diagnostic of free radical or cationic reaction: this monomer does
not polymerize by anionic initiators. Although many details of N-vinyl-
carbazole polymerization by charge transfer initiation remain obscure,
Shirota and Mikawa's research (Natsuume et al., 1970; Tada et al., 1973a;
Shirota et al., 1978a, b) provides valuable insight into basic aspects, for
example, thermal and photosensitized charge transfer initiation and the
effect of solvent basicity on the reaction path. For example, polymerization
of N-vinylcarbazole in the presence of fumaronitrile, that is, an electron
acceptor capable of undergoing radical copolymerization with N-vinyl-
carbazole, produced homopolymer of N-vinylcarbazole (cationic polymerization); however, in conjunction with AIBN a 1:1 alternating
copolymer was obtained (radical polymerization):

Owing to the highly basic nature of N-vinylcarbazole and to the highly stabilized N-vinylcarbazole cation, thermal polymerizations of this monomer induced by electron acceptors are generally cationic in nature. However, the fate of the radical during the cationic polymerization remains largely obscure.

Subsequent to electron transfer and to the formation of the cation–radical anion–radical pair $D^{\cdot \oplus} A^{\cdot \ominus}$ initiation may occur either by a zwitterion or by HCl produced *in situ*. This problem has been studied by Natsuume et al. (1970) in the N-vinylcarbazole/chloranil system where initiation was proposed to involve either cationation by the zwitterion or elimination of HCl, which then functions as the *de facto* initiating entity (-N⟨ indicates a carbazole rest):

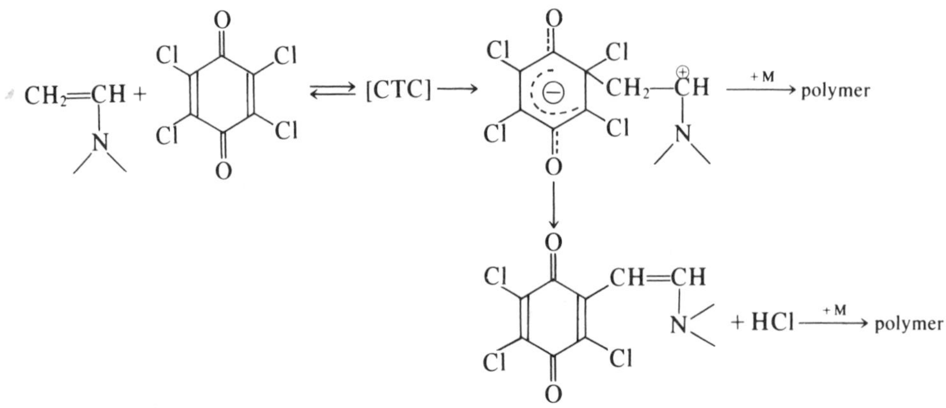

The questions (1) whether initiation is due to direct cationation by the zwitterion or to HCl elimination, and (2) if both processes occur simultaneously what is their relative importance in terms of polymer formed, may be elucidated by appropriate grafting experiments in which the chloranil moiety is attached to an inert flexible macromolecule, that is,

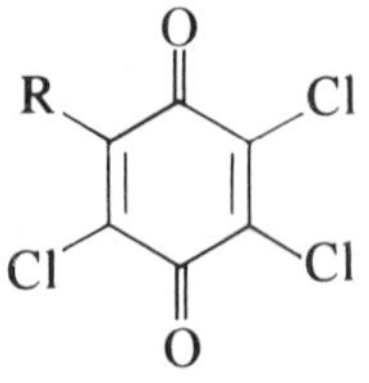

where R = polyisobutylene. In this instance initiation by direct cationation would result in graft copolymer, which could be readily separated from homopolymer of *N*-vinylcarbazole. If chain transfer to monomer occurs during grafting, graft copolymer formation would only prove the presence of direct cationation but could not be used to quantize the relative importance of initiation by direct cationation versus initiation by HCl.

Systematic and careful research by Stille and co-workers (Tarvin et al., 1972) also indicates coupling between the radical cation donor and radical anion acceptor and the formation of zwitterions in the initiation of alkyl vinyl ethers by electron acceptors. For example, initiation in the polymerization of alkyl vinyl ethers by the acceptor 2, 3-dichloro-5, 6-dicyano-*p*-benzoquinone in acetonitrile solvent involves rapid one-electron transfer from the monomer to the acceptor to form a charge transfer complex, followed by slow coupling of the radical cation and radical anion (Tarvin et al., 1972):

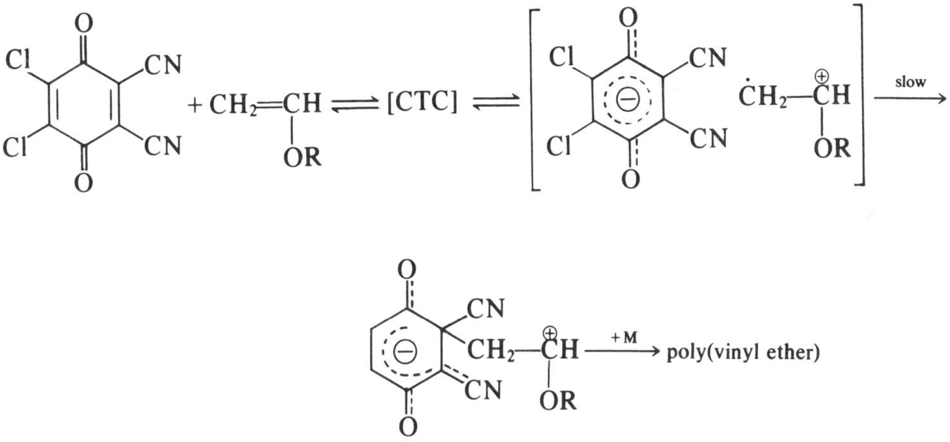

An important side reaction which in many systems interferes with polymerization initiation is 2 + 2 cycloadduct formation (Oguni et al., 1974):

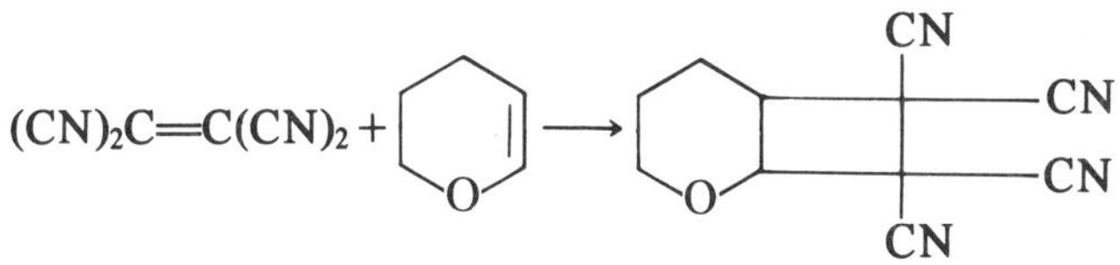

An interesting cationic anionic alternating copolymerization involving ion generation by charge transfer followed by zwitterion formation has been proposed by Butler et al. (1971). According to these authors the first events in the polymerization of ethyl vinyl ether/4-phenyl-1, 2, 4-triazoline-3, 5-dione systems involve charge transfer immediately followed by zwitterion formation, which then leads to alternating copolymerization:

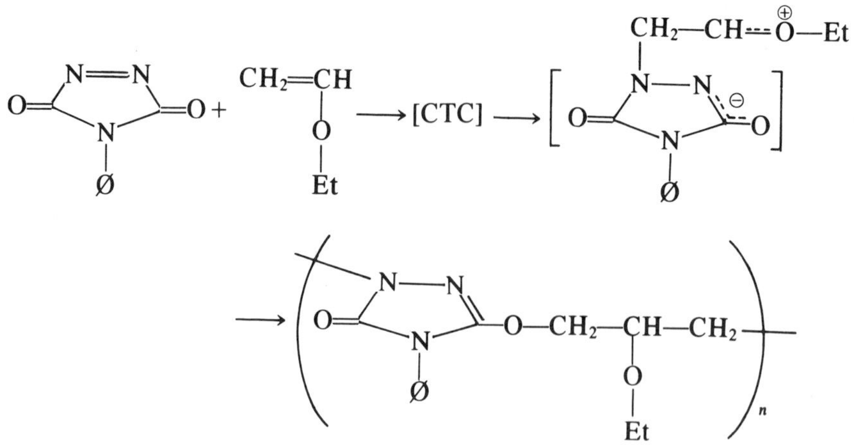

A series of interesting experiments have been carried out by Szwarc and co-workers (Fleichfresser et al., 1968) using the nonpolymerizable olefin 1, 1-diphenylethylene as model compound and $SbCl_5$ as electron acceptor and methylene chloride solvent at $-80°C$. On the basis of UV spectrophotometry and quenching experiments the authors propose the following sequence of events:

$$CH_2=C(C_6H_5)_2 + SbCl_5 \rightleftharpoons [CTC] \rightleftharpoons \dot{C}H_2-\overset{\oplus}{C}(C_6H_5)_2SbCl_5^{\ominus}$$

$$2\dot{C}H_2-\overset{\oplus}{C}(C_6H_5)_2 \rightarrow (C_6H_5)_2\overset{\oplus}{C}-CH_2CH_2-\overset{\oplus}{C}(C_6H_5)_2$$

In a similar series of investigations (Penczek et al., 1968) with 1, 1-diphenylethylene and tetranitromethane carbocation formation was visualized to occur by "dissociative electron transfer":

$$CH_2=C(C_6H_5)_2 + C(NO_2)_4 \rightleftharpoons [CTC] \rightarrow CH_2-\overset{\oplus}{\underset{NO_2}{C}}(C_6H_5)_2 + \overset{\ominus}{C}(NO_2)_3$$

$$CH_2-\overset{\oplus}{\underset{NO_2}{C}}(C_6H_5)_2 + \overset{\ominus}{C}(NO_2)_3 \rightarrow CH=C(C_6H_5)_2 + HC(NO_2)_3$$

The absence of dimer cation $(-CH_2\overset{\oplus}{C}(Ph)_2)_2$ was explained by proposing concerted reactions and/or proton transfer in the ion pair.

Yamada et al. (1966, 1967) reported interesting observations using the less reactive electron donor isobutylene. According to these investigators isobutylene polymerization cannot be initiated by $VOCl_3$ alone; however, polymerization occurs with combinations of $VOCl_3$/naphthalene, anthracene or similar aromatic compounds, amines, and ethers, for example; Table 4.5 shows the data. The originally yellow $VOCl_3$ in n-heptane solution became bluish green on naphthalene or anthracene addition and ultimately turned to dark red on introduction of isobutylene. The red color faded when isobutylene was consumed, but returned when fresh monomer was added and polymerization resumed. Polymerizations were slow but respectable molecular weights were obtained. The authors suggest charge transfer initiation by a charge transfer complex:

$$VOCl_3 + C_{10}H_8 \rightleftharpoons [CTC] \rightleftharpoons VOCl_3^{\ominus} + C_{10}H_8^{\cdot\oplus}$$

$$C_{10}H_8^{\cdot\oplus} + CH_2{=}C(CH_3)_2 \rightarrow C_{10}H_8 + \overset{\cdot}{C}H_2{-}\overset{\oplus}{C}(CH_3)_2$$

$$\overset{\cdot}{C}H_2{-}\overset{\oplus}{C}(CH_3)_2 \rightarrow (CH_3)_2\overset{\oplus}{C}{-}CH_2CH_2{-}\overset{\oplus}{C}(CH_3)_2 \xrightarrow{+M} polymer$$

Unfortunately these interesting observations have not been followed up and no hard evidence to substantiate these claims has become available.

Recently Cesca et al. (1976) described that a large number of compounds, for example, chloranil, p-benzoquinone, 1,3,5-trinitrobenzene, and pyromellitic dianhydride, in the presence of Me_2AlCl and Et_2AlCl efficiently initiate the polymerization of isobutylene and copolymerization of isobutylene–isoprene mixtures to high molecular weight products at relatively high ($-40°C$) temperatures. The chloranil/Et_2AlCl system was investigated in some detail by a variety of techniques, mainly ESR, and initiation by electron transfer was proposed:

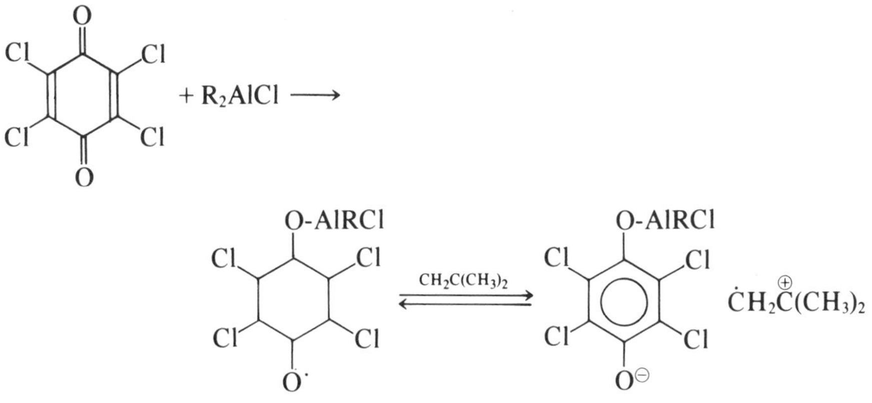

$$2\overset{\cdot}{C}H_2\overset{\oplus}{C}C(CH_3)_2 \rightarrow (CH_3)_2\overset{\oplus}{C}{-}CH_2{-}CH_2{-}\overset{\oplus}{C}(CH_3)_2 \xrightarrow{M} polymer$$

In view of the very high ionization potential of isobutylene, Williams (1976) questioned the possibility of electron transfer to chloranil and

Table 4.5 □ Polymerization of Isobutylene with VOCl₃/Activator in _n_-Heptane[a]

Activator	Activator/VOCl₃ (mole)	Temperature (°C)	Time (hr)	Yield (%)	[η]	$\bar{M}_v \times 10^{-3}$
Naphthalene	1.0	0	20	98.7	—	
Naphthalene	0.5	0	20	96.0	0.15	13.5
Naphthalene	0.05	0	20	81.4	0.46	78.0
Naphthalene	0.5	−78	40	93.5	0.42	67.7
Naphthalene	0.5	−78	40	75	0.65	134
Anthracene	0.2	0	2	94.5	Rubbery polymer	
Fluorene	0.1	−30	10	77.7	Rubbery polymer	
Phenanthrene	0.1	0	10	85.7	Rubbery polymer	
β-Naphthylamine	0.2	0	12	96.4	Rubbery polymer	
Aniline	0.5	−78	20	74.0	Rubbery polymer	
CS₂	1.0	−78	20	31.1	Rubbery polymer	
Phenol	0.2	0	20	51.0	Rubbery polymer	

[a] After Yamada et al. (1966, 1967).

Sigwalt and Szwarc (1976) wondered whether the ESR signals detected in the system are related to the polymerization.

Probably for the same reasons researchers have used stable cation salts for initiation (see section on stable carbenium ion salts above), Oberrauch et al. (1978) have recently prepared a stable cation radical salt, perylene cation radical perchlorate, and used it to initiate carbocationic polymerization of isobutylene, styrene, α-methylstyrene, isobutyl vinyl ether, and N-vinylcarbazole using a variety of common solvents:

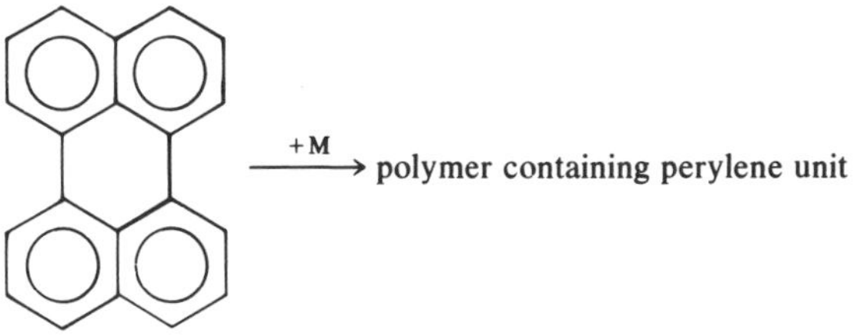

$\xrightarrow{+M}$ polymer containing perylene unit

Unfortunately not many details concerning this system have become available. It is hoped that further research with stable cation radical salts will help illuminate initiation details of carbocationic polymerizations.

Photoinduced Charge Transfer Polymerization

Combinations of strong donor monomers with strong electron acceptors readily induce polymerization even in the dark. With less efficient electron acceptors photoactivation may be necessary for initiation. The reactive intermediates involved in light-induced charge transfer polymerizations, so-called exciplexes or heteropolar excimers, are charge transfer complexes that are stable only in the electronically excited state but dissociate in the ground state. Exciplexes of strong donor–acceptor pairs are usually very polar species and readily lead via electron transfer in polar solvents to solvated ion radicals. They can be characterized by fluorescence spectra (Leonhardt and Weller, 1963, Potashnik et al., 1971; Nagakura, 1975).

Hayashi and co-workers (Irie et al., 1970, 1972a, b, 1975; Suzuki et al., 1976) importantly contributed to the mechanistic understanding of this area. In a series of significant experiments Irie et al. (1972a) illuminated (high-pressure mercury lamp, using filters to cut off wavelengths shorter than 300 nm) α-methylstyrene in the presence of tetracyanobenzene acceptor in 1,2-dichloroethane at −30°C and obtained highly syndiotactic polymer, $\bar{M}_v \sim 77{,}600$. Polymerization did not occur in the dark. Scavenging experiments and copolymerization with styrene proved the mechanism to be cationic. Charge transfer between α-methylstyrene and tetracyanobenzene was indicated by spectroscopy (charge transfer band at 363 nm). The findings led to the conclusion that only photoexcited charge transfer complexes are active, whereas ground state complexes are inactive. Further work by these authors (Irie et al., 1972a) showed that the rate of α-methylstyrene polymerization was proportional to light intensity and the polymer was highly syndiotactic. The mechanism of initiation was studied by ESR and the α-methylstyrene cation radical was proposed to be the initiating species. Pyromellitic anhydride, a strong electron acceptor, was also found useful for photoinduced polymerization of α-methylstyrene in chlorinated solvents at low temperatures (Irie et al., 1972b). The various events during the initiation process have been schematized as follows (Irie et al., 1975):

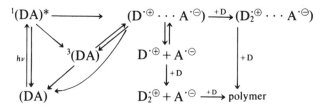

where D = α-methylstyrene donor, A = tetracyanobenzene acceptor, subscripts 1 and 3 denote singlet and triplet states, respectively, the asterisk indicates excited state, and $D_2^{\cdot\oplus}$ is a dimer cation radical. For a detailed discussion of the individual species the reader should consult Irie et al. (1975).

Recently these investigators (Suzuki et al., 1976) extended their studies to styrene and found that this monomer readily polymerizes on illumination at a wavelength longer than 350 nm in the presence of pyromellitic dianhydride, where only the charge transfer band exists. On the basis of molecular weight distribution and other evidence the authors propose the simultaneous occurrence of radical and cationic mechanisms and that the cationic process involves free cations and cation radical–anion radical pairs:

$$D + A \rightleftharpoons (D^{\cdot}A) \xrightarrow{h\nu} (D^{\cdot}A)^* \longrightarrow (D^{\cdot\oplus} \cdots A^{\cdot\ominus}) \rightleftharpoons D^{\cdot\oplus} + A^{\cdot\ominus}$$

polymer radical pzn.	polymer ion pair pzn.	polymer free ion pzn.

where D = styrene and A = pyromellitic anhydride. Increasing solvent polarity increases the contribution of free cationic process.

Photoinduced charge transfer initiation of isobutylene polymerization has recently been proposed by Marek and Toman (1973). The key observation by these authors was that VCl$_4$, TiCl$_4$, or TiBr$_3$ did not induce polymerization of isobutylene in n-heptane solution in the range 0 to $-78°C$; however, rapid polymerization occurred when the mixture was illuminated by visible, UV, or IR light and stopped when illumination was discontinued. Over the intervening years a series of publications have appeared concerning extension and details of the original observation (Toman et al., 1974, 1976; Marek et al., 1975; Marek, 1977; Pilar et al., 1976).

With regard to initiation details, the first step was proposed to occur by photon absorption by the isobutylene–VCl$_4$ donor–acceptor complex which decomposes by homopolar cleavage into ion radicals (Pilar et al., 1976):

$$i\text{-}C_4H_8 + VCl_4 \rightleftharpoons i\text{-}C_4H_8 \cdot VCl_4 \xrightarrow{h\nu} M^{\cdot\oplus} + VCl_4^{\cdot\ominus}$$

Isobutylene cation radicals subsequently recombine and rapidly form dications:

$$2\dot{C}H_2\text{—}\overset{\oplus}{C}(CH_3)_2 \rightarrow (CH_3)_2\overset{\oplus}{C}\text{—}CH_2CH_2\text{—}\overset{\oplus}{C}(CH_3)_2$$

which then initiate conventional cationic chain growth.

Model experiments by Diem et al. (1978) in which mixtures of 2, 4, 4-trimethyl-1-pentane/TiCl$_4$ and VCl$_4$ were illuminated by various light sources failed to produce evidence for the formation of tail-to-tail structures (contiguous–CH$_2$–groups) compatible with the isobutylene cation radical recombination step. Further evidence, preferably chemical, is needed before this initiation mechanism can be accepted.

The principle of photoinduced charge transfer polymerization was used by Butler and Ferree (1976) for photoinitiated cationic crosslinking. These authors first prepared by free radical technique a high molecular weight $\bar{M}_n \sim 110,000$ polymer with pendant vinyl ether groups, that is, poly-m-(vinyloxyethoxy)styrene, mixed this donor "monomer" with acceptors, such as chloranil, and excited this charge transfer system by illuminating with a mercury–xenon lamp at 366 nm in methylene chloride solution at 3°C. Rapid crosslinking ensued and crosslinked precipitate formed on the walls of the tube where the light beam hit the systems. Crosslinking efficiency, however, was rather low. The results of experiments with various quenching agents, oxygen, water, and other additives led the authors to propose photoinitiated cationic crosslinking schematized as follows:

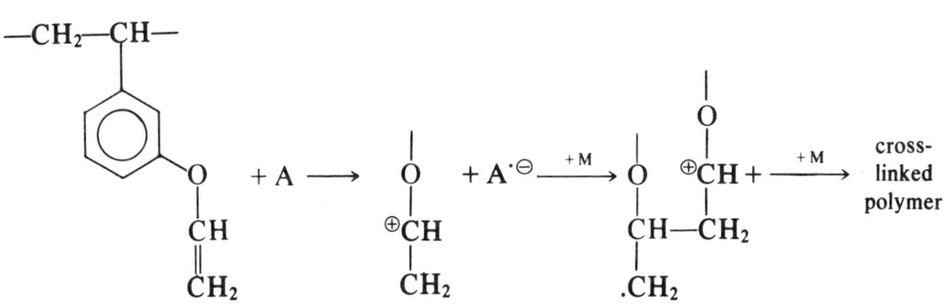

Researches by Shirota et al. (1978a, b) significantly advanced under-standing of mechanisms of light-induced charge transfer polymerizations. According to these authors solvent basicity determines the direction of reaction pathways. They examined in detail light-induced polymerizations (free radical and cationic) and cyclodimerizations of N-vinylcarbazole in the presence of various electron acceptors, for example, chloranil, bromanil, pyromellitic dianhydride, phthalic anhydride, and trinitroben-zene, using a great variety of solvents of various basicities, such as benzene, methylene chloride, nitrobenzene, 1, 2-dichloroethane, acetoni-trile, acetone, methanol, N-methyl-2-pyrrolidone, dimethylformamide, and hexamethylphosphoramide. The polymerization is initiated by the N-vinylcarbazole cation radical NVC$^{\cdot\oplus}$ formed via N-vinylcarbazole-ac-ceptor complexes; the ensuing mechanism is determined by the nature of the medium. The fate of NVC$^\oplus$ depends on the nature and extent of solvation. In strongly basic solvents cationic polymerization is completely inhibited and only free radical mechanism can proceed. In weakly basic media cationic polymerization predominates. In solvents of intermediate character both cationic and free radical polymerizations may occur. Table 4.6 summarizes the results of a large number of studies.

Tada et al. (1973a, b) and Shirota et al. (1970) found that dissolved oxygen may also function as an electron acceptor toward N-vinylcarbazole

Table 4.6 ☐ **Effect of Solvent Basicity on Photoinduced Polymerization of N-Vinylcarbazole in the Presence of Electron Acceptor**

Solvent Basicity	Nature of Reaction in Presence of Small Amounts of Organic Electron Acceptors	Nature of Reaction in Presence of Electron Acceptor Monomers Capable of Radical Copzn. (Fumaro-nitrile, Diethyl Fumarate)
Low (benzene, CH$_2$Cl$_2$; nitro-benzene; 1, 2-dichloro-ethane)	Cationic homopolymeri-zation	Cationic homopzn. of NVC (in benzene, simultaneous radical copzn.)
Medium (acetone; CH$_3$CN; CH$_3$OH)	Cyclodimerization of NVC	Radical copzn. of NVC with electron acceptor monomer; plus cyclo-dimerization for NVC
High (DMF; Me$_2$SO; N-methylpyrrolidone)	Radical homopolymeri-zation + cyclodimeri-zation	Radical copzn. of NVC with electron-accepting monomer
Very high (HMPA)	Radical polymerization	

and cationic polymerization takes place in less basic solvents:

$$NVC \xrightarrow{h\nu} {}^1NVC^* \xrightarrow[\substack{CH_2Cl_2 \\ PhNO_2}]{O_2} NVC^{\cdot \oplus} O^{\cdot \ominus} \rightarrow \text{cationic pzn.}$$

In more basic media cyclodimerization and/or radical polymerization proceed. In the absence of oxygen only radical polymerization occurs in any (nonelectron accepting) solvents.

In addition to these charge transfer polymerization systems in which the donor was invariably a cationically polymerizable monomer, initiating $D \cdot A$ combinations in which D was not monomer have also been described. For example, combinations of propionaldehyde/maleic anhydride polymerized isobutyl vinyl ether only under illumination (Yamaoka et al., 1966) and, similarly, ethers, for example, diethyl ether, dioxane, and bis(chloromethyl)oxetane/maleic anhydride mixtures rapidly, sometimes even explosively, induced polymerization of isobutyl vinyl ether and N-vinylcarbazole only under illumination (Takakura et al., 1965). As to the mechanism of initiation the authors speculate that in the latter systems irradiation creates radicals from the ethers, which then transfer an electron to the maleic anhydride MA acceptor, thus producing the initiating cation radical–anion radical pair (Takakura et al., 1965):

$$R—O—\dot{C}H—R + MA \rightarrow R—O—\overset{\oplus}{C}H—R + MA^{\cdot \ominus}$$

Conclusions: Initiation by One-Electron Transpositions

One-electron transpositions are important, relatively new techniques for initiation of carbocationic polymerizations and lead to interesting new polymer synthesis options. These methods hold great promise for the theory and practice of cationic reactions in general and polymerizations in particular; however, they also have serious fundamental limitations.

Both one-electron transposition techniques, direct radical oxidation and charge transfer polymerization, differ from the more conventional two-electron transposition techniques in details of ion generation. In two-electron transposition methods, cations (protons or carbenium ions) arise by heterolytic bond cleavage and involve interaction between two neutral species: the initiator, that is, cation source, and coinitiator, that is, Friedel–Crafts acid or other species necessary for cation generation. In contrast, ion generation by one-electron transpositions involves either radical generation followed by oxidation or ion radical generation via charge transfer complexes.

Thus, consistent with the terminology used in this book, initiating systems in direct radical oxidation methods are radical precursor/oxidizing agent combinations, such as $C_6H_5COC(OCH_3)_2C_6H_5/AgPF_6$, and the initiating species is the cation produced via the radical intermediate $\dot{C}(OCH_3)_2C_6H_5 \xrightarrow{-e} \overset{\oplus}{C}(OCH_3)_2C_6H_5$. In charge transfer polymerizations $D \cdot A$

complexes control initiation and a typical initiating systems is, for example, N-vinylcarbazole/chloranil. Ion generation involves partial or complete electron transfer (the degree of electron transfer being determined by the nature of the reactants, solvent polarity, and temperature). Excitation may occur thermally or photolytically (exciplexes) and may involve either the donor or the acceptor separately or the D · A complex.

The second step that completes initiation, that is, cationation, is akin to olefin alkylation in all chemical initiation methods.

A caveat in regard to cationation concerns cation radicals: there is evidence that in many systems direct cationation of olefin does not occur (Natsuume et al., 1970); rather, elimination of HX followed by protonation completes initiation:

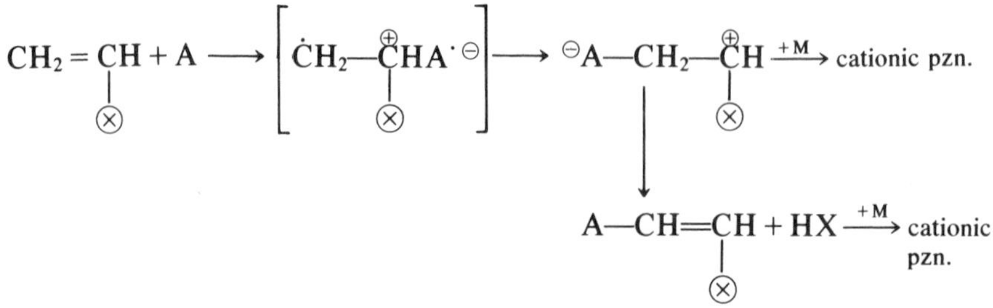

$$CH_2 = CH + A \longrightarrow \left[\dot{C}H_2 - \overset{\oplus}{C}HA^{\cdot\ominus} \right] \longrightarrow \ ^{\ominus}A - CH_2 - \overset{\oplus}{C}H \xrightarrow{+M} \text{cationic pzn.}$$

$$A - CH = CH + HX \xrightarrow{+M} \text{cationic pzn.}$$

where ⊗ = a strongly electron-donating group such as carbazole, alkoxy, or p-methoxyphenyl, and A = electron acceptor such as chloranil. As discussed above, graft copolymerizations may go a long way to elucidate this particular problem.

Detailed consideration of the concept of initiation by one-electron transpositions leads to the proposition that electrophilic polymerizations in general may involve charge transfer intermediates. According to this proposal electrophilic bond formation in general, and cationations during initiation or propagation in particular,

$$C^{\oplus} + C = C \rightarrow C - C - C^{\oplus}$$

may occur in two steps, (1) electron transfer and (2) combination:

$$C^{\oplus} + C = C \rightarrow [\dot{C} + \dot{C} - \overset{\oplus}{C}] \rightarrow C - C - C^{\oplus}$$

Thus charge transfer complexes may in principle always be involved in electrophilic polymerizations though their physical existence would be very difficult to demonstrate. The two-step mechanism would preferentially occur with strong electron donors such as N-vinylcarbazole or alkyl vinyl ethers. Indeed, the presence of radicals has been detected by ESR spectroscopy during cationic polymerization of N-vinylcarbazole (Tazuke et al., 1967). Significantly, cation radicals were trapped by 2, 4, 6-tri-tert-butylnitrosobenzene during styrene and α-methylstyrene polymerization induced by C_6H_5OH/BF_3 or "H_2O"/$BF_3\cdot OEt_2$ or "H_2O"/$AlCl_3$ in benzene solution (Yamada et al., 1976). ESR indicated the formation of two radical species during α-methylstyrene polymerization induced by C_6H_5OH/BF_3:

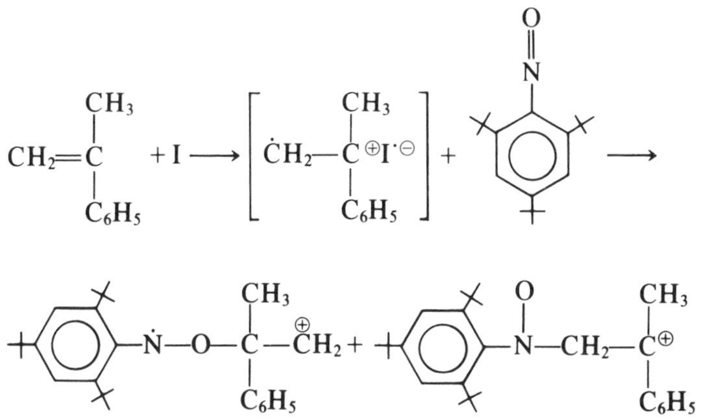

The persistent problem in elucidating initiation details not only in one-electron transpositions but in any kind of initiation methods is the elusive nature of the reactive intermediates. The possibility unfortunately persists that species whose existence in the reaction medium has been proved are not directly involved in initiation but are initiator precursors or have been formed in parasitic reactions. In the absence of direct observation methods, such as exist with stable cation salts, the only way to substantiate the existence of elusive reactive intermediates is to develop overwhelming and therefore convincing circumstantial evidence.

Exploration of carbocationic polymerization initiation by one-electron transposition has been somewhat limited by the need for readily oxidizable radicals or strong electron donors for reaction. Since the number of such species is rather modest and industrially exploitable systems have not yet been developed (although indications of applications in photocurable coatings and image forming exist), this class of initiating systems has inspired only a handful of dedicated researchers within the last few years. Key findings, concerning, for example, reagent purity (Natsuume et al., 1969; Aoki et al., 1970), the effect of solvent basicity on the outcome of reactions (Shirota et al., 1970; Tada et al., 1970, 1972, 1973a, b, 1974), and initiation by direct radical oxidation (Abdul-Rasoul et al., 1978a, b) became available only very recently. It is projected that progress in one-electron transposition polymerization will increasingly accelerate as mechanistic details of these intricate reactions become appreciated.

Physical Methods

In addition to the well-known chemical techniques discussed in the first part of this chapter, important physical methods have also been developed for the initiation of carbocationic polymerizations. In most of these physical methods monomer is converted into a variety of rather ill-defined active species including cations and/or cation radicals by energy transfer from external sources to the monomer M, for example, high-energy or ionizing radiation, photolytic methods, high electric fields, electrolytic processes.

High-Energy or Ionizing Radiation

γ-Ray-Initiated Carbocationic Polymerization

High-energy- or ionizing radiation-induced carbocationic polymerizations not only provide a useful method for the preparation of many high polymers but are also important in the fundamental mechanistic elucidation of these processes. Although many details of initiation of olefin polymerizations by ionizing radiation remain obscure, important aspects have been elucidated and warrant a brief discussion. Excellent introductions to the fundamentals of radiation chemistry for polymer scientists together with critical reviews of the field up to ~ 1961 and subsequently to ~ 1968 have been written by Pinner (1963), Chapiro (1962), and Williams (1968), respectively.

With regard to initiation it is useful to contrast chemical and high-energy irradiation methods. Chemical initiation usually provides specific carbenium ion intermediates and the cationic mechanism of ensuing reactions is not much in doubt. In contrast, when ionizing radiation impinges on olefins a wide variety of active species are generated—, ions, ion radicals, excited species, radicals—and the nature of subsequent reactions is by no means obvious. The novice may presume that in such an ill-defined heterogeneous system many mechanisms initiated by many active species operate simultaneously, leading to a complete lack of polymerization specificity. A closer examination of available information, however, suggests that in spite of the variety of active species formed, polymerization is usually due to one species and most of the product is formed by one predominating mechanism. Even when two mechanisms may proceed side-by-side, for example, in bulk polymerization of almost dry styrene, one of the polymerizations (cationic) is so much faster than the other (free radical) that the final polymer would contain only an insignificant amount of product formed by the slower process. In wet styrene cationic polymerization cannot proceed and, again, only one process, a free radical-induced polymerization, occurs.

For many years it was believed that high-energy radiations induce only radical polymerization of vinyl compounds. The discovery of γ-ray-induced cationic polymerizations of isobutylene (Pinner, 1963; Chapiro, 1962; Williams, 1968; Rogers, 1957; Gromangin, 1957; Davison et al., 1957, 1959; Hoffman, 1959; Worrall and Charlesby, 1959a,; Worrall and Pinner, 1959b; Collinson et al., 1959; Metz, 1969; Hayashi and Okamura, 1967;

Bates et al., 1960) started a new period in synthetic polymer chemistry. Initiation of isobutylene polymerization is rationalized to occur by the following events (Pinner, 1963):

$$i\text{-}C_4H_8 \rightarrow i\text{-}C_6H_8^{\cdot\oplus} + e$$

$$i\text{-}C_4H_8^{\cdot\oplus} + i\text{-}C_4H_8 \rightarrow i\text{-}C_4H_9^{\oplus} + i\text{-}\dot{C}_4H_7$$

$$i\text{-}C_4H_9^{\oplus} \xrightarrow{+M} \text{polymer}$$

$$i\text{-}C_4H_9^{\oplus} + i\text{-}C_4H_7^{\cdot} \rightarrow (CH_3)\dot{C}\text{—}CH_2CH_2\text{—}\overset{\oplus}{C}(CH_3)_2$$

Although isobutylene polymerizations are diagnostic for cationic reactions (this monomer polymerizes *only* by cationic techniques), high-energy-induced styrene polymerizations presented a more complex picture since this monomer readily polymerized by radical and ionic mechanisms. Systematic research using vigorous techniques particularly by Metz (1969) and Hayashi and Okamura (1967) showed that depending on conditions, particularly the dryness of the system, styrene can be polymerized to high molecular weight product at measurable rates by cationic or radical mechanisms. Table 4.7 summarizes the kinetic behavior of this monomer under wet and dry conditions (Metz, 1969).

The presence of cationic polymerizations in high-energy-induced polymerizations have been corroborated by a variety of methods, for example, scavenging experiments (Bates et al., 1960; Bonin et al., 1965), mass

Table 4.7 □ Kinetic Behavior of Styrene Polymerization Initiated by Radiation under Wet (Radical) and Dry (Ionic) Conditions[a]

Parameter	Wet (Radical)	Dry (Ionic)
$R_P [A]^\beta$	$\beta = 0,59$	$0,62 \leqslant \beta \leqslant 1$
DP $[I]^\gamma$	$\gamma = -0,50$	$\gamma \simeq 0$
E_a (rate)	$E_a \simeq 7\,\text{kcal}$	$-6 < E_a < +2\,\text{kcal}$
E_a (DP)	$E_a' \simeq 5\,\text{kcal}$	$E_a \simeq 0$
Copolymerization with	$r_1 = 0.22 \pm 0.2$	$r_1 = 0.25 \pm 0.25$
α-methylstyrene (M2)	$r_2 = 0.6 \pm 0.2$	$r_2 = 8.5 \pm 4.0$
Sensitivity to		
$\quad H_2O$	None	Very high
$\quad O_2$	High	High
$\quad$ DPPH	High	High
$\quad N_2O$		Slight
$\quad CO_2$		Slight
NH_3, amines		Very high

[a] After Metz (1969).

spectrometry (Aquilanti et al., 1967), ion cyclotron resonance (Davison et al, 1959; Viswanathan and Kevan, 1967), electrical conductivity (Hayashi et al., 1968), and electron spin resonance (Yoshida and Hayashi, 1969).

In addition to isobutylene and styrene, many monomers have been shown to polymerize by carbocationic mechanism upon ionizing radiation, for instance, β-pinene (Bates et al., 1960, 1962), cyclopentadiene (Bonin et al., 1965), α-methylstyrene (Hayashi et al., 1977), and vinyl ethers (Hayashi et al., 1977).

An important characteristic, beneficial for theoretical studies, of high-energy-initiated carbocationic polymerizations is the absence of counteranions in these reactions. Electroneutrality in these polymerizations is maintained by electrons trapped by solvation or in cavities in liquids. The kinetic chains thus propagate by free ions and the lifetime of these cations is higher than $\sim 10^{-10}$ sec even in liquids of low dielectric constant (Williams, 1964). Williams estimated the distance between the positive ion and trapped electron to be between 280 and 19Å.

Another advantage of high-energy radiation-induced carbocationic polymerizations is the very high molecular weights obtainable by these techniques. The effect of temperature on the molecular weights of polyisobutylenes obtained by chemical coinitiators and irradiation-induced polymerization has been investigated (Kennedy et al., 1971c).

Pulse Radiolysis

Pulse radiolysis is a relatively recent technique useful for studying fundamentals of polymerizations. Monomers are irradiated for very short periods ($\sim \mu$sec or shorter) using Van de Graaff accelerators or various pulse generators producing pulses with intensity of several meV (Metz et al., 1967; Tabata, 1976). Similar to high energy irradiation, the products of pulse radiolysis are ion radicals, some of which, for example, N-vinylcarbazole, styrene, and α-methylstyrene, can be used to induce cationic polymerizations (Tabata, 1976). For example, using this technique and kinetic data, Tabata (1976) proposes the following ion generation and initiation sequence for styrene, St[$-165°$C, isopentane/n-butyl chloride (4:1 v/v) solvent]:

$$\text{St} \xrightarrow{\text{solvent}^{\oplus}} \text{St}^{\cdot\oplus} \xrightarrow{+M} \text{St}\!-\!\text{St}^{\cdot\oplus} \xrightarrow{+M} \cdot\text{St}\!-\!\text{St}\!-\!\text{St}^{\oplus} \xrightarrow{+M} \text{polymer}$$

According to this view ion generation is the formation of a solvent cation radical which transfers its positive charge to styrene to form a styrene cation radical.

UV Radiation

Direct Techniques Including Ion Injection

Since almost all monomers have ionization potentials in excess of ~ 7 eV, near-UV or visible light, representing only 6 eV, are unable to produce ions and cannot be expected to induce ionic polymerizations.

Until Vermeil (1964a) showed that isobutylene (ethyl chloride or isopentane solvents, $-153°C$) can be polymerized by energetic photons (vacuum UV of 8.4 and 10 eV), it was held that photopolymerization by UV light can proceed only by radical mechanism. Further, Vermeil (1964b) also showed that the yield of polymer can be increased some 2.5-fold by applying an electric field ($\sim 10^4$ V/cm) during irradiation.

Just about the same time, Schlag and Sparapany (1964) and Sparapany (1966) induced isobutylene polymerization and obtained high molecular weight product by injecting *tert*-butyl cations generated by photoionization (~ 10 eV) of pure monomer in the gas phase and directing the cations by means of an electric field into liquid monomer stirred at -115 to $-145°C$. These experiments have been repeated and extended by Viswanathan and Kevan (1967, 1968). In these "ion injection" techniques ions are formed by photolysis in the gas phase and injected into the liquid monomer by means of an electric field. Under suitable conditions only the t-Bu$^{\oplus}$ reaches the liquid phase and the products of the t-Bu$^{\oplus} + i$-C$_4$H$_8$ reaction are readily analyzable. Viswanathan and Kevan (1968) studied the selectivity of *tert*-butylation of isobutylene and found that t-Bu$^{\oplus}$ reacts nonselectively at both the primary and tertiary carbons of isobutylene. When t-Bu$^{\oplus}$ was deexcited in the gas phase by collisions with some inert gas, for example, Ar or Ne, *tert*-butylation selectivity increased markedly.

Indirect Techniques

A fundamentally different type of photoinitiated cationic polymerization has been discovered by Crivello et al. (1976, 1977a, b). According to these authors diaryliodonium salts having stable complex metal halide counteranions, for example, BF$_4^{\ominus}$ and PF$_6^{\ominus}$, are efficient photoinitiators for the cationic polymerization of various monomers such as olefins, N-vinylcarbazole, alkyl vinyl ethers, epoxides, and lactams. Unsubstituted and various substituted diaryl iodonium salts have been prepared and their photodecomposition studied:

where R = H, CH$_3$, t-C$_4$H$_9$; X$^{\ominus}$ = BF$_4^{\ominus}$, PF$_6^{\ominus}$, AsF$_6^{\ominus}$.

Photopolymerizations were carried out by mixing these salts and the monomer in a polar solvent, for example, methylene chloride, at 25°C and illuminating the mixture using a medium-pressure Hanovia lamp from 200 to 300 nm and from 300 to 400 nm. For example, styrene in the presence of p-CH$_3$O–C$_6$H$_4$–$\overset{\oplus}{\text{I}}$–C$_6H_5BF_4^{\ominus}$ within 6 min yielded 89% of polymer of $[\eta] = 0.06$ dl/g. α-Methylstyrene gave similar results.

As to the initiation mechanism, the authors propose the following major sequence of events (Crivello and Lam, 1976, 1977a, b):

$$Ar_2I^{\oplus}X^{\ominus} \xrightarrow{h\nu} Ar_2I^{\oplus}X^{\ominus*} \rightarrow ArI^{\cdot\,\oplus} + Ar^{\cdot} + X^{\ominus}$$

$$ArI^{\cdot\,\oplus} + SH \rightarrow Ar\overset{\oplus}{I}H + S^{\cdot}$$

$$\longrightarrow ArI + H^{\oplus}$$

The photoexcited diaryliodonium salt decomposes to an aryliodo radical cation, aryl radical, and counteranion. The aryl radical may disappear by dimerization to give biphenyl derivatives or by hydrogen atom abstraction from any proton-containing species SH in the medium (H_2O, solvent). More importantly for the initiation process, the aryliodo radical cation interacts with SH to generate a protonated iodoaromatic compound which rapidly loses a proton and thus provides the true initiating entity, the $H^{\oplus}$. Thus the overall material balance of the major path leading to initiation is

$$Ar_2I^{\oplus}X^{\ominus} + SH \rightarrow Ar^{\cdot} + ArI + S^{\cdot} + HX$$

The protic acid formed in this manner can then protonate monomer ($HX + M \rightarrow HM^{\oplus}X^{\ominus}$) and initiate polymerization (Crivello and Lam, 1977a).

Although the feasibility of inducing carbocationic polymerizations has been demonstrated, according to available information the authors must have been more interested in curing epoxy compounds for coating applications (Crivello et al., 1977a). No doubt under suitable conditions high molecular weight polyolefins could also be prepared by this ingenious method.

The fundamental difference between Crivello and Lam's method and charge transfer initiation is noteworthy: although cation radicals are intermediates in both mechanisms, in initiation by charge transfer the cation radical directly attacks olefin $\dot{C}-C^{\oplus} + M \rightarrow \dot{C}-C-M^{\oplus}$, whereas by the Crivello–Lam method the cation radical is an intermediary helping to generate the true initiating entity, the proton from the solvent SH. According to the terminology proposed, then, formally, SH and $Ar_2I^{\oplus}$ are the initiators in these systems.

High Electric Fields: Field Emission and Field Ionization

During the last decade two methods have been developed for the liquid phase generation of polymerization-active radical cations by the use of high $> 10^7$ V/cm electric fields. In field emission polymerizations just as in field emission mass spectroscopy, electrons are emitted from a metal anode of suitable geometry (sharp needle point, edge) by application of a high electric field, and collisions between energetic electrons and monomer produce cation radicals. Conversely, in field ionization, the polarity of electrodes is reversed and no electron current is emitted; rather, neutral molecules in the vicinity of a suitably shaped (needle, thin blade) positive electrode (cathode) lose an electron and are oxidized to radical cations. The following scheme contrasts basic events involved in initiation (M = olefin, e.g., styrene):

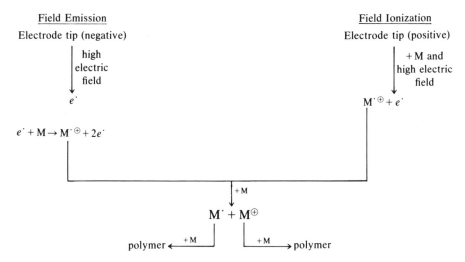

Both techniques have been employed for the polymerization olefins such as styrene, α-methylstyrene, and isobutyl vinyl ether.

Historically, the first field ionization polymerization may have been reported by mass spectroscopists (Migahead and Beckey, 1971), who observed oligomerization of benzonitrile and crotonaldehyde, two by no means conventional cationic monomers. Ion generation and initiation with benzonitrile was visualized to occur as follows:

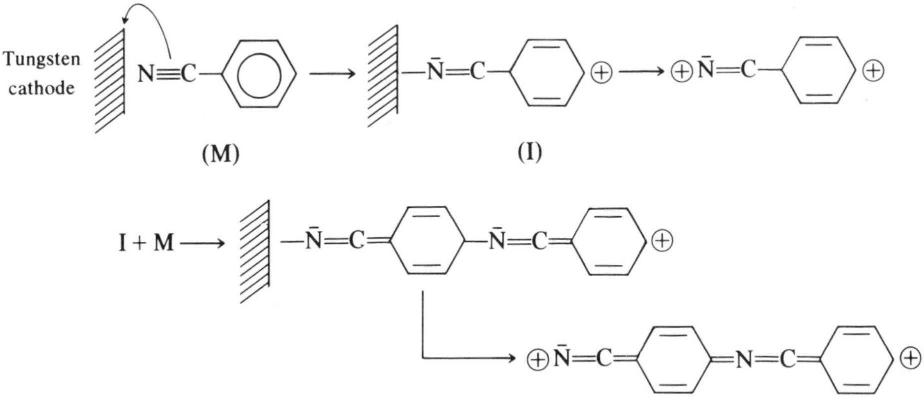

High electric field-induced carbocationic polymerizations have been systematically explored. Brendle and Ilvoas, (1976) recently reinvestigated by improved methods their pioneering studies on field ionization polymerization of styrene, α-methylstyrene, and isobutyl vinyl ether. In addition to substantiating their earlier proposition according to which these polymerizations are induced by radical cations (Brendle, 1971; Brendle and Ilvoas, 1972; Lambla et al., 1970, 1972), they demonstrated the great sensitivity of these systems to impurities, in particular to water. Schnabel and Schmidt, (1971, 1973), and Wablat et al. (1974) studied the fundamentals of styrene and α-methylstyrene polymerization by both field emission and field ionization techniques. Both methods were proposed to yield radical cations capable of initiating polymerizations (Schnabel and Schmidt, 1973). Field emission polymerization of α-methylstyrene was inhibited by both cation scavengers (H_2O, Et_3N) and electron scavengers (SF_6, N_2O) (Wablat et al., 1974). Inhibition by cation scavengers was attributed to interference of propagation by the cation radicals, whereas inhibition by radical traps was proposed to be due to inhibition of initiation. The scheme at the beginning of this section helps to visualize these results: since field emission originally produces only electrons that subsequently collide with neutral monomer to produce cation radicals, electron scavengers should inhibit cation radical generation and thus ensuing polymerization. In contrast, field ionization directly produces radical cations so that electron scavengers should not affect (inhibit) cationic polymerization.

These investigators (Schnabel and Schmidt, 1973; Wablat et al., 1974) inquired as to the kinetics and thermodynamics of initiation by field emission. Only a small fraction of the cation radicals was effective in producing polymer; this was attributed to rapid neutralization of cations at the electrodes. To explain that electrons that possess only ~ 5 eV kinetic energy when they emerge from the anode nonetheless are able to ionize olefins whose ionization potential in the gas phase is much in excess of this value, (e.g., the ionization potential of α-methylstyrene is 8.6 eV, and in the liquid phase should also be much higher than ~ 5.0 eV), the authors propose strong ionization potential reduction by polarization by the high electric field.

In this context it is of interest that isobutylene has also been polymerized to high molecular weight products ($\bar{M}_v = 100,000–400,000$) by glow discharge at $-100°C$ (Pasman et al., 1965). Monomer purity was 99.7% and the carrier gas was hydrogen. Although there is little doubt that the overall polymerization was cationic, details as to initiation mechanism are completely obscure.

Electroinitiation

Significant Contributions

Breitenbach et al. in 1960 hinted and two years later (Breitenbach and Srna, 1962) disclosed that some highly reactive cationic monomers for example, styrene, isobutyl vinyl ether, and N-vinylcarbazole, can be readily pol-merized by electrochemical techniques. The pioneering electroylsis experiment was carried out by disolving $AgClO_4$ in pure monomers or in nitrobenzene solution and passing current through the system at room temperature. For example, a current of 2×10^{-7} farad caused a strongly exothermic polymerization of isobutyl vinyl ether and a rubbery, semisolid polymer was obtained. Similar experiments have been carried out with the other monomers as well and fundamentally similar results have been obtained. That the nature of the polymerization was cationic was indicated by copolymer compositions obtained in a few copolymerization experi-ments using styrene and acrylonitrile monomers (Breitenbach et al., 1960). For the mechanism of initiation the authors suggested anodic oxidation of monomer:

$$ClO_4^{\ominus} \rightarrow ClO_4^{\cdot} + e$$
$$ClO_4^{\cdot} + M \rightarrow ClO_4^{\ominus} + M^{\oplus}$$

These pioneering studies lay dormant until in 1969 Sommer and Brei-tenbach, published a brief account of their work in which they proposed involvement of cation radicals as initiating species:

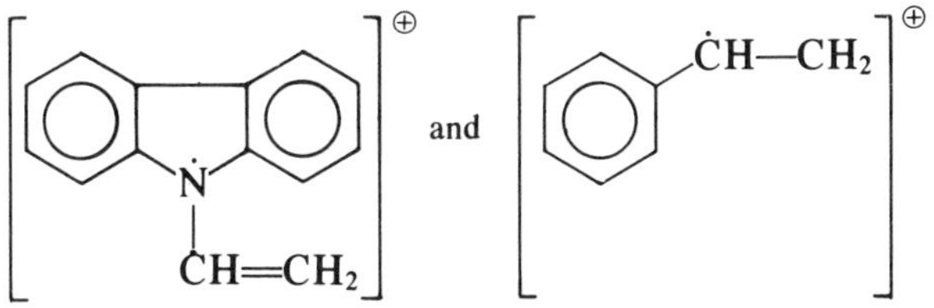

After this communication (Sommer and Breitenbach, 1969) several research groups turned their attention toward the detailed elucidation of the mechanism of electropolymerizations in general and cationic elec-tropolymerizations in particular. However, the basic formalism of anodic monomer oxidation or initiation via radical cations as first postulated by Breitenbach survived, and in different disguises has subsequently been adopted by essentially all investigators in this field.

From the point of view of initiation, perhaps the simplest system analyzed was that by Akbulut et al. (1975). According to these authors the polymerization of styrene in the presence of tetrabutylammonium fluoborate in acetonitrile solution by electrolysis is cationic in nature, and initiation is due to radical cations formed by direct anodic oxidation of the monomer on the surface of the electrode:

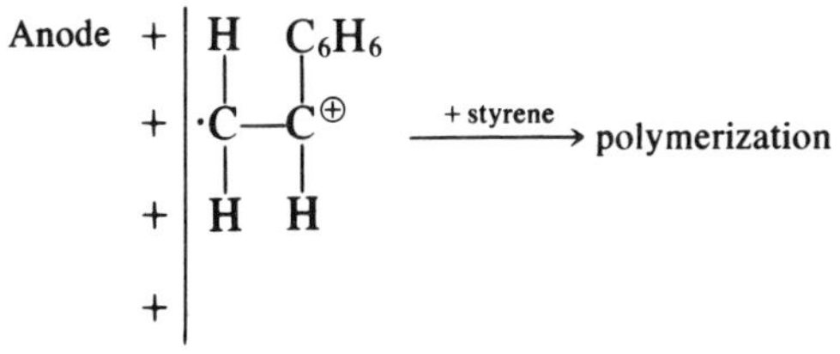

An important body of information relative to the mechanism of electroinitiated olefin polymerization has been generated by Funt and his group (Funt et al., 1970a, b, 1971; Turcot et al., 1974); however, details of initiation remain obscure. For example, when a controlled current was passed through a solution of styrene in methylene chloride in the presence of tetrabutylammonium perchlorate supporting electrolyte, polymer formed only in the anode compartment (Funt and Blain, 1970a). Similar observations have been obtained using tetrabutylammonium hexafluorophosphate and tetrafluoroborate electrolytes. The authors consider two possibilities for initiation: oxidation of $ClO_4^{\ominus}$ to $ClO_4^{\cdot}$, which in turn oxidizes the monomer, or direct anodic oxidation of styrene to a radical cation. Kinetic and other evidence indicated the involvement of cation radicals in initiation. The polymerization of isobutyl vinyl ether was discussed in essentially similar terms (Funt and Blain, 1971). It was proposed that the counteranion, say, $BF_4^{\ominus}$, of the supporting electrolyte is anodically oxidized to the radical $BF_3^{\cdot}$, which in turn either oxidizes the monomer to $\overset{\cdot}{C}H_2$—$\overset{\oplus}{C}HOR$ or abstracts a hydrogen atom from a suitable donor to form $HF + BF_3$, a most reactive initiating system (see discussion above).

Essentially similar conclusions have been reached by several groups of workers. Tidswell and Doughty (1971) electrolyzed a solution of styrene and $NaBF_4$ in sulfolane and obtained low molecular weight products at the anode. According to these authors the polymerization is initiated by electrolytically produced BF_3 and HF and/or H_2O. Ghose and Bhadani (1975) used tetraethylammonium hexachloroantimonate in nitrobenzene solution and produced low molecular weight (5000–8000) polystyrene. Initiation was visualized to occur by anodic oxidation of $SbCl_6^{\ominus}$ to $SbCl_6^{\cdot}$, followed by hydrogen atom abstraction and subsequent formation of $HCl + SbCl_5$.

In several instances the counteranion of the supporting electrolyte was $ClO_4^{\ominus}$ and the *in situ* formation of $HClO_4$ was proposed to explain cationic polymerization. This was the case, for example, with the electrochemically initiated polymerization of styrene in $LiClO_4$–propylene carbonate solutions (Pistoia, 1974a, b), and that of acenaphthylene in n-Bu_4NClO_4–nitrobenzene (Kikuchi and Matsumo, 1975). Pistoia (1974a, b) pointed out the similarity between electroinitiated polymerization in the presence of $LiClO_4$ and conventional $HClO_4$-initiated polymerizations. In contrast, when the polymerization of styrene was induced by passing a current through a system containing n-Bu_4NClO_4 in dimethylsulfate solution, that is, a medium in which polystyrene is insoluble, initiation was visualized to occur by oxidation of styrene to styrene radical cations (Pistoia and Scrosati, 1974c).

The preferential oxidation of styrene to styrene cation radical in the presence of $ClO_4^{\ominus}$ has also been invoked, (Sommer and Breitenbach, 1969; Mengoli and Vidotto, 1972a). According to Mengoli and Vidotto the electroinitiated polymerization of n-butyl vinyl ether in the presence of

$NaB(C_6H_5)_4$ in nitrobenzene may proceed by direct oxidation to radical cations (Mengoli and Vidotto, 1970, 1972b) or by electrolytic processes involving $(C_6H_5)_4B^{\ominus}$ (Mengoli and Vidotto, 1970).

Electrochemically generated radical cations, for example, that of 9, 10-diphenylanthracene, DPA, have been found to be efficient intermediates for the electroinitiated polymerizations of styrene in methylene chloride solution (Turcot et al., 1974). Initiation was proposed to involve the generation of $DPA^{\cdot\oplus}$, which subsequently accepts electrons from styrene to produce a styrene radical cation (the original publication must contain a mistake for it specifies electron transfer from $DPA^{\cdot\oplus}$ to styrene). The fate of the styrene radical cation (coupling to form the dication) remains to be elucidated. Molecular weights were inversely proportional to $DPA^{\cdot\oplus}$ concentration and the highest molecular weight polystyrene obtained was $\sim 10,200$.

Cerrai et al. (1975) investigated the fundamentals of electroinitiated isobutyl vinyl ether polymerization using tetrabutylammonium triiodide $n\text{-Bu}_4NI_3$ as supporting electrolyte in 1, 2-dichloroethane solution. Preliminary studies by the same group of workers (Cerrai et al., 1974) involved the electrolysis of $n\text{-Bu}_4NI$ and $n\text{-Bu}_4NI_3$ in the absence of monomer and showed oxidation of anions to iodine at the anode. By this evidence, together with UV–visible spectrophotometry and kinetic experiments, these authors were able to show that the polymerization was initiated by anodically formed iodine (see discussion above):

$$2\,I_3^{\ominus} \xrightarrow[-2e]{\text{anode}} 3\,I_2$$

Conclusions on Electroinitiated Carbocationic Polymerizations

It has been amply demonstrated by a variety of investigators that cationic polymerizations can be induced by electrochemical methods; however, the detailed mechanism of polymerization in general and initiation in particular are still largely a matter of conjecture.

A survey of available literature leads to the proposition that electrochemical cationic polymerization may be induced by at least two routes: direct or indirect initiation. Direct initiation involves ion generation by anodic oxidation of monomer to a cation radical

$$M \xrightarrow[-e]{\text{anode}} M^{\cdot\oplus}$$

In these instances the supporting electrolyte does not get involved in the process of ion generation or initiation; its sole function is to provide a flow of current. According to the convention followed in this work, then, in direct electrochemical initiation systems the monomer itself is the initiator and the anode may be regarded as the "coinitiator." In direct initiation the counteranion of the supporting electrolyte is first oxidized to a radical, which in turn oxidizes the monomer to the first initiating entity, which can be a cation or cation radical, for example:

$$BF_4^{\ominus} \xrightarrow[-e]{\text{anode}} BF_4^{\cdot}$$

$$BF_4^{\cdot} + CH_2\!\!=\!\!\underset{\underset{OR}{|}}{CH} \rightarrow BF_4^{\ominus} + \dot{C}H_2\!\!-\!\!\underset{\underset{OR}{|}}{\overset{\oplus}{CH}}$$

Or the radical may abstract a hydrogen atom from a hydrogen donor to produce a Brønsted acid–Friedel Crafts acid initiating system (see discussion above):

$$BF_4^{\cdot} + -\overset{|}{\underset{|}{C}}-H \rightarrow -\overset{|}{\underset{|}{C}}{}^{\cdot} + HF + BF_3$$

In these instances, formally, the $-\overset{|}{C}-H$ (which may be the monomer itself) is the initiator (cationogen) and the anode may again be regarded as the coinitiator that is, the agency assisting in the generation of the first polymerization-active cation.

In the presence of supporting electrolytes containing $I^{\ominus}$ or $ClO_4^{\ominus}$ counteranions, anodic oxidation may produce I_2 or $HClO_4$, respectively, which may subsequently induce cationic polymerization of certain monomers for example, vinyl ethers and styrene (see above).

In summary, electroinitiated cationic polymerizations are a highly specialized field of research endeavor which led to the synthesis of some rather low molecular weight products of a few highly reactive monomers, for example, styrene, vinyl ethers, N-vinylcarbazole, and acenaphthylene. Important pioneering breakthroughs are needed to render this field attractive to polymer chemists.

Conclusions: Initiation by Physical Methods

A wide variety of physical methods have been tried and developed to initiate carbocationic polymerizations. The common characteristic of these techniques is energy transfer to the monomer from external sources:

$$M \xrightarrow[\text{energy}]{-e(\text{or}+e)} M^{\oplus}, M^{\cdot\oplus}, \text{ other species}$$

The ionization potential of olefins is ~ 7 eV, which largely determines the energy demand of initiation. In ionizing radiation methods γ-rays, generated usually by cobalt-60 isotope sources, are used. A fraction of the energetic beam collides with matter and expels electrons from the olefin (or other chemicals in the system). In addition to a variety of other species cation radicals, the true initiating entities, are also generated. The structures of transient species responsible for radiation-induced polymerization are largely unknown and the detailed chemistry of ion generation is obscure. Similarly, the nature and position of "counteranions" relative to the propagating cations in γ-ray-induced carbocationic polymerizations is a matter of conjecture. After irradiation the majority of ions and electrons rapidly recombine.

γ-Ray induced polymerizations are regarded to propagate by "free" cations. Strongest chemical evidence for the presence of free ions in these polymerizations are ionic polymerization of, for example, isobutylene, which may reach kinetic chain lengths as high as 10^6 in rigorously dry systems.

In direct UV radiation methods energetic photons (in excess of $\sim 8 \, eV$) are used to expel electrons from compounds containing C–H bonds in the liquid phase and the ionic fragments formed may induce polymerization. In ion injection methods cations produced by UV light in the vapor phase are directed into the liquid phase containing the monomer by an electric field. UV irradiation of certain diaryl iodonium salts has been shown to result in the formation of Brønsted acids. In this manner UV radiation can be used indirectly to induce carbocationic polymerizations.

In addition to irradiation by γ-rays or UV light, carbocationic polymerizations can also be induced by various electric methods. Thus the necessary energy for initiation can be transferred to the monomer by application of strong electric fields. In field emission emission, monomer interacts with electrons emerging from a metal cathode by application of a suitable electric field. In field ionization the monomer is directly oxidized by the anode. Both methods generate radical cations, conceivably the true initiating entities.

Field ionization polymerization is quite similar to electroinitiated polymerizations, a method in which the electric current is sent through a liquid containing the monomer and a supporting electrolyte. The fundamental, cation radical generating processes are believed to be anodic oxidation in both instances. Electroinitiated polymerizations may or may not involve the anodic oxidation of the supporting electrolyte.

It appears that all physical initiation methods are but various ways for the generation of radical cations. The only exception to this generalization may be the indirect UV method in which the initiating electrophile is produced by a circuitous photolysis of a diaryl iodonium salt precursor. In spite of the great interest in these methods, detailed structure of cation radicals is largely obscure and the fate of these reactive transients is unknown. It has often been postulated but never proved that these species may recombine to initiating dications, for example,

Table 4.8 □ Comprehensive Overview of Carbocationic Initiation

A. Two-Electron(Heterolytic) Transpositions

Chemical Methods

Classification of Initiating System	Example of an Initiating System	Initiator	Initiating Species	Coinitiator	Counteranion
Brønsted or protic acid	H_2SO_4	H_2SO_4	$H^{\oplus}$	—	$HSO_4^{\ominus}$
Pseudocationic pzn. init.	$HClO_4$	$HClO_4$	$HClO_4$	—	—
Stable cation salt	$(C_6H_5)_3C^{\oplus}SbCl_6^{\ominus}$	$(C_6H_5)_3CCl$	$(C_6H_5)_3C^{\oplus}$	$SbCl_5$	$SbCl_6^{\ominus}$
Brønsted acid/Friedel–Crafts acid	H_2O/BF_3	H_2O	$H^{\oplus}$	BF_3	$BF_3OH^{\ominus}$
	"H_2O"/$AlCl_3$	"H_2O"	$H^{\oplus}$	$AlCl_3$	$AlCl_3OH^{\ominus}$
	"H_2O"/$BF_3 \cdot OEt_2$	"H_2O"	$H^{\oplus}$	BF_3	$BF_3OH^{\ominus}$
Organic halide/Friedel–Crafts acid	$t\text{-}BuCl/Et_2AlCl$	$t\text{-}BuCl$	$t\text{-}Bu^{\oplus}$	Et_2AlCl	$Et_2AlCl_2^{\ominus}$
Organic halide/silver salt	$(C_6H_5)_2CHCl/AgSbF_6$	$(C_6H_5)_2CHCl$	$(C_6H_5)_2CH^{\oplus}$	$AgSbF_6$	$SbF_6^{\ominus}$
Halogen/Friedel–Crafts acid	Cl_2/BCl_3	Cl_2	$Cl^{\oplus}$	BCl_3	$BCl_4^{\ominus}$
Direct initiation by Friedel–Crafts acid	$(AlBr_3)_2$	$AlBr_3$	$AlBr_2^{\oplus}$	$AlBr_3$	$AlBr_4^{\ominus}$
	$AlBr_3 \cdot TiCl_4$	$AlBr_3$	$AlBr_2^{\oplus}$	$TiCl_4$	$TiCl_4Br^{\ominus}$
Polymerization by condensation	$TiCl_4/M$	M	$H^{\oplus}$ or $\mathbf{R}^{\oplus}$	$TiCl_4$	$TiCl_5^{\ominus}$
Iodine	I_2/M	M	$IM^{\oplus}$	I_2	$I_3^{\ominus}$
Misc. acidic solids, clays, $MgCl_2$, etc.	Acidic zeolite, clay	?	$H^{\oplus}(?)$	?	?

B. One-Electron(Homolytic) Transpositions

	PhC(OMe)₂COPh/AgPF₆/hν $D\cdot A/\Delta$ or $h\nu$	PhC(OMe)₂COPh D	$\overset{\oplus}{P}hC(OMe)_2$ $D'^{\oplus}$ or $^{\oplus}D\cdot A^{\ominus}$	AgPF₆ A	PF₆$^{\ominus}$ A'$^{\ominus}$ or A$^{\ominus}$
Direct radical oxidation Charge transfer initiation by D·A complexes					
Physical Methods					
High-energy or ionizing radiation	γ-Rays/M	M	M'$^{\oplus}$ or M$^{\oplus}$	—	—
Direct UV irradiation— ion injection	UV light/M	M	M$^{\oplus}$	—	—
Indirect UV irradiation— Ar₂I$^{\oplus}$ salts	Ar₂I$^{\oplus}$PF₆$^{\ominus}$/SH/hν	SH	H$^{\oplus}$	Ar₂I$^{\oplus}$PF₆$^{\ominus}$/hν	PF₆$^{\ominus}$
High electric field Field emission	Electrode tip(−)/M	M	M'$^{\oplus}$	—	—
Field ionization	Electrode tip(+)/M	M	M'$^{\oplus}$	—	—
Electroinitiation— direct or indirect	Anode/M	M	M'$^{\oplus}$	(Anode)	X$^{\ominus}$ from supporting electrolyte

The preparative significance of dications would be considerable because they could lead to bifunctional macromolecules. In the absence of this development, physical methods of carbocationic polymerization initiation remain an interesting, theoretically challenging, highly specialized field of endeavor.

Conclusions: Toward a Comprehensive View of Initiation in Carbocationic Polymerization

Carbocationic polymerizations are initiated by acids, either Brønsted acids or Lewis acids. The initiating electrophile may be introduced into the polymerization charge as such (chemical initiation), or generated by a variety of external energy sources (physical initiation). Brønsted or protic acids provide protons, and initiation involves protonation. Stable cations and Friedel–Crafts acids are Lewis acids, and initiate by cationation. Initiating charge transfer complexes usually exist in the form of cation radicals that is, Lewis acids, and as such also initiate by cationation.

The foregoing detailed review of chemical and physical methods of initiation in carbocationic polymerization leads to valuable insight and generalization relative to the chemistry of these processes. Table 4.8 has been compiled by examining and organizing initiation methods used to advantage by a large variety of investigators. The compilation is comprehensive in the sense that it includes representatives of all classes of initiating systems currently available to synthetic polymer scientists. This table helps to perceive relationships between seemingly disparate initiating systems and holds clues for future research by revealing gaps of information.

The first and second columns in Table 4.8 list classes of initiating systems identified and a representative example for every class, respectively. The third column shows the initiator, that is, the chemical that provides the initiating electrophile, and the fourth the actual initiating species. The fifth and sixth columns list the coinitiator, that is, the agent helping to produce the active electrophile, and the counteranion, respectively.

Initiation is formally subdivided into ion formation or priming, and cationation. These processes have been defined and illustrated with examples. In spite of the great dissimilarity between chemical and physical initiation methods this subdivision is valid, although clear distinction between these two events, particularly in physical initiation methods, is not immediately apparent.

Although this subdivision is an aid in systematizing the chemistry of initiation, it must be emphasized that both priming and cationation consist of intricate sequences of only partly understood physical–chemical transformations, for instance, solvation and complexation. Little quantitative information useful to the polymer scientist exists relative to the effect of solvent polarity or nature of the counteranion, or monomer·coinitiator and initiator·coinitiator, complexes, for example, on initiation.

The chemistry of ion formation is far more complex than that of cationation. Indeed the organization of Table 4.8 is based on differences between details of ion formation. For example, the justification for subdividing initiating systems into chemical and physical methods is the fundamentally different ion generation processes, not differences in cationation. Among chemical methods Brønsted acids and organic halide/Friedel–Crafts acid initiating systems belong to different classes on account of differences in ion generation.

First and foremost, Table 4.8 conveys that a large variety (perhaps all) electrophilic species may initiate polymerizations under suitable conditions. Besides protons, the bulk of cations is provided by carbenium ions, that is, aliphatic, cycloaliphatic, or alkyl–aromatic cations. In addition to these species halonium ions ($Cl^{\oplus}$, $Br^{\oplus}$) have been found to be effective initiating species and were investigated in some detail. Significantly, several important electrophiles well-known to organic chemists, for example, $NO^{\oplus}$ and $RSO_2^{\oplus}$, are missing from Table 4.8. It is predicted that these species, under suitable conditions, could initiate and would lead to valuable head-derivatized polymers, for example,

$$NO^{\oplus} + C{=}C \rightarrow ON{-}C{-}C^{\oplus} \xrightarrow{\ M\ } \text{polymer}$$
$$\underset{C_6H_5}{|} \qquad\qquad \underset{C_6H_5}{|}$$

Organization and Classes of Initiating Systems

Distinction between chemical and physical initiation methods is obvious. Chemical initiation involves the spontaneous generation of electrophiles by mixing monomer plus other ingredients whereas physical methods require energy input by radiation or electric current. Chemical initiation usually proceeds by interaction between two ingredients, the initiator and coinitiator, in addition to the monomer, which results in the formation of the initiating species, which carries a full positive charge. As pointed out repeatedly, the role of the coinitiator is to assist ion generation. In contrast, physical initiation methods require interaction between energy and monomer and usually involve cation radicals as initiating species. Since external sources of energy help in generating the initiating species, these energy sources may be regarded in this sense as "coinitiators" of physical initiation methods.

Chemical methods are conveniently discussed in terms of two-electron (heterolytic) and one-electron (homolytic) transpositions. In the former, ion generation necessarily includes a heterolytic bond-breaking step and the initiating species carries a full positive charge; in the latter ion generation proceeds by homolytic cleavage and somewhere along the reaction path a radical or cation radical is formed. In direct radical oxidation the initiating species is a cation produced by thermal or photochemical oxidation of a radical, and in charge transfer polymerizations ion radicals are visualized to be intermediates.

In protic (Brønsted) acid-initiated polymerizations ion generation involves ionization of the acid: $HA \rightleftharpoons H^{\oplus} + A^{\ominus}$. The presence of complex anions $[A \cdots H \cdots A]^{\ominus}$ has been postulated on the basis of results by conductometry and vapor phase osmometry. In cases where complex anions assist ion generation, the Brønsted acid HA may be regarded as providing its own coinitiator. In the presence of highly nucleophilic counteranions $A^{\ominus}$ it may be necessary to use complexing agents (metals, salts, oxides, Friedel–Crafts halides) to immobilize $A^{\ominus}$; these complexors are *de facto* coinitiators. Initiation is completed by protonation and the birth of the first carbenium ion.

Initiation in pseudocationic polymerizations is fundamentally similar to conventional initiation with protic acids; these systems are rather unique, for propagation is thought to occur via covalent species or largely covalent species in equilibrium with carbenium ions.

Initiation by stable cation salts is conceptually the simplest initiating method since in these instances ion generation becomes unnecessary; it has in fact been accomplished prior to assembling the polymerization system. The inherent advantage in using stable cations for initiation resides in their well-defined structures which renders them ideal for kinetic investigations. For example, it is rather easy to follow the rate of disappearance of trityl ion absorption. However, the price for cation stability is limited usefulness: stable cations can be used to initiate the polymerization only of highly reactive olefins, for example, N-vinylcarbazole and α-methylstyrene.

The most important initiating systems both from the scientific and technological point of view are those including Friedel–Crafts acids. Although these coinitiators may in the purest/dryest systems induce polymerization in the absence of purposefully added cationogen (see the section on direct initiation), initiation in the presence of moisture or suitable carbenium ion sources occurs with the greatest of ease. Unless special precautions are taken (Stopping Experiments, see above), it may be safely assumed that polymerizations occurring in the presence of Friedel–Crafts acids have in fact been initiated by traces of cationogenic ingredients. Suspicion that moisture impurity is the operational initiator is indicated by the quotation marks, "H_2O" (see also Section 2.2).

Diorganoaluminum halides and triorganoaluminum compounds are a distinguished class of Friedel–Crafts coinitiators. These materials are much less sensitive toward adventitious cationogenic impurities than conventional Friedel–Crafts acids, such as $AlCl_3$, BF_3, and organoaluminum dihalides. For example, Et_2AlCl can be mixed with rather reactive monomers such as styrene or isobutylene under common laboratory conditions in a nitrogen atmosphere without polymerization. However, as soon as relatively large quantities of suitable cationogens are purposefully added to such quiescent mixtures, rapid polymerization ensues and high molecular weight polymers are obtained. These "controlled" initiation systems are of great value for the preparation of novel block and graft copolymers (see Chapter 8).

For example, the addition of Et_2AlCl or Et_3Al to solutions of chlorinated ethylene–propylene copolymers and styrene rapidly and efficiently yields graft copolymers:

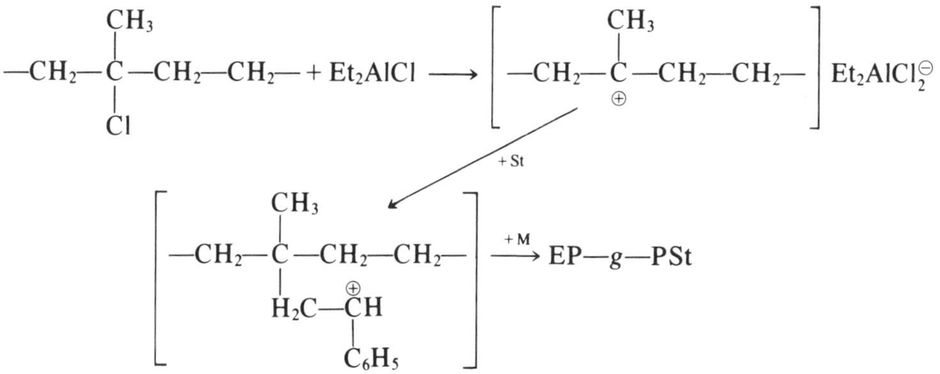

$$
-CH_2-\underset{\underset{Cl}{|}}{\overset{\overset{CH_3}{|}}{C}}-CH_2-CH_2- + Et_2AlCl \longrightarrow \left[-CH_2-\underset{\overset{|}{\oplus}}{\overset{\overset{CH_3}{|}}{C}}-CH_2-CH_2- \right] Et_2AlCl_2^{\ominus}
$$

$$
\left[-CH_2-\underset{\underset{H_2C-\overset{\oplus}{C}H}{\underset{|}{|}}}{\overset{\overset{CH_3}{|}}{C}}-CH_2-CH_2- \right] \xrightarrow{+M} EP-g-PSt
$$

$$+St$$

$$C_6H_5$$

Initiation with halogens in conjunction with these alkylaluminum compounds produces transient halonium ions and thus a route toward α-halogenated polymers:

$$Cl_2/Me_3Al \rightarrow [Cl^{\oplus}Me_3AlCl^{\ominus}] \xrightarrow{+M} Cl\text{\small∿∿}M\text{\small∿∿}^{\oplus} + Me_3AlCl^{\ominus}$$

Under extremely pure conditions in the rigorous absence of cationogens, olefin polymerization can be initiated by certain Friedel–Crafts acids alone or in combination, for example, $AlBr_3/TiCl_4$. Initiation in these systems has been proposed to occur by self-dissociation, direct metalation (hence the misnomer direct initiation) (see above), or hydride abstraction from monomer (see Section 4.3).

Polymerization by condensation denotes a rather unusual, still controversial initiation phenomenon when polymerization of isobutylene and cyclopentadiene by purest $TiCl_4$ occurs only on the cocondensation of ingredients from the vapor phase. Since these experiments have been carried out in the complete absence of exogenous cationogens, it is possible that the monomers themselves fulfill this function; that is, they are *de facto* initiators.

Among initiating systems operating by one-electron transposition very little information is yet available on direct radical oxidation. However, charge transfer initiation by donor–acceptor complexes represents a promising relatively new initiation method. Important aspects of donor–acceptor interactions have been elucidated and exploitation for many useful initiations became possible. In charge transfer systems initiation is determined by $D \cdot A$ complexes excited by either heat or light (exciplexes). Ion generation involves one-electron transfer from the donor (usually monomer) to an acceptor and cation radical formation. The cation radical may subsequently cationate monomer or couple with anion radical to zwitterion, the actual initiating species. The fate of the radicals and/or anions in these polymerizations is largely obscure. In line with the convention used, the acceptor that assists in the generation of the initiating species may be regarded as the coinitiator.

A unique charge transfer initiating system is the alkyl vinyl ether/vinylidene cyanide D·A pair (see above). The charge transfer complex comprises an alkyl vinyl ether radical cation and vinylidene cyanide anion radical which initiate cationic and anionic polymerization, respectively:

$$[CH_2\!\!=\!\!CH\!\!-\!\!OR]^{\cdot\oplus}[CH_2\!\!=\!\!CH(CN)_2]^{\cdot\ominus}$$

$$\begin{array}{cc} \downarrow & \downarrow \\ \text{cationic} & \text{anionic} \\ \text{pzn.} & \text{pzn.} \end{array}$$

In this case the alkyl vinyl ether is the initiator and the vinylidene cyanide the coinitiator for the cationic polymerization. Conversely, the alkyl vinyl ether may be viewed as the coinitiator and the vinylidene cyanide the initiator for the anionic process.

A serious limitation of charge transfer initiations is that this process is useful only for very reactive monomers, for example, N-vinylcarbazole and alkyl vinyl ethers, monomers with relatively low ionization potentials.

In physical initiation methods some form of external energy is absorbed by one (or more) ingredient of the system which is then converted into the initiating species, usually cation radicals. By definition, the species that gives rise to the initiating species, usually the monomer, is the initiator. Formally, the energy or the energy source that assists this process may be regarded as the coinitiator in physical initiation methods.

A Simplified View of Initiation

Initiation is tantamount to the generation of cations of sufficient reactivity for olefin cationation. This process requires a net input of energy E which is needed to convert the first monomer molecule M into cations or cation radicals; schematically,

$$M \xrightarrow{\ E\ } M^{\oplus} \text{ or } M^{\cdot\oplus}$$

where $M^{\oplus}$ stands for any electrophile obtained by two-electron transposition and $M^{\cdot\oplus}$ denotes cation radicals produced by one electron transpositions or physical methods.

In contrast, cationation is usually exoenergetic; that is, ΔG is negative:

$$M^{\oplus} + \overset{|}{\underset{|}{C}}\!\!=\!\!\overset{|}{\underset{|}{C}} \rightarrow M\!\!-\!\!\overset{|}{\underset{|}{C}}\!\!-\!\!\overset{|}{\underset{|}{C}}{}^{\oplus}$$

Table 4.9 □ Ion Generation (Priming) in Carbocationic Polymerization[a]

Method	Energy	Energy Carrier
Chemical		
Two-electron transpositions	Chemical, reaction between initiator and coinitiator	"Energized" species $H^{\oplus}$, $R^{\oplus}$
One-electron transpositions		
Direct radical oxidation	Chemical, oxidation of radical	"Energized" species $R^{\oplus}$
Charge transfer	Chemical, D·A complexation followed by heat or light excitation	Excited species $D \cdot^{\oplus}$, $^{\oplus}D \cdot A^{\ominus}$
Physical		
Ionizing radiation	Electromagnetic, $h\nu$	Photon, γ-rays
Photochemical (direct)	Electromagnetic, $h\nu$	Photon
Field emission	Electric	Current at cathode
Field ionization	Electric	Current at anode
Electrochemical	Electric	Current at anode (direct method) or transient radicals (indirect method)

[a]Fundamental process: $M \xrightarrow{E} M^{\oplus}$ or $M^{\oplus}$.

The energy needed for ion generation may come from a variety of sources and may be in the form of chemical, radiation, or electrical energy. In this broad sense the initiating electrophile may be viewed to be an "energy carrier." Table 4.9 summarizes these thoughts.

4.2 ☐ THE CHEMISTRY OF PROPAGATION

Overview

The understanding and control of propagation is of paramount importance from both scientific and practical points of view since this event affects yields (conversions) and product molecular weight. At first glance carbo-cationic propagation appears to be a simple rapid exothermic repetitive polyalkylation process:

$$\sim C^{\oplus} + C{=}C \rightarrow [\sim \overset{\delta\oplus}{C} \cdots C \overset{\delta\oplus}{\relbar\relbar} C]^{\ddagger} \rightarrow \ \sim C{-}C{-}C^{\oplus}$$

On closer inspection, however, propagation emerges as a rather complex series of subtle events, the control of which is exceedingly difficult to achieve.

The overall energetics of propagation is favorable because it involves the conversion of a π double bond system into a new covalent bond. Since the reaction involves ion molecule interactions, activation energies of carbo-cationic propagations are expected to be low, for example, 0.2 to 2.0 kcal/mole. In the transition state for propagation the charge is distributed over at least three atoms and therefore the process is favored in poorly ionizing (nonpolar) solvents.

In principle the more reactive monomer should yield the more stable cation and the rate-determining step is the formation of the covalent bond between the growing cation and β-carbon of the incoming monomer. Kanoh et al. (1963, 1965) observed that this simple rule is not followed in certain styrene + styrene derivative copolymerization systems (e.g., styrene + p-chlorostyrene/HClO$_4$/ethylene chloride/25°C) and suggested that the rate of propagation is determined by the electrostatic interaction between the cation counteranion pair. According to this author monomer incorporation is controlled by either step 1 or 2:

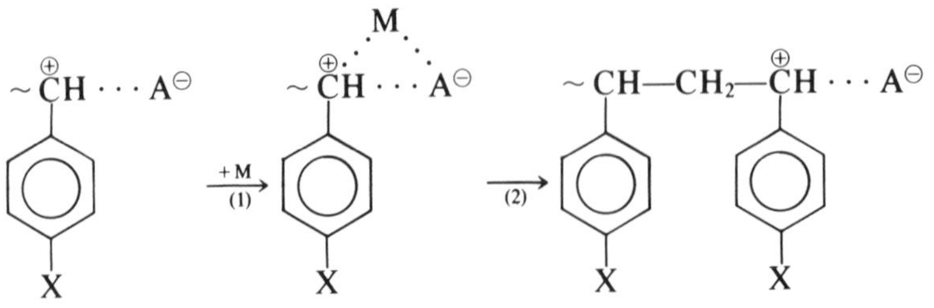

For example, the degree of dissociation of the more stable cation, that is, the one carrying an electron-donating substituent (–CH₃), is higher than that carrying an electron-withdrawing substituent (–Cl) and therefore it appears to be more reactive than the latter.

Complexation between monomer and propagating cation has also been proposed by Fontana (Fontana and Kidder, 1948; Fontana et al., 1952a, b; Fontana, 1959, 1963). According to this author propagation of 1-olefin polymerizations initiated by HBr/AlBr₃ in alkanes in the -20 to $-78°C$ range involves an equilibrium between the propagating ion pair and monomer, followed by rate-determining incorporation of complexed monomer into the ion pair:

$$
\begin{array}{c}
\sim CH_2 \\
|\\
HC^{\oplus} \quad AlBr_4^{\ominus} \; +
\end{array}
\begin{array}{c}
CH_2 \\
\|\\
C \\
|\\
CH_3
\end{array}
\underset{}{\overset{K}{\rightleftharpoons}}
\begin{array}{c}
\sim CH_2 \\
|\\
HC^{\oplus} \quad AlBr_4^{\ominus} \\
|\\
CH_3
\end{array}
\begin{array}{c}
CH_2 \\
\|\\
CH \\
|\\
CH_3
\end{array}
\overset{k}{\longrightarrow}
\begin{array}{c}
\sim CH_2 \\
|\\
HC-----CH_2 \\
|\qquad\ \ |\\
CH_3 \ HC^{\oplus} \ AlBr_4^{\ominus} \\
|\\
CH_3
\end{array}
$$

The key intermediate is the relatively long-lived ion pair–monomer complex that slowly rearranges during propagation. The rate of propagation by this mechanism is a complex function of monomer concentration $(-d[M]/dt = kK[C][M]/(1+K[M])$ where C is AlBr₃ or HBr, whichever is limiting) (in regard to the Fontana mechanism see also Section 4.4).

Among factors that determine chemical details of propagation (for kinetic aspects see Chapter 5) are the nature of the counteranion and solvent, ionicity of the growing species, desolvation–solvation equilibria, the possibility of isomerization prior to propagation, stereoregular placement of incoming monomer, and medium viscosity. The balance of this section concerns an examination of some of these factors, in particular the effect of ionicity, electron acceptors, isomerization, and stereochemistry on carbocationic propagation.

Ionicity of the Propagating Species

Static aspects of carbocations of interest to polymer scientists have been discussed in Section 3.1 including evidence for the existence of free ions, ion pairs, and aggregates. All these species are known to play a role in anionic polymerization; however, their identification in cationic polymerization currently is a largely unresolved problem. Circumstantial results increasingly seem to indicate the simultaneous existence of several propagating centers during propagation (Plesch, 1968; DeSorgo et al., 1973).

Although it is now generally accepted that propagating anion/cation pairs may be dissociated (free ions) or nondissociated (ion pairs), very little is known about the ionicity of propagating carbocations. Higashimura and co-workers (Higashimura et al., 1974, 1976a, 1977d, 1978, 1979a, b; Masuda and Higashimura, 1971a, b, 1976; Sawamoto et al. 1976, 1977) have recently carried out important investigations relative to the problem of ionicity in cationic aromatic olefin polymerizations. Significantly, these workers observed bimodal molecular weight distributions (MWD) in certain cationic styrene polymerization systems and on the basis of this evidence proposed the coexistence of dissociated and nondissociated chain carriers. They also found significant differences between polymerizations induced by Friedel–Crafts acids and "nonmetal halide" initiators such as oxo acids and iodine (Higashimura and Hirokawa, 1977a).

The first evidence for the coexistence of and independent propagation by dissociated and nondissociated carbocations was obtained by finding bimodal MWD's with the acetyl perchlorate/styrene/CH_2Cl_2/0°C system, shown in Figure 4.2. Number average molecular weights of the high and low polymers were 20,000 and 2000, respectively. Molecular weights (position of the peaks) were independent of solvent; however, relative polymer amount was governed by solvent polarity: high polymer was mainly obtained using polar solvents whereas low polymer predominantly formed in less polar media. High polymer formation was suppressed whereas low polymer formation remained unaffected in the presence of common ion salt, that is, n-Bu_4NClO_4 (Higashimura and Kishiro, 1974). Similar bimodal MWD's have been obtained with polystyrenes obtained with various oxo acid initiators, such as $HClO_4$ (Higashimura and Kishiro, 1974; Pepper, 1974a, b, 1975), CF_3SO_3H (Masuda et al., 1976), and CF_3COOH (Sawamoto et al., 1977).

These results demonstrate the coexistence of two propagating carbocations of different ionicities (dissociation states) and the existence of a

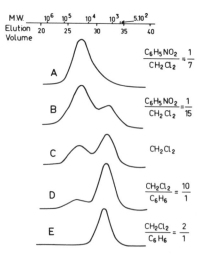

Figure 4.2 □   MWD of polystyrene obtained by $AcClO_4$ at 0°C ([M]$_0$ = 1.0 M) (Higashimura and Kishiro, 1974; Hisgashimura et al. 1976a).

dynamic relative to propagation slow equilibrium between these species. The effects of solvents and added common ion salts indicate that the propagating species leading to high polymer is highly "dissociated" and should be free ionic in nature. This aspect of Higashimura's proposition has been questioned by Szwarc and Litt (see general discussion and comment in Higashimura, 1976b).

Significantly, a close correlation between polymer MWD and overall polymerization rate was found. For example, in the styrene/CF_3SO_3H system (Sawamoto et al., 1976) illustrated in Figure 4.3, the rate of polymerization and the relative amount of high molecular weight material H increased with solvent dielectric constant. Over a rather narrow dielectric constant range bimodal MWD's were obtained, whereas with nonpolar medium only low molecular weight product L formed. Also, the rate decreased and the amount of high polymer decreased on introduction of common ion salt (Masuda and Higashimura, 1971b; Higashimura et al., 1978). On the basis of these results it was concluded that, similarly to the observations made with anionic polymerization systems, the dissociated propagating species leading to high polymer is of higher reactivity than the nondissociated chain carrier producing the low polymer.

The coexistence of a dissociated and nondissociated chain carrier does not necessarily yield bimodal MWD. For example, the polymerization of p-methoxystyrene (p-MeOSt) with CH_3COClO_4 irrespective of solvent fails to give polymer with bimodal MWD; however, bimodality appears in the presence of iodine initiator which yields $I_3^{\ominus}$, a more nucleophilic counteranion than $ClO_4^{\ominus}$. Similarly, polystyrene produced by conventional Friedel–Crafts acid metal halides exhibits monomodal MWD, although the position of the peak in the GPC trace shifts toward higher molecular weights with increasing solvent polarity, as shown in Figure 4.4.

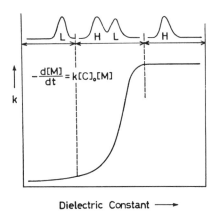

Figure 4.3 □ Scheme of correlation between rate constant k, solvent dielectric constant, and MWD (Sawamoto et al., 1976). (Reprinted by permission of Hüthig and Wepf Verlag, Basel.)

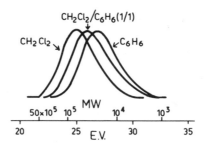

Figure 4.4 □ MWD of polystyrene obtained by BF₃·OEt₂ at 0°C ([M]₀ = 1.0 M).

According to Higashimura (Higashimura, 1976b) polymer MWD is affected by counteranion nucleophilicity and propagating carbocation stability. Scheme 4.1 summarizes observations with regard to four aromatic monomers and three initiating systems. The single arrows indicate unimodal MWD and the double arrows signify the appearance of bimodality.

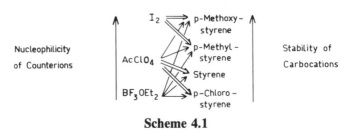

Scheme 4.1

Strong interaction between growing carbocations and highly nucleophilic counteranions may prevent rapid interconversion of propagating species having different ionicities and thus propagation by two independent species may occur. Counteranions derived from Friedel–Crafts acid metal halides seem to interact weakly with growing cations even in nondissociated form. Higashimura (1976b) concludes that cationic olefin polymerizations involve two independent dissociated and nondissociated propagating species and that two types of nondissociated species exist, one with strong cation–anion interaction and another with weak interaction.

Since low polymer predominates in nonpolar solvents and its formation is not suppressed by common anion salt addition in polar media, the propagating species responsible for low polymer is a nondissociated species and should be an ion pair or ester. Sawamoto et al. (1977) feel that esters are probably not propagating entities, at least in certain systems. For example, polymer with bimodal MWD is obtained in the CF₃COOH/styrene/(CH₂Cl)₂/50°C system; however, the styryl ester of the initiating acid, CH₃CH(C₆H₅)OCOCF₃, that is, a model for the hypothetical ester propagating species, does not initiate polymerization under identical conditions. Evidently the nondissociated species in these systems are ion pairs or ion pair aggregates (Sawamoto et al., 1977).

Nondissociated species, that is, a stable carbocation and a strongly interacting counteranion, may propagate in the absence of undesirable side reactions since the active site is protected by the nucleophilic counteranion in its proximity. Results in line with this expectation have been obtained in the I_2/p-MeOSt/CCl$_4$/0 to $-15°C$ system (Higashimura and Kishiro, 1977d; Higashimura, 1979a, b). The molecular weight of the polymer increased linearly with conversion, and addition of a second batch of monomer after completion of the first polymerization resulted in resumption of linear molecular weight increase with conversion. Or, introduction of isobutyl vinyl ether (IBVE) as second monomer yielded a novel block copolymer poly(p-MeOSt-b-IBVE) (Higashimura et al., 1979b). These results indicate the presence of long-lived carbocationic propagating species.

Effect of Electron Acceptors on Propagation

Panayotov (Panayotov et al., 1969, 1975; Dimitrov and Panayotov, 1971; Toncheva et al., 1974; Velichkova and Panayotov (1970) and Heublein (Heublein and Adelt, 1972; Heublein and Spange, 1974a; Heublein and Barth, 1974b; Heublein et al., 1975a, Heublein and Schutz, 1975b) separately and jointly (Panayotov and Heublein, 1977) have studied the effect of electron acceptors (EA) on the propagation of cationic polymerizations. According to these investigators EA's affect the overall rate of polymerization. For example, the rate of styrene polymerization in the presence of tetracyanoethylene (TCNE) exhibits a maximum, as shown in Figure 4.5 EA's also affect molecular weights, though the effect is much less than that on yields.

The effect of TCNE on the overall rate indicates that this EA can affect the rate of one or more elementary steps, which in turn may be due to complexation with either monomer or counteranion or both.

The increase of molecular weights with TCNE indicates that the rate of propagation may be the molecular weight determining step. It would be of interest to examine the effect of an EA on propagation. Panayotov et al. (1974) have analyzed reactivity ratios obtained in the presence of EAs; these investigations are discussed in Chapter 6.

The counteranion or monomer may form 1:1 complexes with EA's (Toncheva et al., 1974):

$$M + EA \rightleftharpoons [M \cdots EA] \qquad (4.1)$$

Such complexes may form in nonelectron-donating or highly polar solvents; thus benzene or nitrobenzene are unsuitable solvents. The stability of the complex increases with the ionization potential of the monomer as illustrated by the data in Table 4.10.

The EA may also form complexes with counteranions $A^\ominus$:

$$C^\oplus A^\ominus + M \cdots EA \rightleftharpoons C^\oplus M[A^\ominus \cdots EA] \qquad (4.2)$$

and the formation of these complexes would lead to the decomposition of

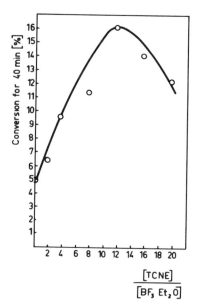

Figure 4.5 ☐ Polymerization of styrene in the presence of TCNE ([St] = 1 M, [BF$_3$·Et$_2$O] = 0.01 M, ClCH$_2$CH$_2$Cl, 20°C, 40 min) (from Panayotov et al., 1969).

Table 4.10 ☐ **Stability Constants K_s of TCNE Complexes with Monomers and Counteranions in 1,2-Dichloroethane**[a]

Donor	T (°C)	K_s(1/mole)
4-Vinylcyclohexene	20	0.11
Styrene	20	0.80
Indene	20	1.12
p-Methylstyrene	20	2.00
α-Methylstyrene	20	3.18
SbF$_6^{\ominus}$	25	6.7
AsF$_6^{\ominus}$	25	7.0
SbCl$_6^{\ominus}$	25	26.0
TiCl$_5^{\ominus}$	25	38.0

[a]The K_s values of monomers have been obtained by spectroscopy (Toncheva et al., 1974) and those of counteranions by calorimetry (Heublein and Spange, 1974a; Heublein and Barth, 1974b).

Table 4.11 □ K_s Values for Styrene Polymerization in Presence and Absence of EA^a

Polymerization Conditions	$K_s \times 10^3$
In absence of EA	0.13 (K_1)
Presence of TNB^b	0.25 (K_2)
Presence of TCNE	0.34 (K_2)

^{a}[St] = 1M, [BF$_3$·OEt$_2$] = [EA] = 2.5 × 10^{-3} M,
1,2-dichloroethane, 20°.
^{b}1,3,5-Trinitrobenzene.

M···EA complexes. The displacement reaction (equation 4.2) occurs since K_s values of counteranions are usually larger than those of monomers (see Table 4.10).

Thus two situations may arise. In the absence of EA,

$$\sim C^{\oplus}A^{\ominus} \underset{}{\overset{K_1}{\rightleftharpoons}} \sim C^{\oplus} + A^{\ominus} \tag{4.3}$$

and in the presence of EA,

$$\sim C^{\oplus}[A^{\ominus} \cdots EA] \underset{}{\overset{K_2}{\rightleftharpoons}} \sim C^{\oplus} + [A \cdots EA]^{\ominus} \tag{4.4}$$

According to the values in Table 4.11 $K_2 > K_1$. Thus the stability constant increases with the nature of the acceptor and changes electron delocalization in the counteranion. The rate of polymerization may increase and then decrease with increasing [EA] in the system. In the presence of relatively small [EA], the contribution of free ions would increase, which in turn would enhance overall rate.

With increasing [EA] the separation between the ion and counteranion increases so that "contact ion pairs" may be transformed into "EA-separated ion pairs." After the solvate shell of the ion pair is saturated the excess may form EA complex with monomer. Since M ··· EA complexes are less active during polymerization than uncomplexed monomer, the rate of polymerization tends to decrease with increasing [EA].

Isomerization Polymerization

Carbocations are inherently reactive species and as such are prone to undergo a great variety of rearrangements. The sometimes facile and predictable rearrangements of propagating carbenium ions have been exploited for the synthesis of many unique polymer structures and led to the development of a subdiscipline, cationic isomerization polymerizations.

Isomerization polymerizations are polyaddition (chain growth) reactions in which the propagating species rearranges to energetically preferred structures prior to propagation so that the repeat unit of the polymer does not possess the structure of the original monomer employed (Kennedy, 1967i). The defining equation of isomerization polymerization is

$$nA \rightarrow +B+_n$$

Isomerization polymerizations often lead to unique structures available only by this route, for example,

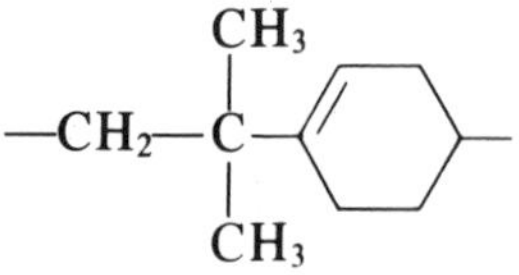

poly(β-pinene) or perfectly alternating isobutylene-cyclohexene copolymer,

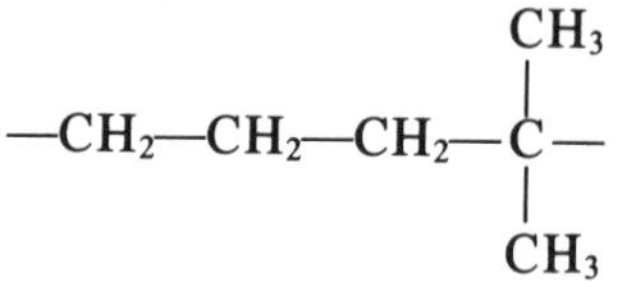

poly(4-methyl-1-pentene) or perfectly alternating ethylene–isobutylene copolymer,

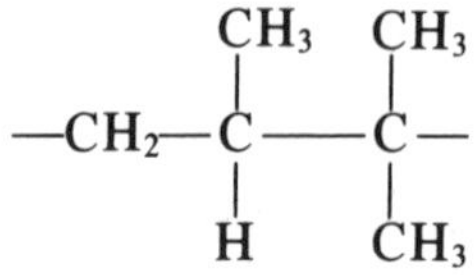

and poly(3, 3-dimethyl-1-butene). Since the repeat units of these and similar polymers do not correspond to monomers in the conventional sense, the term "phantom polymers" has been coined to describe such systems (Plesch, 1963).

Since the field of cationic isomerization polymerizations has been comprehensively surveyed and critically examined (Kennedy, 1967i), this section focuses on developments published after the earlier review.

A useful classification of isomerization polymerizations is based on the type of rearrangement occurring during propagation: isomerizations by bond (electronic) rearrangements or by material transport (Kennedy, 1967i).

Isomerizations by Bond (Electron) Rearrangement

Intra-Intermolecular Polymerization

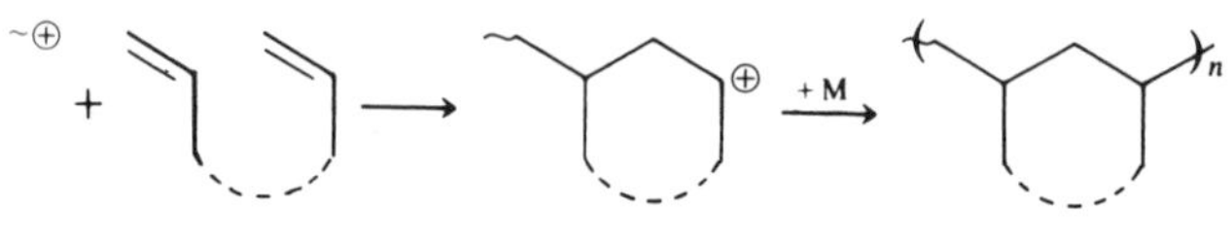

Pioneering research (Butler and Miles, 1966; Butler et al., 1965) in cationic intra–intermolecular polymerizations, schematized by has been followed up, for example, by Aso et al. (1972), who obtained condensed rings from *cis*-1,2-divinylcyclohexane:

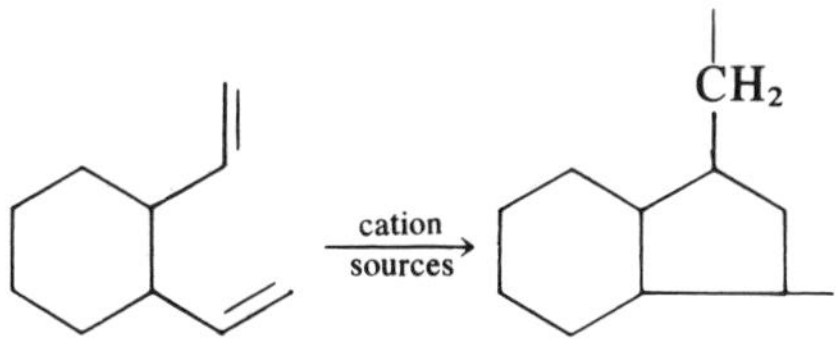

An interesting carbocationic cyclopolymerization has recently been described (Furukawa and Nishimura, 1976). Thus α, ω-bis(4-vinylphenyl)alkanes,

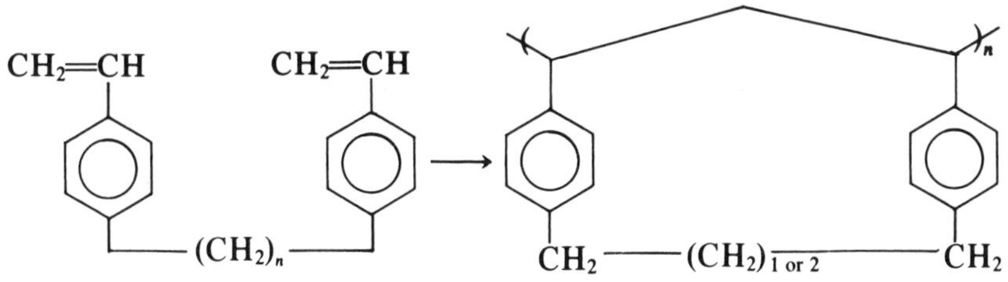

with $n = 3$ and 4 have been claimed to yield in the presence of CH_3COClO_4 initiator macro-ring polymers of reasonably high (30,000 to 40,000) molecular weights.

Transannular Polymerization

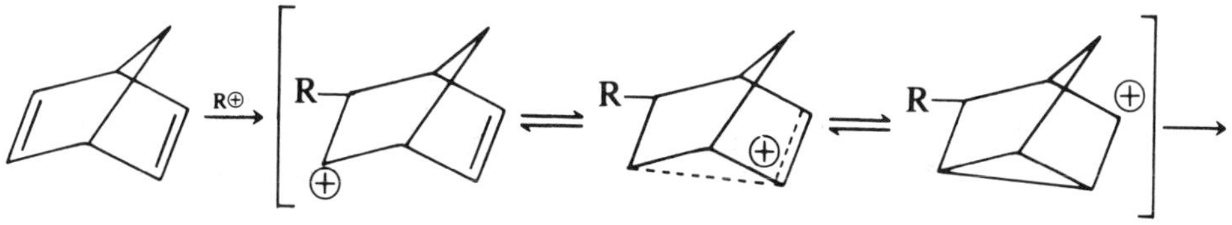

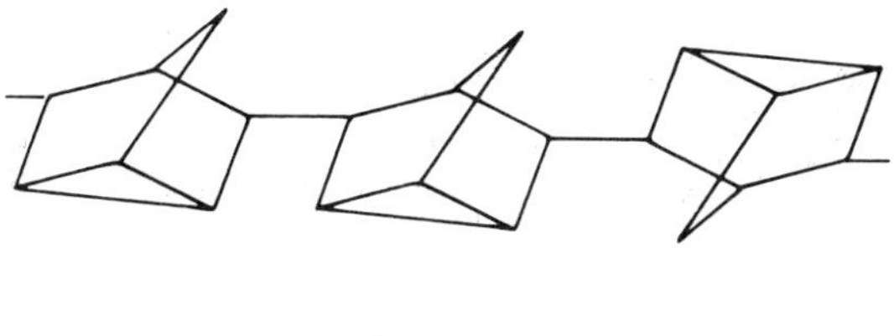

The field of transannular polymerizations opened by investigations concerning the polymerization of 2, 5-norbornadiene (Kennedy and Hinlicky, 1965a), and 2-methylene-5-norbornene (Sartori et al., 1963),

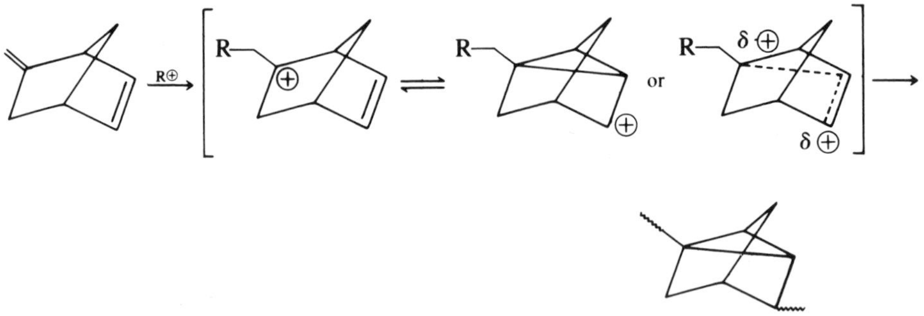

has been further expanded by elucidating the polymerization behavior and structure of various other bicyclo[2, 2, 1]heptane derivatives, that is, norbornene, 2-vinyl-5-norbornene, and 2-isopropenyl-5-norbornene (Kennedy and Makowski, 1967j, 1968a). The T_g of polynorbornadiene 320°C (at less than 1 cps) is probably the highest known T_g for a linear soluble polyhydrocarbon (Roller et al., 1973).

Polymerization by Strain Relief and Ring Opening

The common mechanistic feature of polymerizations in this group are the opening of strained rings which may in addition be followed by rearrangements. Simplest representatives of these polymerizations are cyclopropane and 1, 1-dimethylcyclopropane, which under cationic conditions lead to low molecular weight, branched (isomerized) structures (Tipper and Walker, 1959) and rather well-defined linear products (Ketley, 1963), respectively:

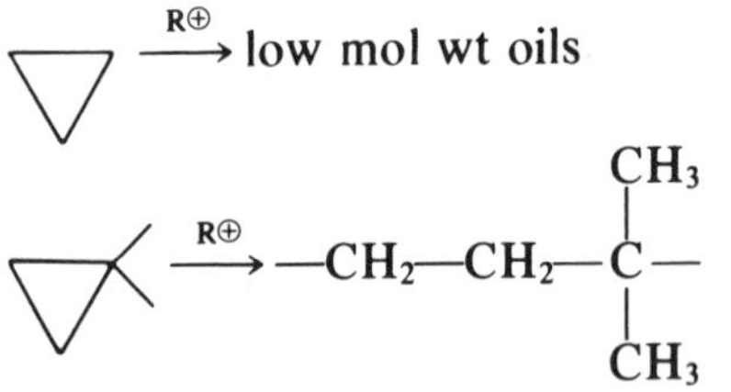

Polymerization of vinylcyclopropane (Takashashi and Yamashita, 1965) and its derivatives have been shown to yield interesting repeat structures suggesting the involvement of propagating species as follows (Ketley et al., 1967; Kennedy et al., 1968b):

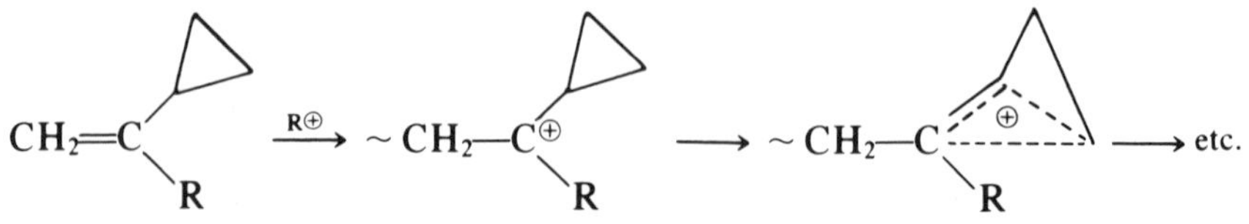

The polymer chemistry of terpene derivatives, particularly those of β- and α-pinene, continue to occupy several research groups (Kennedy and Chou, 1976a; Huet and Maréchal, 1970). The structure of poly(β-pinene) has been elucidated (Ruckel et al., 1975; Snyder et al., 1977; Sivola and Harva, 1970; Huet and Maréchal, 1970; Roberts and Day, 1950) and a generally acceptable propagation mechanism has been proposed (Roberts and Day, 1950; Kennedy, 1967i).

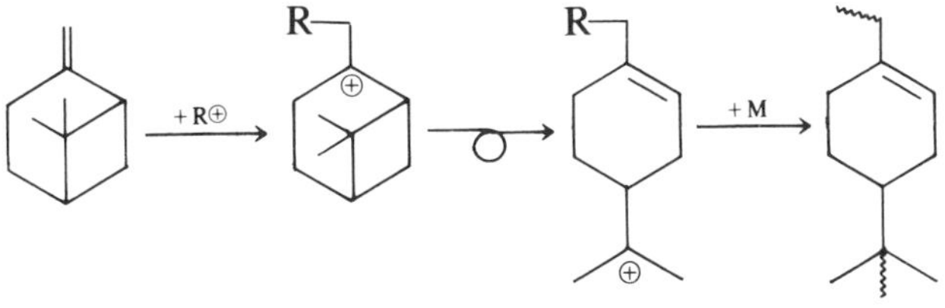

Isomerization copolymerization of β-pinene with isobutylene has been demonstrated to occur readily (Kennedy and Chou, 1976) and the random copolymers were found to yield completely ozone-resistant high molecular weight sulfur-vulcanizable elastomers:

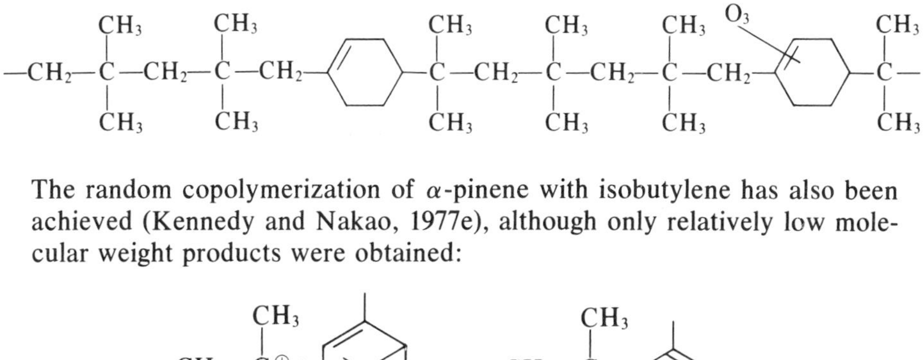

The random copolymerization of α-pinene with isobutylene has also been achieved (Kennedy and Nakao, 1977e), although only relatively low molecular weight products were obtained:

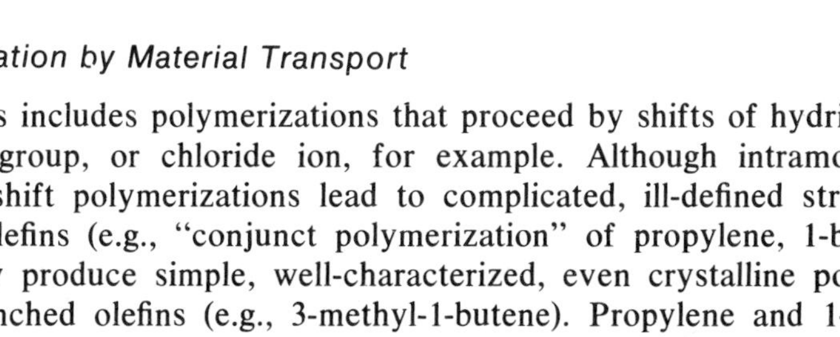

Isomerization by Material Transport

This class includes polymerizations that proceed by shifts of hydride ion, methide group, or chloride ion, for example. Although intramolecular hydride shift polymerizations lead to complicated, ill-defined structures with α-olefins (e.g., "conjunct polymerization" of propylene, 1-butene), they may produce simple, well-characterized, even crystalline polymers with branched olefins (e.g., 3-methyl-1-butene). Propylene and 1-butene under cationic polymerization conditions yield rather low molecular weight products, the structures of which suggest rapid 1, 2 hydride shifts (Ketley and Harvey, 1961).

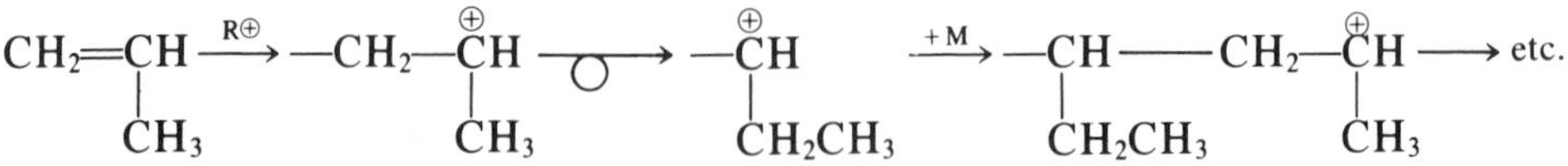

$$CH_2{=}CH \xrightarrow{R\oplus} {-}CH_2{-}\overset{\oplus}{C}H \xrightarrow{} \overset{\oplus}{C}H \xrightarrow{+M} {-}CH{-\!\!\!-\!\!\!-}CH_2{-}\overset{\oplus}{C}H \longrightarrow etc.$$

with substituents: CH_3 ; CH_3 ; CH_2CH_3 ; CH_2CH_3 ; CH_3

Recent work using high-resolution spectroscopy has helped to elucidate structural details of cationically prepared polymers of normal olefins (Puskas et al., 1976; Corno et al., 1977; Watson and Rysselberge, 1978). In confirmation of the prognostications of earlier authors (Fontana, 1963; Kennedy, 1967i), these materials are hopelessly complicated mixtures of various contributing repeat units and sequences. The best-known example of intramolecular hydride shift polymerization is that of 3-methyl-1-butene, a monomer that at very low temperatures ($> -100°C$) produces crystalline high polymer, but at moderately low temperatures ($< -100°C$) leads to rubbery products, the structures of which comprise rearranged and unrearranged units (Kennedy et al., 1963, 1964a–e, 1965c; Hudson, 1966):

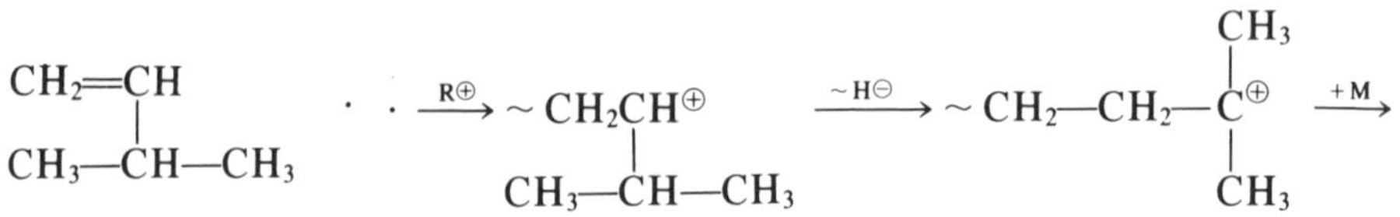

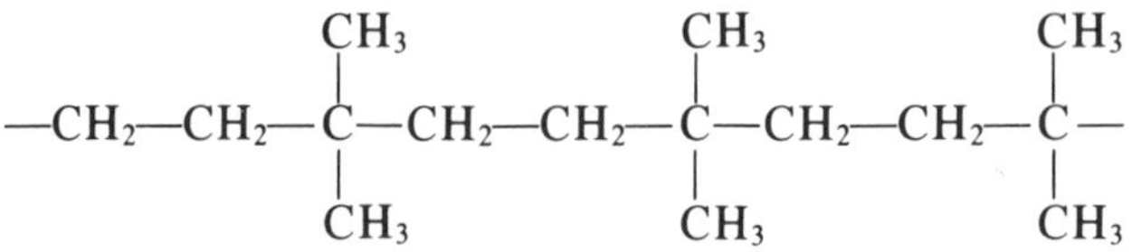

$$CH_2{=}CH \quad \cdots \xrightarrow{R\oplus} {\sim}CH_2CH^{\oplus} \xrightarrow{\sim H\ominus} {\sim}CH_2{-}CH_2{-}\overset{\oplus}{\underset{CH_3}{C}} \xrightarrow{+M}$$

with: $CH_3{-}\overset{|}{C}H{-}CH_3$ (first), $CH_3{-}\overset{|}{C}H{-}CH_3$ (second), CH_3 (top of third)

$$-CH_2{-}CH_2{-}\underset{\underset{CH_3}{|}}{\overset{\overset{CH_3}{|}}{C}}{-}CH_2{-}CH_2{-}\underset{\underset{CH_3}{|}}{\overset{\overset{CH_3}{|}}{C}}{-}CH_2{-}CH_2{-}\underset{\underset{CH_3}{|}}{\overset{\overset{CH_3}{|}}{C}}-$$

Crystalline structure obtainable at very low temperatures

$${\sim}CH_2{-}CH{-\!\!\!-\!\!\!-}CH_2{-}CH_2{-}\overset{\overset{CH_3}{|}}{C}{-}CH_2{-}CH{-\!\!\!-\!\!\!-}CH_2{-}CH\,{\sim}$$

with substituents: $CH_3{-}\overset{|}{\underset{H}{C}}{-}CH_3$; CH_3 ; $CH_3{-}\overset{|}{\underset{H}{C}}{-}CH_3$; $CH_3{-}\overset{|}{\underset{H}{C}}{-}CH_3$

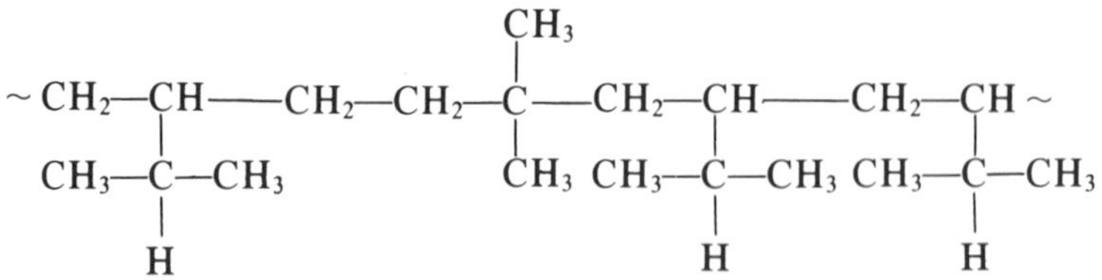

Rubbery structure obtainable at moderately low temperature

In contrast, Ziegler–Natta coordination catalysts produce crystalline isotactic poly(3-methyl-1-butene) of conventional 1, 2 enchainment. Thus depending on the temperature and initiating system employed, 3-methyl-1-butene can be polymerized to three fundamentally dissimilar structures yielding three fundamentally different materials:

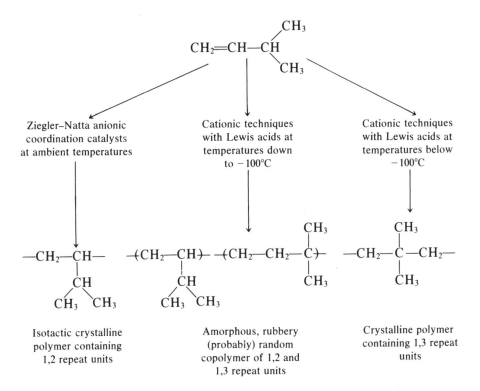

A comprehensive survey of the cationic polymerization behavior of 3-methyl-1-butene has been published (Kennedy, 1967i). Recent research carried out mainly with carbon (Tanaka and Sato, 1976; Kennedy and Johnston, 1975a) and high resolution proton NMR (Kennedy and Johnston, 1975a) is in complete agreement with the earlier findings.

The polymerizations of 3-methyl-1-pentene (Bacskai, 1967), 3-ethyl-1-pentene (Kennedy, 1967i), and vinylcyclohexane (Kennedy et al., 1964c; Ketley et al., 1964b) also proceed by intramolecular hydride shift, for example:

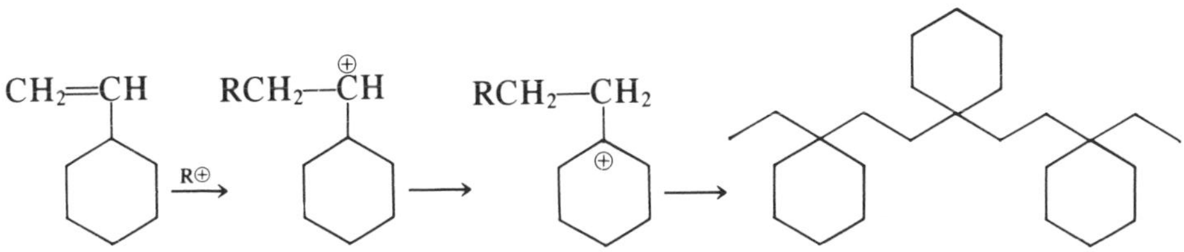

In addition to these 1, 2 hydride shifts several examples of longer-range, multiple hydride migration polymerizations have also been described.

The cationic polymerization behavior of 4-methyl-1-pentene and the structure of poly(4-methyl-1-pentene) has been examined in detail by several research groups (Edwards and Chamberlain, 1963; Kennedy et al., 1964f; Goodrich and Porter, 1964; Wanless and Kennedy, 1965; Kennedy and Johnston, 1975a; Ferraris et al., 1977), who are in general agreement that the product is a copolymer comprising at least three repeat units:

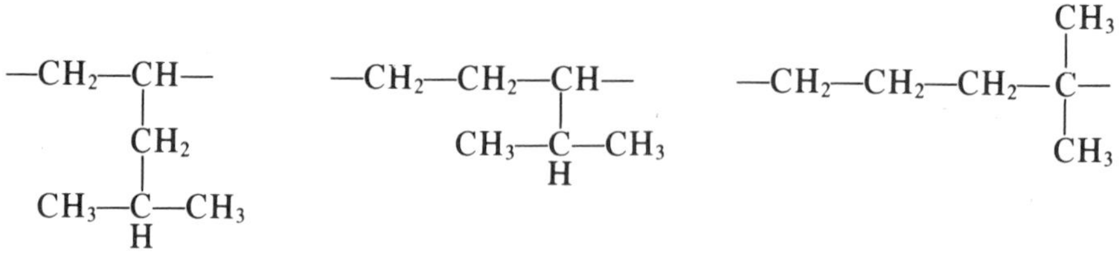

1,2 Enchainment   1,3 Enchainment   1,4 Enchainment

On the basis of structural information the following propagation mechanism has been proposed to account for the various repeat units (Wanless and Kennedy, 1965).

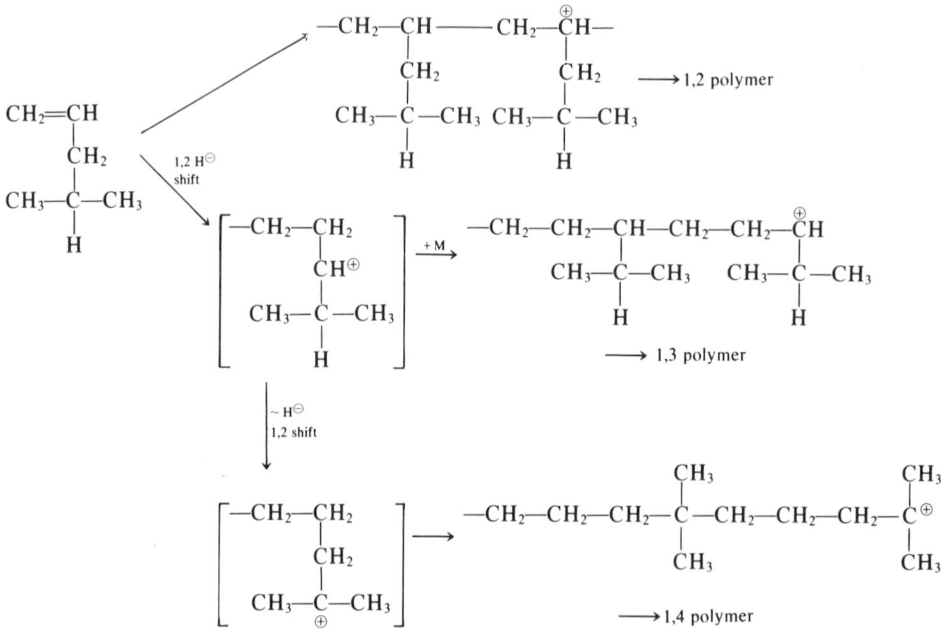

The question has been raised whether the 1,4 enchainment is due to a 1,3 hydride jump or to two consecutive 1,2 shifts. Experiments with the deuterated monomer CH_2=CH–CH_2–$CD(CH_3)_2$ indicate that rearrangement occurs by two consecutive 1,2 hydride shifts (Wanless and Kennedy,

1965). The 1,4 structure is of particular interest since it is formally equivalent to a perfectly alternating ethylene–isobutylene copolymer unavailable by any other known polymerization mechanism.

Italian investigators (Ferraris et al., 1977) on the basis of extensive ^{13}C-NMR spectroscopy recently proposed two additional contributory structures:

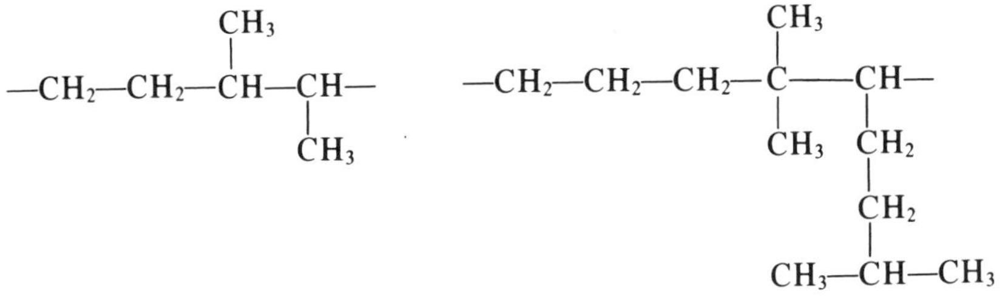

$$
\begin{array}{cc}
\overset{\displaystyle CH_3}{\underset{\displaystyle CH_3}{\overset{|}{-CH_2-CH_2-CH-CH-}}} & \overset{\displaystyle CH_3}{\underset{\displaystyle \underset{\displaystyle CH_3-CH-CH_3}{\overset{|}{\underset{\displaystyle CH_2}{\overset{|}{CH_2}}}}}{\overset{|}{-CH_2-CH_2-CH_2-\overset{\displaystyle CH-}{\underset{\displaystyle CH_3}{\overset{|}{C}}}}}
\end{array}
$$

and suggested reasonable sequential hydride and methide shifts to account for them.

Various other longer-range hydride shift polymerizations leading to some quite complicated structures have also been described, for example, 4-methyl-1-hexene (Kennedy et al., 1965b; Bácskai, 1967), 5-methyl-1-hexene (Edwards and Chamberlain, 1963; Ketley, 1964a), 5-methyl-1-heptene (Bácskai, 1967), and 6-methyl-1-heptene (Ketley, 1964a).

Mechanistic details of hydride shifts have been investigated in an elegant series of experiments (Bacskai, 1967) by the use of optically active monomers in which isomerization polymerization was designed to destroy optical activity, for example:

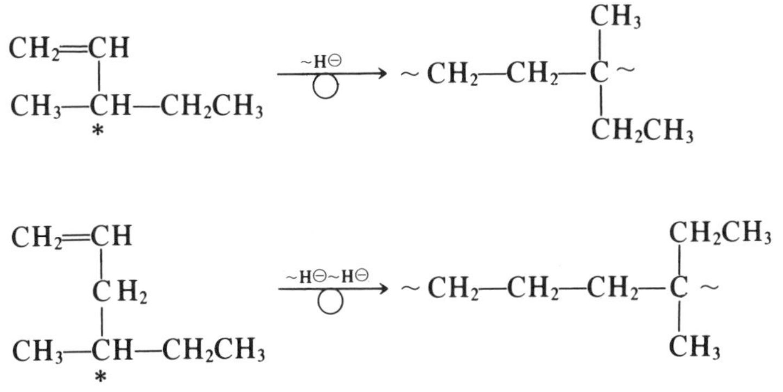

$$
\underset{*}{\overset{\displaystyle CH_2{=}CH}{\underset{\displaystyle CH_3-CH-CH_2CH_3}{|}}} \quad \xrightarrow[\bigcirc]{\sim H^{\ominus}} \quad \sim CH_2-CH_2-\overset{\displaystyle CH_3}{\underset{\displaystyle CH_2CH_3}{\overset{|}{\underset{|}{C}}}} \sim
$$

$$
\underset{*}{\overset{\displaystyle CH_2{=}CH}{\underset{\displaystyle CH_3-CH-CH_2CH_3}{\overset{|}{\underset{\displaystyle }{\overset{\displaystyle CH_2}{|}}}}}} \quad \xrightarrow[\bigcirc]{\sim H^{\ominus}\sim H^{\ominus}} \quad \sim CH_2-CH_2-CH_2-\overset{\displaystyle CH_2CH_3}{\underset{\displaystyle CH_3}{\overset{|}{\underset{|}{C}}}} \sim
$$

By determining quantitatively the loss of optical activity the extent of isomerization could be obtained. Polymers of (+)3-methyl-1-pentene and (−) 4-methyl-1-hexene were found to be completely optically inactive after low-temperature polymerization with "H_2O"/$AlCl_3$ in CH_2Cl_2, indicating essentially complete isomerization polymerization. In contrast, 14 to 20% of the original optical activity remained after polymerization of (+)5-methyl-1-heptene, which was attributed to the increased length of the path over which consecutive hydride shifts occur:

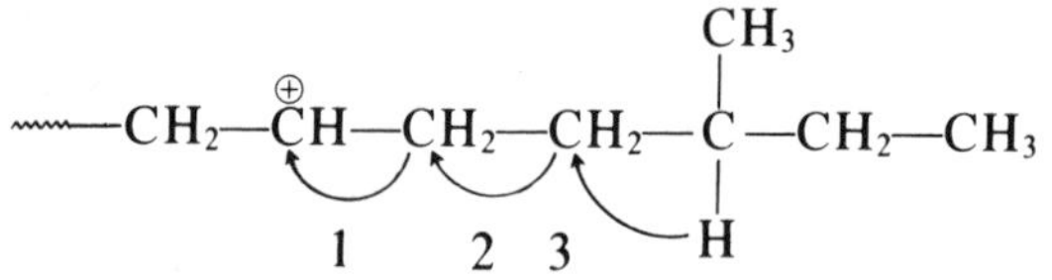

$$\text{\textasciitilde\textasciitilde\textasciitilde}-CH_2-\overset{\oplus}{C}H-CH_2-CH_2-\underset{\underset{H}{|}}{\overset{\overset{CH_3}{|}}{C}}-CH_2-CH_3$$

$$1 \quad\quad 2 \quad 3$$

The structure of the propagating carbocation formed from 3-methyl-1-pentene and 2-methyl-1-butene are similar:

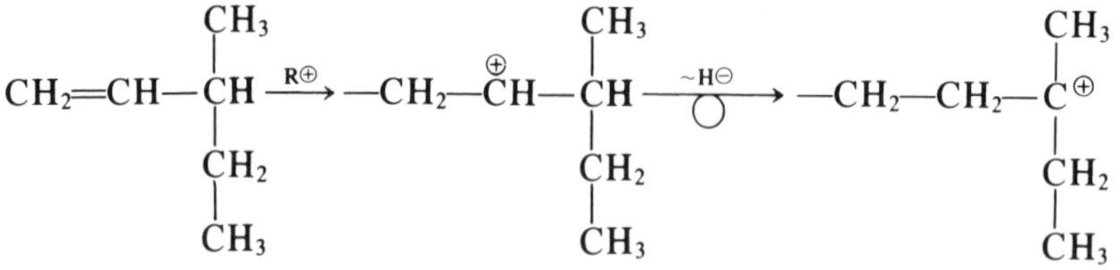

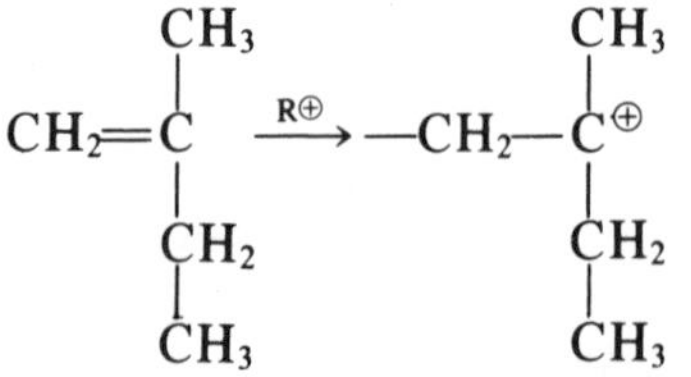

$$CH_2=CH-\underset{\underset{CH_3}{\underset{|}{CH_2}}}{\overset{\overset{CH_3}{|}}{CH}} \xrightarrow{R\oplus} -CH_2-\underset{\underset{CH_3}{\underset{|}{CH_2}}}{\overset{\oplus}{C}H}-\underset{}{\overset{\overset{CH_3}{|}}{CH}} \xrightarrow{\sim H\ominus} -CH_2-CH_2-\underset{\underset{CH_3}{\underset{|}{CH_2}}}{\overset{\overset{CH_3}{|}}{C}}^{\oplus}$$

$$CH_2=\underset{\underset{CH_3}{\underset{|}{CH_2}}}{\overset{\overset{CH_3}{|}}{C}} \xrightarrow{R\oplus} -CH_2-\underset{\underset{CH_3}{\underset{|}{CH_2}}}{\overset{\overset{CH_3}{|}}{C}}^{\oplus}$$

Differences in polymerization behavior between these two monomers (Kennedy, 1967a) are probably due to the larger steric inhibition of propagation of the latter isoolefin.

Intramolecular hydride shift polymerization helps to explain differences in polymerization behavior between cyclopentene or cyclohexene and 3-methylcyclopentene or 3-methylcyclohexene. Although the parent cycloolefins cannot be polymerized by "H$_2$O"/AlCl$_3$, the 3-substituted derivatives polymerize under essentially identical conditions (Boor et al., 1966). Inspection of structures suggests an explanation:

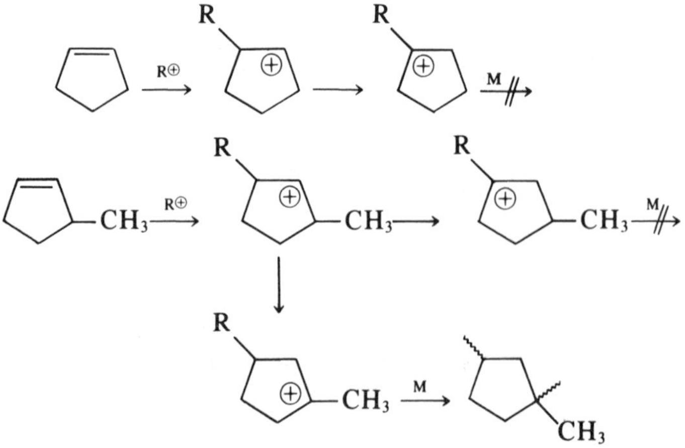

Though the secondary carbenium ion formed with the unsubstituted cycloalkenes immediately rearranges to a sterically "buried" tertiary cation unable to sustain propagation, the 3-methyl derivatives may produce by hydride shift a less sterically compressed tertiary cation, permitting the approach of monomer.

With 3-phenyl-1-butene $CH_2=CH-CH(C_6H_5)CH_3$ cationic propagation may involve competition between $H^\ominus$, $CH_3^\ominus$, or $C_6H_5^\ominus$ shifts; however, product analysis produced only ambiguous results (Kennedy et al., 1964g; Davidson, 1966).

The cationic polymerization of 2-vinyl-1,3-dioxolane may proceed partially (2 to 15 mole %) by intramolecular hydride shift coupled with ring opening (Tada et al., 1967). The following equations illustrate the rearrangements involved (Kennedy, 1967i):

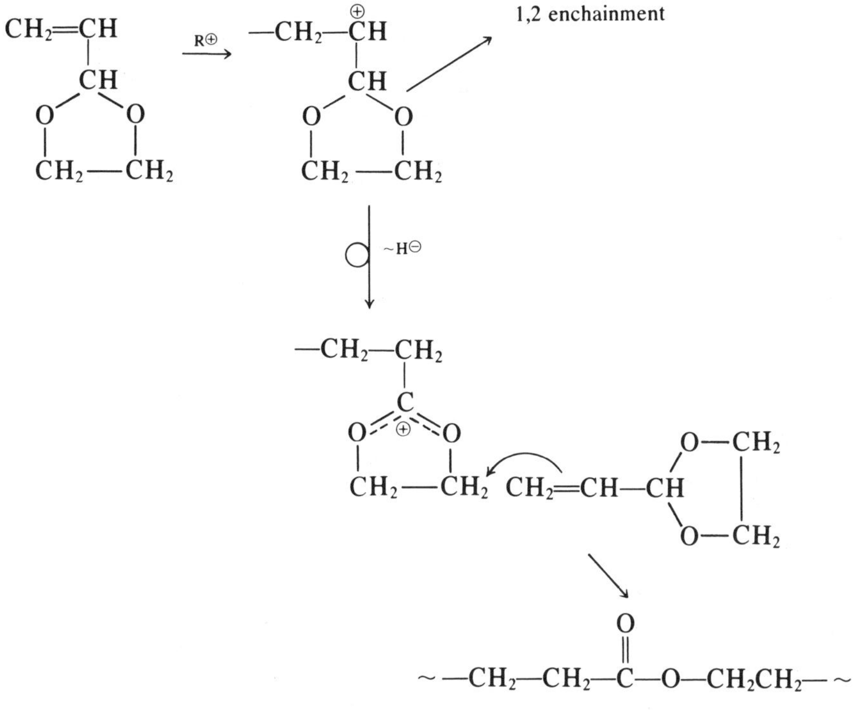

The cationic polymerization of 9-vinylanthracene (Michel, 1964),

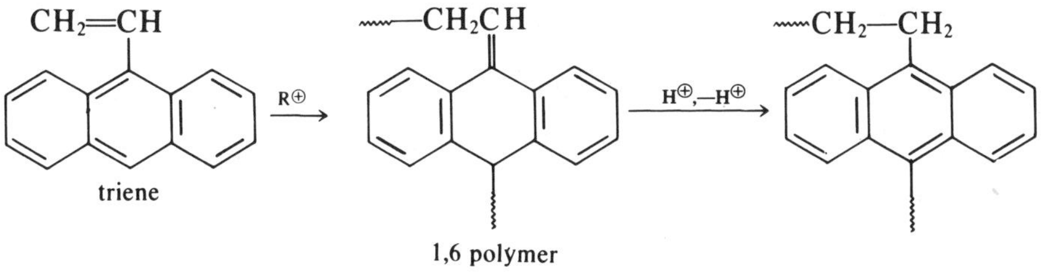

triene

1,6 polymer

was further investigated by Laguerre and Maréchal (1974), who showed that methyl substitution in the 10 position prevents isomerization polymerization. According to Coudane et al. (1979) the interpretation of spectroscopic results (Michel, 1964) may need revision. These authors also point out that unless the poly(9-vinylanthracene) is stored in the absence of oxygen in the dark the polymer rapidly degrades.

Besides intramolecular hydride shift polymerizations, interesting examples of group migration and combinations of hydride + alkyl group migration polymerizations have also been described. A good documentation of group migration polymerization concerns the low-temperature polymerization of 3,3-dimethyl-1-butene (Edwards and Chamberlain, 1963; Kennedy et al., 1964h; Maltsev et al., 1969). This monomer produces only low molecular oligomers at moderately low temperatures, probably because proton elimination (i.e., chain transfer to monomer) is faster than propagation at the sterically "buried" secondary carbenium ion. At very low temperatures, such as − 130°C, energetically favorable methide shift occurs, which facilitates propagation by formation of less compressed tertiary cation (Kennedy et al., 1964h):

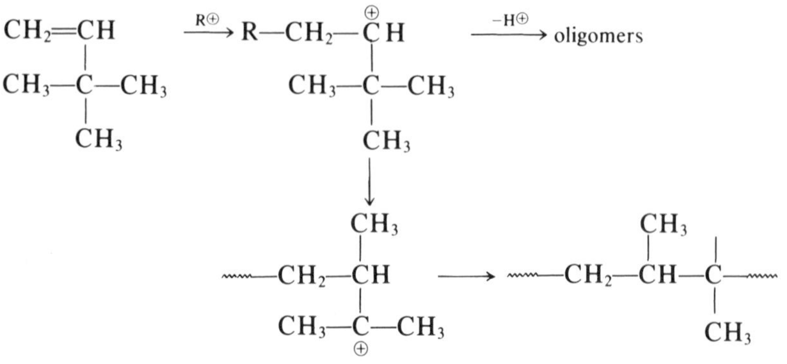

1,3 Polymer

Polymerization in the presence of counteranions derived from TiCl$_4$ has been claimed to produce essentially pure 1,3 enchainment (Maltsev et al., 1969).

In this context it is rather interesting to contemplate that both 3,3-dimethyl-1-butene and 2,3-dimethyl-1-butene produce polymer of the same repeating unit, the former by methide migration, the latter by hydride shift (Van Lohuizen and de Vries, 1968).

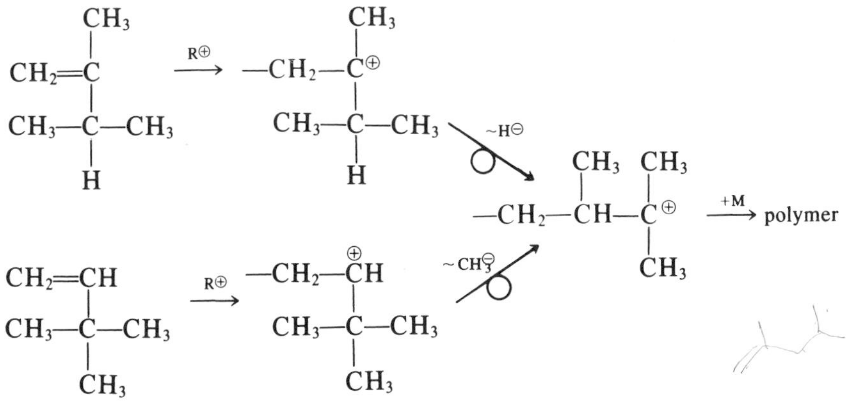

$$
\begin{array}{ccc}
& & CH_3 \\
& & | \\
CH_2{=}C & \xrightarrow{\ R^{\oplus}\ } & -CH_2-C^{\oplus} \\
| & & | \\
CH_3-\overset{\displaystyle H}{\underset{\displaystyle |}{C}}-CH_3 & & CH_3-\overset{\displaystyle H}{\underset{\displaystyle |}{C}}-CH_3 \xrightarrow{\ \sim H^{\ominus}\ }
\end{array}
$$

$$
\begin{array}{cc}
& CH_3\ \ CH_3 \\
& |\ \ \ \ | \\
-CH_2-CH-C^{\oplus} & \xrightarrow{\ +M\ } \ \text{polymer} \\
& | \\
& CH_3
\end{array}
$$

$$
\begin{array}{ccc}
CH_2{=}CH & \xrightarrow{\ R^{\oplus}\ } & -CH_2-\overset{\oplus}{CH} \\
| & & | \\
CH_3-\overset{\displaystyle CH_3}{\underset{\displaystyle |}{C}}-CH_3 & & CH_3-\overset{\displaystyle CH_3}{\underset{\displaystyle |}{C}}-CH_3 \xrightarrow{\ \sim CH_3^{\ominus}\ }
\end{array}
$$

In contrast, under essentially the same conditions, 2,4,4-trimethyl-1-butene did not polymerize (Van Lohuizen and de Vries, 1968). This may be due to steric compression to propagation of the initially formed tertiary cation and the inability to form by methide migration a strained cation:

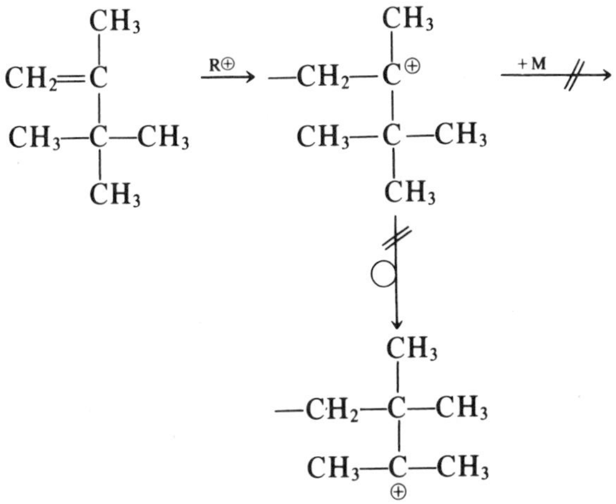

$$
\begin{array}{ccc}
CH_3 & & CH_3 \\
| & & | \\
CH_2{=}C & \xrightarrow{\ R^{\oplus}\ } & -CH_2-C^{\oplus} \xrightarrow{\ +M\ } \not\rightarrow \\
| & & | \\
CH_3-\overset{\displaystyle CH_3}{\underset{\displaystyle |}{C}}-CH_3 & & CH_3-\overset{\displaystyle CH_3}{\underset{\displaystyle |}{C}}-CH_3
\end{array}
$$

$$
\begin{array}{c}
CH_3 \\
| \\
-CH_2-C-CH_3 \\
| \\
CH_3-\underset{\oplus}{C}-CH_3
\end{array}
$$

A graceful example of a consecutive hydride + methide shift polymerization has been described in the 4,4-dimethyl-1-pentene system (ethyl chloride diluent, "H₂O"/AlCl₃, −78 to −130°C (Sartori et al., 1971).

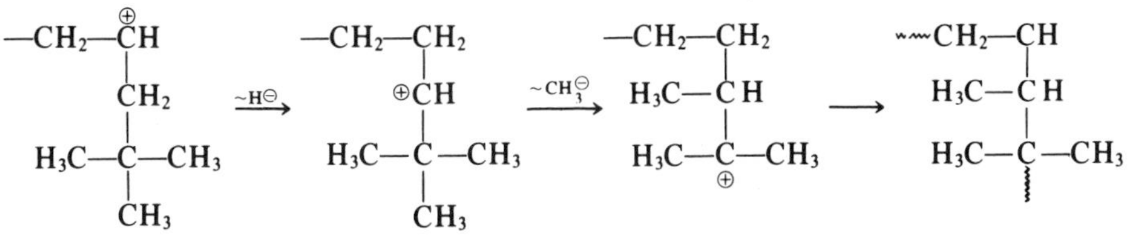

$$
\begin{array}{ccccc}
-CH_2-\overset{\oplus}{CH} & & -CH_2-CH_2 & & -CH_2-CH_2 \\
| & \xrightarrow{\ \sim H^{\ominus}\ } & | & \xrightarrow{\ \sim CH_3^{\ominus}\ } & | \\
CH_2 & & \overset{\oplus}{CH} & & H_3C-\overset{\displaystyle }{C}H \\
| & & | & & | \\
H_3C-\overset{\displaystyle CH_3}{\underset{\displaystyle |}{C}}-CH_3 & & H_3C-\overset{\displaystyle CH_3}{\underset{\displaystyle |}{C}}-CH_3 & & H_3C-\underset{\oplus}{C}-CH_3
\end{array}
$$

$$
\begin{array}{c}
\sim\!\sim\!CH_2-CH \\
| \\
H_3C-CH \\
| \\
H_3C-C-CH_3
\end{array}
$$

Finally one intramolecular chloride shift polymerization has also been described (Kennedy et al., 1966c; Kennedy and Squires, 1967l). Surprisingly, the polymerization of 3-chloro-1-butene in the presence of "H_2O"/$AlCl_3$ over the temperature range from -30 to $-130°C$ yields a copolymer composed of a mixture of about equal portion of 1,2 and 1,3 chloride shift units:

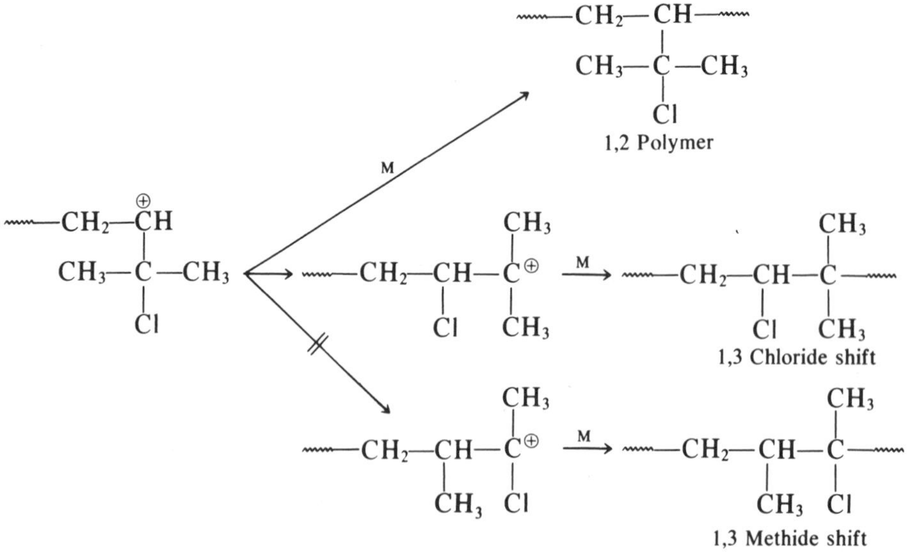

$$\text{\sim\sim}CH_2-CH-\text{\sim\sim}$$
$$CH_3-\underset{|}{C}-CH_3$$
$$Cl$$
1,2 Polymer

$$\overset{\oplus}{\text{\sim\sim}CH_2-CH}$$
$$CH_3-\underset{|}{C}-CH_3 \quad \longleftrightarrow \quad \text{\sim\sim}CH_2-CH-\overset{\oplus}{C} \xrightarrow{M} \text{\sim\sim}CH_2-CH-C\text{\sim\sim}$$
$$Cl \qquad\qquad\qquad Cl \quad CH_3 \qquad\qquad Cl \quad CH_3$$
1,3 Chloride shift

$$\text{\sim\sim}CH_2-CH-\overset{\oplus}{C} \xrightarrow{M} \text{\sim\sim}CH_2-CH-C\text{\sim\sim}$$
$$CH_3 \quad Cl \qquad\qquad CH_3 \quad Cl$$
1,3 Methide shift

No evidence for the presence of 1,3 methide shift unit was found. The overall structure was only slightly affected by changing the temperature in the range from -30 to $-130°C$.

Controversial Ill-Supported Claims in the Field of Isomerization Polymerizations

The literature of isomerization polymerizations contains a number of controversial reports, some of which have already been refuted and some of which remain unchallenged and are in need of confirmation.

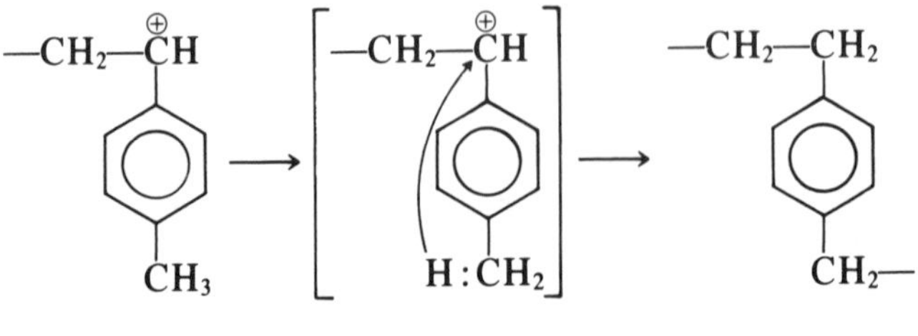

Claims by Krentsel et al. (Prishepa et al., 1956) and Disselhof and Braun, 1964) that p-methylstyrene contains "anomalous structures" which arise by a nebulous hydrogen migration, have been laid to rest by Kennedy et al. (1971b). Nor do hydride jumps occur during the cationic polymerization of o-methylstyrene (Magagnini et al., 1971b). Korshak et al.'s mechanism (1963) concerning the poly-merization of p-methylphenyldiazomethane by $BF_3 \cdot Et_2O$ has been strongly criticized (Kennedy, 1967i)

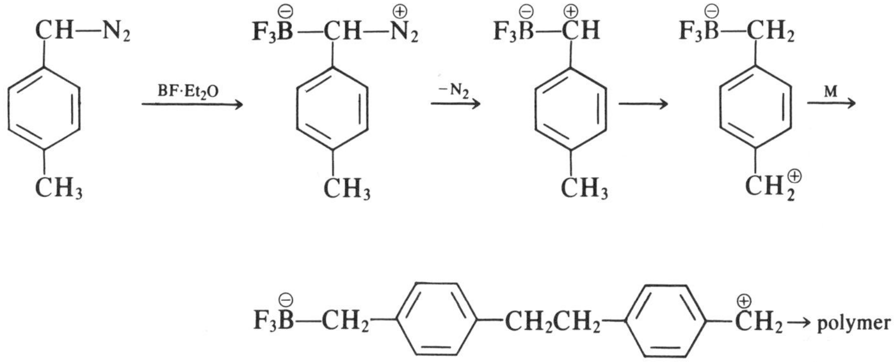

and still remains unconfirmed. The claim (Disselhof and Braun, 1964) that the growing p-isopropylstyrene carbenium ion undergoes rearrangement and yields p-substituted repeat units,

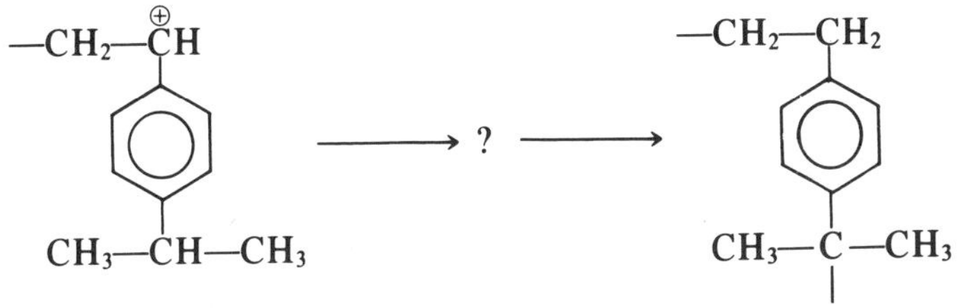

has been reinvestigated and refuted (Kennedy, 1967i; Magagnini and Plesch, 1971a). Kennedy's· suggestion (1967i) that the structure of o-isopropylstyrene might be more conducive for hydride migration,

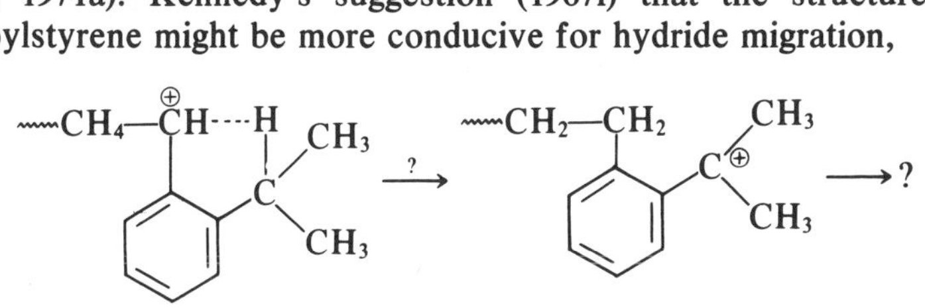

was followed up by Aso et al. (1969), who believed they had found supporting evidence for such a trans-spatial hydride jump. Close scrutiny of the theory revealed it to be seriously in error (Kennedy et al., 1971b), and direct experimental proof was obtained according to which the struc-ture of poly(o-isopropylstyrene) is conventional 1,2 enchainment (Ken-nedy et al., 1971b).

Controversial claims concerning anomalous structures of polyisobutylene and poly(isopropyl vinyl ether) have been strongly criticized (Kennedy, 1967i, 1975b) and remain unconfirmed to date.

Stereochemistry of Propagation

Although they are overshadowed by the glamour of stereoregular polymerizations induced by Ziegler–Natta coordination catalysts, usually but by no means always propagating by anionic coordinated mechanism, a considerable amount of research has also been committed to cationic stereoregular polymerizations. It is a matter of record that the very first synthetic stereoregular high polymer poly(isobutyl vinyl ether) was prepared by carbocationic polymerization using $BF_3 \cdot OEt_2$ at $-78°C$ (Schildknecht et al., 1948). Unfortunately the physical–mechanical properties of the isotactic poly(isobutyl vinyl ether) have not been attractive for commercial development and the field of stereoregular polymerizations lingered for about a decade until Ziegler and Natta's discovery of crystalline polyolefins by stereospecific catalysts.

This section concerns an examination of various experimental parameters (nature of monomer, temperature, etc.) on the stereochemistry of carbocationic propagation and in turn on polymer structure.

Vinyl Ethers

Numerous studies concern the stereochemistry of alkyl vinyl ether polymerizations (Higashimura et al., 1967b; Dombroski and Schuerch, 1971; Matsuzaki et al., 1967; Ohsumi et al., 1967a, b; Okamure et al., 1959; Goodman and Fam, 1964) and propenyl ethers (Higashimura et al., 1967b, 1968a, 1969; Higashimura and Hoshino, 1972; Higashimura and Hirokawa, 1977a; Dombroski and Schuerch, 1971; Ohsumi et al., 1967a, b, 1968; Matsuzaki et al., 1967; Okamura et al., 1959; Goodman and Fan, 1964). Considerable attention has been devoted to examine stereoregularity of polymers obtained exclusively by carbocationic techniques (Higashimura et al., 1967a, b, 1972; Dombroski and Schuerch, 1971; Matsuzaki et al., 1967; Ohsumi, 1968). Care should be exercised when studying the literature concerning the stereochemistry of carbocationic polymerizations because Higashimura et al. first employed Natta's definition for threo and erythro diisotactic dyads (Ohsumi et al., 1967a, b, 1968) and later switched to Bovey's notation Higashimura et al., 1969).

Influence of Monomer Geometry on Stereochemistry

The structures of polymers obtained from vinyl or propenyl ethers are largely determined by monomer geometry. According to Natta et al. (1960, 1962), trans alkenyl ethers and β-chloroalkyl vinyl ethers undergo cis

Table 4.12 □ Nature and Extent of Double Bond Opening of Methyl Propenyl Ether at Various Temperatures[a]

Polymerization Temperature (°C)	Cis Isomer		Trans Isomer	
	Cis Opening (%)	Trans Opening (%)	Cis Opening (%)	Trans Opening (%)
0	47	53	80	20
− 40	50	50	89	11
− 78	60	40	≃ 100	≃ 0

[a]After Ohsumi et al. (1967b).

opening of the double bond during cationic polymerization. Ohsumi et al. (1967a) studied diisotacticity of methyl propenyl ether and showed that polymer obtained by $BF_3 \cdot OEt_2$ from a monomer mixture rich in trans isomer is highly crystalline whereas that from a cis-rich monomer mixture is amorphous. Since cis alkenyl ethers are generally more reactive than trans isomers, polymerization of isomer mixtures should give cis-rich polymers and leave behind trans-rich monomer charges. The steric structure and amount of units in the polymer can be determined by analyzing the product and the residual monomers, and the relative contribution of cis or trans openings can be quantitatively obtained. Table 4.12 shows for methyl propenyl ether the nature and extent of double bond openings at various temperatures using $BF_3 \cdot OEt_2$ (Ohsumi et al., 1968).

According to these data and in conformity with Natta's analysis (Natta, 1960), the trans isomer undergoes cis opening. Conversely, however, with the cis isomer significant amounts of both cis and trans openings are observed even at − 78°C and amorphous polymer is obtained by using homogeneous initiating systems. Evidently the type of opening is strongly influenced by the nature of the geometric isomer.

Higashimura et al. (1969) also studied the polymerization of mixtures of cis and trans methyl propenyl ethers. Figure 4.6 shows the effect of methyl propenyl ether geometric structure on the steric structure of the polymer. The stereochemical structure of the polymer is a linear function of isomer composition in the charge, indicating that the stereochemistry of monomer addition is the same for growing chain ends derived from cis or trans isomers. Copolymerization experiments with cis and trans propenyl and butenyl ethers did not show differences in chain end reactivities derived from either isomer (Higashimura et al., 1968b, c).

Effect of the Nature and Concentration of Coinitiator and Solvent on Stereochemistry

Ohsumi et al. (1967a) studied the polymerization of methyl vinyl ether by various Friedel–Crafts acid-based initiating systems under various

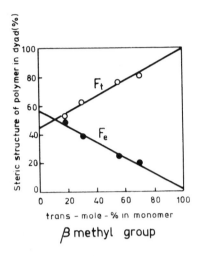
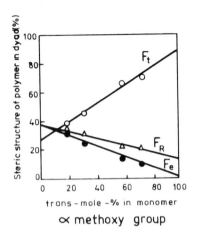

Figure 4.6 □ Effect of geometric structure of methyl propenyl ether on steric structures of the β-methyl and α-methoxy groups in the polymer (from Higashimura et al., 1969) (F_t, F_r, F_e refer to configurations relative to the β carbon according to Bovey (Fowells et al., 1967; Schuerch et al., 1964):

$$
\begin{array}{ccc}
\text{OCH}_3 & \text{H} & \text{OCH}_3 \\
| & | & | \\
\sim\text{C}\!\!-\!\!\!-\!\!\!-\!\!\!-\!\!\text{C}\!\!-\!\!\!-\!\!\text{C}\!\!-\!\!\!-\!\!\!-\!\!\sim \\
| & | & | \\
\text{H} & \text{CH}_3 & \text{H}
\end{array}
\qquad
\begin{array}{ccc}
\text{OCH}_3 & \text{H} & \text{H} \\
| & | & | \\
\sim\text{C}\!\!-\!\!\!-\!\!\!-\!\!\text{C}\!\!-\!\!\!-\!\!\text{C}\!\!\sim \\
| & | & | \\
\text{H} & \text{CH}_3 & \text{OCH}_3
\end{array}
\qquad
\begin{array}{ccc}
\text{H} & \text{H} & \text{H} \\
| & | & | \\
\sim\text{C}\!\!-\!\!\!-\!\!\!-\!\!\text{C}\!\!-\!\!\!-\!\!\text{C}\!\!\sim \\
| & | & | \\
\text{OCH}_3 & \text{CH}_3 & \text{OCH}_3
\end{array}
$$

$$\quad F_t \qquad\qquad\qquad F_r \qquad\qquad\qquad F_e$$

Threo–meso   Racemic   Erythro-meso

conditions. Figure 4.7 shows the triad tacticity of poly(methyl vinyl ether) obtained in the presence of $BF_3 \cdot OEt_2$ as a function of solvent composition (Ohsumi et al., 1967a). Similar results have also been obtained with the $CCl_3COOH \cdot SnCl_4$ initiating system. According to the findings the isotactic component decreases with the polarity (amount of CH_2Cl_2) of the solvent (Ohsumi et al., 1967a). These results confirm earlier observations by the senior author (Higashimura et al., 1968c) that isotactic poly(alkyl vinyl ethers) are preferentially obtained in homogeneous systems using solvents of low dielectric constant. The same observation has been made by Matsuzaki et al. (1967) with α-methylvinyl isobutyl and methyl ether. For example, isotactic poly(α-methylvinyl isobutyl ether) is obtained with $BF_3 \cdot OEt_2$ or Et_2AlCl in toluene and the addition of polar solvent causes an increase of syndiotactic material. In contrast, Higashimura et al. (1969) found that for methyl propenyl ether the steric structure of the polymer is almost independent of solvent polarity.

The polymerization of alkenyl propenyl ethers has similarly been studied using solvents of various polarities and homogeneous (e.g., $BF_3 \cdot OEt_2$) or

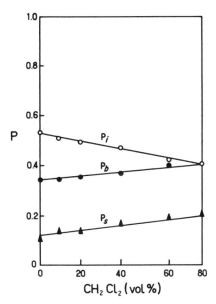

Figure 4.7 □ Tacticity of poly(methyl vinyl ether) as a function of solvent composition ([M] = 20 vol %, [BF$_3$·OEt$_2$] = 1 M, solvent = toluene–methylene chloride, − 78°C. P$_i$, P$_h$, P$_s$ = isotactic, heterotactic, syndiotactic triads (from Ohsumi et al., 1967a).

heterogeneous (e.g., Al$_2$(SO$_4$)$_3$·H$_2$SO$_4$ complexes) initiating systems (Higashimura et al., 1968a). Except with the *tert*-butyl propenyl ether, crystalline polymers were obtained only with trans isomers. These results are in agreement with those published by Natta (1960). Crystalline polymers were, however, produced in both polar (CH$_2$Cl$_2$) and nonpolar (C$_6$H$_5$CH$_3$) media and the degrees of crystallinities were about equal. In this regard the behavior of propenyl ethers and vinyl ethers is different: whereas the former give crystalline material in polar and nonpolar media, the latter produce crystalline polymer only in nonpolar solvent. The behavior of *tert*-butyl propenyl ether represents an exception; this monomer leads to highly crystalline polymer in toluene which gives rise to amorphous material in methylene chloride.

In contrast to these results, the cis isomer is converted to highly crystalline polymers whereas the trans isomer produces only oily products using the heterogeneous Al$_2$(SO$_4$)$_3$·H$_2$SO$_4$ complex initiating system.

One cannot generalize the effect of system heterogeneity or homogeneity on stereoregularity of polymerization on account of Okamura et al.'s results (Okamura et al., 1959; Ohsumi et al., 1968a). These authors studied the effect of experimental conditions on stereoregularity of isobutyl vinyl ether polymers (Okamura et al., 1959) and found at low monomer concentration larger amounts of isotactic product under homogeneous than heterogeneous conditions, although the rate of the homogeneous polymerization was rapid.

Effect of Temperature on Stereochemistry

Because system polarity affects polymer stereochemistry, it is rather difficult to assess uniquely the effect of temperature on stereochemistry of cationic polymerizations because by changing the temperature the polarity of the medium also changes. Unfortunately for this analysis many authors used polar solvents at various temperatures in their investigations.

Ohsumi et al. (1967a) studied the influence of temperature on stereoregularity of poly(methyl vinyl ether) using methylene chloride and toluene/n-hexane (1:1) mixtures. In both solvents the isotactic fraction decreased whereas the syndio-and heterotactic fractions increased with increasing temperatures. Using Bovey's definition (Fowells et al., 1967; Schuerch et al., 1964) of steric structures, Higashimura et al. (1969) studied the effect of temperature on erythro-meso and threo-meso fractions of poly(methyl propenyl ether); the results are shown in Figure 4.8. Similarly to earlier results (Ohsumi et al., 1968), the erythro-meso form increases at both α-and β-carbons obtained with cis-rich monomer charges by lowering the temperature. In contrast, the steric structure of polymers obtained with equimolar cis/trans monomer charges is independent of temperature, con-

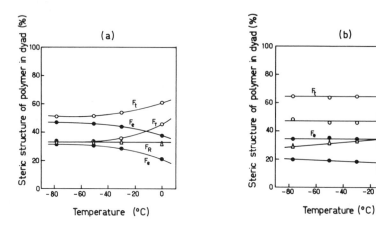

Figure 4.8 □ Effect of temperature on steric structure of β-methyl and α-methoxy groups in poly(methyl propenyl ether). ([M_0] = 10 vol %, [$BF_3 \cdot OEt_2$] = 2 mM, toluene solvent. (a) cis/trans in monomer 0.784/0.216, (b) cis/trans in monomer 0.490/0.510, F_t, F_e, F_r as in Figure 4.6. F_T, F_R, F_E according to Bovey (Fowells et al., 1967; Schuerch et al., 1964) are configurations relative to the α-carbon, respectively: threo-meso, racemic, erythro-meso. (Higashimura, 1969).

ceivably because the threo-meso form from the trans isomer increases at lower temperatures.

The Penultimate Effect

According to Ohsumi et al. (1967a), in some cases the stereochemistry of addition of an incoming monomer is determined not only by the ultimate unit of the growing chain but also by the penultimate unit. These authors studied the steric course of methyl vinyl ether (MVE) polymerization and calculated differences of activation enthalpies and entropies for syndiotactic and isotactic additions independent of the steric structure of the end unit by the following equation:

$$\ln\left(\frac{P_s}{P_i}\right) = \frac{-(\Delta H_s^{\ddagger} - \Delta H_i^{\ddagger})}{RT} + \frac{(\Delta S_s^{\ddagger} - \Delta S_i^{\ddagger})}{R}$$

where P_i and P_s are dyad isotacticity and syndiotacticity, respectively. Table 4.13 shows representative values.

$\Delta H_s^{\ddagger} - \Delta H_i^{\ddagger}$ values for MVE are constant irrespective of the nature of solvent. Values derived from NMR measurements were found to be smaller than those estimated from softening points (Kodama et al., 1961). Absolute values of enthalpy differences were not much different from those obtained in the radical polymerization of vinyl chloride (Fordham et al., 1959) and methyl methacrylate (Bovey, 1960) except that the sign of $\Delta H_s - \Delta H_i$ was negative.

Table 4.13 □ **Activation Enthalpy and Entropy Differences of Dyad Addition**[a]

Monomer	Polymerization Conditions	$\Delta H_s^{\ddagger} - \Delta H_i^{\ddagger}$ (cal/mole)	$\Delta S_s^{\ddagger} - \Delta S_i^{\ddagger}$ (e.u.)
MVE	BF$_3$·OEt$_2$ n-hexane—toluene (50/50, v/v)	450 ± 20	0.0 ± 0.05
MVE	BF$_3$·OEt$_2$; toluene	450 ± 20	0.6 ± 0.05
MVE	BF$_3$·OEt$_2$; methylene chloride	450 ± 20	0.8 ± 0.05
Vinyl chloride[b]	Radical polymerization	-600	—
Methyl methacrylate[c]	Radical polymerization	-775 ± 75	0.0 ± 0

[a] After Ohsumi et al. (1967a).
[b] Fordham et al. (1959).
[c] Bovey (1960).

Let P_{ii} be the probability of l addition of a monomer to an ll growing chain end of a d addition of a monomer to a dd growing chain end. In the same way P_{si} is l addition to a dl growing chain end or d addition to a growing ld chain end.

The coefficient $\alpha = P_{si}/P_{ii}$ was shown to be (Ohsumi et al., 1965) a good measure for a penultimate effect. Because $\alpha > 1$, the stereochemistry of the last monomer unit appears to be determined not only by the ultimate unit in the chain but also by other factors including the penultimate unit. The coefficient α increased slightly with solvent polarity. Chujo's method (1966) was used to ascertain the existence of the penultimate effect; that is, the penultimate effect on the energy differences between tactic placements of the last monomer unit in the chain (see above) was calculated by the following equations:

$$\ln\left(\frac{P_{is}}{P_{ii}}\right) = -\frac{\Delta H_{is}^{\ddagger} - \Delta H_{ii}^{\ddagger}}{RT} + \frac{\Delta S_{is}^{\ddagger} - \Delta S_{ii}^{\ddagger}}{R}$$

$$\ln\left(\frac{P_{ss}}{P_{si}}\right) = -\frac{\Delta H_{ss}^{\ddagger} - \Delta H_{si}^{\ddagger}}{RT} + \frac{\Delta S_{ss}^{\ddagger} - \Delta S_{si}^{\ddagger}}{R}$$

The results of calculations are shown in Table 4.14.

The monomer adds preferentially in an isotactic fashion in both solvents irrespective of the penultimate unit. Isotactic addition to an isotactic–isotactic end unit seems to be energetically most favorable.

Stereoselective Polymerization of Racemic Monomer Mixture

Chiellini (1970) has shown that the cationic copolymerization of racemic 1-methylpropyl vinyl ether with (R)- or (S)-1-phenylethyl vinyl ether is stereoselective in the presence of the heterogeneous initiating system $Al(O\text{-}i\text{-}Pr)_3 \cdot H_2SO_4$.

Stereoselective polymerization of a racemic monomer mixture in the presence of an asymmetric cationic initiating system was reported for the first time by Higashimura and Hirokawa (1977a). These authors studied the

Table 4.14 □ **Activation Enthalpy and Entropy of Triad Addition**[a]

Polymerization Conditions	$\Delta H_{ii}^{\neq} - \Delta H_{is}^{\neq}$ (cal/mole)	$\Delta H_{ss}^{\neq} - \Delta H_{si}^{\neq}$ (cal/mole)	$\Delta S_{ii}^{\neq} - \Delta S_{is}^{\neq}$ (e.u.)	$\Delta S_{ss}^{\neq} - \Delta S_{si}^{\neq}$ (e.u.)
$BF_3 \cdot OEt_2$; toluene	-430 ± 20	60 ± 20	0.0 ± 0.05	0.0 ± 0.05
$BF_3 \cdot OEt_2$; methylene chloride	-550 ± 20	240 ± 20	0.9 ± 0.05	0.9 ± 0.05

[a] After Ohsumi et al. (1967a).

polymerization of racemic cis and trans 1-methylpropyl propenyl ethers and of racemic 1-methylpropyl vinyl ether by asymmetric alkoxyaluminum dichlorides. Stereoselectivity has been observed only with the racemic cis-1-methylpropyl propenyl ether using (−) menthoxyaluminum dichloride in toluene at − 78°C. The polymer showed positive optical rotation and the residual monomers were converted by $BF_3 \cdot OEt_2$ to a polymer having negative optical rotation. These observations lead to several conclusions. (1) Only monomers with bulky side groups give optically active polymer and only one of the geometric isomers (cis) leads to this result. (2) The optical activity of the resulting polymer shows that the S monomer antipode is polymerized faster than the R antipode by (−)menthoxy-aluminum dichloride. (3) The absence of stereoselectivity when the same monomer is used with (S)-1-methylpropoxyaluminum dichloride or (S)-2-methylbutoxyaluminum dichloride initiator indicates involvement by the counteranion in stereoselection. (4) The counterion from menthoxyalu-minum dichloride is probably bulkier than those from the other coinitiators.

α-Methylstyrene

Stereoregularity of poly(α-methylstyrene) can readily be analyzed by high-resolution NMR spectroscopy (Brownstein et al., 1961; Sakurada et al., 1963; Braun et al., 1964; Ohsumi et al., 1965) since the α-methyl protons are sensitive to stereochemical configuration. Ohsumi et al. (1966) in-vestigated the stereospecific polymerization of α-methylstyrene obtained by Friedel–Crafts acid coinitiators under a variety of conditions and concluded that the physical state of the system (i.e., homogeneity or heterogeneity) and temperature determine polymer stereoregularity. Highly isotactic polymer is obtained in homogeneous systems (good solvent) and at low temperature.

In strong contrast with results obtained with alkyl vinyl ethers, solvent polarity and coinitiator nature affect the steric structure of poly(α-methyl-styrene) only slightly.

According to Ohsumi et al. (1966) in homogeneous systems the inter-action between the counterion, the substituents of the growing end, and those of the attacking monomer determine polymer stereoregularity. In homogeneous systems the monomer attacks the growing end from the least hindered direction. α-Methylstyrene carries two α-substituents so that, particularly in good solvents, stereoregularity is determined mainly by repulsive interactions between the α-methyl groups of the growing ion and incoming monomer. In poor solvents atactic product is formed because the polymer is forced out of solution and the incoming monomer cannot attack from the least hindered direction owing to interference by chain segments in the vicinity of the growing site. Isotactic polymer is formed pref-erentially when a small amount of strongly solvating polar solvent (nitro-benzene) is added to a poor solvent (n-hexane). Evidently the polar solvent which solvates the growing ion impedes free entry of incoming monomer and thus enhances polymer stereoregularity.

With regard to the effect of temperature, increased stereoregularity of poly(α-methylstyrene) at lower temperatures was attributed to increased interaction of α-substituents.

Kunitake and Aso (1970) studied stereoregularity of poly(α-methylstyrene) obtained by Friedel–Crafts coinitiators in toluene–methylcyclohexane mixtures. In contrast to Ohsumi et al.'s findings (1966), these authors maintain that in some cases polymer stereoregularity is affected by temperature, solvent, and coinitiator even in homogeneous systems. Thus the 97% syndiotactic content of polymer obtained with AlCl$_3$ decreases to 86% in the presence of SnCl$_4$.

This result, together with the observation that the syndiotactic content gradually decreases with increase of the methylcyclohexane content in the solvent, would indicate that steric structure is influenced by the nature of solvation rather than by the physical state of the system.

The nature of the coinitiator did not affect polymer stereoregularity in pure toluene; however, a noticeable effect was obtained by increasing the methylcyclohexane content in the medium. Apparently the effect of counteranion becomes more pronounced by decreasing solvation. The syndiotactic content increases along the following series: AlCl$_3$ > TiCl$_4$ ~ BF$_3$·OEt$_2$ > SnCl$_4$.

Heilbrunn and Maréchal (1973) obtained partially crystalline poly(α-methylstyrene) by the use of a Ziegler–Natta initiator system TiCl$_4$/Et$_3$Al and CH$_2$Cl$_2$ solvent and showed unambiguously that the polymerization mechanism was cationic in nature.

Stereochemical Mechanism of Propagation

Models for isotactic propagation in homogeneous systems have been developed mainly for vinyl ethers. Thus Cram and Kopecky (1959) and Bawn and Ledwith (1962) proposed the formation of a six-membered ring at the growing vinyl ether chain end. For example, Cram and Kopecky (1959) suggested the following scheme to explain the formation of isotactic poly(isobutyl vinyl ether) in the presence of BF$_3$ coinitiator:

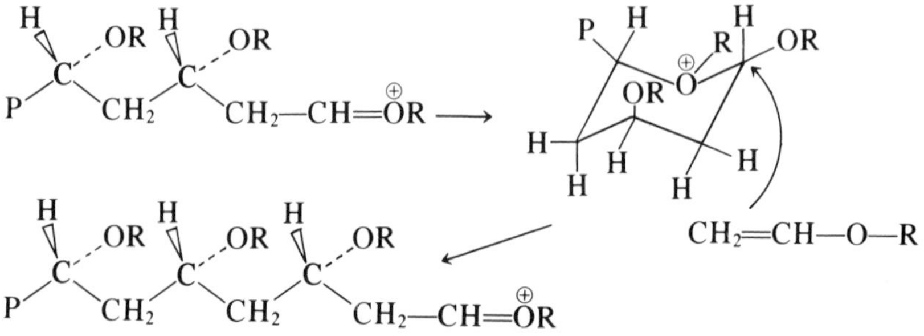

Substituents P (polymer chain) and OR (R = isobutyl) are visualized to occupy the least compressed equatorial positions in the transition state and the carbon carrying the OR group is inverted during ring opening. The relative configurations of the first two asymmetric centers formed determine the overall configurations of the chain. If the configurations are identical isotactic polymer is formed, whereas if they are opposite syndiotactic polymer arises.

This mechanism was adopted by Goodman and Fan (1964) to explain syndiotactic propagation in methyl propenyl ether polymerization. The authors propose a growing oxonium ion in the form of a pseudo six-membered ring formed by interaction of the terminal carbenium ion with the ether oxygen of the penultimate unit.

The overall configuration is determined by the relative configurations of the two asymmetric carbons in the penultimate and prepenultimate positions. If their configurations are identical the polymer chain becomes isotactic; if they are opposite (as the authors found) the polymer chain is syndiotactic.

Higashimura et al. (1959) visualize for the growing site an sp^3-hybridized carbenium ion and attribute hybridization to interaction with the counterion. Particularly in the absence of appreciable solvation hybridization of the growing carbenium ion might be due to deformation of the empty orbital as a result of electrostatic interaction with the counterion. In addition the substituents of the attacking monomer and the ultimate and penultimate monomer units of the polymer repel each other so that the conformation of substituents of the last two units in solution is nearly equal to that in a solid isotactic polymer. In such a "gauche" conformation repulsive interactions similar to that prevailing in 1,2-dichloroethane arise. The latter postulate is in agreement with statistical considerations relative to conformation of stereoregular polymers in solution. Figure 4.9 helps to visualize these postulates (Higashimura et al., 1959).

The vacant sp^3 hybrid orbital of the cation is nearly parallel to the plane of the incoming monomer so that overlap is maximum between the sp^3 atomic orbital of the growing carbenium ion and the p orbital of the β-carbon of the attacking monomer. The double bond of the attacking monomer is predominantly oriented in the C_1, C_2, and G (counterion) plane for two reasons. (1) This conformation is preferred in the presence of repulsive interactions between R_0 and R_1 and between R_0 and R_2. The C_1–G distance is about 2.2 Å with isobutylene and $BF_3OH^{\ominus}$ (Plesch, 1955), so that this conformation of the attacking monomer is geometrically possible. (2) The coulomb interaction between electric charges on C_1, G, C_α, and C_β requires G to be on the C_1, C_α, and C_β plane (Figure 4.9b).

According to Yonezawa et al.'s (1957) quantum chemical calculations the preferred head-to-tail (as contrasted to the head-to-head) configuration in isotactic poly(alkyl vinyl ether) may be due to a term akin to stabilization energy of conjugation.

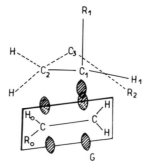

a. Most probable configuration of attacking monomer

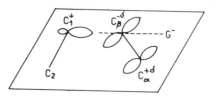

b. Orientation of attacking monomer modified by Coulomb interaction

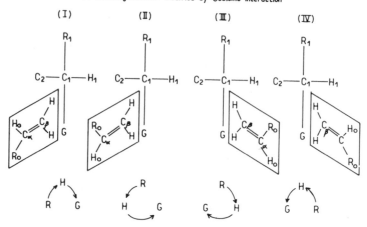

C. Four possible modes of configuration of attacking monomer. Relative positions
of R,H,and G at the end of the polymer after addition of monomer, viewing from C_α
to C_β direction: (I) clockwise; (II) counterclockwisse; III clockwise; (IV) counterclockwisse

Figure 4.9 ☐   Representation of postulates for propagation mechanism (Higashimura et al.,
1959). (Reprinted by permission of Hüthig and Wepf Verlag, Basel.)

    Kunitake and Aso (1970) criticized these mechanisms and suggested an
alternate route for stereoregular cationic propagation. According to these
authors the steric course of propagation is primarily determined by repulsion between the propagating chain, the counteranion, and the incoming
monomer as well as the direction of monomer attack, which in turn is

determined by the ionicity (tightness) of the growing ion pair. The growing carbenium ion is assumed to be sp^2-hybridized, leading to the conformation of least steric repulsion shown by structure **I** in Figure 4.10.

The substituents at the terminal and penultimate units are far apart and steric crowding is minimal. Molecular models indicate that this is especially the case for large substituents (e.g., *tert*-butoxy). The counterion is assumed to be at the side of the carbenium ion away from the penultimate unit. The stability of this conformation must be affected strongly by temperature and bulk of substituents. Stereoregular polymers, be they isotactic or syndiotactic, are expected to form only at low temperatures in homogeneous systems.

In polar solvents interaction between the growing cation and counteranion is weak or absent and the incoming monomer is expected to attack the carbenium ion from the least hindered side (front-side attack) leading to syndiotactic placement (**II** in Figure 4.10).

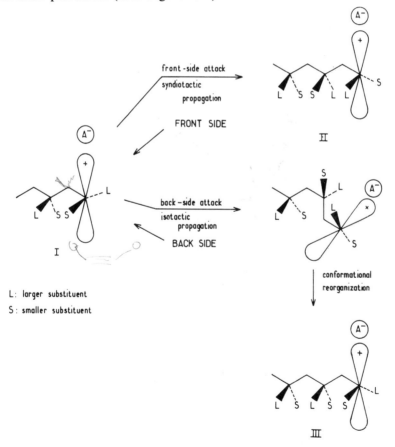

Figure 4.10 □ Stereochemical scheme of propagation in the homogeneous cationic polymerization of vinyl and related monomers (from Kunitake and Aso, 1970).

Although the incoming monomer approaches the carbenium ion so as to minimize steric repulsion between substituents of the terminal carbon and monomer, the direction of monomer approach does not determine stereoregularity since the terminal carbenium ion is capable of free rotation in vinyl monomers. The stereochemistry of the terminal carbon in **I** is locked in only after addition is complete.

In nonpolar solvents coulombic interaction between the carbenium ion and the counteranion is stronger than in polar solvents and the free energy of ion pair separation necessary for monomer insertion could reach several kilocalories per mole. The free energy of ion pair separation is expected to be compensated by the heat of polymerization (ΔH) more efficiently in the case of back-side attack than front-side attack, because ion pair separation and bond formation can occur simultaneously in the former case. Therefore with tight growing ion pairs the incoming monomer is expected to attack the carbenium ion from the back side, leading to isotactic placement (**III** in Figure 4.10). However, in the presence of large penultimate substituents steric hindrance to back-side attack increases and front-side attack occurs even in nonpolar media. Thus depending on the ionicity (tightness) of the growing ion pair and extent of steric hindrance between the substituents on the growing carbenium ion and incoming monomer, front-side or back-side monomer attack occurs, preferentially, which in turn determines isotactic or syndiotactic placement.

According to Kunitake and Aso (1970) interactions between counteranion and incoming monomer can be neglected because electron-rich monomers, for example, alkyl vinyl ethers and α-methylstyrene, would interact only weakly, if at all, with the anion.

This model has been extended to β-substituted vinyl monomers as well (Kunitake and Aso, 1970).

4.3 □ THE CHEMISTRY OF CHAIN TRANSFER

Introduction and Terminology

Chain transfer, or charge transfer as it is sometimes called, is the process in which the growing carbenium ion transfers its charge and with it its propagating ability to a monomer M either directly;

$$\sim C^{\oplus} + M \rightarrow [\text{transition states}] \rightarrow \sim C^* + HM^{\oplus}$$

or indirectly by the intervention of a nucleophilic transfer agent:

$$\sim C^{\oplus} + Nu \rightarrow [\text{transition states}] \xrightarrow{+M} \sim C^* + -M^{\oplus}$$

where $\sim C^{\oplus}$ is the growing carbenium ion, $\sim C^*$ is a nonpropagating olefinic or cyclic or –CH terminus, Nu is a charged or neutral nucleophile (transfer agent), and $-M^{\oplus}$ stands for the new cationated monomer chain carrier. Phenomenologically the most important consequence of chain transfer is the appearance of a new propagating unit the structure of which is identical to the original chain carrier, that is, $\sim C^{\oplus} = -M^{\oplus}$. In this sense all chain-transfer events may be viewed as chain transfer to monomer processes.

The processes of chain transfer and termination (see Section 4.4) are different and should be carefully differentiated. Whereas during chain transfer the kinetic chain survives and propagation continues, termination consists of irreversible loss of propagating ability. As a result of chain transfer the kinetic chain remains operational and unless the rate of cationation by the new chain carrier is markedly different from the original carbenium ion, the overall rate of polymerization should not change. Although the rate may be insensitive to chain transfer, molecular weights are greatly affected by it.

The elucidation and understanding of the chemistry of chain transfer in cationic polymerizations is of paramount importance for both molecular weight control and molecular engineering. In view of a large variety of efficient initiating systems (see Section 4.1) and usually favorable kinetics and thermodynamics of propagation, and in view of usually less important termination reactions, the only factor that stands between the experimentalist and high molecular weight products is chain transfer.

Chain transfer processes are the most complex and consequently least understood events in the course of carbocationic polymerizations. They involve a series of profound reorganization of bonds and groups not usually found in the other elementary events (initiation, propagation, termination). Our understanding as to the details of these reactions does not improve by disregarding the poorly understood secondary effects of solvation, counteranion structure, and aggregation phenomena, for example, and focusing only on the primary events.

Owing to some superficial similarities between chain transfer and termination, regrettably, even knowledgeable writers often use sloppy terminology and in spite of repeated warnings by thoughtful authors (Plesch 1955, 1968; Sawamoto et al., 1977) fail to distinguish between these two processes. Unfortunately, this confusion is evident even in polymer chemistry textbooks.

A process schematized by

$$\sim \overset{|}{\underset{\underset{H}{|}}{C}} - \overset{|}{\underset{|}{C}}{}^{\oplus} + \overset{|}{\underset{|}{C}} = \overset{|}{\underset{|}{C}} \rightarrow \sim \overset{|}{\underset{|}{C}} = \overset{|}{\underset{|}{C}} + H\overset{|}{\underset{|}{C}} - \overset{|}{\underset{|}{C}}{}^{\oplus}$$

should not be termed termination. Although the molecular weight is determined by such a chain transfer event, the kinetic chain is still operational and polymerization has not terminated. The case in which the newly formed carbenium ion $H\overset{|}{\underset{|}{C}}-\overset{|}{\underset{|}{C}}{}^{\oplus}$ is unable to propagate, say, proton transfer to a highly branched olefin leading to a sterically buried carbenium ion,

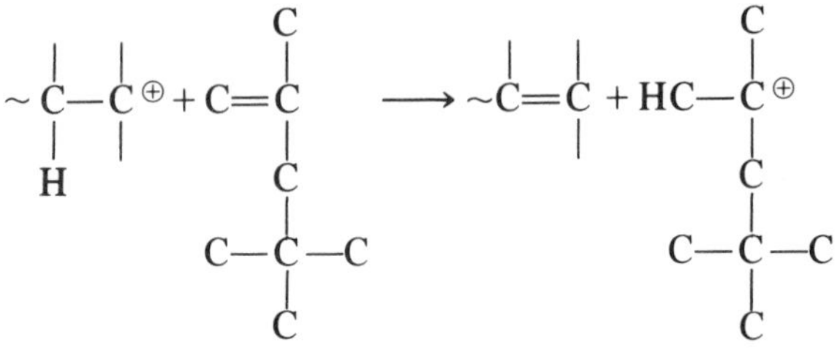

$$\sim \overset{|}{\underset{H}{C}}-\overset{|}{\underset{}{C}}{}^{\oplus}+ C{=}\overset{|}{\underset{\underset{\underset{|}{C}}{C{-}\overset{|}{\underset{}{C}}{-}C}}{C}} \longrightarrow \sim \overset{|}{C}{=}\overset{|}{\underset{}{C}} + HC{-}\overset{C}{\underset{\underset{\underset{|}{C}}{C{-}\overset{|}{\underset{}{C}}{-}C}}{\underset{|}{C}}}{}^{\oplus}$$

Sterically
inaccessible
carbenium ion

is a challenge to the terminologist. However, it still should not be termed termination because the propagating ability has not been lost; the carbenium ion is operationally still active.

Our insistence on accurate terminology is motivated not only by pedagogy but more importantly by conviction that careful differentiation and separation between the chemistries of termination and chain transfer processes is necessary for developing concepts of macromolecular engineering, that is, tailoring sequential copolymers, telechelic polymers, and functional polymers.

The term "chain breaking" embraces both chain transfer and termination processes. This term is well chosen (Plesch, 1968) because it combines the two fundamental events that interrupt, though not necessarily terminate, the kinetic chain.

The chemistry of chain transfer can conveniently be subdivided by the nature of the nucleophile involved: chain transfer by nucleophiles carrying a net negative charge, that is, $Nu^{\ominus}$ = counteranion, those involving unshared electron pairs (Nu = RX), π electrons (Nu = olefins or aromatic compounds), and σ electrons (Nu = alkanes, polymers with aliphatic C–H bonds).

Chain Transfer Reactions

Reactions discussed in this section are arranged by the nucleophiles involved and are presented in the sequence of decreasing nucleophilicities, that is, counteranion > unshared electron pairs > π electron systems (olefins, aromatics) > σ electrons (*n*-bases).

Chain Transfer by Counteranion

In this type of chain transfer the counteranion may be regarded as assisting in transferring propagating ability, that is, proton, to the monomer. Schematically;

$$\sim C^{\oplus}G^{\ominus} \rightarrow [\sim C^{\delta\oplus} \cdots H^{\delta\oplus} \cdots G^{\ominus} \leftrightarrow \sim C^* + H^{\oplus}G^{\ominus}] \xrightarrow{+M} \sim C^* + HM^{\oplus}G^{\ominus}$$

where the asterisk indicates a terminal C atom. This reaction may be unimolecular or bimolecular. In the unimolecular process the rate-determining step is the (re)generation of a proton–counteranion pair, $\sim C^{\oplus}G^{\ominus} \rightarrow$ $\sim C^* + H^{\oplus}G^{\ominus}$, followed by rapid protonation of incoming monomer, whereas in the bimolecular process the slow step is direct proton transfer to monomer $\sim C^{\oplus} + M \rightarrow C^* + HM^{\oplus}$. The unimolecular process is usually termed chain transfer to the counteranion, and the bimolecular is often called spontaneous or direct proton transfer to monomer (see below). In practice the demarcation between these reactions is hazy and in the absence of specific kinetic information separation of these events is almost impossible. Since chain transfer reactions are classified by nucleophiles in this book, chain transfer by counteranion $(Nu = G^{\ominus})$ and that by monomer $(Nu = olefin)$ are in separate sections; however, owing to the great similarity between these processes these sections should be considered jointly.

Comparison between chain transfer to counteranion and monomer (discussed below) is instructive. In chain transfer to counteranion the counteranion "assists" proton transfer, whereas in chain transfer to monomer direct or spontaneous proton transfer is visualized to occur in the absence of counteranion. The corresponding transition states may be formulated as in Scheme 4.2.

Scheme 4.2 ☐ Possible Chain Transfer Mechanisms

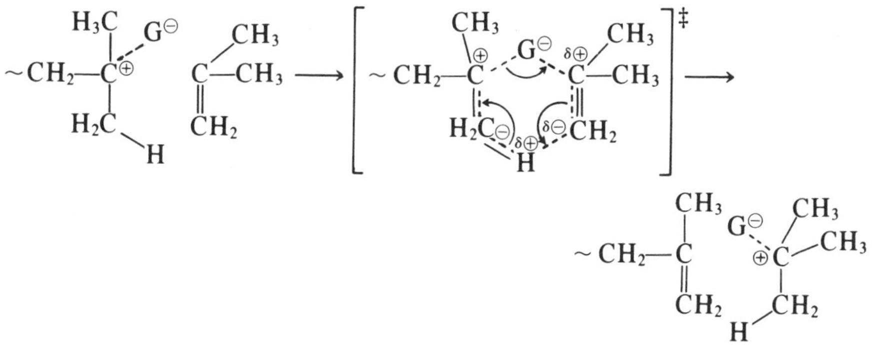

Counteranion assisted chain transfer: chain transfer by counteranion

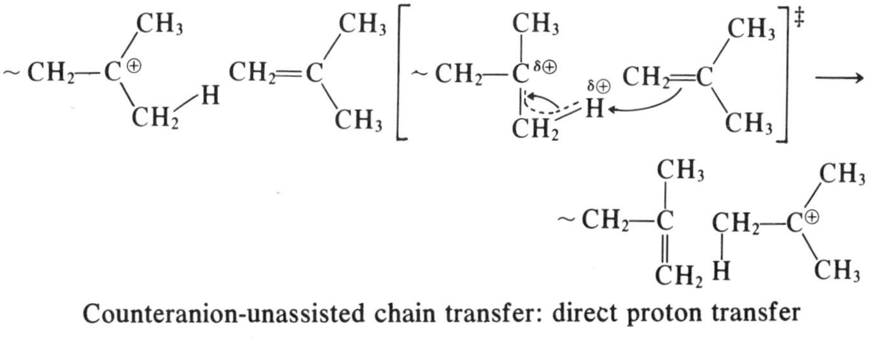

Counteranion-unassisted chain transfer: direct proton transfer

Chain transfer by counteranion may proceed with any cationically polymerizable monomer. In the case of alkenes the proton originates β to the propagating carbenium ion center and the nonpropagating end group $\sim C^*$ is olefinic. For example, in the $H_2O/BF_3/i\text{-}C_4H_8$ system,

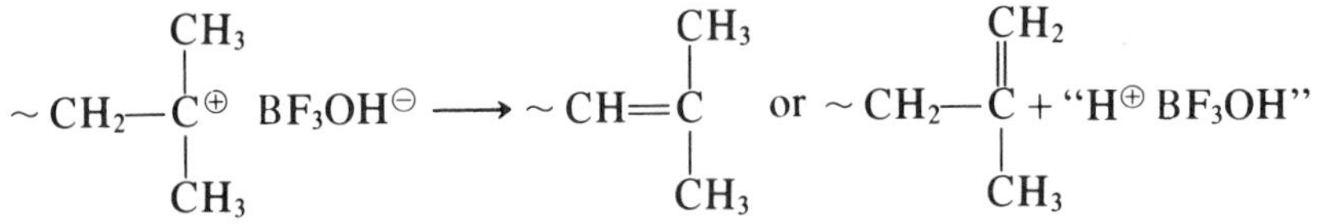

$$\sim CH_2-\overset{\overset{\displaystyle CH_3}{|}}{\underset{\underset{\displaystyle CH_3}{|}}{C}}{}^{\oplus}\ BF_3OH^{\ominus} \longrightarrow \sim CH=\overset{\overset{\displaystyle CH_3}{|}}{\underset{\underset{\displaystyle CH_3}{|}}{C}}\quad \text{or}\ \sim CH_2-\overset{\overset{\displaystyle CH_2}{\|}}{\underset{\underset{\displaystyle CH_3}{|}}{C}} + \text{``}H^{\oplus}\ BF_3OH\text{''}$$

(see discussion below). The quotation marks again indicate the hypothetical nature of the species. Direct evidence for the presence of such end groups has been obtained (Flett and Plesch, 1952, 1953; Mannatt et al., 1977; Dainton and Sutherland, 1949). With aromatic monomers, the proton may arise by β-elimination or intramolecular alkylation (see also Section on chain transfer by aromatic group, below); that is, $\sim C^*$ may indicate terminal unsaturation or cyclic, indane, unit. For example, with styrene, under nonpseudocationic conditions, an indane derivative is formed preferentially:

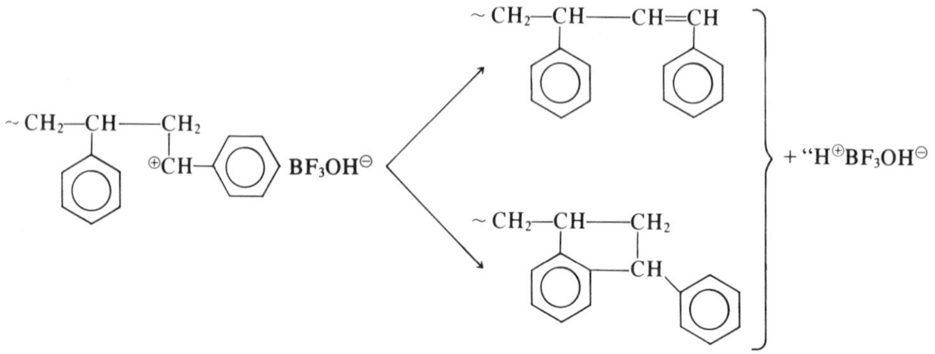

Evidence for both unsaturated and indane end groups has been presented (Pepper and Reilly, 1961, 1966). Simultaneously with the nonpropagating olefinic or indane terminus a protonating species also arises which may or may not be identical to the original initiating system employed. If (re)protonation of monomer is slow for some reason, chain transfer to counteranion may become kinetic termination (terminative chain transfer; see below).

Insight into details of competition between the paths leading to linear olefin terminus by β-hydrogen elimination or indane end group by self-alkylation may be obtained from studies concerning styrene dimerization. The dimerization of styrene has been investigated by numerous authors using sulfuric acid (Stoermer and Kootz, 1928; Rosen, 1953; Spoerri and Rosen, 1950; Calder et al., 1969; Corson et al., 1962), phosphoric acid

(Corson et al., 1954), and silica (Corson et al., 1954), and the products were always mixtures of a linear unsaturated dimer 1,3-diphenyl-1-butene (**I**) and a cyclic dimer 1-methyl-3-phenylindane (**II**) plus oligomer:

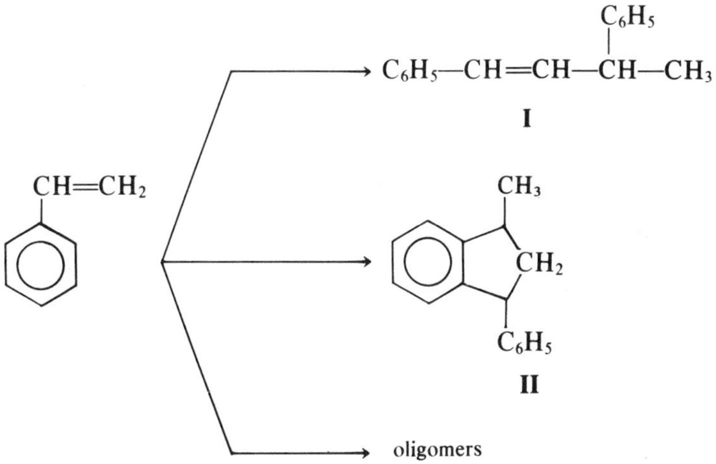

According to Higashimura et al. (Higashimura and Nishii, 1977c; Sawamoto et al., 1975) CH_3COClO_4 or CF_3SO_3H produce only **I** whereas Friedel–Crafts acid-based initiators in benzene give rise only to polymers or oligomers. Selective formation of linear dimer requires the presence of a multidentate anion, for example, $ClO_4^{\ominus}$ (see also Section 4.1).

This argument is in line with the fact that in solvolytic eliminations in poorly solvating media the leaving group is determined by the counteranion (Cram and Sahyum, 1963; Cocivera and Winstein, 1963; Skell and Hall, 1963).

Monomer transfer involves complicated reorganization of bonds, atoms, or groups of atoms and as such requires considerable activation energy. Owing to this circumstance, cooling to low temperatures often leads to high molecular weight products. Indeed, the reciprocal effect of temperature on product molecular weight is one of the very few general phenomena, seldom violated, in cationic polymerizations.

The temperature coefficient of molecular weights has been shown to hold important clues for the polymerization mechanism. The effect of temperature on molecular weights is usually expressed by Arrhenius plots, that is, plots of log $\bar{M}$ versus $1/T$. Flory (1953) was first to apply Arrhenius plots to molecular weights published by Thomas et al. (1940) and since that time many authors have employed such plots to describe the effect of temperature on molecular weights. Slopes of Arrhenius plots yield $\Delta H_{\bar{M}}$, the "activation enthalpy of molecular weight," a complex composite quantity from the fundamental point of view but one with great diagnostic significance. The activation enthalpies of propagation, transfers, termination, and their relation to each other determine $\Delta H_{\bar{M}}$ values. A linear

Arrhenius plot over a temperature range suggests that the contribution of individual activation enthalpy differences of elementary processes remains constant and the molecular weight-determining mechanism does not change in this range. If and when the relative contributions change with temperature, the linear plot exhibits curvature or a slope change. According to an exhaustive analysis of published data (see Table 5.4), Arrhenius plots for polyisobutylenes, irrespective of conditions employed for synthesis, exhibit only two slopes and a transition region (a representative plot is shown in Figure 4.11 (Kennedy and Thomas, 1962) and the higher slope corresponding to $\Delta H_{\bar{M}} = -6.6 \pm 1.0$ kcal/mole (-27.6 ± 4.2 kJ/mole) is diagnostic for temperature regimes over which molecular weight control is by chain transfer,

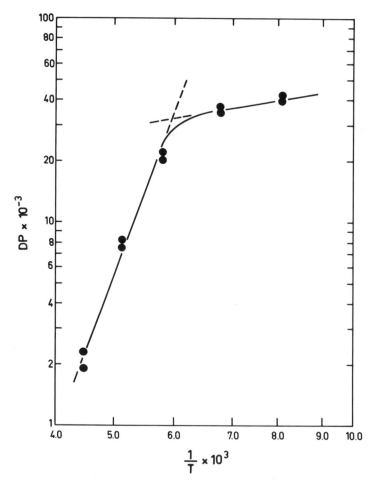

Figure 4.11 □ Temperature dependence of DP of polyisobutylene in propane solvent (Kennedy and Thomas, 1962).

probably by counteranion. Further discussion concerning the diagnostic significance of $\Delta H_{\bar{M}}$ values can be found in Sections 4.4 and 5.5 and in the detailed analysis by Kennedy and Trivedi (1978d).

Important insight into the mechanism of cationic polymerizations in general and chain transfer processes in particular has recently been obtained by the use of "proton traps" (Kennedy et al., unpublished results). Proton traps are sterically hindered, highly basic amines that are able specifically to react with proton but not with any other electrophilic species. Brown and Kanner (1953) and Brown (1956) have shown that, for example, 2,6-di-*tert*-butylpyridine DTBP can react with HCl but not with BF$_3$. Evidently, the two bulky *tert*-butyl

groups effectively block the approach of electrophiles except that of the proton. Even protons bonded by hydrogen bridges are reluctant to react with proton traps. Since Brown's and Kanner's pioneering research (1953) several highly basic proton traps have been explored (Alder et al., 1968; Olah et al., 1975; Stang and Anderson, 1978).

According to recent findings (Kennedy and Chou, 1979e), polymerization of α-methylstyrene by "H$_2$O"/BCl$_3$ in CH$_2$Cl$_2$ in the -20 to $-60°C$ range is characteristically affected by the presence of DTBP. Table 4.15 and Figure 4.12 show representative results. Under identical conditions except in the absence of a proton trap α-methylstyrene conversions were 100%, whereas in the presence of DTBP, conversions were much reduced (depending on temperature, 3.4 to 58.3%). Polymerizations were rapid and could not be stopped at low conversions in the absence of DTBP. Further, molecular weights of poly(α-methylstyrenes) obtained in control experiments were much lower than those produced in the presence of proton traps over the whole temperature range. Molecular weights in excess of 5×10^5 can readily be obtained at $\sim -60°C$. Significantly, the molecular weight distribution $\bar{M}_w/\bar{M}_n$ of poly(α-methylstyrenes) obtained in the absence of proton trap was in the range of 2.5 to 5.0, whereas those harvested in experiments with DTBP showed $\bar{M}_w/\bar{M}_n = 1.4$ to 1.8.

Conclusions of general significance are as follows. The low α-methylstyrene conversions in the presence of proton traps indicate the appearance and immediate trapping of protons somewhere along the reaction path. Evidently the kinetic chain cannot progress beyond the first chain transfer step because the chain carrier, a species akin to proton momentarily emerging during chain transfer, is immediately deactivated by the strongly basic proton sponge. The exact point at which "trapping" occurs is unknown. The

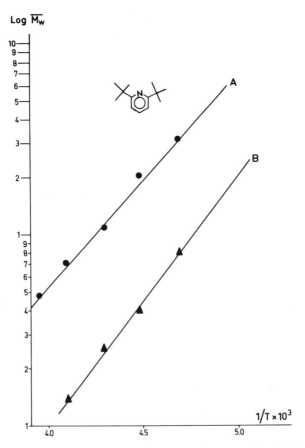

Figure 4.12 □ The effect of temperature on the molecular weight of poly(α-methylstyrene) obtained in the presence and absence of proton trap: A, with trap; B, without trap (Kennedy and Chow, 1979).

quaternized amine cannot sustain propagation and polymerization ceases. This process may be termed terminative chain transfer or terminative proton entrapment.

The observations outlined above raise a fundamental question: considering similarities between cationation and chain transfer to counteranion, say, with the "H_2O"/BCl_3/α-methylstyrene system,

$H^{\oplus}$ from initiation

$H^{\oplus}$ from chain-transfer
  by counteranion

$$\text{"}H^{\oplus}BCl_3OH^{\ominus}\text{"} + CH_2{=}\underset{\underset{\bigcirc}{|}}{\overset{\overset{CH_3}{|}}{C}} \longrightarrow HCH_2{-}\underset{\underset{\bigcirc}{|}}{\overset{\overset{CH_3}{|}}{C}}^{\oplus} \quad BCl_3OH^{\ominus}$$

Table 4.15 □ **The effect of a Proton Trap on α-Methylstyrene Polymerization**[a]

Sample	T (°C)	Conv. (%)	$\bar{M}_n$	$\bar{M}_w$	$\bar{M}_w/\bar{M}_n$	Yield/$\bar{M}_n$ (mole)	No. of Chain Transfers per Kinetic Chain
1	−20	100	2×10^3	5.2×10^3	2.7	9.1×10^{-4}	162
1[b]	−20	3.4	2.9×10^4	4.8×10^4	1.65	5.6×10^{-6}	
2	−30	100	4.0×10^3	1.4×10^4	3.5	4.3×10^{-4}	82
2*[b]	−30	12	4.4×10^4	7.1×10^4	1.00	5.2×10^{-6}	
3	−40	100	6.8×10^3	2.5×10^4	3.7	2.8×10^{-4}	40
3*[b]	−40	22.5	6.2×10^4	1.1×10^5	1.78	7×10^{-6}	
4	−50	100	1.1×10^4	4.1×10^4	3.9	1.7×10^{-4}	27
4*[b]	−50	42	1.3×10^5	2.1×10^5	1.6	6.2×10^{-6}	
5	−60	100	2.1×10^4	7.8×10^4	3.6	9×10^{-5}	15
5*[b]	−60	58.3	1.9×10^5	3.1×10^5	1.6	6.0×10^{-6}	

[a] $[M] = 0.62\ M$, $[BCl_3] = 1 \times 10^{-2}\ M$, CH_2Cl_2 solvent.
[b] $[DTBP] = 2.6 \times 10^{-2}\ M$.

how is the proton trap able to trap the proton arising from chain transfer but not the one arising by initiation? Or, formulated in a different way, how is it that initiation can occur but chain transfer cannot go to completion in the presence of a proton trap? One conclusion is to assume the complete absence of free protons during initiation (otherwise the DTBP would trap them and polymerization would not occur) and the presence of trappable protons during chain transfer.

An interesting application of proton traps is to increase grafting efficiencies. As discussed in Chapter 8, many interesting carbocationic graftings have been carried out by exploiting the principle of controlled initiation, however, the possibility of homopolymer formation exists if chain transfer to monomer is faster than termination. In the presence of proton traps terminative proton entrapment occurs and the formation of homopolymer is avoided. Thus recent experiments by Kennedy and Guhaniyogi (1980) have shown that the grafting efficiency of poly(α-methylstyrene) branches from polychloroprene and chlorinated butyl rubber backbones in the presence of BCl$_3$ and SnCl$_4$ coinitiators and DtBP can be increased close to 100%.

Research with hindered amines promises to be a rewarding new technique in the elucidation of cationic polymerization mechanisms.

Chain Transfer by Unshared Electron Pair

Chain transfer involving unshared electron pairs may proceed in the presence of, say, halogen-, oxygen-, or sulfur-containing compounds. With alkyl halides, often used as solvents in cationic polymerizations, this process is termed "solvent transfer." Schematically,

$$\sim C^\oplus + RX \rightleftharpoons [\overset{\delta\oplus}{C} \cdots \overset{\delta\oplus}{X} \cdots \overset{\delta\oplus}{R}] \xrightarrow{+M} \sim CX + RM^\oplus$$

where R = organic group (or H if X is not halogen) and X = usually halogen; for X = O or N, for example, the equation must be slightly recast to account for the multivalent nature of these atoms (see below). (The above symbolism is not intended to imply a necessarily linear transition state.)

A glance at this equation reveals the great similarity between this process and initiation with hydrogen halide or alkyl halide/Friedel–Crafts acid systems: in both processes a Lewis acid (carbenium ion or Friedel–Crafts acid) interacts with a hydrogen or alkyl halide. Thus it is not very surprising that the same hydrogen or alkyl halides can function as initiators and/or transfer agents. This circumstance can be exploited for molecular weight control. For example, in isobutylene polymerization addition of HCl or t-BuCl (Kennedy et al., 1967f) increases the rate of reaction and decreases molecular weights in proportion to the amount of halide added. It is thought that this process is employed in the production of polyisobutylene and butyl rubber.

Polystyrenes obtained by $TiCl_4$ coinitiator in $ClCH_2CH_2Cl$ solvent in the presence of EtBr, i-PrBr, and t-BuBr contain fragments of these materials (Plesch, 1953a). Since $ClCH_2CH_2Cl$ is not an initiator in conjunction with $TiCl_4$ (Longworth et al., 1960), fragments of these materials in the polymer are probably incorporated by solvent transfer. Similarly, there is evidence for chain transfer with t-BuCl and i-PrCl in styrene polymerization coinitiated by $SnCl_4$ in $ClCH_2CH_2Cl$ solution (Colclough and Dainton, 1958a, b). The fact that α-methylstyrene polymerization by Friedel–Crafts acids produces lower molecular weight products in ethyl chloride than in CS_2 may also be due to chain transfer to the halide solvent (Hersberger et al., 1945). Experiments with $^{14}CH_3Cl$ diluent in the "H_2O"/$AlCl_3$/isobutylene/CH_3Cl/ $-78°C$ system showed significant radiocarbon incorporation into the carefully purified polymer (Kennedy and Thomas, 1960a). The findings indicate methylation by $CH_3^{\oplus}AlCl_3^{\ominus}$ or chain transfer by $\sim C^{\oplus} + CH_3Cl \rightarrow \sim CCl + Me^{\oplus}$ or both.

The effect of HCl (Kennedy and Squires, 1967g) and numerous alkyl and allyl halides on isobutylene polymerization with "H_2O"/$AlCl_3$ has been investigated (Kennedy et al., 1967f). In agreement with expectations, HCl increases the yield (initiating action) but decreases the molecular weights (transfer agent). Differences and similarities between the effects of olefins and organic halides have been analyzed in terms of their poison and transfer coefficients (Kennedy et al., 1967f).

Although the mechanism of poisoning by olefins and allyl halides may have a common basis (hydride abstraction; see section on chain transfer by hydride transfer below), their chain transfer activity must be due to dissimilar processes, that is, cross-transfer with olefins and solvent transfer with the halides (Kennedy et al., 1967f).

That alkyl halides may simultaneously function as initiators *and* transfer agents has recently been used to great advantage for the synthesis of desirable biterminally functional—telechelic—polymers (Kennedy and Smith, 1979h). Thus α,ω-bifunctional polyisobutylenes, for example.

have been prepared by the use of bifunctional initiator–transfer agents, "inifers" (term coined by combination of *ini*tiator–trans*fer* agent).

The structure of this telechelic molecule has been carefully determined by a variety of analytical techniques (Kennedy and Smith, 1979g). The presence *and* position of terminal tertiary chlorines have been established by the synthesis of PαMeSt-b-PIB-b-PαMeSt (see also Sections 8.3 and 9.2).

Inifers X–R–X must fulfill two functions simultaneously: as bifunctional initiators they must initiate two-kinetic chains $^{\oplus}C$–R–$C^{\oplus}$ and as bifunctional transfer agents they must effect the following chain transfer sequence:

$$\sim C^{\oplus} + X{-}R{-}X \rightarrow \sim CX + X{-}R^{\oplus}$$

$$X{-}R^{\oplus} + M \rightarrow X{-}RM^{\oplus} \xrightarrow{+nM} X{-}R$$

The above α,ω-dichloropolyisobutylene has been obtained by carefully balancing the relative stability of the initiating–transferring cation with that of the propagating cation, specifically, by using the following system:

$$\text{Cl}(CH_3)_2{-}CC_6H_4C{-}(CH_3)_2Cl/BCl_3/\text{isobutylene}/CH_2Cl_2/n\text{-}C_5H_{12}/-50°C$$

The inifer, p-dicumylbenzene, initiates two kinetic chains by the cumyl cation (it is not necessary that the two initiations be simultaneous), and effects chain transfer by essentially the same mechanism. Termination is by chlorination (see Section 4.4) and leads to the same terminus as chain transfer by inifer in the particular system assembled. The reactions involved in the synthesis are summarized as follows (ClRØRCl = inifer, M = isobutylene; H atoms and counteranion, when obvious, omitted):

1 □ *Ion generation*

$$\text{ClRØRCl} + BCl_3 \rightleftharpoons \text{ClRØR}^{\oplus} BCl_4^{\ominus}$$

2 □ *Cationation and propagation*

$$\text{ClRØR}^{\oplus} \xrightarrow{+nM} \text{ClRØR} \sim\sim M^{\oplus}$$

3 □ *Chain transfer to inifer*

$$\text{ClRØR} \sim\sim M^{\oplus} + \text{ClRØRCl}$$

$$\text{ClRØR} \sim\sim MCl + \text{ClRØR}^{\oplus}$$

4 □ *Termination*

$$\sim\sim M^{\oplus} + BCl_4^{\ominus} \rightarrow \sim\sim MCl + BCl_3$$

The detailed mechanism is shown in Scheme 4.3. Thus p-dicumylchloride and BCl_3 yield the initiating ion **II**, which cationates isobutylene and produces the growing chain **III**. Chain **III** reacts with the inifer **I** and produces the "half-terminated" chain **IV** plus ion **II**. The formation, consumption, and re-formation of **II** represents an "inifer loop." Species **IV** contains a benzylic chlorine which in the presence of BCl_3 ionizes to **V** and polymerizes isobutylene **VI**. The growing chain reacts by chain transfer with **I** or **IV** to give the end product **VII** plus **II** or the end product plus **V**, respectively. The formation of species **II** and **V** is part of two additional inifer loops involving **I** and **IV**. The solid arrows indicate inifer loops.

Scheme 4.3 □ Synthesis of Biterminally Chlorinated PIB by Inifer Method

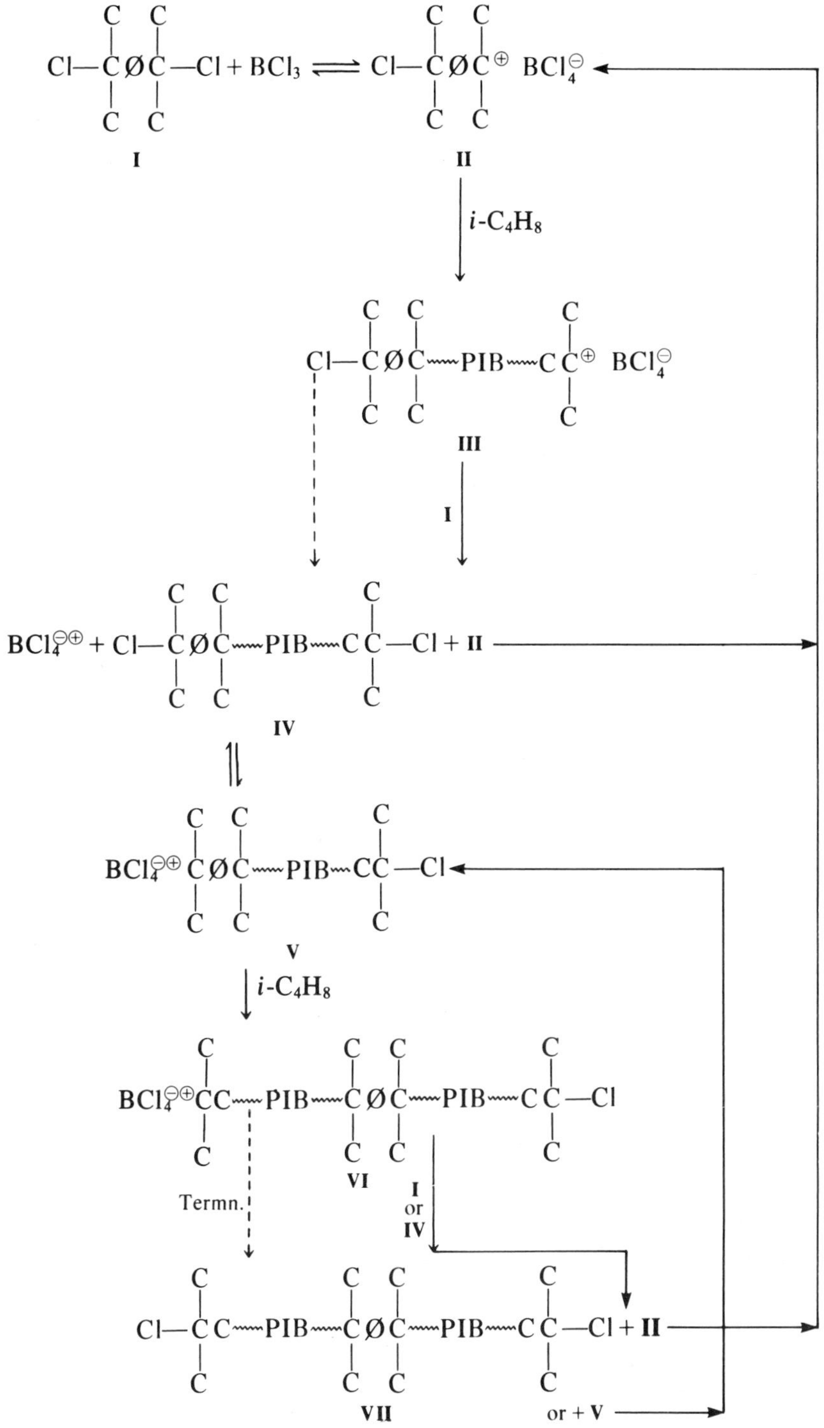

205

The two broken arrows indicate irreversible termination, that is, the collapse of the $PIB^{\oplus}BCl_4^{\ominus}$ ion pair. The frequency of these events is low compared to that of transfer to inifer.

In the final analysis all routes must lead to **VII**; that is, the only product is Cl–PIB–Cl.

In addition to the numerous solvent transfer reactions involving alkyl halides, many types of oxygen-containing compounds for example, CH_3OH, CH_3COOH, and $(CH_3CO)_2O$, have been found to function as chain transfer agents in styrene polymerizations coinitiated by BF_3, $SnCl_4$ (Higashimura and Okamura, 1956a, 1958). These chain transfer reactions may be formulated as substitutive transfer via tertiary oxonium ion:

$$\sim C^{\oplus} + R\text{---}O\text{---}R \rightarrow [\sim C\text{---}\overset{\oplus}{\underset{\underset{R}{|}}{O}}\text{---}R] \xrightarrow{+M} \sim C\text{---}OR + RM^{\oplus}$$

or proton transfer via secondary oxonium ion:

$$\sim C\text{---}C^{\oplus} + R\text{---}O\text{---}R \rightarrow \sim C\text{=\!=}C + R\text{---}\overset{\oplus}{\underset{\underset{H}{|}}{O}}\text{---}R \xrightarrow{+M} R\text{---}OH + RM^{\oplus}$$

or

$$R\text{---}O\text{---}R + HM^{\oplus}$$

It would be instructive to determine the nature of end groups.

Chain Transfer by π Electron Systems

This class of chain transfer processes includes two important subclasses determined by the nature of π systems involved, that is, olefins or aromatic compounds. Chain transfer to olefins, in the form of chain transfer to monomer or simply monomer transfer, is often molecular weight-determining and chain transfer to aromatic compounds is responsible for ring alkylation and branching in polystyrene.

Chain Transfer by Olefin

Schematically this process may be depicted as follows:

$$\sim C^{\oplus} + M \rightleftharpoons [\overset{\delta\oplus}{C} \cdots M^{\delta\oplus}] \rightarrow \sim C^* + HM^{\oplus}$$

where M = olefin, most commonly monomer, and $\sim C^*$ stands for terminal olefinic or indane group. The similarity between chain transfer to counteranion and direct chain transfer to monomer has been discussed above.

Chain transfer to counteranion is more likely to occur in nonpolar media (associated cation/counteranion pair) whereas chain transfer to monomer proceeds preferentially with free ions in polar solvents or necessarily in polymerizations induced by irradiation which occur in the absence of counteranions as such (see Section 4.1). At which point on the polarity scale the mechanism switches from preferentially unimolecular counterion transfer to bimolecular monomer transfer cannot be answered at present.

It is instructive to examine factors that determine the site of the leaving proton in monomer transfer. Because monomer transfer most likely involves El elimination, the leaving proton must be situated β to the carbenium center. For example, in the case of propagating polyisobutylene cations proton elimination occurs from the CH_3- or $-CH_2-$ groups to form internal or external terminal unsaturation (Manatt et al., 1977; Dainton and Sutherland, 1949; Flett and Plesch, 1953):

$$\sim CH_2-\underset{\underset{CH_3}{|}}{\overset{\overset{CH_3}{|}}{C}}\oplus \xrightarrow{-H\oplus} \sim CH_2-\underset{\underset{CH_3}{|}}{\overset{\overset{CH_2}{\|}}{C}} \ or \ \sim CH=\underset{\underset{CH_3}{|}}{\overset{\overset{CH_3}{|}}{C}}$$

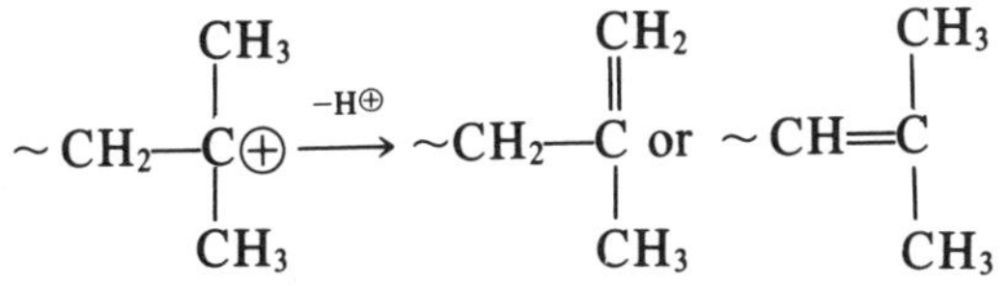

According to physical organic–chemical principles El eliminations usually favor the formation of internal olefins and the structures of the products usually follow Saytzeff's rule; that is, the olefin formed carries the largest possible number of substituents. This rule is clearly not followed by the polyisobutylene carbocation, for analytical evidence indicates the presence of significant quantities of external unsaturation in the polymer (Manatt et al., 1977; Dainton and Sutherland, 1949; Flett and Plesch, 1953). An explanation for preferential proton elimination from the CH_3 group could be formulated along Brown's suggestion (Brown, 1956), according to which in certain branched carbenium ions the transition state of proton elimination leading to the internal olefin would be unfavorable owing to unavoidable steric hindrance between the methyl and *tert*-butyl groups:

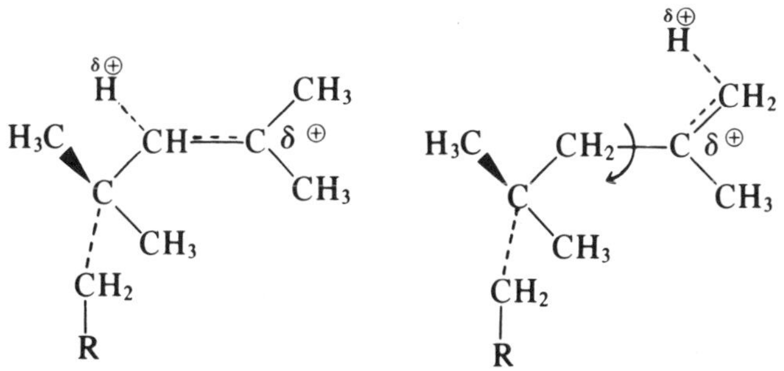

Restricted rotation,
unfavorable
transition state

Freely rotating, favorable
transition state

A detailed study concerning chain transfer by various olefins in isobutylene polymerization has been made (Kennedy et al., 1967a–h). Isobutylene polymerizations under set conditions ("H_2O"/$AlCl_3$/n-pentane/ $-78°C$) in the absence and presence of many olefinic compounds were conducted and the molecular weight-depressing effect of these materials was quantitatively correlated by so-called transfer coefficients (representative data are shown in Table 4.17; see Section 4.4). (The yield-depressing effect of olefins was attributed to termination by hydride transfer and expressed by poison coefficients; see Section 4.4). Molecular weight reduction of polyisobutylenes by olefins was explained by chain transfer by olefin, for example, with 2-octene:

$$\sim \overset{|}{\underset{|}{C}}{}^{\oplus} + CH_3{-}CH{=}CH{-}CH_2{-}CH_2{-}C_3H_7 \rightarrow$$

$$CH_3{-}\overset{\delta\oplus}{C}H{-}CH{-}\overset{\delta\oplus}{C}H{-}CH_2{-}C_3H_7 \xrightarrow{-H^{\oplus}+M}$$

$$CH_3{-}CH{=}CH{-}CH{=}CH{-}C_3H_7 + HM^{\oplus}$$

The molecular weight-reducing effect of 2-olefins is more pronounced than that of 1-olefins, which has been attributed to differences in the stabilities of the allyl cation intermediates and conjugated diolefin products (Kennedy and Squires, 1967b).

Chain transfer by highly branched olefins is particularly interesting. Experience has shown that 1,1-disubstituted olefins, for example, 2,4,4-trimethyl-1-pentene, cause a precipitous drop in polyisobutylene molecular weight (Thomas et al., 1940; Horrex and Perkins, 1949). This effect was explained by assuming that although this olefin can readily be incorporated in a growing polyisobutylene chain, the new cation is unable to sustain propagation because of steric hindrance; consequently proton elimination, that is, chain transfer, results (Kennedy and Squires, 1967c):

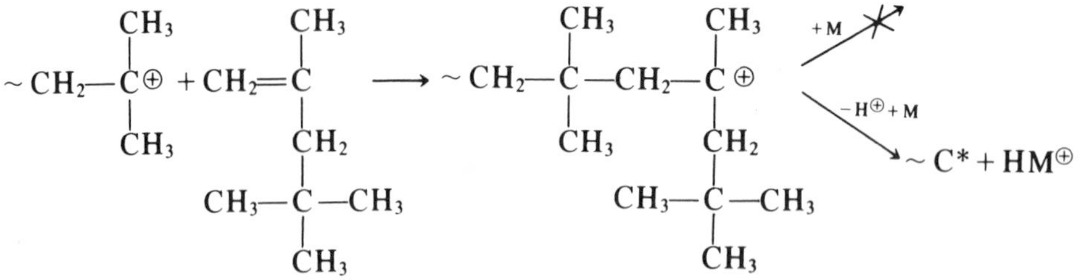

The 2,4,4-trimethyl-2-pentene isomer is unable to add to the growing polyisobutylene cation because of sterically inaccessible π systems (Kennedy, 1967c). This isomer effects modest molecular weight reduction, conceivably by a different mechanism as discussed in connection with termination by these molecules (see Section 4.4).

It is thought that diisobutylene, a commercially available mixture of 2,4,4-trimethyl-1-pentene and 2,4,4-trimethyl-2-pentene, is used as a molecular weight control agent in butyl rubber manufacture.

As a model compound for isobutylene 2,4,4-trimethyl-1-pentene has provided important insight into the polymerization mechanism of this
monomer (Kennedy et al., 1972e, 1974b, 1976b; Priola et al., 1975a, b; Kriz
and Marek, 1973).

According to Ledwith (Bawn et al., 1971), protor transfer commonly
assumed to operate in alkyl vinyl ethers, that is,

$$\sim CH_2\text{—}CH + CH_2\text{=}CH \rightarrow \sim CH\text{=}CH + CH_3\text{—}CH$$
$$\underset{RO\oplus}{\|} \qquad \underset{OR}{|} \qquad \underset{OR}{|} \qquad \underset{RO\oplus}{\|}$$

is insufficient to explain the tremendous efficiency of monomer transfer
(for example, in *i*-BuVE polymerization monomer transfer is only 30 to 50
times slower than that of propagation at 0°C), particularly since propenyl alkyl
ethers may be more reactive than vinyl alkyl ethers. The authors offered two
ingenious alternative monomer transfer mechanisms, involving cyclopropyl
or cyclic intermediates:

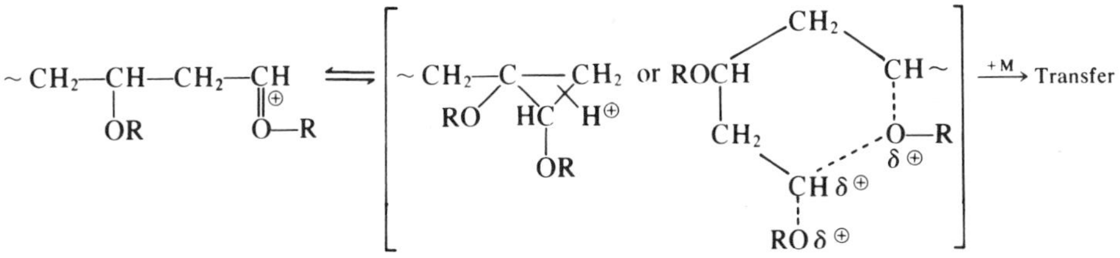

Chain transfer by olefin is responsible for the low molecular weights of
copolymers obtained by carbocationic mechanisms. In random copolymerizations chain transfer by olefin is usually termed "cross-transfer" and
is discussed in Chapter 6.

Chain Transfer by Aromatic Group

In the presence of aromatic groups substitutive aromatic alkylation by the
growing chain, termed chain transfer by aromatic groups, may take place.
Schematically,

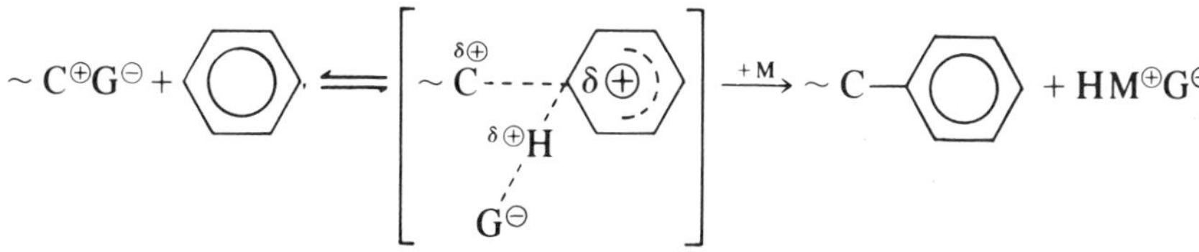

Characteristically, the displaced proton is taken up by monomer to start a
new chain. A number of rate-determining events may be envisioned, for
example, the reaction $\sim C^\oplus + \text{⟨◎⟩}$ leading to a σ complex, the reaction

between the σ complex and monomer $\sim \overset{H}{C}\,\text{◁}\!\!\text{○}\,\text{▷} + M$, and protonation of monomer $H^{\oplus}G^{\ominus} + M$. The $H^{\oplus}G^{\ominus}$ may or may not be identical to the original initiating system which started the kinetic chain. The formation of indane end groups in polystyrene arising by intramolecular aromatic alkylation presented in the preceeding section is an important example for chain transfer by aromatic groups.

Several groups of authors have studied various aspects of chain transfer by aromatic groups (Endres and Overberger, 1955; Higashimura and Okamura, 1956c; Penfold and Plesch, 1961; Anton and Maréchal, 1971; Garreau and Maréchal, 1973). According to Plesch (1953c) the markedly lower molecular weight polystyrenes formed in toluene than in CCl$_4$ solvent reported by Williams (1940) indicate alkylative chain transfer. (Overberger and Endres, 1953, 1955; Endres and Overberger, 1955) studied in detail the polymerization of styrene using C$_6$H$_5$NO$_2$/CCl$_4$ solvent mixtures in the presence of various aromatic compounds and found incorporation of one aromatic transfer agent per macromolecule. The k_{tr}/k_p ratio decreased along the sequence:

$$\text{anisole} > \text{thiophene} > p\text{-xylene} > p\text{-chloroanisole}$$

which follows the order of expected electrophilic substitution of these compounds. Endres et al. (1962) also presented evidence for branching in polystyrene obtained at high conversions and ascribed this to chain transfer to polymer. Whether chain transfer by aromatic group or by hydride transfer (see next section) was meant has not been discussed. Chain transfer by anisole (CH$_3$OC$_6$H$_5$) in the H$_2$O/TiCl$_4$/i-C$_4$H$_8$/CH$_2$Cl$_2$/ -9 to $-89°C$ system has been formulated to occur by alkylative chain transfer and to result in $\sim$ C$_6$H$_4$OMe end groups (Penfold et al., 1961). In the CCl$_3$COOH/TiCl$_4$/stilbene/toluene/25°C system chain transfer by aromatic group sustains the kinetic chain (Brackman and Plesch, 1955):

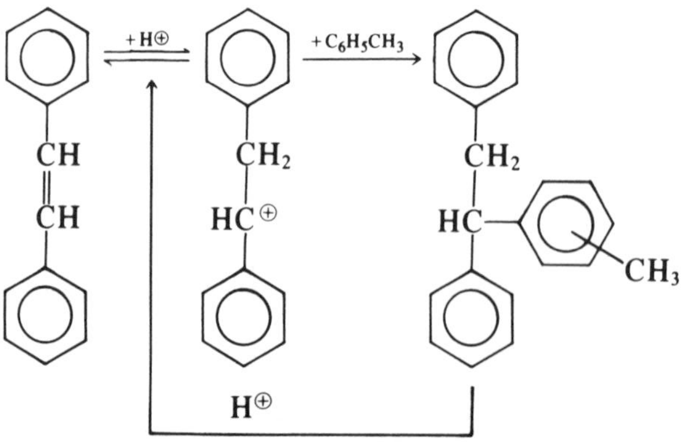

Russel et al. (Bauer et al., 1970, 1971; Russel and Vail, 1976) thoroughly studied isobutylene polymerization initiated by various phenols in conjunction with SnCl$_4$ using EtCl at -78 and, for example, with 2,6-di-*tert*-butylphenol (Russel and Vail, 1976) and found strong evidence for chain breaking by ring alkylation:

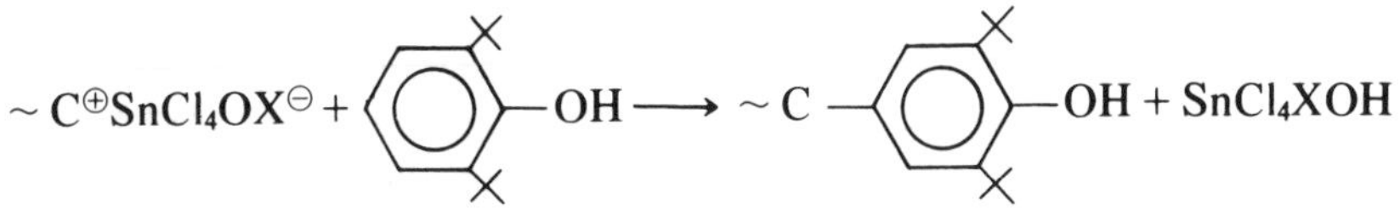

$$\sim C^{\oplus}SnCl_4OX^{\ominus} + \underset{X}{\overset{X}{\bigcirc}}-OH \longrightarrow \sim C -\underset{X}{\overset{X}{\bigcirc}}-OH + SnCl_4XOH$$

Significantly reduced rates and product molecular weights of course were obtained in the presence of this hindered phenol.

Higashimura and Okamura, 1956b carried out LCAO calculations with regard to alkylative transfer; however, their results and interpretations were severely criticized (Mathieson, 1963) and remain in doubt.

Some time ago efforts were made to employ chain transfer to aromatic groups as a means of synthesizing graft copolymers; however, preparative success was at best modest. The general idea was to carry out cationic polymerization of isobutylene (Penfold and Plesch, 1961) or styrene (Haas et al., 1957; Overberger and Burns, 1969) in the presence of pre-formed polystyrene or polystyrene derivatives in the charge; it was hoped that chain transfer by aromatic group would produce graft copolymer by "grafting onto" (see Chapter 8):

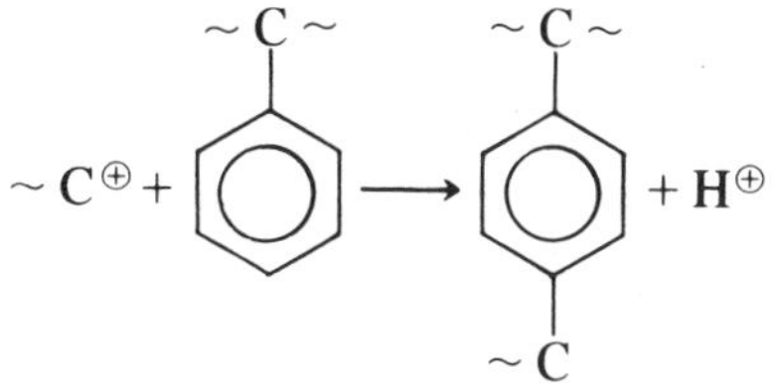

$$\sim C^{\oplus} + \underset{\sim C}{\overset{\sim C \sim}{\bigcirc}} \longrightarrow \underset{}{\overset{\sim C \sim}{\bigcirc}} + H^{\oplus}$$

In view of recent advances in graft copolymer syntheses (Chapter 8) these efforts are of diminished significance; however, they forcefully demonstrate the presence of chain transfer to aromatic groups.

Polyarylations leading to polyphenyls $\{\bigcirc\}_n$ or polybenzyls $\{\bigcirc-CH_2\}_n$ (Chapter 7) and chain transfer by aromatic group are similar processes, since in all these reactions aromatic substitutions occur. What distinguishes chain transfer from other Friedel–Crafts ring alkylation processes is the presence of excess olefin in the charge which kinetically forces the reaction toward chain growth.

Chain Transfer by Hydride Transfer

In this type of chain transfer C–H bonds (*n*-bases) are the nucleophiles; schematically,

$$\sim C^{\oplus} + -\overset{|}{\underset{|}{C}}-H \rightleftharpoons [\sim \overset{\delta\oplus}{C} \cdots \overset{\delta\ominus}{H} \cdots -\overset{|}{C}{}^{\delta\oplus}] \xrightarrow{+M} \sim \overset{|}{C}H + -\overset{|}{\underset{|}{C}}M^{\oplus}$$

The rate of the forward reaction would be increased by factors that increase the stability of the $\sim C^{\oplus}$, for example, electron-releasing groups in the hydride donor; however, if this cation is too stable to add monomer kinetic termination would result (Section 4.4).

Hydride transfer between alkyl cations is extremely rapid, as first shown by Bartlett et al.'s now classical reaction sequence; hydride transfer from isopentane to the t-Bu$^{\oplus}$ cation is over in less than 0.002 sec (Bartlett et al., 1944):

$$(CH_3)_3CCl + AlBr_3 \rightleftharpoons (CH_3)_3C^{\oplus} + AlBr_3Cl^{\ominus}$$

$$(CH_3)_3C^{\oplus} + CH_3CH_2(CH_3)_2CH \rightleftharpoons (CH_3)_3CH + CH_3CH_2(CH_3)_2C^{\oplus}$$

$$CH_3CH_2(CH_3)_2C^{\oplus} + AlBr_3Cl^{\ominus} \rightleftharpoons CH_3CH_2(CH_3)_2CBr + AlBr_2Cl$$

Hydride transfer has been invoked to explain a variety of observations in cationic polymerizations. For example, short-chain branching during the polymerization of linear olefins, for example, 1-butene 2-butene, pentenes, octenes, and decenes, has been attributed (Fontana et al., 1948, 1952a, b) to hydride transfer (Plesch, 1953c):

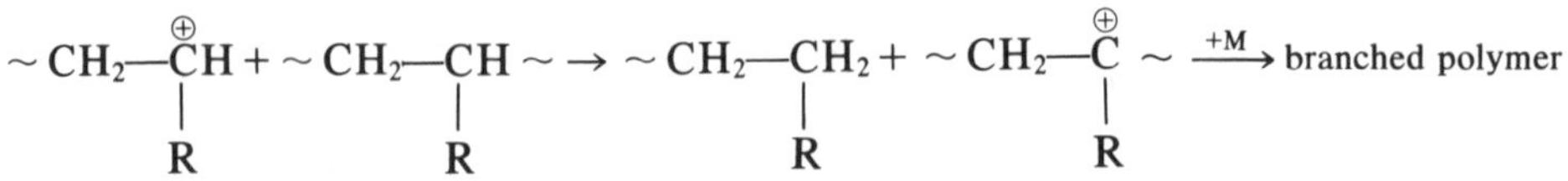

Similarly, long-chain branching in polystyrene was explained in terms of

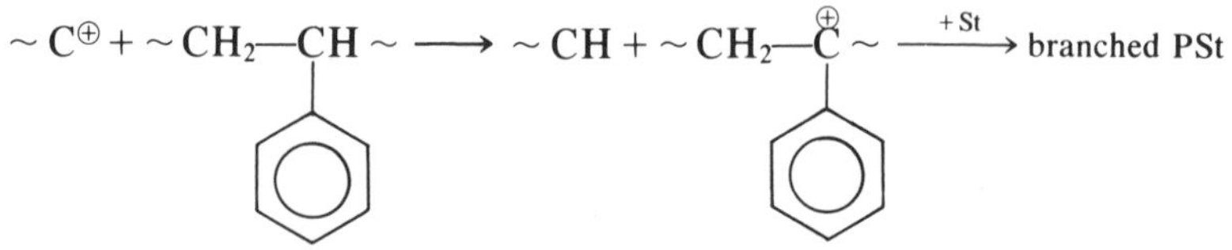

Although various grafting-onto reactions have been attributed to chain transfer by aromatic group, for example, polymerization of polystyrene onto preformed poly(2,6-dimethoxystyrene) (Overberger and Burns, 1969) or poly(p-methoxystyrene) (Haas et al., 1957), it is conceivable that addition was at least partly due to chain transfer by hydride ion. The cationic polymerization of conjugated olefins frequently leads to crosslinking (Kennedy, 1975c; Cooper, 1963), which may be due to hydride transfer, for example, in isoprene,

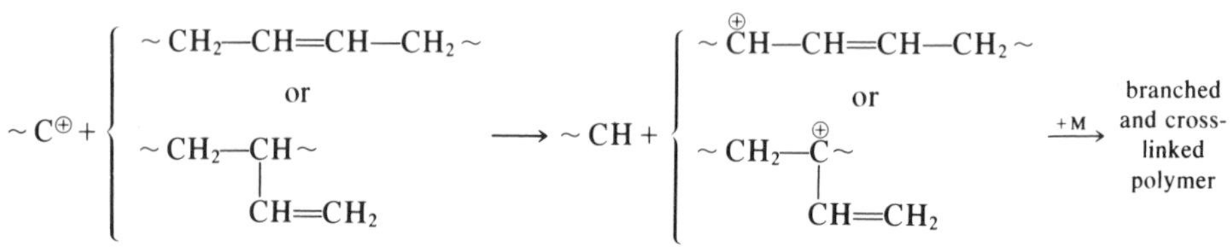

Although among n-bases only C–H bonds have been postulated to enter chain transfer reactions, chain transfer by methide ion $CH_3^{\ominus}$ may be energetically more favorable. According to Plesch (1955) the heterolytic dissociation energy for $R–CH_3$ is some 30 kcal lower than those for R–H. Experimental corroboration of this possibility would be desirable.

Conclusions

Chain transfer is the process during which the charge, and with it the propagating ability, of the chain carrier is transferred either directly or indirectly via a nucleophile to monomer. In this manner a molecule of "dead" polymer is formed; however, the kinetic chain remains operational and further growth may occur. In contrast, termination is the irreversible destruction of chain-propagating ability, which may occur before or after chain transfer takes place (see Section 4.4). Although chain transfer and termination are both chain-breaking processes, they should be carefully differentiated.

Terminative chain transfer occurs when a proton is eliminated from the growing carbenium ion and is captured by a base, for example, by a hindered amine, so that further chain propagation is rendered impossible. Terminative chain transfer is in fact incomplete chain transfer (to monomer) and arises when the chain-carrying proton is intercepted by proton traps. Degradative chain transfer, a term widely used in free radical polymerization, and terminative chain transfer are similar concepts; both concern transfer reactions in the course of which the kinetic chain is short-stopped (stable allyl radical formation in the former case or proton trapping in the latter). Chain transfer is of great influence on molecular weights; indeed more often than not, chain transfer processes are *the* molecular weight-determining factors (or more precisely, the ratio of rate of propagation to that of chain transfer). In contrast, termination determines the ultimate yield of a polymerization and controls molecular weights only in the absence of chain transfer.

The particular nucleophile determines the chemistry of the chain transfer process. Table 4.16 is a subdivision of chain transfer reactions by nucleophiles and serves to facilitate the summary of a large amount of information.

Characteristically, all chain transfer processes except terminative chain transfers produce a nonpropagating "dead" polymer molecule and a new propagating carbenium ion. Phenomenologically, then, all chain transfer reactions may be viewed as chain transfer to monomer processes.

Chain transfer by counteranion profoundly affects molecular weights in probably all chemically initiated cationic polymerizations. In spontaneous or direct transfer to monomer the counteranion does not formally participate. The latter reaction is probably molecular weight-determining in polymerizations in which counteranions are necessarily absent (e.g.,

Table 4.16 □ System of Chain Transfer Reactions

Nucleophile	Name and Schematic Defining Equation	Remarks
Counteranion $G^\ominus$	Chain transfer by counteranion $\sim\!C^\oplus G^\ominus \rightleftharpoons [\sim\!C^{\delta\oplus}\cdots H^{\delta\oplus}\cdots G^\ominus] \xrightarrow{+M} \sim\!C* + HM^\oplus G^\ominus$	Unimportant only in absence of $G^\ominus$, e.g., irradiation induced pzn.
Unshared electron pair. Mainly RX but also ROH, etc.	Chain transfer by unshared electron pair; solvent transfer $\sim\!C^\oplus + RX \rightleftharpoons [\sim\!C^{\delta\oplus}\cdots X^{\delta\ominus}\cdots R^{\delta\oplus}]^{+M} \longrightarrow \sim\!CX + RM^\oplus$	May be useful in mol wt control. Inifer concept important for synthesis of terminally functional polymers.
π-Base, olefin. Mainly monomer but also other olefins	Chain transfer by olefin; monomer transfer $\sim\!C^\oplus + M \rightleftharpoons [\sim\!C^{\delta\oplus}\cdots M^{\delta\oplus}] \rightarrow \sim\!C* + HM^\oplus$	Unimportant only with contact ion pairs (nonpolar conds.) Mol wt control via hindered olefins.
π-Base, aromatic group	Chain transfer by aromatic group 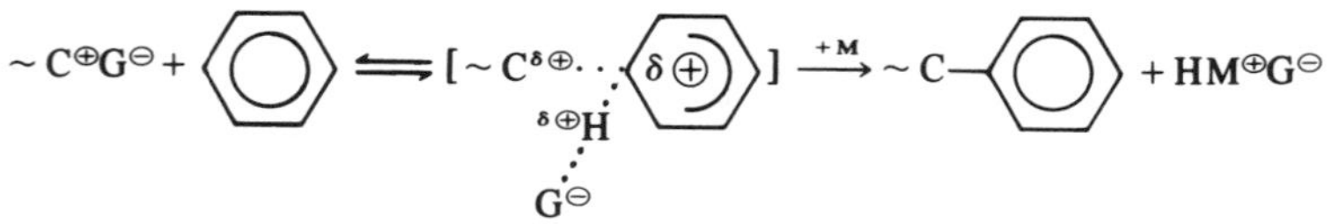 	Branching in polystyrene. Grafting onto.
n-Base, σ-bond in C–H	Chain transfer by hydride transfer $\sim\!C^\oplus + {-}CH \rightleftharpoons [\sim\!C^{\delta\oplus}\cdots H^{\delta\ominus}\cdots C^{\delta\oplus}]^{+M} \longrightarrow \sim\!CH + {-}C M^\oplus$	May effect short- and long-chain branching, crosslinking. Grafting method.

irradiation-induced polymerizations; see Section 4.1) or not readily available (counteranion in solid phase).

Chain transfer by unshared electron pairs is important in the form of solvent transfer. The potential usefulness of this reaction has recently been demonstrated by the synthesis of telechelic polymers using inifers.

Chain transfer by aromatic groups is a special case of Friedel–Crafts ring alkylation and readily occurs in the presence of excess olefin.

Chain transfer by hydride ion has often been invoked but never rigorously proved to explain branch formation under cationic polymerization conditions.

A few remarks relative to molecular engineering by cationic processes are in order (see also conclusion in Section 4.4). Chain transfer is undesirable not only from the point of view of molecular weight buildup, but also for the synthesis of functional polymers, block and graft copolymers, and telechelic polymers. Indeed, unless the objective is the synthesis of oligomers, chain transfer is the bane of the molecular engineer.

Examination of chain transfer reactions encountered in α-olefin polymerizations leads to the conclusion that except for chain transfer by inifer, these reactions yield "sterile" CH_3 head groups unsuitable for subsequent functionalizations, together with a variety of undesirable tail groups that are, for example, saturated or cyclized. Similarly, chain transfer during blocking or grafting reactions leads to undesirable homopolymers or other by-products (see Chapter 8). Thus chain transfer processes are in general unacceptable if the objective is the preparation of well-defined structures or starting materials for subsequent use. Unfortunately, owing to the inherently low thermodynamic stability of useful, propagating carbocations, chain transfer is a "built-in" feature of carbocationic polymerizations.

However, the scene is by no means as bleak as it appears by examining the older literature (in this context older means about 5 years old or more). Recently important breakthroughs have been made in molecular engineering by carbocationic polymerizations. First, it has been recognized that chain transfer can be suppressed by controlled termination and end group engineering has been demonstrated in several cases. These matters are discussed in Section 4.4. Second, the inherent problem of chain transfer has been turned to advantage by the inifer concept and synthesis of biterminally functional polymers by inifers has recently been demonstrated. Down stream applications of new α, ω-bifunctional prepolymers promise to yield many heretofore unavailable cheap and potentially useful materials. Experimentation with trifunctional inifers has already shown the way toward star-shaped polymers (Kennedy et al., 1981). Third, proton traps have been found to intercept certain kinds of chain transfer reactions that may lead to new possibilities in molecular engineering, at present only dimly perceived. One such possibility, molecular weight distribution control, has already been demonstrated.

4.4 □ THE CHEMISTRY OF TERMINATION

Introduction

Termination of carbocationic polymerizations, or of all polymerizations for that matter, consists of the irreversible loss/destruction of propagating ability of a kinetic chain.

Termination and chain transfer are different processes and must be carefully distinguished. Unfortunately termination and chain transfer are often confused and misused even by knowledgable authors (see the introduction to Section 4.3).

Termination, necessarily, is a rare event; it can occur only once per kinetic chain. Several systems have been described in which termination is absent (Plesch, 1954).

The existence of termination in a polymerization system can be readily demonstrated by determining the amount of monomer converted. Less than 100% conversion in a pure system indicates the presence of termination. For example, Figure 4.13 (Kennedy and Thomas, 1960b) shows the results of an experiment in which aliquots of a solution of $AlCl_3$ in CH_3Cl (i.e., a "H_2O"/$AlCl_3$/CH_3Cl initiating system) were added into a stirred charge of isobutylene/n-pentene at $-78°C$. Each $AlCl_3$ addition resulted in a burst of polymerization and produced amounts of polyisobutylene proportional to the quantity of $AlCl_3$ introduced.

In view of the obvious scientific and technological importance of the understanding of termination it is truly amazing how little solid information on this subject is available. Only very recently have systematic investigations been directed toward the elucidation of the chemistry and kinetics of termination. A comprehensive review of termination in carbocationic polymerization has not yet been made.

Very few investigators have asked the question, why do carbocationic polymerizations stop? In the majority of cases—and this is particularly true for industrially or technologically oriented investigations—the problem of increasing the yield of a polymerization process was solved by adding increasing amounts of initiating systems (coinitiator) to the monomer charge. Scientific publications are replete with incomplete, sometimes very low conversion data, without an inquiry into the reason(s) for premature polymerization cessation.

The situation is similar with regards to catalyst efficiency, that is, the amount of polymer produced per unit of initiating system, which is largely controlled by termination. Although this parameter is of great importance for the overall profitability of commercial-scale polymerization processes, there is little indication for its quantitative or mechanistic exploration by either industrial or academic researchers.

This lack of inquiry into the chemistry of termination becomes even more puzzling in view of the plethora of kinetic schemes proposed for

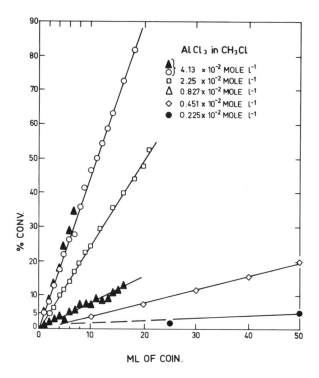

Figure 4.13 □ Dependence of monomer conversion on coinitiator concentration. The concentrations of AlCl₃ in CH₃Cl solutions used are shown in the figure (Kennedy and Thomas, 1960b).

polymerizing systems and the fact that many kinetic expressions cannot be derived without some consideration of the chemical events occurring during termination. As a rule, authors who had fitted conversion and molecular weight data to rate equations did not proceed to unravel details of the chemistry of termination. Admittedly research into chemical changes occurring as a consequence of termination is extremely difficult, not in the least because of the analysis of end groups, that is, structures present at extremely low concentrations. Mainly because of this experimental difficulty, model investigations proved to be extremely useful in helping to elucidate mechanistic details of termination.

The understanding and subsequent control of termination processes holds the key to many exciting polymer synthetic possibilities. Terminal functionalization by reactive groups of inexpensive polymers such as polyisobutylene and polystyrene may lead to a host of new products of desirable combination of physical, mechanical, and chemical properties. Possibilities have already been demonstrated, for example, in the synthesis of block copolymers (see Chapter 8).

Termination Reactions

Mechanistically, termination is a chemical reaction in the course of which the propagating ability of the kinetic chain is irreversibly destroyed. A thorough survey of the literature results in the following classification of termination reactions.

1 □ *Termination by propagating ion pair collapse or neutralization* □ Schematically

$$\sim C^{\oplus} + ZG^{\ominus} \underset{\searrow}{\overset{\nearrow}{}} \begin{array}{l} \sim C\text{---}ZG \\[4pt] \sim CZ + G \end{array}$$

where $\sim C^{\oplus}$ and $\sim CZ$ are the propagating and dead chain ends, $ZG^{\ominus}$ and G are the counteranion and the entity into which the counteranion is converted upon termination, respectively, and $\sim C\text{---}ZG$ indicates a chain end formed by the addition of $ZG^{\ominus}$ to $\sim C^{\oplus}$.

2 □ *Termination by destruction of the propagating cation with the simultaneous formation of carbocation too stable to sustain propagation* □ Schematically,

$$\sim C^{\oplus} + ZR \rightarrow \sim CZ + R^{\oplus}$$

where ZR is an organic molecule carrying an (acidic) group Z, usually hydrogen, that can be transferred to the growing cation and $R^{\oplus}$ is a carbenium ion, usually allylic, unable (too stable) to propagate.

3 □ *Quenching or killing, that is, purposefully terminating the polymerization by introducing a nucleophile* □ Schematically

$$\sim C^{\oplus} + Nu \rightarrow \sim CNu^{\oplus}$$

where Nu is a neutral nucleophile, for example, NH_3, ROH, or H_2O, purposefully introduced into the polymerization system to stop propagation by producing a stable cation.

The balance of this section concerns a detailed examination of these reactions.

Termination by Neutralization

Termination by neutralization involves chemical interaction between the propagating carbocation and counteranion. These entities may collapse to a new electrically neutral species $\sim C^{\oplus} + G^{\ominus} \rightarrow \sim CG$, for example, macroester, or they may give rise to two neutral entities $\sim C^{\oplus} + ZG^{\ominus} \rightarrow \sim CZ + G$.

Neutralization by Reversal of Ionization (Macroester Formation)

Termination in which the growing ion pair collapses to a terminus incorporating all the components of the ion pair that is, $\sim C^{\oplus} + G^{\ominus} \rightarrow$

~ C–G, is, in fact, a reverse of ionization. This kind of termination should necessarily be very rare under polymerization conditions, which, after all, are assembled to yield high polymer.

Several systems have been described in which initiation was affected by protic acids having rather weakly nucleophilic conjugate bases, for example, CF_3COOH (Sawamoto et al., 1977), and termination occurred by reversal of ionization of the growing carbocation/counteranion pair to a macroester:

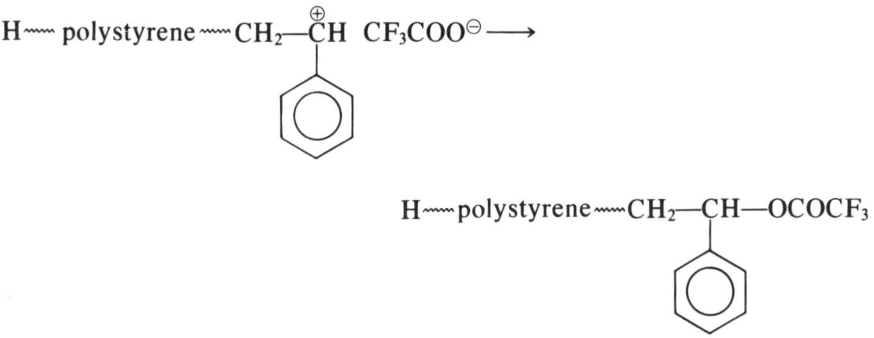

$$H \text{---} \text{polystyrene} \text{---} CH_2 \overset{\oplus}{-} CH \quad CF_3COO^{\ominus} \longrightarrow$$

$$H \text{---} \text{polystyrene} \text{---} CH_2 - CH - OCOCF_3$$

That perchlorate groups may in fact be end groups in polystyrene obtained with $HClO_4$ has been demonstrated elegantly by Goethals, Pepper et al. (Bossaer et al., 1977). The authors initiated the block polymerization of aziridine derivatives by "dormant" polystyryl perchlorate obtained in nonstationary stage styrene polymerization at $-97°C$ (see Chapter 8):

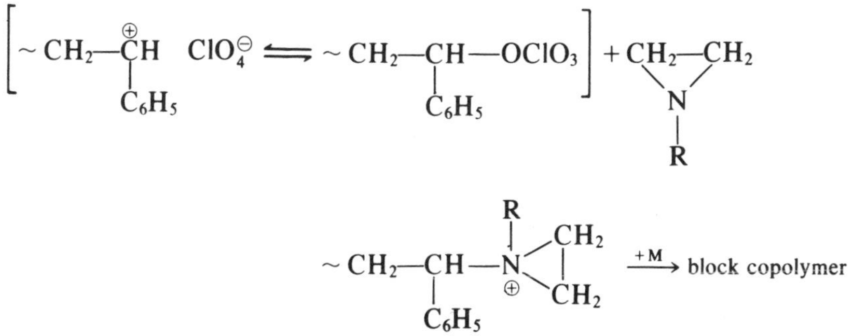

$$\left[\sim CH_2 \overset{\oplus}{-} CH \quad ClO_4^{\ominus} \rightleftharpoons \sim CH_2 - CH - OClO_3 \right] + CH_2 - CH_2$$
$$\overset{|}{C_6H_5} \qquad\qquad\qquad \overset{|}{C_6H_5} \qquad\qquad \underset{\underset{R}{|}}{N}$$

$$\sim CH_2 - CH - \overset{\overset{R}{|}}{\underset{\oplus}{N}} \overset{CH_2}{\underset{CH_2}{<}} \quad \xrightarrow{+M} \text{block copolymer}$$
$$\overset{|}{C_6H_5}$$

Polymerizations initiated by H_2SO_4 or H_3PO_4 may produce the corresponding macroesters; however, other possibilities may also be envisioned and experiments to settle this point have not yet been carried out.

Termination of styrene polymerization induced by HCl may be slowed down and some low molecular weight product is obtained by carrying out the reaction in CH_3NO_2. However, in less solvating hydrocarbon solvents polymerization is absent either because protonation is too slow or because the ion pair would immediately collapse (Pepper, 1954). Polymerization of styrene initiated by CF_3COOH also indicates the determining role of solvation (Throssel et al., 1956). Polymerization is absent when CF_3COOH is added to styrene; however, the addition of this monomer to the acid rapidly yields high molecular weight (20,000 to 30,000) product (see also Section 4.1).

There is a suggestion and follow-up report in the literature (Plesch, 1950; Flett and Plesch, 1952) according to which termination in the $CCl_3COOH/TiCl_4/isobutylene/n$-hexane$/-90$ to $0°C$ system may take place by

$$\sim CH_2\!-\!\overset{\overset{\textstyle CH_3}{|}}{\underset{\underset{\textstyle CH_3}{|}}{C^{\oplus}}}TiCl_4CCl_3COO^{\ominus} \rightarrow \sim CH_2\!-\!\overset{\overset{\textstyle CH_3}{|}}{\underset{\underset{\textstyle CH_3}{|}}{C}}\!-\!OCOCCl_3$$
$$\downarrow$$
$$TiCl_4$$

Evidence for this proposition was infrared spectroscopy and the presence of chlorine in the polymer (Flett and Plesch, 1952).

Neutralization with the Formation of Two Species

Termination by $\sim C^{\oplus} + ZG^{\ominus} \rightarrow \sim CZ + G$ has often been postulated but seldom experimentally substantiated. For example, the reaction

$$\sim \overset{\overset{\textstyle CH_3}{|}}{\underset{\underset{\textstyle CH_3}{|}}{C}}\!-\!C^{\oplus}BF_3OH^{\ominus} \rightarrow \sim CH_2\!-\!\overset{\overset{\textstyle CH_3}{|}}{\underset{\underset{\textstyle CH_3}{|}}{C}}\!-\!OH + BF_3$$

has been shown without reference or independent verification in otherwise meritorious textbooks (Heublein, 1975c) as termination in the $H_2O/BF_3/i$-C_4H_8 system; however, because $(CH_3)_3COH$ is an initiator in conjunction with BF_3 coinitiator for isobutylene polymerization (Evans and Polanyi, 1947a), chances for this termination to occur are slim. Indeed, it appears that kinetic termination is absent in this (Plesch, 1954) as well as in the $H_2O/TiCl_4/styrene/CH_2Cl_2$ system (Plesch, 1953b). Details of termination in industrially important polymerizations of isobutylene or copolymerization of isobutylene with isoprene by $AlCl_3$ in methyl chloride diluent are still obscure. It is possible that $AlCl_3$ losses are due to encapsulation during polymer precipitation in the heterogeneous polymerization system, or, according to Plesch's theory (Plesch, 1953a), to the formation of strong inactive $CH_2=C(CH_3)_2 \cdot AlCl_3$ complexes.

Since the synthetic significance of functional termini for polymer derivatization and block copolymer preparation has been recognized, several groups of workers have turned their attention to the detailed elucidation of termination by neutralization with the simultaneous formation of two entities since this type of termination was shown to lead to desirable, functional end groups. As suggested by the schematic equation

$$\sim C^{\oplus} + ZG^{\ominus} \rightarrow \sim CZ + G$$

terminal functionality can be introduced into the growing polymer by appropriately designed counteranions $ZG^{\ominus}$. Because the nature of the counteranions is mainly determined by the nature of the coinitiator, counteranion design is equivalent to understanding the function and role of coinitiators in cationic polymerizations.

To date, three kinds of such terminations have been studied in some detail that is, when Z in the above equation is organic group, hydrogen, or halogen.

Alkylations and Arylations of Growing Cation (Z = Organic Group)

Organoaluminum compounds were found to be very useful in building counteranions able to transfer organic groups to the growing cation and thus bring about controlled termination by alkylation or arylation, for example.

Termination by alkylation occurs in isobutylene polymerizations coinitiated by certain alkylaluminum compounds. Among the best-characterized systems is the t-BuCl/Me$_3$Al/i-C$_4$H$_8$/CH$_3$Cl (Kennedy et al., 1970, 1972a, b, 1973a, 1974b), where termination is visualized to occur by methylation:

$$\sim CH_2-\underset{\underset{CH_3}{|}}{\overset{\overset{CH_3}{|}}{C^{\oplus}}} + Me_3AlCl^{\ominus} \rightarrow \sim CH_2-\underset{\underset{CH_3}{|}}{\overset{\overset{CH_3}{|}}{C}}-Me + Me_2AlCl$$

Model experiments carried out with t-BuCl + Me$_3$Al in the absence of isobutylene (Kennedy et al., 1970, 1973a, b) demonstrated the formation of neopentane in 100% yield:

$$CH_3-\underset{\underset{CH_3}{|}}{\overset{\overset{CH_3}{|}}{C}}-Cl + Me_3Al \rightleftharpoons CH_3-\underset{\underset{CH_3}{|}}{\overset{\overset{CH_3}{|}}{C^{\oplus}}}Me_3AlCl^{\ominus} \rightarrow CH_3-\underset{\underset{CH_3}{|}}{\overset{\overset{CH_3}{|}}{C}}-Me + Me_2AlCl$$

This reaction may be regarded to be an initiation followed by immediate termination ("polymerization without propagation") (Kennedy, 1973a). In addition to methylation of t-BuCl, Me$_3$Al turned out to be an efficient methylating agent for a large variety of organic halides, in particular for tertiary chlorides. Indeed methylation by Me$_3$Al has become a new synthetic method for the preparation of quaternary carbon-containing compounds (Kennedy, 1970; Kennedy and Sivaram, 1973a).

Interestingly, the rate of methylation of t-BuBr and t-BuI was found to become increasingly slower than that of t-BuCl. Also, methylation was strongly solvent dependent, proceeding much faster in CH_3Cl, than in CH_3Br, which in turn was much faster than in CH_3I, and it was slowest in cyclopentane solvent. These phenomena have been discussed in detail (Kennedy and Sivaram, 1973a).

Similarly to methylation, model experiments have also been carried out with Et_3Al and i-Bu_3Al (Kennedy and Sivaram, 1973a, b; Kennedy and Rengachary 1974b); however, these studies showed that instead of ethylation or isobutylation, hydridation occurred, that is, hydrogen transfer from the counteranion to the growing chain (see next section).

Organoaluminum coinitiators turned out to be uncommonly valuable for the introduction of a variety of organic end groups in polyolefins. Thus in a series of publications Kennedy et al. (Mandal et al., 1978a, b) explored synthetic possibilities offered by trivinylaluminum $(CH_2=CH)_3Al$. First, by model experiments these authors showed that $(CH_2=CH)_3Al$ is an efficient vinylating agent, particularly for tertiary halides (Mandal et al., 1978a), and subsequently that $(CH_2=CH)_3Al$ in conjunction with tertiary or allylic halides is an efficient coinitiator for isobutylene polymerization (Mandal et al., 1978b). For example, a termination-dominated (essentially transferless polymerization of isobutylene induced by the $CH_2=CHCH(CH_3)Cl/(CH_2=CH)_3Al$ initiating system yielded an α,ω-diene-polyisobutylene containing 1.8 ± 0.1 terminal unsaturation:

$$CH_3CH=CHCH_2\text{\small w}PIB\text{\small w}CH=CH_2$$

Termination by cyclopentadienylation has recently been demonstrated (Kennedy and Castner, 1979b, c). First the model reaction between t-BuCl and dimethylcyclopentadienylaluminum Me_2CPDAl showed that the main product was t-butylcyclopentadiene:

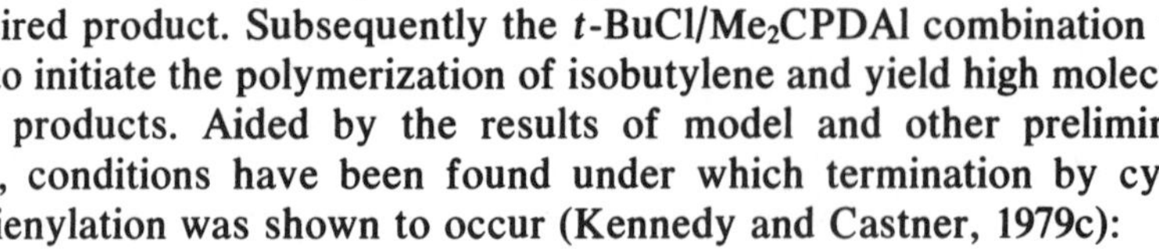

Extensive kinetic studies were carried out at various Cl/Al ratios using various solvents in the temperature range from -40 to $+20°C$ to maximize the desired product. Subsequently the t-BuCl/Me_2CPDAl combination was found to initiate the polymerization of isobutylene and yield high molecular weight products. Aided by the results of model and other preliminary studies, conditions have been found under which termination by cyclopentadienylation was shown to occur (Kennedy and Castner, 1979c):

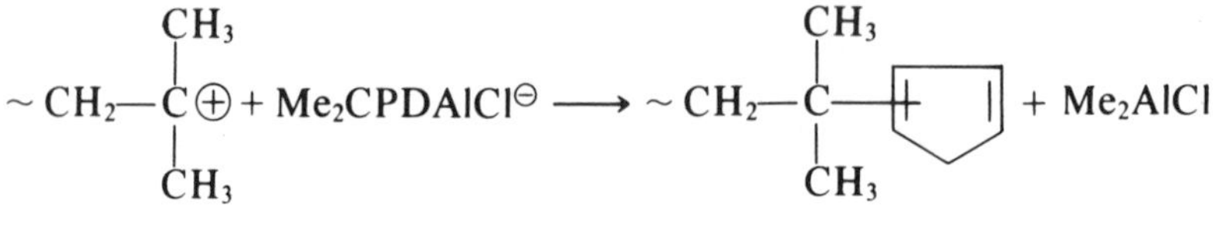

The terminal cyclopentadiene groups have been demonstrated by spectroscopy (UV and NMR) and by reaction with maleic anhydride.

Current work indicates the possibility of synthesis of α,ω-biterminally cyclopentadienylated polymers:

Methylation, vinylation, phenylation, and cyclopentadienylation of macromolecular carbenium ions generated from macromolecular halides have been achieved by techniques similar to those used to obtain terminally functional polymers discussed above (Kennedy and Mandal, 1977b; Kennedy and Chung, 1980; Kennedy and Castner, 1979c) and series of interesting functionalized polymers have been obtained.

Hydridation of Growing Cation (Z = H)

Termination by hydridation (reduction) occurs in olefin polymerizations coinitiated by organoaluminum compounds containing hydrogen to the aluminum atom in the counteranion (Kennedy and Rengachary, 1974b). Simultaneously with the reduction of the growing carbocation, olefin is eliminated from the counteranion. For example, in the presence of the counteranion formed from t-BuCl/Et$_3$Al (Kennedy and Rengachary, 1974b):

$$\sim \overset{|}{\underset{|}{C}}{}^{\oplus}\ Et_3AlCl^{\ominus} \rightleftharpoons \left[\begin{array}{c} \overset{Et}{\diagup} \\ \overset{\ominus}{.}\overset{..}{Cl}-Al-Et \\ \sim \overset{|}{\underset{|}{C}}{}^{\oplus}\ \overset{\frown}{\underset{H-CH_2}{\big(}}CH_2 \end{array}\right] \longrightarrow\ \sim CH + CH_2{=}CH_2 + Et_2AlCl$$

Evidently, hydridation is much faster than ethylation (with Et$_3$Al) or isobutylation (with i-Bu$_3$Al) (Kennedy and Rengachary, 1974b) and this method may be developed into an efficient and clean reduction of active (tertiary, allylic, benzylic) halides. Indeed, hydridation is extremely fast even at very low temperatures. For example, the p-CH$_3$C$_6$H$_4$CH$_2$Cl/Et$_3$Al/i-C$_4$H$_8$/CH$_3$Cl/$-50°$C system is rather inefficient for polymerization presumably because rapid hydridation of the p-methylbenzyl carbocation to p-xylene occurs (Reibel et al., 1977). In contrast, with Et$_2$AlCl, efficient polymerization proceeds most likely because the lifetime of the p-CH$_3$C$_6$H$_4\overset{\oplus}{C}H_2$ cation is relatively long in the presence of the less nucleophilic Et$_2$AlCl$_2^{\ominus}$ and cationation of isobutylene is faster than ion pair collapse.

A conceptually similar reductive termination method was found to be useful for the synthesis of graft copolymer. Kennedy and Delvaux (1981) initiated the polymerization of styrene using t-BuCl/Et$_2$AlCl in the presence of polybutadiene and were able to obtain by this grafting onto method high grafting efficiencies. Subsequent analysis and model experiments indicated that grafting-onto was accompanied by rapid termination (see also Chapter 8):

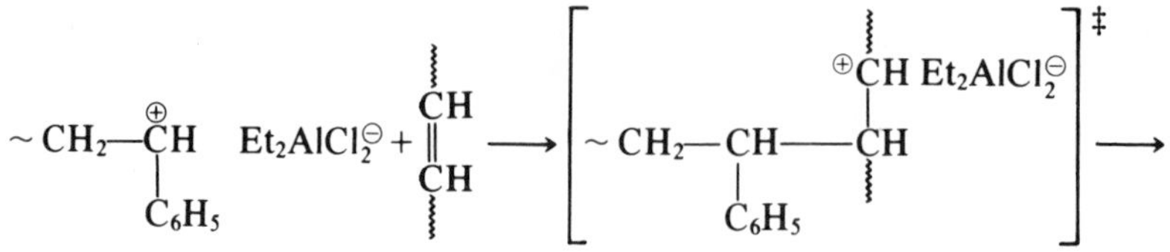

$$\sim CH_2-\overset{\oplus}{\underset{\underset{C_6H_5}{|}}{C}H} \; Et_2AlCl_2^{\ominus} + \overset{\vdots}{\underset{\vdots}{\underset{CH}{\overset{CH}{\|}}}} \longrightarrow \left[\sim CH_2-\underset{\underset{C_6H_5}{|}}{C}H-\overset{\overset{\vdots}{\oplus}}{\underset{\vdots}{C}H} \; Et_2AlCl_2^{\ominus} \right]^{\ddagger} \longrightarrow \begin{array}{l} \text{grafting by} \\ \text{termination by} \\ \text{hydridation} \end{array}$$

Evidently hydridation was faster than chain transfer to monomer and/or reinitiation from the sterically highly hindered cationic site. Although termination by ethylation was postulated to occur in isobutylene polymerizations coinitiated by Et_2AlI (Giusti et al., 1975), in view of the above, hydridation appears to be more likely:

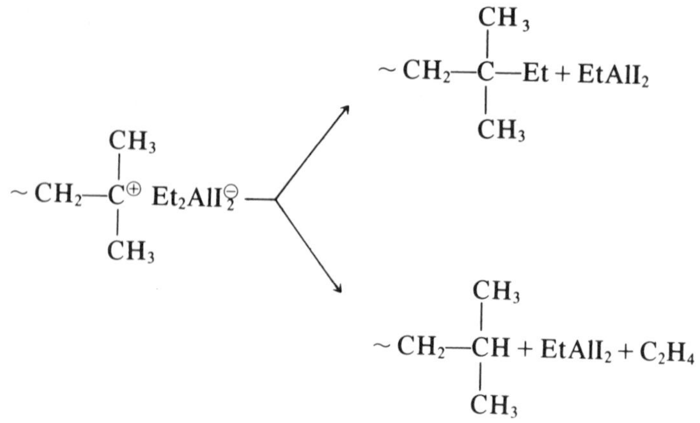

$$\sim CH_2-\overset{CH_3}{\underset{CH_3}{\overset{|}{\underset{|}{C}^\oplus}}} \; Et_2AlI_2^{\ominus} \longrightarrow \begin{array}{l} \sim CH_2-\overset{CH_3}{\underset{CH_3}{\overset{|}{\underset{|}{C}}}}-Et + EtAlI_2 \\[2em] \sim CH_2-\overset{CH_3}{\underset{CH_3}{\overset{|}{\underset{|}{C}H}}} + EtAlI_2 + C_2H_4 \end{array}$$

Halogenation of Growing Cation (Z = Cl, Br)

Boron-containing counteranions, particularly those with B–Cl bonds, were found to be valuable terminating entities. Extensive studies centered on elucidating the chemistry of termination in BCl_3-coinitiated olefin polymerizations. In a series of publications including model experiments, chemical characterizations, and kinetic studies, (Kennedy et al., 1976b, 1977a, c; 1978a, b; Feinberg and Kennedy, 1976) have shown that polymerization of isobutylene initiated by H_2O/BCl_3 systems in CH_2Cl_2 diluent terminate by chlorination of the growing carbocation:

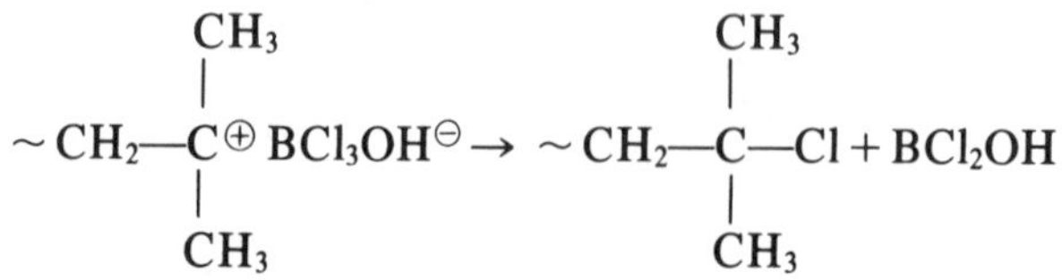

$$\sim CH_2-\overset{CH_3}{\underset{CH_3}{\overset{|}{\underset{|}{C}^\oplus}}} \; BCl_3OH^{\ominus} \rightarrow \sim CH_2-\overset{CH_3}{\underset{CH_3}{\overset{|}{\underset{|}{C}}}}-Cl + BCl_2OH$$

A key finding that substantiated this reaction was that BCl_3 was found to be unable to ionize $(CH_3)_3CCl$ or even $(CH_3)_3CCH_2(CH_3)_2CCl$ (Kennedy et al.,

1977c). The characterization of high molecular weight polyisobutylenes carrying a single terminal chlorine atom represented an uncommonly difficult assignment because the presence *and* position of the halogen had to be ascertained. Proof positive for termination by chlorination was obtained by initiating the polymerization of a second monomer (styrene) by the polyisobutylene carrying the tertiary chloride end group in conjunction with Et$_2$AlCl as coinitiator, and isolating and characterizing the isobutylene-styrene block copolymer thus formed (Kennedy et al., 1978a):

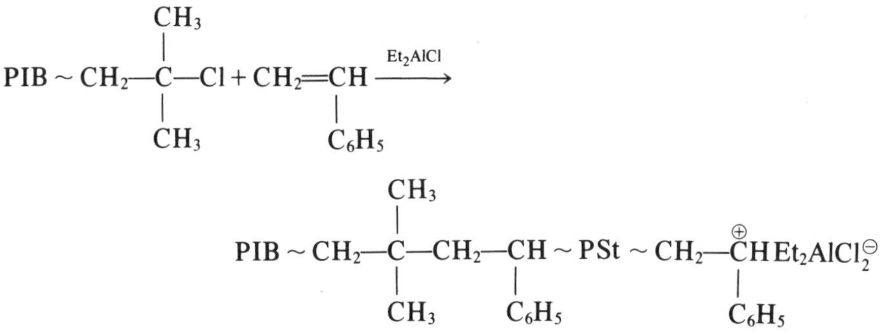

$$PIB \sim CH_2 - \overset{\overset{\displaystyle CH_3}{|}}{\underset{\underset{\displaystyle CH_3}{|}}{C}} - Cl + CH_2 = \overset{}{\underset{\underset{\displaystyle C_6H_5}{|}}{CH}} \xrightarrow{\text{Et}_2\text{AlCl}}$$

$$PIB \sim CH_2 - \overset{\overset{\displaystyle CH_3}{|}}{\underset{\underset{\displaystyle CH_3}{|}}{C}} - CH_2 - \overset{}{\underset{\underset{\displaystyle C_6H_5}{|}}{CH}} \sim PSt \sim CH_2 - \overset{\oplus}{\underset{\underset{\displaystyle C_6H_5}{|}}{CH}} Et_2AlCl_2^{\ominus}$$

Systematic further work showed that termination by chlorination also operates with styrene, and blocking of isobutylene from $-CH_2-\underset{\underset{\displaystyle C_6H_5}{|}}{CH}-Cl$

end groups in polystyrenes obtained with "H$_2$O"/BCl$_3$ can be achieved, although less efficiently than in the former system (Kennedy and Feinberg, 1978b).

It has also been demonstrated that isobutylene polymerization can be induced by RCl/BCl$_3$ systems and in this manner the head group as well as the end group could be tailor-made. For example (Kennedy et al., 1977f), the (CH$_3$)$_2$C=CH–CH$_2$Cl/BCl$_3$ combination gave rise to

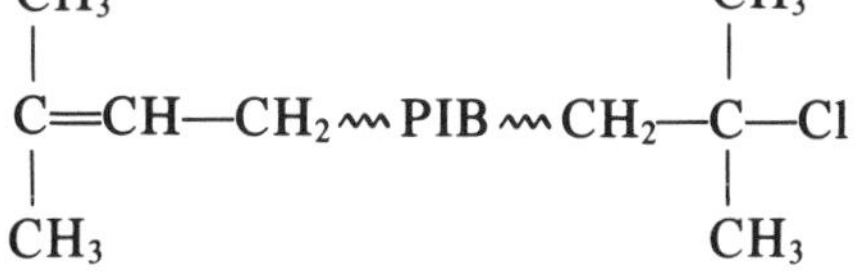

$$\overset{\overset{\displaystyle CH_3}{|}}{\underset{\underset{\displaystyle CH_3}{|}}{C}} = CH - CH_2 \rightsquigarrow PIB \rightsquigarrow CH_2 - \overset{\overset{\displaystyle CH_3}{|}}{\underset{\underset{\displaystyle CH_3}{|}}{C}} - Cl$$

Similarly, telechelic molecules carrying chlorine atoms as head and end groups have been obtained by the use of Cl$_2$/BCl$_3$ combinations. Thus the synthesis of α,ω-dichloropolyisobutylene has recently been described (Kennedy and Chen, 1979d):

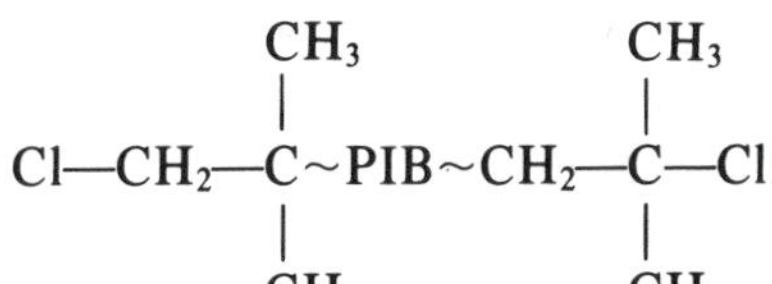

$$Cl - CH_2 - \overset{\overset{\displaystyle CH_3}{|}}{\underset{\underset{\displaystyle CH_3}{|}}{C}} \sim PIB \sim CH_2 - \overset{\overset{\displaystyle CH_3}{|}}{\underset{\underset{\displaystyle CH_3}{|}}{C}} - Cl$$

Blocking of styrene from the tertiary chlorine end group of this molecule by the use of Et$_2$AlCl produced α-chlorine containing block copolymers (Kennedy et al., 1979f):

$$\text{Cl}\longrightarrow\boxed{\text{Polyisobutylene}}\longrightarrow\boxed{\text{Polystyrene}}$$

Termination in "H$_2$O"/BCl$_3$-initiated α-methylstyrene polymerization in CH$_2$Cl$_2$ solvent is absent and polymerizations proceed to 100% monomer conversions (Section 4.3). Evidently, chlorination to

$$\begin{array}{c}\text{CH}_3\\|\\-\text{CH}_2-\text{C}-\text{Cl}\\|\\\text{C}_6\text{H}_5\end{array}$$

is slow either because steric hindrance to chlorination by the BCl$_3$OH$^\ominus$ counteranion is prohibitively large or because the tertiary benzylic chloride would rapidly ionize in the presence of excess BCl$_3$ and thus sustain chain propagation, or both. Nonetheless, new methods exist to stop the polymerizations of α-methylstyrene at low conversions (see discussion of terminative chain transfer, Section 4.3).

In addition to BCl$_3$, experiments have also been carried out with the other boron trihalides; however, these have shown that BF$_3$-coinitiated olefin polymerizations result in transfer-dominated mechanisms and BBr$_3$ or BI$_3$ are rather inefficient coinitiators. The inherent difficulty with BF$_3$ is that RF + BF$_3$ ionizes and gives R$^\oplus$BF$_4^\ominus$ (particularly when R = tertiary or benzylic moiety), so that even if momentary termination would occur by ion pair collapse, the ~RF end group would immediately ionize to ~R$^\oplus$BF$_4^\ominus$ and thus sustain propagation. Under these conditions, then, chain transfer would occur sooner or later, rendering systematic end group design virtually impossible. With BBr$_3$, and even more so with BI$_3$, rapid side reactions with H$_2$O (or RX) occur and lead to the decomposition of these strong Lewis acids.

Termination by halogenation has also been invoked in FSO$_3$H- and ClSO$_3$H-initiated styrene polymerizations (Masuda et al., 1976):

$$\text{H}\sim\text{PSt}\sim\text{CH}_2-\overset{\oplus}{\text{C}}\text{HSO}_3\text{Cl}^\ominus\overset{\text{C}_6\text{H}_6}{\longrightarrow}\text{H}\sim\text{PSt}-\text{CH}_2-\text{CH}-\text{Cl}+\text{SO}_3$$

with C$_6$H$_5$ below the cation carbon on the left and C$_6$H$_5$ below the CH on the right.

Infrared spectroscopy (C–Cl stretching at 620 cm^{-1}) and absence of S in the polymer was cited to substantiate the chemistry. Termination was found to be much less important with CF$_3$SO$_3$H- and CH$_3$SO$_3$H-initiated polymerizations, which was attributed to the S–C bonds in these acids being more covalent than the S–X bonds in the former.

According to Bawn et al. (1971) a probable termination in isobutyl vinyl ether polymerization initiated by stable cation/SbCl$_6^\ominus$ systems is halogenation of the propagating ion by the counteranion:

$$\sim CH_2\!\!-\!\!\underset{\underset{\oplus O\text{-}i\text{-}Bu}{\|}}{CH}\, SbCl_6^\ominus \rightleftharpoons\, \sim CH_2\!\!-\!\!\underset{\underset{O\text{-}i\text{-}Bu}{|}}{CH}\!\!-\!\!Cl + SbCl_5$$

coupled by subsequent reactions in which SbCl$_5$ is consumed.

Termination by chlorination has been postulated to occur in isobutylene polymerizations coinitiated by AlCl$_3$ or EtAlCl$_2$ (Magagnini et al., 1977) as follows:

$$\sim CH_2\!\!-\!\!\underset{\underset{CH_3}{|}}{\overset{\overset{CH_3}{|}}{C}}^{\oplus}\!\!AlCl_4^\ominus \rightleftharpoons\, \sim CH_2\!\!-\!\!\underset{\underset{CH_3}{|}}{\overset{\overset{CH_3}{|}}{C}}\!\!-\!\!Cl\cdot AlCl_3 \rightleftharpoons\, \sim CH_2\!\!-\!\!\underset{\underset{CH_3}{|}}{\overset{\overset{CH_3}{|}}{C}}\!\!-\!\!Cl + AlCl_3$$

$$AlCl_3 + CH_2\!\!=\!\!C(CH_3)_2 \rightarrow complex$$

It is known that both AlCl$_3$ and/or EtAlCl$_2$ rapidly polymerize isobutylene (Kennedy and Thomas, 1960b; di Maina et al., 1977) and in the presence of relatively low amounts of Friedel–Crafts acids, conversions are modest due to rapid termination (see Figure 4.13). According to Italian workers termination occurs by chlorination of the growing carbenium ion site. Further, these authors assume that the $\sim CH_2C(CH_3)_2Cl\cdot AlCl_3$ complex is relatively stable, nonionizing, since t-BuCl$\cdot$AlCl$_3$ complexes are known to be stable at low temperatures (Cesca et al., 1972). Free AlCl$_3$ released in the equilibrium is immediately complexed by isobutylene and, according to Plesch (1973), i-C$_4$H$_8\cdot$AlCl$_3$ complexes are unable to initiate isobutylene polymerization. Thus the above set of equilibria depicts true termination (Magagnini et al., 1977). Direct chemical evidence is necessary to corroborate this interesting suggestion.

Termination Involving Stable Cation Formation

The chain carrier may be destroyed by transferring to it a negative entity from an organic molecule and simultaneously creating a carbenium ion that is too stable to sustain chain growth:

$$\sim C^{\oplus} + ZR \rightarrow \sim CZ + R^{\oplus}$$

Although in principle Z could be hydrogen, organic group, or halogen, for instance, to date only systems in which Z = H and R$^{\oplus}$ = allylic carbocation have been discussed.

Kennedy's concept of "allylic termination" (1967a–h) is summarized by the following equation:

$$\sim \overset{|}{\underset{|}{C}}{}^{\oplus} + \overset{|}{\underset{|}{C}} = \overset{|}{\underset{|}{C}} - \overset{|}{\underset{|}{C}} H \rightarrow \sim \overset{|}{\underset{|}{C}} H + \delta^{\oplus}\overset{|}{\underset{|}{C}} \cdots \overset{|}{\underset{|}{C}} \cdots \overset{|}{\underset{|}{C}} \delta^{\oplus}$$

The driving force is provided by the formation of a covalent C–H bond and a stable substituted allyl carbenium ion from an unsaturated molecule and a relatively less stable carbocation. According to this theory olefins with allylic hydrogen atoms are potential terminators and monomers may function as their own terminating agents ("suicide polymerizations").

The allylic termination theory was tested under a variety of conditions, in particular in the presence of a large number of various potential hydride donors. The terminating efficiency of hydride donors was quantized by "poison coefficients" (P.C.). Poisons were defined as materials that decrease the overall polymer yield, presumably by allylic termination, but do not necessarily affect molecular weight. The P.C. expresses quantitatively yield reduction under specified conditions, that is, "H_2O"/$AlCl_3(CH_3Cl)$/isobutylene/n-pentane/ $-78°C$. Several materials have been found that reduce the yield without affecting molecular weights. Among these so-called pure poisons are 1-alkenes, 4-methyl-1-pentene, and norbornadiene (Kennedy and Squires, 1967g). Another group of chemicals, termed "transfer agents," for example, certain alkyl halides, reduced only molecular weights, and their molecular weight-reducing effect was quantized by "transfer coefficients" (T.C.) (see also Section 4.3). The largest number of materials exhibited both poisoning and transfer activity to a more or less pronounced degree. Table 4.17 shows a representative list of materials together with their experimental P.C. and T.C. values.

1-Alkenes are pure poisons, that is, compounds with T.C. = 0, that diminish conversions but do not affect molecular weights. The P.C. increases from 4.9 for propylene to 11.9 for 1-hexene. According to the allylic termination theory, propylene gives the least stable, unsubstituted allyl cation by allylic termination. With 1-butene, 1-pentene, and 1-hexene, olefins containing two secondary allylic hydrogen atoms, allyl cation formation is facilitated owing to substitution. That 2-octene is a much stronger poison is probably due to the formation of the more stable doubly substituted allyl cation:

$$\overset{\delta\oplus}{\underset{\underset{CH_3}{|}}{CH}} \cdots CH \cdots \overset{\delta\oplus}{\underset{\underset{C_4H_9}{|}}{CH}} > \overset{\delta\oplus}{CH} \cdots CH \cdots \overset{\delta\oplus}{\underset{\underset{C_nH_{2n+1}}{|}}{CH_2}} > \overset{\delta\oplus}{CH_2} \cdots CH \cdots \overset{\delta\oplus}{CH_2}$$

Similar self-consistent arguments have been developed to account for P.C. values of branched olefins (Kennedy and Squires, 1967b, c), conjugated dienes (Kennedy and Squires, 1967d), and cyclic olefins (Kennedy et al., 1967e).

Table 4.17 □ Empirical Poison Coefficients and Transfer Coefficients

Material	P.C.	T.C.
Propylene	4.9	0
1-Butene	6.2	0
1-Pentene	9.7	0
1-Hexene	11.9	0
2-Octene	20.8	12.1
Butadiene	7.6	2.0
2,5-Dimethyl-2,4-hexadiene	26.7	15.2
2,3-Dimethyl-1,3-butadiene	107.0	3.6
Cyclohexadiene	103.0	48.0
Isoprene	140.0	60.0
Piperylene	170.0	327.0
Cyclopentadiene	900.0	
2-Methylcyclopentadiene	685.0	
2-Methyl-1-pentene	3.5	53.8
2-Ethyl-1-hexene	24.7	248.0
2,4,4-Trimethyl-1-pentene	66.7	700.0
2,4,4-Trimethyl-2-pentene	66.7	34.6
3-Methyl-1-butene	0	0
4-Methyl-1-pentene	2.9	0
3,3-Dimethyl-1-butene	0	0
Vinylcyclohexane	3.7	6.6

The results obtained with 2,4,4-trimethyl-1-pentene (2,4,4-TM-1-P) and 2,4,4-trimethyl-2-pentene (2,4,4-TM-2-P) (components of diisobutylene) are particularly revealing (Kennedy and Squires, 1967c). Both materials have P.C. = 66.7 but vastly different T.C.s. The allyl cations expected from 2,4,4-TM-1-P are **I** and **II**, whereas from 2,4,4-TM-2-P only **II** arises. Since cation **I** is much less favored than **II**, the allyl cations arising from the two TMP isomers could conceivably be the same, explaining the identical P.C. values found. For an interpretation of the significantly different T.C. values, (see Section 4.3).

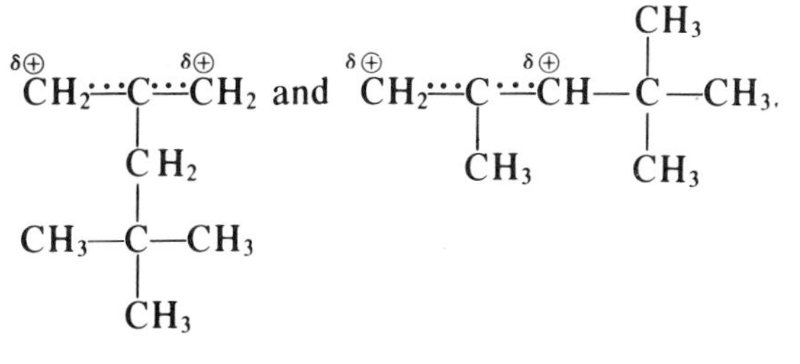

I   II

A by-product of this work on allylic termination was a relative stability order of allyl carbenium ions (Kennedy and Squires, 1967h):

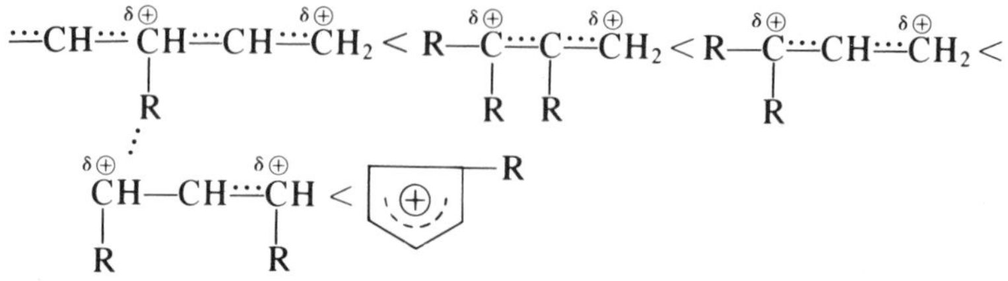

$$\cdots CH \overset{\delta\oplus}{\cdots} \underset{R}{\overset{\cdots}{CH}} \cdots CH \overset{\delta\oplus}{\cdots} CH_2 < R - \underset{R}{\overset{\delta\oplus}{\underset{|}{C}}} \cdots \underset{R}{\overset{|}{C}} \overset{\delta\oplus}{\cdots} CH_2 < R - \underset{R}{\overset{\delta\oplus}{\underset{|}{C}}} \cdots CH \overset{\delta\oplus}{\cdots} CH_2 <$$

$$\underset{R}{\overset{\delta\oplus}{\underset{|}{CH}}} - CH \overset{\delta\oplus}{\cdots} \underset{R}{\overset{|}{CH}} < \boxed{\oplus}^{-R}$$

Hydride abstraction has also been invoked to explain termination in isobutylene polymerization in the presence of allyl halides (Kennedy et al., 1967f). According to the concept of allylic termination allyl halides may terminate cationic chains by the following process:

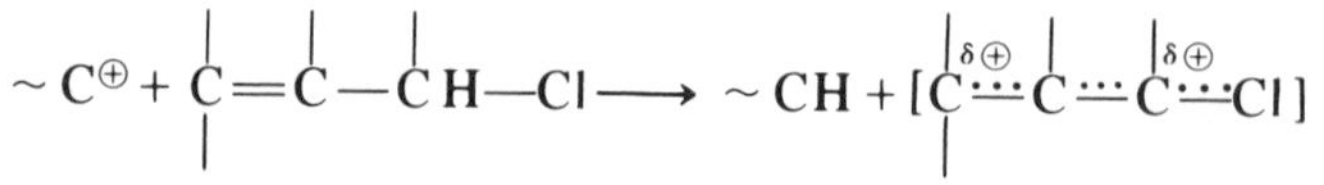

$$\sim C^{\oplus} + \overset{|}{\underset{|}{C}} = \overset{|}{C} - \overset{|}{\underset{|}{C}}H - Cl \longrightarrow \sim CH + [\overset{|}{\underset{|}{C}} \overset{\delta\oplus}{\cdots} \overset{|}{C} \cdots \overset{|}{\underset{|}{C}} \overset{\delta\oplus}{\cdots} Cl]$$

A variety of considerations including molecular orbital calculations indicate that chloride or hydride loss from allyl chlorides may be competitive (Kennedy et al., 1967f).

Fontana's termination mechanism (Fontana et al., 1952b; Fontana, 1953a) arose from an extensive study of 1-alkene, for example, propylene and 1-butene, polymerizations with HBr/AlBr₃. It was observed that the initiating system does not remain active indefinitely and the following reactions were proposed to account for termination:

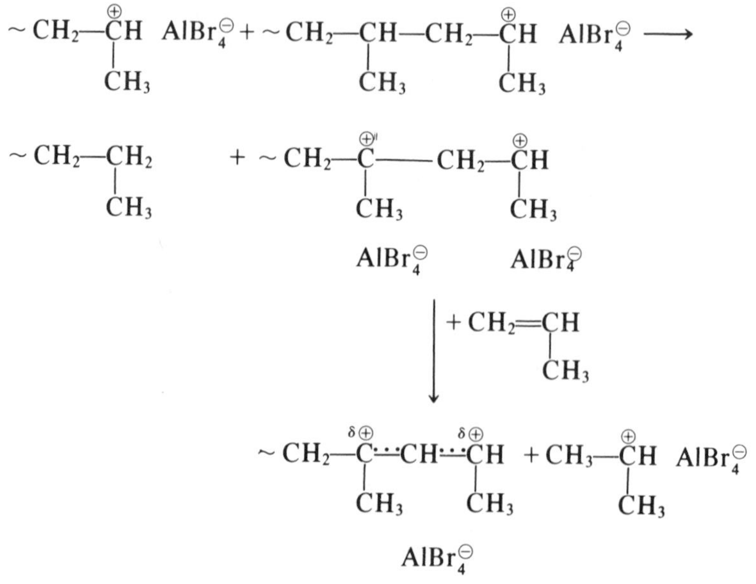

$$\sim CH_2 - \underset{CH_3}{\overset{\oplus}{\underset{|}{CH}}} \ AlBr_4^{\ominus} + \sim CH_2 - \underset{CH_3}{\overset{|}{\underset{|}{CH}}} - CH_2 - \underset{CH_3}{\overset{\oplus}{\underset{|}{CH}}} \ AlBr_4^{\ominus} \longrightarrow$$

$$\sim CH_2 - \underset{CH_3}{\overset{|}{\underset{|}{CH_2}}} \quad + \sim CH_2 - \underset{\underset{AlBr_4^{\ominus}}{CH_3}}{\overset{\oplus}{\underset{|}{C}}} - CH_2 - \underset{\underset{AlBr_4^{\ominus}}{CH_3}}{\overset{\oplus}{\underset{|}{CH}}}$$

$$\Big\downarrow \overset{+ CH_2 = CH}{\underset{\overset{|}{CH_3}}{}}$$

$$\sim CH_2 - \underset{CH_3}{\overset{\delta\oplus}{\underset{|}{C}}} \cdots CH \overset{\delta\oplus}{\cdots} \underset{CH_3}{\overset{|}{\underset{|}{CH}}} + CH_3 - \underset{CH_3}{\overset{\oplus}{\underset{|}{CH}}} \ AlBr_4^{\ominus}$$

$$AlBr_4^{\ominus}$$

This reaction consists of hydride transfer followed by proton transfer which yields a resonance-stabilized allyl carbocation. Both Kennedy's and Fontana's termination mechanisms are in principle valid for most α-olefins.

Subtle differences in hydride transfer may account for some as yet unexplained findings relative to 3-methyl-1-butene and 4-methyl-1-pentene polymerization. It has repeatedly been reported (Kennedy et al., 1964a, b) that the molecular weights of poly(3-methyl-1-butene) were markedly lower than those of poly(4-methyl-1-pentene) obtained under essentially identical conditions, that is, "H_2O"/$AlCl_3$/CH_3Cl or CH_2Cl_2/$-78°C$ (Kennedy and Johnston, 1975a; Kennedy, 1975b). Conceivably termination by hydride transfer in the 3-methyl-1-butene system,

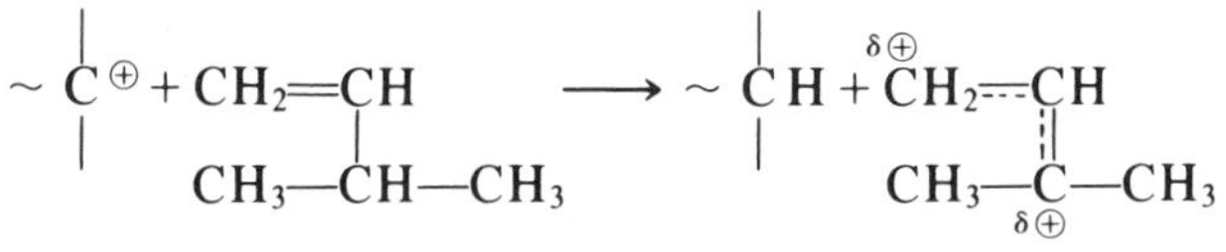

$$\sim \overset{|}{\underset{|}{C}}{}^{\oplus} + \underset{CH_3-CH-CH_3}{CH_2{=}\overset{|}{C}H} \longrightarrow \sim \overset{|}{\underset{|}{C}}H + \underset{CH_3-\underset{\delta\oplus}{C}-CH_3}{\overset{\delta\oplus}{C}H_2{\cdots}\overset{}{C}H}$$

is faster due to the formation of a relatively stable primary–tertiary allylcarbenium ion than that in the 4-methyl-1-pentene system:

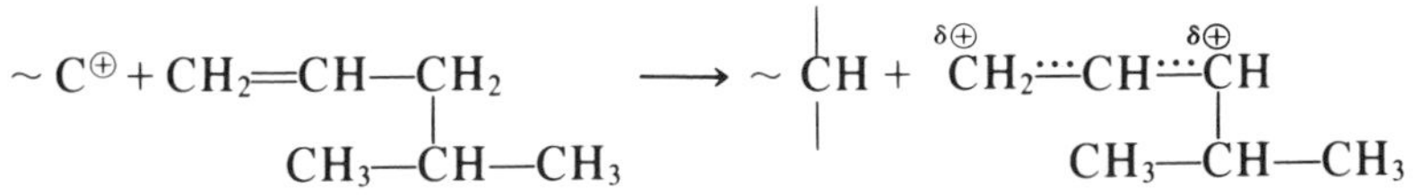

$$\sim C^{\oplus} + \underset{CH_3-CH-CH_3}{CH_2{=}CH{-}CH_2} \longrightarrow \sim \overset{|}{C}H + \underset{CH_3-CH-CH_3}{\overset{\delta\oplus}{C}H_2{\cdots}CH{\cdots}\overset{\delta\oplus}{C}H}$$

which yields a somewhat less stable primary–secondary allyl cation. Hydride transfer from the tertiary site in 4-methyl-1-pentene is still less favored. Because the poison coefficient for 3-methyl-1-butene is zero (Kennedy and Squires, 1967c), hydride transfer probably involves the unrearranged $-CH_2-\overset{\oplus}{C}H-CH(CH_3)_2$ cation and not the isomerized $-CH_2-CH_2-\overset{\oplus}{C}(CH_3)_2$ species, whose structure is quite similar to that of the propagating polyisobutylene cation.

Conceptually, one could visualize termination by the above-discussed mechanisms not only by hydride transfer but also by transfer of other groups, particularly CH_3-, because the heterolytic dissociation energy of $R-CH_3$ is some 30 kcal lower than that of $R-H$ (Plesch, 1955). Mechanisms along these lines, however, have not yet been proposed.

Finally, in the context of termination by stable cation formation, termination by "wrong monomer addition" may be mentioned. Propagation of N-vinylcarbazole may be visualized to occur on the vinyl group or on the nitrogen atom, leading to an unreactive quaternary ammonium ion (Pac and Plesch, 1967):

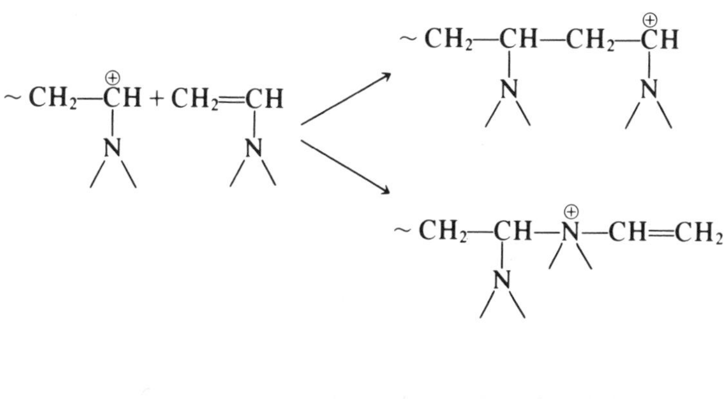

Quenching

The routine quenching or "killing" of polymerizations is carried out at the end of polymerization experiments usually by introducing to the "live" system an excess of a highly nucleophilic quenching reagent, for example, alcohols, ammonia, amines, or KOH in CH_3OH. Although this operation is often referred to as "terminating a polymerization," this terminology may be a misnomer since kinetic termination of a cationically propagating chain is usually complete long before the quenching agent is introduced and the addition of alcohols or bases merely serves to convert Friedel–Crafts acids to conveniently disposable oxides.

In view of the great variety of Friedel–Crafts acids used for polymerizations and other petrochemical operations, and considering that virtually all laboratory experiments or industrial processes employing Friedel–Crafts acids are quenched by large excesses of some strong nucleophile, it is truly remarkable that the chemistry of quenching has not been studied systematically. Usually it is tacitly assumed that quenching involves hydrolysis/alcoholysis reactions identical to those occurring between Friedel–Crafts acids and bases in the absence of monomer, polymer, and nonaqueous solvent. This lack of experimental attention is the more surprising in view of the complex reactions between alcohols or water and Friedel–Crafts acids even under well-controlled conditions. Similarly, it is further assumed that carbenium ions in propagating systems and nucleophilic quenching agents interact in a manner identical to that expected of conventional carbenium ions. In one instance this assumption has been examined and found to be borne out (Kennedy and Chou, 1976a).

Quenching by water is expected to proceed by

$$\sim C-\overset{|}{\underset{|}{C}}{}^{\oplus} + (H_2O)_n \rightarrow \sim C{=}\overset{|}{C} + H_3O^{\oplus}(H_2O)_{n-1}$$

or

$$\sim \overset{|}{\underset{|}{C}}{}^{\oplus} + (H_2O)_n \rightarrow \sim \overset{|}{\underset{|}{C}}-OH + H_3O^{\oplus}(H_2O)_{n-2}$$

and thermochemical calculations indicate the reasonableness of these overall equations. For example, when the $H_3O^{\oplus}$ ion is fully solvated by H_2O the latter reaction is exothermic by > 100 kcal/mole owing to the heat of solvation (Bell, 1959).

Quenching with ammonia or amines is visualized to result in quaternization of nitrogen (George et al., 1950a, b):

$$\sim \overset{|}{\underset{|}{C}}{}^{\oplus} + NH_3 \rightarrow \sim \overset{|}{\underset{|}{C}}NH_3^{\oplus}$$

and KOH in methanol that is, $CH_3O^{\ominus}$, is expected to yield terminal methyl ether:

$$\sim \overset{|}{\underset{|}{C}}{}^{\oplus} + CH_3O^{\ominus} \rightarrow \sim \overset{|}{\underset{|}{C}}-OCH_3$$

However, it should again be stressed that most of these very plausible expectations have not been experimentally scrutinized or verified under polymerization conditions.

Conclusions

The question "Why do cationic polymerizations stop?" has seldom been asked and even more seldom answered. That termination occurs in many cationic polymerization systems can readily be and has often been demonstrated (see, for example, Figure 4.13). A review of pertinent information reveals three fundamentally different types of termination: (1) collapse of the propagating carbenium ion/counteranion pair or neutralization, (2) transfer of hydride ion to the chain carrier with simultaneous formation of stable allylic carbenium ions, and (3) deliberate destruction of the chain carrier by purposeful addition of nucleophiles or quenching. Although advances in the elucidation of the chemistries of the latter two types were rather modest during the last decade, significant progress has been made in the understanding and exploitation of the first and by far the most important type, that is, termination by neutralization.

In regard to termination by neutralization in olefin polymerizations coinitiated by well-known Friedel–Crafts acids, for example, BF_3, $TiCl_4$, $AlBr_3$, $AlCl_3$, and $SnCl_4$, very little can be added to Plesch's analysis written some 25 years ago (Plesch, 1954): "termination reactions...may be of many different kinds,...they may differ for apparently closely related systems, and...they may even be entirely absent." Regrettably, systematic investigation in this field lay dormant until it was recently recognized that desirable. functional end groups can be introduced into cationically obtained polymers by "controlled termination" under "transferless" polymerization conditions. It is not surprising that until the concept of controlled termination was fully appreciated and its usefulness for the synthesis of unique terminally and biterminally functionalized polymers had been demonstrated, research on termination had little justification other than academic.

Macromolecular engineering by the concept of controlled termination requires the coexistence of two mechanistic components: transferless polymerization and end group control by counteranion. The prerequisite for controlled termination is the assembly of a polymerization system in which chain transfer is absent and in which the only chain-breaking step is termination. Such systems are not "living" since termination is operative; they are transferless. The second requirement is the control of the chemistry of termination by neutralization. The latter is effected by the use of suitably designed counteranions. Transferless polymerizations with controlled termination are of great synthetic potential; they complement and in certain respects may even surpass the utility of true "living," that is, transferless and terminationless, systems.

The term transferless cationic polymerizations needs some elaboration. Operationally this term means that reaction conditions must be found under which termination by neutralization is faster than proton elimination. Except for a very few isolated instances in which carbocationic polymerizations exhibited some kinetic behavior akin to those of "living" systems, that is, ranges over which conversion and molecular weight increased in unison (Higashimura and Kishiro, 1977d; Kennedy et al., 1964b; Pepper, 1974b), truly living carbocationic polymerizations have not yet been devised (Kennedy et al., 1974a; Pepper, 1975), because they cannot be devised given the fundamentally unstable nature of propagating cations. (However, see remarks concerning quasiliving carbocationic polymerizations, Chapter 9).

Olah's technique (1973) developed to stabilize carbocations by the use of super acids, that is, solutions of "HF"/SbF$_5$ in nonnucleophilic solvents like SO$_2$, failed to produce "living" polyisobutylene or polystyrene and yielded only low molecular weight, ill-defined products (Kennedy et al., 1974a). Higashimura et al.'s most interesting lead (1977d) with the I$_2$/p-methoxystyrene/CCl$_4$ or CHCl$_3$/0°C system produces only a partially long-lived system (only the low molecular weight fraction as characterized by GPC exhibits linear molecular weight–conversion trend) and, disturbingly, all the molecular weight-conversion plots (Figure 2, 4, 5, 7, and 8 in Higashimura et al., 1977d) exhibit unexplained intercepts at 0% conversion, indicating the presence of unknown factor(s).

A truly transferless system would demand the existence of a propagating carbenium ion unable to eliminate protons and to maintain its integrity for long periods. Although such very desirable cationic systems for the synthesis of functional and block copolymers are presently unavailable, operationally just as valuable conditions can be achieved with systems in which elimination is suppressed by rapid termination. During the last decade several transferless cationic polymerization systems have been found witness the finding of cationic graft copolymerizations exhibiting virtually 100% grafting efficiencies (Kennedy and Smith, 1974c; Kennedy and·Charles, 1977d; Oziomek and Kennedy, 1977) and their synthetic potential demonstrated.

In practical terms, finding transferless systems is equivalent to finding propagating cations with suitable life-spans. Long-lived propagating carbenium ions are prone to eliminate protons and thus may lead to unacceptable chain transfer. Too short-lived propagating cations are obviously also unacceptable because they do not permit molecular weight buildup. According to this analysis, then, desirable life-spans of cation carriers are between these extremes.

The lifetime of propagating carbenium ions is determined by the rate of termination, which in turn can be controlled by the structure of counteranion and careful selection of experimental conditions in terms of initiating system, temperature range, and medium polarity. According to experimental evidence, both highly stable or highly nonnucleophilic and unstable or highly nucleophilic counteranions are counterproductive; rather, counteranions of intermediate stability or nucleophilicity are desirable. In the presence of highly stable counteranions the life-span of carbocations is prolonged and from the point of view of end group control unacceptable proton elimination may occur. On the other end of the scale, highly nucleophilic counteranions are also undesirable because in their presence chain propagation is short-stopped or oligomers may at best be formed.

Table 4.18 is a compilation of polymerization systems in which controlled termination is possible. The end groups shown have been substantiated by various evidences. Highest confidence may be placed in BCl_3 and Me_2CPDAl-based systems where direct evidence for the presence *and* position of terminal $-Cl$ and $-CPD$ groups has been obtained. Although transferless polymerizations seem to exist in a variety of isobutylene systems, styrene polymerizations are not completely transferless.

Analysis of the systems shown in Table 4.18 and considerations of chemical principles discussed in this section lead to the compilation in Table 4.19, which lists counteranions by nucleophilicities for which sufficient data in isobutylene polymerization systems exist. Those in the high nucleophilicity group are most likely unsuitable for controlled termination because in their presence ion pair collapse is too fast for (efficient) polymerization. For example, HCl and isobutylene produce t-BuCl and not polymer; in the presence of $Et_3AlCl^\ominus$ or i-$Bu_3AlCl^\ominus$, that is, counteranions containing a hydrogen β to aluminum, hydridation (see there) is prohibitively rapid.

Counteranions in the medium nucleophilicity group may be suitable for controlled termination. For example, tertiary chlorine end groups have been obtained by chlorinative termination in transferless polymerizations in the presence of $BCl_4^\ominus$ or $BCl_3OH^\ominus$. Methylation can be controlled with $Me_3AlCl^\ominus$, $Me_2AlCl_2^\ominus$ in transferless polymerizations. Similarly, hydridation is less rapid with $Et_2AlCl_2^\ominus$ than with $Et_3AlCl^\ominus$, and the t-$BuCl/Et_2AlCl$ initiating system produces transferless polymerizations. The coinitiators $(C_6H_5)_3Al$, $(CH_2=CH)_3Al$, and Me_2CPDAl have been shown to introduce phenyl, vinyl, and cyclopentadienyl groups at the terminus of polyisobutylene chains in transferless polymerizations.

The jury is still out with regard to $TiCl_4$. According to Cheradame (Cheradame, 1978; Sigwalt et al., 1973) $TiCl_4$ is "inert" with t-BuCl and the yield of polyisobutylene increases with decreasing temperatures in $TiCl_4$-coinitiated polymerizations. These two key facts can be readily explained assuming termination by chlorination by the $TiCl_5^\ominus$ or $TiCl_4OH^\ominus$ coun-

Table 4.18 □ **Polymerization Systems and Probable End Groups**

System	Terminating Counteranion	End Group	Reference
CCl3COOH/TiCl4/isobutylene/n-hexane/ -90 to 0°C[a]	TiCl4CCl3COO⊖	⁻C(CH3)2CCl3COO	Plesch (1968), Flett and Plesch, (1952)
In situ formation of AlCl3/isobutylene/CH3Cl/ < -78°C[a]	AlCl4⊖	—C(CH3)2Cl	Magagnini et al. (1977)
H2O/BCl3/styrene/CH3Cl2/ -20 to -78°C[a]	BCl3OH⊖	—CH(C6H5)Cl	Kennedy and Feinberg, (1978b)
H2O/BCl3/isobutylene/CH2Cl2/ -20 to -78°C	BCl3OH⊖	—C(CH3)2Cl	Kennedy et al. (1977a)
(CH3)2CH—CH=CHCH2Cl/BCl3/isobutylene/ CH2Cl2/ -20 to -78°C	BCl4⊖	—C(CH3)2Cl	Kennedy et al. (1979f)
Cl2/BCl3/isobutylene/CH2Cl2/ -20 to -78°C	BCl4⊖	—C(CH3)2Cl	Kennedy and Chen (1979d)
t-BuCl/AlMe3/isobutylene/CH3Cl/ -40 to -100°C	AlMe3Cl⊖	—C(CH3)3	Kennedy et al. (1969a, 1970, 1973b, 1974b)
t-BuCl/AlEt3/isobutylene/CH3Cl/ -20 to -100°C	AlEt3Cl⊖	—C(CH3)2H and —C(CH3)Et	Kennedy et al. (1969a, 1970, 1974b)
t-BuCl/Al(i-Bu)3/isobutylene/CH2Cl/ -20 to 100°C	Al(i-Bu)3Cl⊖	—C(CH3)2H	Kennedy and Milliman, (1969a), nedy (1970)
t-BuCl/Al(C6H5)3/isobutylene/C6H5Cl/ -20 to -40°C	Al(C6H5)3Cl⊖	—C(CH3)2C6H5	Kennedy and Chung (1980)
Cl(CH3)2C(-CH2-)3C(CH3)2Cl/Al(C6H5)3/ isobutylene/C6H5Cl/ -20 to -40°C	Al(C6H5)3Cl⊖	—C(CH3)2C6H5	Kennedy and Chung (1980)
t-BuCl/Me2AlC5H5/isobutylene/(CH2Cl)2/ 19 to -40°C	AlMe2C5H5Cl⊖	—C(CH3)2C5H5	Kennedy and Castner (1979b, c)
t-BuCl/Al(CHCH2)3/isobutylene/CH3Cl/ -10 to -78°C	Al(CHCH2)3Cl⊖	—C(CH3)2CH=CH2	Mandal and Kennedy (1978b)
ClSO3H/styrene/benzene/0°C	SO3Cl⊖	—CH(C6H5)Cl	Masuda et al. (1976)
HClO4/styrene/CH2Cl2/ -97°C	ClO4⊖	—CH(C6H5)OClO3	Bossaer et al. (1977)

[a]System probably not transferless, particularly at higher temperatures.

Table 4.19 □ Groups of Counteranion Nucleophilicities for Isobutylene Polymerization

High Nucleophilicity: Controlled Termination Unlikely due to too short $C^{\oplus}$ Lifetime (Rapid Neutralization)	Medium Nucleophilicity: Controlled Termination Possible	Low Nucleophilicity Controlled Termination Unlikely due to Prolonged $C^{\oplus}$ Lifetime (Propensity for Proton Elimination)
$F^{\ominus}$	$BCl_4^{\ominus}$	$BF_4^{\ominus}$
$Cl^{\ominus}$	$BCl_3OH^{\ominus}$	$BF_3OH^{\ominus}$
$Br^{\ominus}$	$AlCl_4^{\ominus}$	$AlBr_4^{\ominus b}$
$Et_3AlCl^{\ominus a}$	$AlCl_3OH^{\ominus}$	$AlBr_3OH^{\ominus}$
$i\text{-}Bu_3AlCl^{\ominus a}$	$Me_3AlCl^{\ominus}$	$MeAlCl_3^{\ominus}$
	$Et_2AlCl_2^{\ominus}$	$SbF_6^{\ominus}$
	$Et_2AlI_2^{\ominus}$	$SbF_5OH^{\ominus}$
	$(C_6H_5)_3AlCl^{\ominus}$	
	$(CH_2{=}CH)_3AlCl^{\ominus}$	
	$Me_2C_5H_5AlCl^{\ominus}$	
	$TiCl_5^{\ominus b}$	
	$TiCl_4OH^{\ominus b}$	
	$ClO_4^{\ominus}$ at $-97°C$	

[a] Rapid hydridation.
[b] Probable, unverified.

teranion. A blocking experiment (see Chapter 8) could substantiate this possibility. In contrast, some of Plesch's experiments (Biddulph and Plesch, 1960) with $TiCl_4$-initiating systems, that is, with CCl_3COOH or CF_3COOH initiators in *n*-hexane, indicate significant chain transfer to monomer. In a later publication (Biddulph et al., 1965), however, this worker reported the presence of measurable amounts of chlorine in low molecular weight polyisobutylenes obtained with $H_2O/TiCl_4/i\text{-}C_4H_8/CH_2Cl/$from $+18$ to $-12°C$ and attributed these to $ClCH_2{-}$ head groups arising by solvent transfer. Another likely explanation would be to assume the presence of $-CH_2C(CH_3)Cl$ end groups and the absence of ionization of these groups by $TiCl_4$.

Finally, counteranions in the low nucleophilicity group are deemed unsuitable for controlled termination under conventional conditions because in the presence of these species the life-span of propagating carbocations may be sufficiently long for proton elimination. Proton elimination necessarily leads to chain transfer which obviates end group control. For example, the initiating systems (RF/BF_3 or H_2O/BF_3) that give rise to $BF_4^{\ominus}$ or $BF_3OH^{\ominus}$ counteranions are useful for rapid initiation; however, they are so highly nonnucleophilic that in their presence neutralization by ion pair collapse is virtually absent. Thus the life-span of the propagating end $\sim CH_2-\overset{\oplus}{C}(CH_3)_2$ is extended and proton elimination occurs. Further, even if termination by fluorination were to occur momentarily, the fluorinated end group $\sim CH_2C(CH_3)_2F$ would immediately ionize in the presence of BF_3 in the system and propagation would continue. In fact, kinetic termination seems to be absent in reasonably pure $H_2O/BF_3/i\text{-}C_4H_8$ systems (Plesch, 1954). In the presence of relatively large amounts of impurities or under conventional commercial/industrial conditions this analysis of course cannot hold and multiple BF_3 additions are necessary to reach 100% conversion. Termination in $AlCl_3$-coinitiated polymerizations has not yet been completely settled, although Beard et al. (1964) have shown that in pure $H_2O/AlCl_3/i\text{-}C_4H_8/CH_2Cl_2/$from 0 to $-60°C$ systems polymerizations go to completion; that is, termination is absent, whereas chain transfer is significant. According to Kennedy and Thomas (1962) in a number of $AlCl_3$-based systems, for example, in "H_2O"$/AlCl_3/i\text{-}C_4H_8/CH_3Cl/$from -20 to $-150°C$, chain transfer dominates in the higher temperature range down to $\sim -80°C$. Below this point chain transfer is essentially frozen out and chain breaking only by termination takes over. In contrast, Magagnini et al. (1975) have evidence for termination by chlorination by the $AlCl_4^{\ominus}$ counteranion.

It was to be expected, and recently it has been demonstrated, that in addition to the nature of the counteranion, temperature also greatly influences conditions under which transferless polymerizations can exist. According to Kennedy and Trivedi (1978c, d) the slope of the log $\bar{M}_n$ of polyisobutylene versus $1/T$ plot—see, for example, Figure 4.11—($\Delta H_{\bar{M}} =$ activation enthalpy difference of molecular weight) is diagnostic for transfer-dominated and/or termination-dominated, that is, essentially transferless, polymerization conditions. These authors found that independent of the nature of the initiating system the slope of the log $\bar{M}_n$ versus $1/T$ plot (Arrhenius line) yields only three $\Delta H_{\bar{M}}$ values (in kcal/mole): -6.6 ± 1.0, -4.6 ± 1.0, and -1.8 ± 1.0. The temperature ranges over which these values hold are determined by the nature of the coinitiator. The temperature range over which $\Delta H_{\bar{M}} = -1.8 \pm 1.0$ kcal/mole indicates termination-dominated, transferless mechanism. In contrast $\Delta H_{\bar{M}} = -6.6 \pm 1.0$ kcal/mole suggests transfer-dominated conditions. The intermediate value $\Delta H_{\bar{M}} = -4.6 \pm 1.0$ kcal/mole indicates a transition range over which transfer domination gradually changes to termination domination. Table 5.4 is a compilation of data that illustrates these findings (Kennedy and Trivedi, 1978d) and these matters are further discussed in Section 5.6.

These data are of great help to the molecular engineer because they indicate whether or not controlled termination can be achieved with a particular initiating system and, if so, over what temperature range the experiments have to be carried out to achieve it. Further, in the case of new or little-tested initiator systems, $\Delta H_{\bar{M}_n}$ values provide valuable guidance toward the temperature range of transferless conditions.

As discussed in this section transferless polymerizations with controlled termination are now important synthetic options. By the use of the techniques described in this chapter the synthesis of various unique functional and telechelic polymers has been demonstrated. By skillful combination of controlled initiation and termination the preparation of many linear polymers with tailor-made termini can be envisioned. Several polymers carrying different and/or identical termini have been prepared and techniques for permutations are available. The cationic synthesis of polymers carrying one polymerizable or copolymerizable terminus, for example, $CH_2=CH-R$ where R = polymer, so-called macromers, has also become feasible. Indeed, the development of controlled initiation by recently developed boron- and aluminum-containing coinitiators in combination with transferless mechanisms and controlled termination leads to molecular engineering unheard of a decade ago and may herald a renaissance in the science of cationic polymerization.

□ REFERENCES

Abdul-Rasoul, F. A. M., Ledwith, A., and Yagci, Y. (1978a), *Polymer* **19**, 1219.

Abdul-Rasoul, F. A. M., Ledwith, A., and Yagci, Y. (1978b), *Polym. Bull.* **1**, 1.

Addecott, K. S. B., Mayor, L., and Turton, C. N. (1967), *Eur. Polym. J.* **3**, 601.

Akbulut, U., Fernandez, J. E., and Birke, R. L. (1975), *J. Polym. Sci. Polym. Chem. Ed.* **13**, 133.

Albert, A. A., and Pepper, D. C. (1951), *Proc. Roy. Soc. London Ser. A* **263**, 75.

Alder, R. W., Bowman, P. S., Steele, W. R. S., and Winterman, D. R. (1968), *Chem. Commun.*, 723.

Andrews, L. J., and Keefer, R. M. (1964), *Molecular Complexes in Organic Chemistry*, Holden-Day, San Francisco.

Anonymous (1964), *Hydrocarbon Proc. Petrol.* **43**, 183.

Anton, A., and Maréchal, E. (1971), *Bull. Soc. Chim. Fr.*, 2669.

Anton, A., and Maréchal, E. (1971), *Bull. Soc. Chim. Fr.*, 3753.

Aoki, S., Nakamura, H., and Otsu, T. (1968), *Makromol. Chem.* **115**, 282.

Aoki, S., Tarvin, R. F., Stille, J. K. (1970), *Macromolecules* **3**, 472.

Aoki, S., Shirafuji, C., and Otsu, T. (1971), *Polym. J.* **2**, 257.

Aquilanti, V., Galli, A., Giardini-Guidoni, A. and Volpi, G. G. (1967), *Trans. Faraday Soc.* **63**, 926.

Armstrong, V. C., Katovic, Z., and Eastham, A. M. (1971), *Can. J. Chem.* **49**, 2119.

Aso, C., Kunitake, T. and Shinkai, S. (1969), *Kobunshi Kagaku* **26**, 280.

Aso, C., Kunitake, T. and Tagami, S. (1972), *Prog. Polym. Sci. Jap.* **1**, 149.

Baccaredda, M., Bruzzone, M., Cesca, S., Dimaina, M., Ferraris, G., Giusti, P., Magagnini, P. L., and Priola, A. (1973), *Chim. Ind.* **55**, 109.

Bacskai, R. (1967), *J. Polym. Sci. Part A1* **5**, 619.

Barrer, R. M., and Oei, A. T. T. (1973), *J. Catal.* **30**, 460.

Bartlett, P. D., Condon, F. E. and Schneider, A. (1944), *J. Am. Chem. Soc.* **66**, 1531.

Bates, T. H., Best, J. V. F. and Williams, T. F. (1960), *Nature* **188**, 469.

Bates, T. H., Best, J. V. F., and Williams, T. F. (1962), *J. Chem. Soc.*, 1531.

Bauer, R. F., LaFlair, R. T., and Russel, K. E. (1970), *Can. J. Chem.* **48**, 1251.

Bauer, R. F., and Russel, K. E. (1971), *J. Polym. Sci: Part A1* **9**, 1951.

Bawn, C. E. H., and Ledwith, A. (1962), *Quart. Rev.* **16**, 361.

Bawn, C. E. H., Fitzsimmons, C., and Ledwith, A. (1964), *Proc. Chem. Soc.* 391

Bawn, E. H., Bell, R. M., and Ledwith, A. (1965a), *Polymer* **6**, 95.

Bawn, E. H., Carruthers, R. and Ledwith, A. (1965b), *Chem. Commun.*, 522.

Bawn, C. E. H., Fitzsimmons, C., Ledwith, A., Penfold, J., Sherrington, D. C., and Weightman, J. A. (1971), *Polymer* **12**, 119.

Beard, J. H., Plesch, P. H., and Rutherford, P. P. (1964), *J. Chem. Soc.*, 2566.

Bell, R. P. (1959), *The Proton in Chemistry*, Methuen, London.

Bergman, E. D., and Katz, D. (1958), *J. Chem. Soc.*, 3216.

Bertoli, V., and Plesch, P. H. (1968), *J. Chem. Soc. B*, 1500.

Bethea, S. R., and Karchmer, J. H. (1956), *Ind. Eng. Chem.* **48**, 370.

Biddulph, R. H. and Plesch, P. H. (1960), *J. Chem. Soc.*, 3913.

Biddulph, P. H., Plesch, P. H., and Rutherford, P. P. (1965), *J. Chem. Soc.*, 275.

Biswas, M., and Mishra, P. K. (1973), *J. Polym. Sci. Polym. Lett. Ed.* **11**, 639.

Biswas, M., and Mishra, P. K. (1975), *Polymer* **16**, 621.

Biswas, M., and Kabir, G. M. A. (1978), *Polymer* **19**, 595.

Bonin, M., Busler, W. R., and Williams, F. (1962), *J. Am. Chem. Soc.* **84**, 2895.

Bonin, M., Busler, W. R., and Williams, F. (1965), *J. Am. Chem. Soc.* **87**, 199.

Bonner, T. G., Clayton, J. M., and Williams, G. (1958), *J. Chem. Soc.*, 1705.

Boor, J., Youngman, E. A., and Dimbat, M. (1966), *Makromol. Chem.* **90**, 26.

Borson, C. A., Knight, J. R., and Robb, J. C. (1972), *Brit. Polym. J.* **4**, 427.

Bossaer, P. K., Goethals, E. J., Hackett, P. J., and Pepper, D. C. (1977), *Eur. Polym. J.* **13**, 489.

Bovey, F. A. (1960), *J. Polym. Sci.* **46**, 59.

Bowyer, P. M., Ledwith, A., and Sherrington, D. C. (1971), *J. Chem. Soc. B*, 1511.

Bowyer, P. M. (1972), Ph.D. Thesis, University of Liverpool.

Brackman, D. S., and Plesch, P. H. (1955), *Chem. Ind.*, 255.

Brake, W., Cheng, W. J., Pearson, J. M., and Szwarc, M. (1969), *J. Am. Chem. Soc.* **91**, 203.

Branchu, R. B., Cheradame, H. and Sigwalt, P. (1969), *C. R. Acad. Sci.* **268**, 1292.

Braun, D., Heufer, G., Johnsen, Y. and Kolbe, K. (1964), *Ber. Bunsenges,* **68**, 959.

Breitenbach, J. W., and Srna, Ch., and Olaj, O. F., (1960), *Makromol. Chem.* **42**, 171.

Breitenbach, J. W., and Srna, Ch. (1962), *Pure Appl. Chem.* **4**, 245.

Brendle, M. C. (1971), *C. R. Acad. Sci. C* **272**, 743.

Brendle, M. C., and Ilvoas-Frèmond, A. M. (1972), *J. Chim. Phys.* **69**, 1748.

Brendle, M. C., and Ilvoas-Frèmond, A. M. (1976), *Brit. Polym. J.*, 11.

Brown, H. C., and Brady, J. D. (1953), *J. Am. Chem. Soc.* **75**, 3570.

Brown, H. C., and Kanner, B. W. (1953), *J. Am. Chem. Soc.* **75**, 3865.

Brown, H. C. (1956), *J. Chem. Soc.*, 1248.

Brown, C. P., and Mathieson, A. R. (1957), *Trans. Faraday Soc.* **53**, 1033.

Brown, C. P., and Mathieson, A. R. (1958), *J. Chem. Soc.*, 3445.

Brownstein, S., Bywater, S., and Worsfold, D. J. (1961), *Makromol. Chem.* **48**, 127.

Burgess, F. J., Cunliffe, A. V., Richards, D. H., and Thomson, P. (1978), *Polymer* **19**, 334.

Butler, G. B., Miles, M. L., and Brey, W. S. (1965), *J. Polym. Sci. Part A1* **3**, 723.

Butler, G. B. and Miles, M. L. (1966), *Polym. Eng. Sci.* 71.

Butler, G. B., Guilbault, L. J., and Turner, S. R. (1971), *J. Polym. Sci. Polym. Lett.* **9**, 115.

Butler, G. B., and Ferree, Jr., W. I. (1976), *J. Polym. Sci. Polym. Symp.*, **56**, 397.

Butlerov, A. M. (1877), *Ann. Chem.* **189**, 47.

Bywater, S. (1963), in Plesch (1963) p. 312.

Bywater, S., and Worsfold, D. J. (1966), *Can. J. Chem.* **44**, 1671.

Calder, I. C., Lee, W. Y., and Treloar, F. E. (1969), *Aust. J. Chem.* **22**, 2689.

Cerrai, P., Giusti, P., Guerra, G., and Tricoli, M. (1974), *Eur. Polym. J.* **10**, 1195.

Cerrai, P., Giusti, P., Guerra, G., and Tricoli, M. (1975), *Eur. Polym. J.* **11**, 101.

Cesca, S., Priola, A., and Ferraris, G. (1972), *Makromol. Chem.* **156**, 325.

Cesca, S., Giusti, P., Magagnini, P., and Priola, A. (1975a), *Makromol. Chem.* **176**, 2319.

Cesca, S., Priola, A., Bruzzone, M., Ferraris, G., and Giusti, P. (1975b), *Makromol. Chem.* **176**, 2339.

Cesca, S., Priola, A., Ferraris, G., Busetto, C. and Bruzzone, M. (1976), Fourth International Symposium on Cationic Polymerization, *J. Polym Sci. Polym. Symp.* **56**, 159.

Chapiro, A. (1962), *Radiation Chemistry of Polymeric Systems*, Interscience, New York, p. 310.

Cheradame, H., and Sigwalt, P. (1964), *C. R. Acad. Sci.* **259**, 4273.

Cheradame, H., and Sigwalt, P. (1965), *C. R. Acad. Sci.* **260**, 159.

Cheradame, H., Hung, N. A., and Sigwalt, P. (1969), *C. R. Acad, Sci.* **268**, 476

Cheradame, H., and Sigwalt, P. (1970), *Bull. Soc. Chim. Fr.*, 843.

Cheradame, H., Mazza, M., Hung, N. A. and Sigwalt, P. (1973), *Europ. Polym. J.* **9**, 375.

Cheradame, H. (1978), personal communication, Paris.

Chiellini, E. (1970), *Macromolecules*, **3**, 527.

Chmelir, M., Marek, M., and Wichterle, O. (1965), *Symp. Macromol. Chem.*, *Prague*, Preprint. 110.

Chmelir, M., and Marek, M. (1967a), *Coll. Czech. Chem. Commun.* **32**, 3047.

Chmelir, M., Marek, M. and Wichterle, O. (1967b), *J. Polym. Sci. Polym. Symp.* **16**, 833.

Chmelir, M., and Marek, M. (1968), *J. Polym. Sci. Polym. Symp.* **22**, 177.

Chujo, R. (1966), *J. Phys. Soc. Jap.* **21**, 2669.

Clark, D. (1953), *Cationic Polymerization and Related Complexes*, P. H. Plesch, Ed., Academic Press, New York.

Clayton, J. M., and Eastham, A. M. (1957), *Can. J. Chem.* **39**, 138.

Clayton, J. M., and Eastham, A. M. (1961), *J. Am. Chem. Soc.* **79**, 6368.

Cocivera, M., and Winstein, S. (1963), *J. Am. Chem. Soc.* **85**, 1702.

Colclough, R. O. and Dainton, F. S. (1958a), *Trans. Faraday Soc.* **54**, 894.

Colclough, R. O. and Dainton, F. S. (1958b), *Trans. Faraday Soc.* **54**, 898.

Collinson, E., Dainton, F. S., and Gillis, H. A. (1959), *J. Phys. Chem.* **63**, 909; *J. Polym. Sci.* **34**, 241.

Cooper, W. (1963), in *The Chemistry of Cationic Polymerization*, P. H. Plesch, Ed., Macmillan, New York, p. 349.

Corno, C., Ferraris, G., and Cesca, S. (1977), *Ital. Macromol. Sci. Meet. Abstr.*, p. 194.

Corson, B. B., Dorsky, J., Nickels, J. E., Kutz, W. M., and Thayer, H. I. (1954), *J. Org. Chem.* **19**, 17.

Corson, B. B., Heintzelman, W. T., Moe, H. and Rousseau, C. R. (1962), *J. Org. Chem.* **27**, 1636.

Cotrel, R., Sauvet, G., Vairon, J. P., and Sigwalt, P. (1976), *Macromolecules* **9**, 931.

Coudane, J., Brigodiot, M. and Maréchal, E. (1979), *J. Macromol. Sci. Chem.* **A13**, 827.

Cowell, G. W., Kocharyan, K., Ledwith, A., and Woods, J. H. (1970), *Eur. Polym. J.*, **6**, 551.

Cram, D. J., and Kopecky, K. R. (1959), *J. Am. Chem. Soc.* **81**, 2748.

Cram, D. J., and Sahyun, M. R. V. (1963), *J. Am. Chem. Soc.* **85**, 1257.

Crivello, J. V., and Lam, J. H. W. (1976), *J. Polym. Sci. Polym. Symp.* **56**, 383.

Crivello, J. V., Lam, J. H. W., and Volante, C. N. (1977a), *J. Radiat. Curing*, 2.

Crivello, J. V., and Lam, J. H. W. (1977b), *Macromolecules* **10**, 1307.

Dainton, F. S., and Sutherland, G. B. (1949), *J. Polym. Sci.* **4**, 37.

Dainton, F. S., and Tomlinson, R. H. (1953), *J. Chem. Soc.*, 151.

Davidson, E. B. (1966), *J. Polym. Sci. Polym. Lett. Ed.* **4**, 175.

Davison, W. H. T., Pinner, S. H., and Worrall, R. (1957), *Chem. Ind. (London)*, 1274.

Davison, W. H. T., Pinner, S. H., and Worrall, R. (1959), *Proc. Roy. Soc. (London) A* **252**, 187.

De Sorgo, M., Pepper, D. C., and Szwarc, M. (1973), *Chem. Commun.*, 419.

Diem, T., and Kennedy, J. P., (1978), *J. Macromol. Sci. Chem.* **A12**, 1359.

Di Maina, M., Cesca, S., Giusti, P., Ferraris, G. and Magagnini, P. L. (1977), *Makromol. Chem.* **178**, 2223.

Dimitrov, I. K., and Panayotov, I. M. (1971), *Commun. Dept. Chem. (Sofia)* **4**, 137; *C. R. Acad. Bulg. Sci.* **24**, 191.

Disselhof, R., and Braun, D. (1964), diploma work by R. Disselhof, Darmstadt.

Dombroski, J. R., and Schuerch, C. (1971), *Macromolecules* **4**, 447.

Drijvers, W., and Goethals, E. J. (1971), *IUPAC Symp. Macromol. Chem., Boston Prepr.*, 663.

Eastham, A. M. (1956), *J. Am. Chem. Soc.* **78**, 6040.

Edwards, W. R., and Chamberlain, N. F. (1963), *J. Polym. Sci., Part A* **1**, 2299.

Eidus, Y. T., and Nefredov, B. K. (1960), *Usp. Khim.* **29**, 833.

Eley, D. D., and Richards, A. W. (1949), *Trans. Faraday Soc.* **45**, 436.

Eley, D. D., and Saunders, J. (1952), *J. Chem. Soc.*, 4167.

Eley, D. D., and Seabrooke, A. (1964), *J. Chem. Soc.*, 2226.

Eley, D. D., Isack, F. L., and Rochester, C. H. (1968), *J. Chem. Soc.*, 872, 1651.

Ellinger, L. P. (1963), *Chem. Ind.*, 1982

Emmet, P. H. (1958), *Catalysis*, Reinhold, New York, Vols. 1–3.

Endres, G. F., and Overberger, C. G. (1955), *J. Am. Chem. Soc.* **77**, 2201.

Endres, G. F., Kamath, V. G., and Overberger, C. G. (1962), *J. Am. Chem. Soc.* **84**, 4813 and previous publications in this series.

Evans, A. G., Holden, D., Plesch, P. H., Polanyi, M., Skinner, H. A. and Weinberger, M. A. (1946a), *Nature* **157**, 102.

Evans, A. G., Meadows, G. W., and Polanyi, M. (1946b), *Nature*, **158**, 94.

Evans, A. G., and Polanyi, M. (1947a), *J. Chem. Soc.* 252.

Evans, A. G., Meadows, G. W., and Polanyi, M. (1947b), *Nature* **160**, 869.

Evans, A. G., and Weinberger, M. A. (1947c), *Nature* **159**, 437.

Evans, A. G., and Meadows, G. W. (1950), *Trans. Faraday Soc.* **46**, 327.

Evans, A. G. (1951), *J. Appl. Chem.* **1**, 240.

Evans, A. G., and Lewis, J. (1957), *J. Chem. Soc.*, 2975.

Evans, A. G., and Owen, E. D. (1959), *J. Chem. Soc.*, 4723.

Evans, A. G., James, E. A., and Owen, E. D. (1961), *J. Chem. Soc.*, 3532.

Fairbrother, F. (1941), *Trans. Faraday Soc.* **37**, 763.

Feinberg, S. C., and Kennedy, J. P. (1976), *Polym. Prepr.* **17**, 797.

Ferraris, G., Corno, C., Priola, A., and Cesca, S. (1977), *Macromolecules* **10**, 188.

Fleichfresser, B. E., Cheng, W. J., Person, J. M., and Szwarc, M. (1968), *J. Am. Chem. Soc.* **90**, 2172.

Flett, M. St. C., and Plesch, P. H. (1952), *J. Chem. Soc.*, 3355.

Flett, M. St. C., and Plesch, B. M. (1953), in Plesch (1953a), p. 119.

Flory, P. J. (1953), *Principles of Polymer Chemistry*, Cornell University Press, Ithaca, N.Y.

Fontana, C. M., and Kidder, G. A. (1948), *J. Am. Chem. Soc.* **70**, 3745.

Fontana, C. M., Kidder, G. A., and Herold, R. J. (1952a), *Ind. Eng. Chem.* **44**, 1688.

Fontana, C. M., Herold, R. J., Kinney, E. J., and Miller, R. C. (1952b), *Ind. Eng. Chem.* **44**, 2955.

Fontana, C. M. (1959), *J. Phys. Chem.* **63**, 1167.

Fontana, C. M., (1963), in Plesch (1963) p. 209, 218.

Fordham, J. W., Burleigh, P. H., and Sturm, C. L. (1959), *J. Polym. Sci.* **41**, 73.

Fowells, W., Schuerch, S., Bovey, F. A., and Hood, E. P. (1967), *J. Am. Chem. Soc.* **89**, 1396.

Franta, E., Rempp, P., and Afshar-Taromi, F. (1977), *Makromol. Chem.* **178**, 2139.

Fujimatsu, M., Natsuume, T., Hirata, H., Shirota, Y., Kusabayashi, S., and Mikawa, H. (1970), *J. Polym. Sci. A1*, **8**, 3349.

Funt, B. L., and Blain, T. J. (1970a), *J. Polym. Sci. Part A1* **8**, 3339.

Funt, B. L., Blain, J., and Young, R. A. (1970b), *Polym. Prepr.* **11**, 2.

Funt, B. L., and Blain, J. (1971), *J. Polym. Sci. Part A1*, **9**, 115.

Furukawa, J. and Nishimura, J. (1976), *J. Polym. Sci., Polym. Symp.* **56**, 437.

Gandini, A., and Plesch, P. H. (1965), *J. Chem. Soc.*, 4826.

Gandini, A., and Plesch, P. H. (1968), *Eur. Polym. J.* **4**, 55.

Gandini, A., Cheradame, H., and Sigwalt, P. (1980), *Polym. Bull.*, **2**, 731.

Garreau, H., and Maréchal, E. (1973), *Eur. Polym. J.* **4**, 9.

Gayer, F. H. (1933), *Ind. Eng. Chem.* **25**, 1122.

Gaylord, N. G., Matyska, B., Mach, K., and Vodehnal (1966), *J. Polym. Sci. Part A1* **4**, 2493.

George, J., Wechsler, H., and Mark, H. (1950a), *J. Am. Chem. Soc.* **72**, 3891.

George, J., Mark, H., and Wechsler, H. (1950b), *J. Am. Chem. Soc.* **72**, 3896.

Ghanem, N. A., and Marek, M. (1972), *Eur. Polym. J.* **8**, 999.

Ghose, A., and Bhadani, S. N., (1975), *Indian J. Technol.* **13**, 172.

Gilbert, H., Miller, F. F., Averill, S. J., Carlson, E. J., Folt, V. L., Heller, H. J., Stewart, F. D., Schmidt, R. F. and Trumbull, H. L. (1956), *J. Am. Chem. Soc.* **78**, 1669.

Gippin, M. (1962), *Ind. Eng. Prod. Res. Dev.* **1**, 32.

Gippin, M. (1965), *Ind. Eng. Prod. Res. Dev.* **4**, 160 (1965).

Giusti, P. and Andruzzi, F. (1966a), *IUPAC Symp. Macromol. Chem. Praha*, Paper A215.

Giusti, P., Puce, G., and Andruzzi, F. (1966b), *Makromol. Chem.* **98**, 170.

Giusti, P., Andruzzi, F., Cerrai, P., and Possanzini, G. L. (1970), *Makromol. Chem.* **136**, 97.

Giusti, P., Priola, A., Magagnini, P. L., and Narducci, P., (1975), *Makromol. Chem.* **176**, 2303.

Goethals, E. J. (1974), *Makromol. Chem.* **175**, 1309.

Goka, A. M., and Sherrington, D. C. (1976), *J. Chem. Soc.*, *Perkin II* 329.

Goodman, M., and Fan, Y. L. (1964), *J. Am. Chem. Soc.* **86**, 4922.

Goodrich, J. E., and Porter, R. S. (1964), *J. Polym. Sci. Polym. Lett.* **2**, 353.

Gromangin, J. (1957), unpublished work quoted by Magat, M. in *Coll. Czech. Chem. Commun.* **22**, 141.

Guterbok, H. (1959), *Polyisobutylene*, Springer, Berlin.

Haas, H. C., Kamath, P. M., and Schuler, N. W. (1957), *J. Polym. Sci.* **24**, 85.

Hall, F. C., and Nash, A. W. (1938), *J. Int. Petrol. Technol.* **24**, 471.

Haugh, M. J., and Dalton, D. R. (1975), *J. Am. Chem. Soc.* **97**, 5674.

Hayashi, Ka., Takagaki, T., Takada, K., Hayashi, K., and Okamura, S. (1958), *Bull. Chem. Soc. Japan* **41**, 1261.

Hayashi, K., and Okamura, S. (1967), *J. Polym. Sci. Polym. Symp.* **22**, 15.

Hayashi, Ka., Yamazawa, Y., Takagaki, T., Williams, F., Hayashi, K., and Okamura, S. (1977), *Trans. Faraday Soc.* **63**, 1489.

Hayes, M. J., and Pepper, D. C. (1961), *Proc. Roy. Soc. (London)* **A263**, 63.

Heilbrunn, A. G., and Maréchal, E. (1973), *Bull. Soc. Chim. Fr.*, 470.

Hersberger, A. B., Reid, J. C., and Heiligmann, R. G. (1945), *Ind. Eng. Chem.* **37**, 1073.

Hersberger, A. B. and Heiligmann, R. C., to Atlantic Refining Company (1947), U.S. Patent 2,429,719.

Heublein, G., and Adelt, B. (1972), *Plaste Kautsch.* **19**, 177, 728.

Heublein, G., and Spange, St. (1974a), *Z. Chem.* **14**, 22.

Heublein, G., and Barth, O. (1974b), *J. Prakt. Chem.* **316**, 649.

Heublein, G., Ludewig, M., and Eschtke, H. D. (1975a), *Z. Chem.* **15**, 150.

Heublein, G., and Schutz, H. (1975b), *Faserforsch. Textiltech.* **26**, 213.

Heublein, G. (1975c), *Zum Ablauf Ionischer Polymerisations Reaktionen*, Akademy-Verlag, Berlin, p. 125.

Higashimura, T., and Okamura, S. (1956a), *Chem. High Polym. Jap.* **13**, 342.

Higashimura, T., and Okamura, S. (1956b), *Chem. High Polym. Jap.* **13**, 431.

Higashimura, T., and Okamura, S. (1956c), *Chem. High Polym. Jap.* **13**, 397.

Higashimura, T., and Okamura, S. (1958), *Chem. High Polym. Jap.* **15**, 702.

Higashimura, T., Yonezawa, T., Okamura, S., and Fukui, K. (1959), *J. Polym. Sci.* **39**, 487.

Higashimura, T., and Okamura, S. (1960), *Kobunski Kagaku* **13**, 57.

Higashimura, T., Kanoh, N., and Okamura, S. (1961), *Makromol. Chem.* **47**, 35.

Higashimura, T., Fukushima, P. T., and Okamura, S. (1967a), *J. Macromol. Sci.* **A1**, 683.

Higashimura, T., Ohsumi, Y., Kuroda, K., and Okamura, S. (1967b), *J. Polym. Sci.* **A1 3**, 863.

Higashimura, T., Kusudo, S., Ohsumi, Y., Mizote, A., and Okamura, S. (1968a), *J. Polym. Sci. Part A1* **6**, 2511.

Higashimura, T., Kusudo, S., Ohsumi, Y., and Okamura, S. (1968b), *J. Polym. Sci. Part A1* **6**, 2523.

Higashimura, T., Kusudo, S., and Okamura, S. (1968c), *Chem. High Polym. Tokyo* **25**, 694.

Higashimura, T., Ohsumi, Y., Okamura, S., Chujyo, R., and Kuroda, T., (1969), *Makromol. Chem.* **126**, 87.

Higashimura, T., and Hoshino, M. (1972), *J. Polym. Sci. Polym. Lett.* **10**, 269.

Higashimura, T., and Kishiro, O. (1974), *J. Polym. Sci.* **12**, 967.

Higashimura, T., Kishiro, O., and Takeda, T. (1976a), *J. Polym. Sci.* **14**, 1089.

Higashimura, T. (1976b), *J. Polym. Sci., Polym. Symp.* **56**, 71.

Higashimura, T., and Hirokawa, Y. (1977a), *J. Polym. Sci.* **15**, 1137.

Higashimura, T., Sawamoto, M. and Masuda, T. (1977b), *Prepr. Polym. Colloq. Kyoto, Sept.* 26.

Higashimura, T., and Nishii, H. (1977c), *J. Polym. Sci.* **15**, 329.

Higashimura, T., and Kishiro, O. (1977d), *Polym. J.* **9**, 87.

Higashimura, T., Takeda, T., Sawamoto, M., Matsuzaki, K. and Uryu, T. (1978), *J. Polym. Sci., Polym. Chem. Ed.* **16**, 503.

Higashimura, T. (1979a), *Polym. Prepr. (ACS)* **20**(1), 161.

Higashimura, T., Mitsuhashi, M., and Sawamoto, M. (1979b), *Macromolecules* **12**, 178.

Hoffman, A. S. H. (1959), *J. Polym. Sci.* **34**, 241.

Horrex, C., and Perkins, F. T. (1949), *Nature* **163**, 486.

Houdry, E. J. (1940), U.S. Patent 2,226,562.

Hudson, B. E., Jr. (1966), *Makromol. Chem.* **94**, 172.

Huet, J. M., and Maréchal, E. (1970), *C. R. Acad. Sci.*, **271**, 1058.

Hunter, W. H., and Yohe, R. V. (1933), *J. Am. Chem. Soc.* **55**, 1248.

Hyde, P., and Ledwith, A. (1974), in *Molecular Complexes*, R. Foster, Ed., Vol. 2, Elek Science, London, p. 174.

Imanishi, Y., Yamamoto, R., Higashimura, T., Kennedy, J. P., and Okamura, S. (1967), *J. Macromol. Sci. (Chem.)* **A1**, 877.

Ipatieff, V. N., and Grosse, A. V. (1936), *J. Am. Chem. Soc.* **58**, 915.

Irie, M., Tomimoto, S., and Hayashi, K. (1970), *J. Polym. Sci. Lett.* **8**, 585.

Irie, M., Tomimoto, S., and Hayashi, K. (1972a), *J. Polym. Sci.* **10**, 3235.

Irie, M., Tomimoto, S., and Hayashi, K. (1972b), *J. Polym. Sci.* **10**, 3243.

Irie, M., and Hayashi, K. (1975), *Prog. Polym. Sci. Jap.* **8**, 105.

Johnson, A. F., and Young, R. N. (1976a), *J. Polym. Sci.* **56**, 211.

Johnson, A. F., and Pearce, D. A. (1976b), *J. Polym. Sci. Polym. Symp.* **56**, 57.

Jones, E. K. (1956), *Adv. Catal.* **8**, 219.

Jones, G. D. (1963), in Plesch (1963), p. 542.

Jones, J. F. (1958), *J. Polym. Sci.* **33**, 513.

Jordan, D. O., and Treloar, F. E. (1961), *J. Chem. Soc.*, 737.

Kagiya, T., Izu, M., Maruyama, H., and Fukui, K. (1969), *J. Polym. Sci. Part A1* **7**, 917.

Kalfoglou, N., and Szwarc, M. (1968), *J. Phys. Chem.* **72**, 2233.

Kampmeier, J. A., Geer, R. P., Meskin, A. J., and D'Silva, R. M. (1966), *J. Am. Chem. Soc.* **88**, 1257.

Kanoh, N., Higashimura, T., and Okamura, S. (1962a), *Kobunski Kagaku* **19**, 181.

Kanoh, N., Higashimura, T., and Okamura, S. (1962b), *Makromol. Chem.* **56**, 65.

Kanoh, N., Gotoh, A., Higashimura, T., and Okamura, S. (1963), *Makromol. Chem.* **63**, 115.

Kanoh, N., Ikeda, K., Gotoh, A., Higashimura, T., and Okamura, S. (1965), *Makromol. Chem.* **86**, 200.

Kennedy, J. P. (1959), *J. Macromol. Sci., Chem.* **A3**, 861.

Kennedy, J. P., and Thomas, R. M. (1960a), *J. Polym. Sci.* **45**, 227.

Kennedy, J. P., and Thomas, R. M. (1960b), *J. Polym. Sci.* **46**, 233.

Kennedy, J. P., and Thomas, R. M. (1961), *J. Polym. Sci.* **49**, 189.

Kennedy, J. P., and Thomas, R. M. (1962), *Adv. Chem. Ser.* **34**, 111.

Kennedy, J. P., Minckler, L. S., Wanless, G. G., and Thomas, R. M. (1963), *J. Polym. Sci. Polym. Symp.* **4**, 289.

Kennedy, J. P., Minckler, L. S. Jr., Wanless, G. G., and Thomas, R. M. (1964a), *J. Polym. Sci., Part A* **2**, 1441, 2093.

Kennedy, J. P., Elliott, J. J. and Groten, B. (1964b), *Makromol. Chem.*, **77**, 26.

Kennedy, J. P., Minckler, L. S., Jr., and Thomas, R. M. (1964c), *J. Polym. Sci., Part A* **2**, 367.

Kennedy, J. P. (1964d), *J. Polym. Sci., Part A* **2**, 381.

Kennedy, J. P., Elliott, J. J., and Naegele, W. (1964e), *J. Polym. Sci., Part A* **2**, 5029.

Kennedy, J. P., Wanless, G. G., and Elliott, J. J. (1964f), *Polym Prepr. A.C.S. Div. Polym. Chem.* **5**, 679.

Kennedy, J. P., Cohen, C. A., and Naegele, W. (1964g), *J. Polym. Sci. Polym. Lett.* **2**, 1159.

Kennedy, J. P., Elliott, J. P. and Hudson, B. E. (1964h), *Makromol. Chem.* **79**, 109.

Kennedy, J. P., and Langer, A. W., Jr. (1964i), *Adv. Polym. Sci.* **3**, 508.

Kennedy, J. P. and Hinlicky. J. A. (1965a), *Polymer* **6**, 133.

Kennedy, J. P., Naegele, W., and Elliott, J. J. (1965b), *J. Polym. Sci. Polym. Lett.* **3**, 729.

Kennedy, J. P., Schulz, W. W., Squires, R. G., and Thomas, R. M. (1965c), *Polymer* **6**, 287.

Kennedy, J. P., and Squires, R. G. (1965d), *Polymer* **6**, 579.

Kennedy, J. P. (1966a), *Int. Symp. Macromol. Chem., Tokyo-Kyoto, Sept.*, Abstr. 2.1.04.

Kennedy, J. P. (1966b), *Polym. Prepr.* **7**, 485; (1968) *J. Polym. Sci. Part A1* **6**, 3139.

Kennedy, J. P., Borzel, P., Naegele, W., and Squires, R. G. (1966c), *Makromol. Chem.* **93**, 191.

Kennedy, J. P., and Squires, R. G. (1967a), *J. Macromol. Sci. Chem.* **A1**, 805.

Kennedy, J. P., and Squires, R. G. (1967b), *J. Macromol. Sci. Chem.* **A1**, 831.

Kennedy, J. P., and Squires, R. G. (1967c), *J. Macromol. Sci. Chem.* **A1**, 847.

Kennedy, J. P., and Squires, R. G. (1967d), *J. Macromol. Sci. Chem.* **A1**, 861.

Kennedy, J. P., Bank, S., and Squires, R. G. (1967e), *J. Macromol. Sci. Chem.* **A1**, 961.

Kennedy, J. P., Bank, S., and Squires, R. G. (1967f), *J. Macromol. Sci. Chem.* **A1**, 977.

Kennedy, J. P., and Squires, R. G. (1967g), *J. Macromol. Sci. Chem.* **A1**, 995.

Kennedy, J. P., and Squires, R. G. (1967h), *Polym. Prepr.* **8**, 460.

Kennedy, J. P. (1967i), *Encyclopedia of Polymer Science and Technology*, Wiley-Interscience, New York, Vol. 7., p. 754.

Kennedy, J. P., and Makowski, H. S. (1967j), *J. Macromol. Sci. Chem.* **A1**, 345.

Kennedy, J. P. (1967k), U.S. Patent 3,349,065.

Kennedy, J. P., and Squires, R. G. (1967l), *J. Polym. Sci., Polym. Symp.* **16**, 1541.

Kennedy, J. P., and Makowski, H. S. (1968a), *J. Polym. Sci., Polym. Symp.* **22**, 247.

Kennedy, J. P., Elliott, J. J. and Butler, P. E. (1968b), *J. Macromol. Sci. Chem.* **A2**, 1415.

Kennedy, J. P., and Tornquist, E. G. M. (1968c), *Polymer Chemistry of Synthetic Elastomers*, Wiley-Interscience, pp. 301–307, 409–413.

Kennedy, J. P. (1968d), discovery on Nov. 19, 1968, filed for patent March 9, 1970; abandoned January 12, 1972; refiled Jan. 28, 1972. J. P. Kennedy, to Exxon Research and Engineering Co. U.S. Patent 4,029,866 (June 14, 1978).

Kennedy, J. P., and Milliman, G. E. (1969a), *Adv. Chem. Ser.* **91**, 287.

Kennedy, J. P. (1969b), *J. Macromol. Sci. Chem.* **A3**, 885.

Kennedy, J. P. (1970), *J. Org. Chem.* **35**, 532.

Kennedy, J. P. (1971a), XXIII *Int. Cong. Pure Appl. Chem., Macromol. Prepr.* **1**, 105.

Kennedy, J. P., Magagnini, P. L., and Plesch, P. H. (1971b), *J. Polym. Sci. Part A1* **9**, 1635, 1647.

Kennedy, J. P., Shinkawa, A., and Williams, F. (1971c), *J. Polym. Sci. Part A1* **9**, 1551.

Kennedy, J. P. (1972a), *Pure Appl. Chem.*, 179.

Kennedy, J. P. (1972b), *J. Macromol. Sci. Chem.* **A6**, 329.

Kennedy, J. P. (1972c), *Int. Union Pure Appl. Chem., Macromol. Chem.* **8**, 179.

Kennedy, J. P. (1972d), *J. Macromol. Sci. Chem.* **A6**, 329.

Kennedy, J. P. and Gillham, J. K. (1972e), *Adv. Polym. Sci.* **10**, 1.

Kennedy, J. P. and Sivaram, S. (1973a), *J. Org. Chem.* **38**, 2262.

Kennedy, J. P., Desai, N. V. and Sivaram, S. (1973b), *J. Am. Chem. Soc.* **95**, 6386.

Kennedy, J. P., and Sivaram, S. (1973c), *J. Macromol. Sci., Chem.* **A7**, 969.

Kennedy, J. P., Melby, E., and Johnston, J. (1974a), *J. Macromol. Sci. Chem.* **A8**, 463.

Kennedy, J. P., and Rengachary, S. (1974b), *Adv. Polym. Sci.* **14**, 1.

Kennedy, J. P., and Smith, R. R. (1974c), in *Recent Advances in Polymer Blends, Grafts and Blocks*, L. H. Sperling, Ed., Plenum, New York.

Kennedy, J. P., and Trivedi, P. D. (1974d), *XXIII IUPAC Meet. Macromol., Madrid* **1**, 198.

Kennedy, J. P., and Johnston, J. E. (1975a), *Adv. Polym. Sci.* **19**, 57.

Kennedy, J. P. (1975b), *Cationic Polymerization of Olefins: A Critical Inventory*, Wiley, New York.

Kennedy, J. P. (1975c), in Kennedy (1975b), p. 154.

Kennedy, J. P., and Chou, T. (1976a), *Adv. Polym. Sci.* **21**, 1.

Kennedy, J. P., Feinberg, S. C., and Huang, S. Y. (1976b), *Polym. Prepr.* **17**, 194.

Kennedy, J. P., and Trivedi, P. D. (1976c), *Polym. Prepr.* **17**, 791.

Kennedy, J. P., Huang, S. Y., and Feinberg, S. C. (1977a), *J. Polym. Sci.* **15**, 2801.

Kennedy, J. P., and Mandal, B. M. (1977b), *Polym. Lett.* **15**, 595.

Kennedy, J. P., Feinberg, S. C., and Huang, S. Y. (1977c), *J. Polym. Sci., Chem. Ed.*, **15**, 2869.

Kennedy, J. P., and Charles, J. J. (1977d), *J. Appl. Polym. Sci., Appl. Polym. Symp.* **30**, 119.

Kennedy, J. P., and Nakao, M. (1977e), *J. Macromol. Sci. Chem.* **A11**, 1621.

Kennedy, J. P., Huang, S. Y. and Feinberg, S. C. (1978a), *J. Polym. Sci., Polym. Chem. Ed.* **16**, 243.

Kennedy, J. P., and Feinberg, S. Y. (1978b), *J. Polym. Sci., Polym. Ed.* **16**, 2191.

Kennedy, J. P., and Trivedi, P. D. (1978c), *Adv. Polym. Sci.* **28**, 83.

Kennedy, J. P., and Trivedi, P. D. (1978d), *Adv. Polym. Sci.* **28**, 112.

Kennedy, J. P., and Diem, T. (1978e), *Polym. Bull.* **1**, 29.

Kennedy, J. P., unpublished results (1979a).

Kennedy, J. P., and Castner, K. F. (1979b), *J. Polym. Sci. Polym. Chem. Ed.* **17**, 20.

Kennedy, J. P., and Castner, K. F. (1979c), *J. Polym. Sci. Polym. Chem. Ed.* **17**, 2055.

Kennedy, J. P., and Chen, F. J.-Y., (1979d) *Polym. Prepr.* **20**, 310.

Kennedy, J. P., and Chou, R. T. (1979e), *Polym. Prepr.*, **20**, 306.

Kennedy, J. P., Huang, S. Y., and Smith, R., (1979f), *Polym. Bull.* **1**, 371.

Kennedy, J. P., and Smith, R. C., (1979g), *Polym. Prepr.* **20**, 316.

Kennedy, J. P., and Plamthottam, S. S. (1980) *J. Macromol. Sci., Chem.*, **A14**, 729.

Kennedy, J. P., and Chung, D. Y. L. (1980), *Polym. Prepr.* **21**, 150.

Kennedy, J. P., and Delvaux, J. M. (1981), *Adv. Polym. Sci.* **38**, 143.

Kennedy, J. P., and Guhaniyogi, S., (1981), to be published.

Kennedy, J. P., Ross, L. R., Lackey, J. E., and Nuyken, O. (1981) *Polym. Bull.* **4**, 67.

Ketley, A. D., and Harvey, M. C. (1961), *J. Org. Chem.* **26**, 4649.

Ketley, A. D. (1963), *J. Polym. Sci. Polym. Lett.* **1**, 313.

Ketley, A. D. (1964a), *J. Polym. Sci. Polym. Lett.* **2**, 827.

Ketley, A. D. and Ehri, R. J. (1964b), *J. Polym. Sci. Part A* **2**, 4461.

Ketley, A. D., Berlin, A. J., and Fisher, A. P. (1967), *J. Polym. Sci. Part A1* **5**, 227.

Kikuchi, Y., and Matsumo, A. (1975), *Makromol. Chem.* **176**, 515.

Kilpatrick, M., and Luborsky, F. E. (1953), *J. Am. Chem. Soc.* **75**, 577.

Kodama, T., Higashimura, T., and Okamura, S. (1961), *Kobunshi Kagaku*, **18**, 267.

Korshak, V., and Lebedev, N. N. (1948), *J. Gen. Chem. U.S.S.R.* **18**, 1766.

Korshak, V. V., Sergeev, V. A., Shitikov, V. K., and Burenko, P. S. (1963), *Vysokomol. Soedin.* **5**, 1597.

Kössler, I., Stolka, M., and Mach, K. (1963), *J. Polym. Sci. Polym. Symp.* **4**, 977.

Kriz, J., and Marek, M. (1973), *Makromol. Chem.* **163**, 155.

Kubisa, P., Penczek, S. (1971), *Makromol. Chem.* **144**, 169.

Kunitake, T. and Aso, C. (1970), *J. Polym. Sci. Part A1* **8**, 665.

Kusabayashi, S., and Mikawa, H. (1969), *Chem. Commun.*, 189.

Kusabayashi, S., and Mikawa, H. (1970), *Polym. J.* **1**, 181.

Laguerre, J. P., and Maréchal, E. (1974), *Ann. Chim.* **9**, 163.

Lambla, M., Scheibling, G., and Banderet, A. (1970), *C. R. Acad. Sci.* **271**, 924.

Lambla, M., Koenig, R. and Banderet, A., (1972), *Eur. Polym. J.* **8**, 1.

Langlois, G. E. (1953), *Ind. Eng. Chem.* **45**, 1470.

Ledwith, A., and Sambhi, M. (1965), *Chem. Commun.*, 64.

Ledwith, A. (1969a), *Ann. N.Y. Acad. Sci.* **155**, 385.

Ledwith, A. (1969b), *Adv. Chem. Ser.* **91**, 317.

Ledwith, A., and Sherrington, D. C. (1971), *Polymer* **12**, 344.

Ledwith, A., and Sherrington, D. C. (1975), *Adv. Polym. Sci.* **19**, 1.

Ledwith, A. (1978), *Polymer* **19**, 1217.

Ledwith, A. (1979) *Pure Appl. Chem.* **51**, 159.

Lee, W. Y. and Treloar, F. E. (1969), *J. Phys. Chem.* **73**, 2458.

Leonhardt, H., and Weller, A. (1963), *Ber. Phys. Chem.* **67**, 791.

Lichin, N. N., and Pappas, P. (1957), *N.Y. Acad. Sci. Ser II* **20**, 143.

Longworth, W. R., and Plesch, P. H. (1958), *J. Chem. Soc.*, 451.

Longworth, W. R., and Plesch, P. H. (1959a), *J. Chem. Soc.*, 1887.

Longworth, W. R., Plesch, P. H., and Rutherford, P. P. (1959b), *Int. Conf. Coordination. London.*

Longworth. W. R., and Plesch, P. H. (1959c), *Symp., Wiesbaden*, p. III-A-11.

Longworth, W. R., Plesch, P. H., and Rutherford, P. P. (1960), *Proc. Chem. Soc.*, 68.

Longworth, W. R., and Mason, C. P. (1966), *J. Chem. Soc. A*, 1164.

Lopour, P., and Marek, M. (1970), *Makromol. Chem.* **134**, 23.

Lyudvig, E. V., Gantmacher, A. R., and Medvedev, S. S. (1968), *Symp. Macromol. Chem. Prague*, p. 324.

Magagnini, P. L., and Plesch, P. H. (1971a), unpublished results in Kennedy et al. (1971b).

Magagnini, P. L., Plesch, P. H. and Kennedy, J. P. (1971b), *Eur. Polym. J.* **7**, 1161.

Magagnini, P. L., and Priola, A. (1973), *Chim. Ind.* **55**, 109.

Magagnini, P. L., Cesca, S., Giusti, P., Priola, A., and di Maina, M. (1977), *Makromol. Chem.* **178**, 2235.

Maltsev, V. V., Plate, N. A., Azimov, T., and Kargin, V. A. (1969), *Vysokomol. Soedin.* **A11**, 220; *Polym. Sci. USSR* **2**, 248.

Manatt, S. L., Ingham, J. D., and Miller, J. A., Jr. (1977), *Org. Magn. Reson.* **10** (*Suppl. Vol.*), 198.

Mandal, B. M., Kennedy, J. P., and Kiesel, R. (1978a), *J. Polym. Sci., Chem. Ed.* **16**, 821.

Mandal, B. M. and Kennedy, J. P. (1978b), *J. Polym. Sci., Chem. Ed.* **16**, 833.

Maréchal, E., Basselier, J. J., and Sigwalt, P. (1964), *Bull. Soc. Chim. Fr.*, 1740.

Marek, M. (1959), Czech. Patent 88,879.

Marek, M., Roosova, M., and Doskocilova, D. (1967), *J. Polym. Sci. Polym. Symp.* **16**, 971.

Marek, M. and Chmelir, M. (1968), *J. Polym. Sci. Polym. Symp.* **22**, 177.

Marek, M. and Toman, L. (1973), *J. Polym. Sci. Polym. Symp.* **42**, 339.

Marek, M., Toman, L., and Pilar, J. (1975), *J. Polym. Sci. Chem. Ed.* **13**, 1565.

Marek, M. (1977), *J. Polym. Sci., Polym. Symp.* **56**, 149.

Marschner, F. (1940), U.S. Patent 2,199,133.

Maslinska-Solich, J., Chmelir, M. and Marek, M. (1969), *Coll. Czech. Chem. Commun.* **34**, 2611.

Mass, O. (1918), *J. Am. Chem. Soc.*, **40**, 1561.

Mass, O. (1921), *J. Am. Chem. Soc.*, **43**, 1227.

Mass, O., Boomer, E. H., and Morrison, D. M. (1923), *J. Am. Chem. Soc.* **45**, 1433.

Mass, O. (1925), *J. Am. Chem. Soc,* **47**, 2883.

Mass, O. (1930), *Can. J. Res.*, **3**, 526.

Masuda, T., and Higashimura, T. (1971a), *J. Polym. Sci. Polym. Lett.* **9**, 783.

Masuda, T., and Higashimura, T. (1971b), *J. Polym. Sci. Part A1*, **9**, (1563).

Masuda, T., Sawamoto, M. and Higashimura, T. (1976), *Makromol. Chem.* **177**, 2981.

Mathieson, A. R. (1963), in *The Chemistry of Cationic Polymerization* P. H. Plesch, Ed., Macmillan, New York, p. 291.

Matsuda, M., and Abe, K. (1968a), *J. Polym. Sci., Part A1* **6**, 1441.

Matsuda, M., and Abe, K. (1968b), *Kogyo Kagaku Zasshi* **71**, 425.

Matsushima, S. (1972a) Ger. Patent 2,163,956 to Sumitomo Chemical Co.

Matsushima, S., and Ueno, K. (1972b), Ger. Patent 2,163,957 to Sumitomo Chemical Co.

Matsuzaki, K., Hamada, M. and Arita, K. (1967), *J. Polym. Sci.* **5**, 1233.

Matyska, M., Mach, K., Vodehnal, J. and Kossler, I. (1965), *Coll. Czech. Chem. Commun.* **30**, 2569.

Matyska, M., Svestka, M., and Mach, K. (1966), *Coll. Czech. Chem. Commun.* **31**, 659.

Mayen, M., and Maréchal, E. (1972), *Bull. Soc. Chim. Fr.*, 4662.

Mayo, F. R., and Katz, J. J. (1947), *J. Am. Chem. Soc.* **69**, 1339.

Mengoli, G., and Vidotto, G. (1970), *Makromol. Chem.* **139**, 293.

Mengoli, G., and Vidotto, G. (1972a), *Eur. Polym. J.* **8**, 661.

Mengoli, G., and Vidotto, G. (1972b), *Eur. Polym. J.* **8**, 671.

Metz, D. J., Potter, R. C., and Thomas, J. K. (1967), *J. Polym. Sci.* **5**, 877.

Metz, D. J. (1969), "Radiation in induced ionic polymerization" *Addition and Condensation Polymerization Process, Adv. Chem. Ser.* **91**, 202.

Michel, R. H. (1964), *J. Polym. Sci., Part A* **2**, 2533.

Migahead, M. D., and Beckey, D. (1971), *Kolloid-Z. Z. Polym.*, **246**, 679.

Miyoshi, M., Uemura, S., Tsuchiya, S. and Kato, O. (1958), U.S. Patent 3,402,145.

Monroe, L. A., and Gilliland, E. R. (1938), *Ind. Eng. Chem.*, **30**, 58.

Mulliken, R. S., and Peason, W. B. (1969), *Molecular Complexes*, Wiley-Interscience, New York.

Nagai, T., Miyazaki, T., Sonoyama, Y., and Tokura, N. (1968), *J. Polym. Sci. Part A1* **6**, 3087.

Nagakura, S. (1975), *Excited States*, E. Lim, Ed., Academic Press, New York, Vol. 2.

Nakane, R., Watanabe, T., and Kurihara, O. (1962), *J. Chem. Soc. Jap.* **35**, 1747.

Natsuume, T., Akana, Y., Tanabe, K., Fujimatsu, M., Shimizu, M., Shirota, Y., Hirata, H., Kusabayashi, S., and Mikawa, H. (1969), Chem. Commun. 189.

Natsuume, T., Nishimura, M., Fujimatsu, M., Shimizu, M., Shirota, Y., Hirata, H., Kusabayashi, S., and Mikawa, H., (1970), *Polymer J.* **1**. 181.

Natta, G. (1960), *J. Polym. Sci.* **48**, 219.

Natta, G., Peraldo, M., Farina, M. and Bressan, G. (1962), *Makromol. Chem.* **55**, 139.

Nippon Oil Co. Ltd. (1965), Brit Patent 1,656,730.

Nippon Petrochemicals Co. Ltd. (1967), Brit. Patent 1,183,118.

Nippon Petrochemicals Co. Ltd. (1971), Brit. Patent 1,233,557.

Norrish, R. G., and Russel, K. E. (1952), *Trans. Faraday Soc.* **48**, 91.

Norton, C. J. (1964), *Ind. Eng. Chem.* **3**, 230.

Oberrauch, E., Salvatori, T., and Cesca, S. (1978), *Polym. Lett.* **16**, 345.

Oguni, N., Kamachi, M. and Stille, J. K. (1974), *Macromol.* **7**, 435.

Ohsumi, Y., Higashimura, T., and Okamura, S. (1965), *J. Polym. Sci., Part A1* **3**, 3729.

Ohsumi, Y., Higashimura, T., and Okamura, S. (1966), *J. Polym. Sci. Part A1* **4**, 923.

Ohsumi, Y., Higashimura, T., and Okamura, S. (1967a), *J. Polym. Sci. Part A1* **5**, 849.

Ohsumi, Y., Higashimura, T., Okamura, S., Chujo, R., and Kuroda, T., (1967b), *J. Polym. Sci. Part A1*, **5**, 3009.

Ohsumi, Y., Higashimura, T., Okamura, S., Chujo, R. and Kuroda, T., (1968), *J. Polym. Sci. Part A1*, **6**, 3015.

Okamura, S., Higashimura, T., and Sakurada, I. (1959), *J. Polym. Sci.* **39**, 507.

Okamura, S., Kanoh, N., and Higashimura, T. (1961), *Makromol. Chem.* **47**, 19.

Okamura, S., Kanoh, N., and Higashimura, T. (1962), *Chem. High Polym. Jap.* **19**, 181.

Olah, G. A., Quinn, H. W., and Kuhn, S. J. (1960), *J. Am. Chem. Soc.* **82**, 426.

Olah, G. A., Baker, E. B., Evans, T. C., Tolgyesi, W. S., McIntyre, J. S., and Bastien, I. J. (1964), *J. Am. Chem. Soc.* **86**, 1360.

Olah, G. A. (1973), *Angew. Chem.* **85**, 173.

Olah, G. A. (1974), *Makromol. Chem.* **175**, 1039.

Olah, G. A., Halpern, Y., and Lin, H. C. (1975), *Synthesis*, 315.

Ottolenghi, M. (1973), *Acc. Chem. Res.* **6**, 153.

Overberger, C. G., and Endres, G. F. (1953), *J. Am. Chem. Soc.* **75**, 6349.

Overberger, C. G., and Endres, G. F. (1955), *J. Polym. Sci.* **16**, 283.

Overberger, C. G., Ehring, R. J., and Marcus, R. A. (1958), *J. Am. Chem. Soc.* **80**, 2456.

Overberger, C. G., and Burns, C. M. (1969), *J. Polym. Sci. Part A1* **7**, 333.

Oziomek, J. and Kennedy, J. P. (1977), in *Cationic Graft Copolymerization*, J. P. Kennedy, Ed., J. Appl. Polym. Sci., Appl. Polym. Symp. **30**, 91.

Pac, J., and Plesch, P. H. (1967), *Polymer* **8**, 237.

Panayotov, I. M., Dimitrov, I. K., and Bakerdjiev, E. I. (1969), *J. Polym. Sci. Part A1* **7**, 2421.

Panayotov, I. M., Velichkova, R. S., and Matev, N. (1974), *C. R. Acad. Bulg. Sci.* **27**, 1679.

Panayotov, I. M., Velichkova, R. S., Dimitrov, I., and Jossifov, H., (1975), *C. R. Acad. Bulg. Sci.* **28**, 339.

Panayotov, I. M., and Heublein, G. (1977), *J. Macromol. Chem.* **A11**, 2065.

Pasman, H. J., Vlugter, J. C., and Breukink, C. J. (1965), *Brennstoff Chem.* **46**, 271.

Penczek, S., Jagur-Grodzinski, J., and Szwarc, M. (1968), *J. Am. Chem. Soc.* **90**, 2174.

Penczek, S., and Kubisa, P. (1973), *Makromol. Chem.* **165**, 121.

Penczek, S. (1974), *Makromol. Chem.* **175**, 1217.

Penfold, J., and Plesch, P. H. (1961), *Proc. Chem. Soc.*, 311.

Pepper, D. C. (1949), *Trans. Faraday Soc.* **45**, 404.

Pepper, D. C. (1954), Quart. Rev. **8**, 88.

Pepper, D. C. (1959), *Int. Symp Macromol. Chem.*, *Wiesbaden*, paper III, A9.

Pepper, D. C., and Reilly, P. J. (1961), *Proc. Chem. Soc.*, 460.

Pepper, D. C. (1963), in Plesch (1963) p. 257.

Pepper, D. C., and Reilly, P. J. (1966), *Proc. R. Soc. (London) Ser. A* **291**, 41.

Pepper, D. C. (1972), *IUPAC Macromol. Symp., Helsinki, Prepr.*, Vol. 1, p. 30.

Pepper, D. C. (1974a), *23rd Int. Symp. Macromol., Madrid.*

Pepper, D. C. (1974b), *Makromol. Chem.* **175**, 1077.

Pepper, D. C. (1975), *J. Polym. Sci., Polym. Symp.* **50**, 51.

Pilar, J., Toman, L., and Marek, M. (1976), *J. Polym. Sci., Chem. Ed.* **14**, 2399.

Pinner, S. H. (1963), in Plesch (1963) p. 611.

Pistoia, G. (1974a), *Eur. Polym. J.* **10**, 279.

Pistoia, G. (1974b), *Eur. Polym. J.* **10**, 285.

Pistoia, G., and Scrosati, B. (1974c), *Eur. Polym. J.* **10**, 1115.

Plesch, P. H., Polanyi, M., and Skinner, H. A. (1947), *J. Chem. Soc.*, 257.

Plesch, P. H. (1950), *J. Chem. Soc.*, 543.

Plesch, P. H. (1953a), *Cationic Polymerizations and Related Complexes*, Heffer, Cambridge, p. 85.

Plesch, P. H. (1953b), *J. Chem. Soc.*, 1653.

Plesch, P. H. (1953c), *J. Chem. Soc.*, 1662.

Plesch, P. H. (1954), *J. Polym. Sci.* **12**, 481.

Plesch, P. H. (1955), *Ric. Sci. Suppl. Simp. Int. Chim. Macromol.* **25A**, 140.

Plesch, P. H. (1963), *The Chemistry of Cationic Polymerization*, Pergamon, New York.

Plesch, P. H. (1966), *Pure Appl. Chem.* **12**, 117.

Plesch, P. H. (1968), *Prog. High Polym.* **2**, 737.

Plesch, P. H. (1973), *Pure Appl. Chem., IUPAC Macromol. Chem.* **8**, 305.

Plesch, P. H. (1974), *Makromol. Chem.*, **175**, 1065.

Polton, A., and Sigwalt, P. (1969), *C. R. Acad. Sci.* **268**, 1214.

Polton, A., and Sigwalt, P. (1970), *Bull. Soc. Chim. Fr.*, 131.

Potashnik, R., Goldschmidt, C. R., Ottolenghi, M., and Weller, A. (1971), *J. Chem. Phys.* **55**, 5344.

Priola, A., Ferraris, G., Di Maina, M., and Giusti, P. (1975a), *Makromol. Chem.* **176**, 2271.

Priola, A., Cesca, S., Ferraris, G., and Di Maina, M., and Giusti, P., (1975b), *Makromol. Chem.* **176**, 2289.

Prishepa, N. D., Goldfarb, Y. Y., and Krentsel, B. A. (1956), *Vysokomol. Soedin.* **8**, 1658.

Puskas, I., Banas, E. M., and Nerheim, A. G. (1976), *J. Polym. Sci. Polym. Symp.* **56**, 191.

Rehner, J., Zapp, R. L., and Sparks, W. J. (1953), *J. Polym. Sci.* **9**, 21.

Reibel, L., Kennedy, J. P., and Chung, Y. L. (1977), *J. Org. Chem.* **42**, 690.

Reibel, L., Kennedy, J. P., and Chung, Y. L. (1979), *J. Polym. Sci., Chem. Ed.*, **17**, 2757.

Roberts, W. J., and Day, A. R. (1950), *J. Am. Chem. Soc.* **72**, 1226.

Rogers, R. L. (1957), Third Biennal Electron Beam Symposium, Milwaukee, summarized in *Nucleonics* **15**, 180.

Roller, M. B., Gillham, J. K., and Kennedy, J. P. (1973), *Appl. Polym. Sci.* **17**, 2223.

Rosen, M. J. (1953), *J. Org. Chem.* **18**, 1701.

Ruckel, E. R., Arlt, H. G., and Wojcik, R. T. (1975), *Adhesion Sci. Technol.* **9A**, 395.

Russel, K. E., and Vail, G. M. C. (1976), *J. Polym. Sci. Polym. Symp.* **56**, 183.

Saegusa, T., Imai, H., and Furukawa, J. (1964), *Makromol. Chem.* **79**, 207.

Saegusa, T. (1969), in *Structure and Mechanism in Vinyl Polymerization*, Tsuruta and O'Driscoll, Eds., Dekker, New York, p. 283.

Sakurada Y., Matsumoto, M., Imai, K., Nishioka, A., and Kato, Y. (1963), *J. Polym. Sci., Polym. Lett.* **1**, 633.

Sambhi, M. S., and Treolar, F. E. (1967), *J. Polym. Sci. Polym. Lett.* **3**, 445.

Sambhi, M. S. (1970), *Macromolecules* **3**, 351.

Sartori, G., Valvassori, A., Turba, V., and Lachi, M. P. (1963), *Chim. Ind. (Milan)* **45**, 1529.

Sartori, G., Lammens, H., Siffert, J., and Bernard, A. (1971), *J. Polym. Sci. Polym. Lett.* **9**, 599.

Sauvet, G., Vairon, J. P., and Sigwalt, P. (1967), *C. R. Acad. Sci.* **265c**, 1090.

Sauvet, G., Vairon, J. P., Sigwalt, P. (1969), *J. Polym. Sci. Part A1* **7**, 983.

Sauvet, G., Vairon, J. P., and Sigwalt, P. (1974), *Eur. Polym. J.* **10**, 501.

Sauvet, G., Vairon, J. P., and Sigwalt, P. (1978), *J. Polym. Sci., Chem. Ed.* **16**, 3047.

Sawada, H. (1972), *J. Macromol. Sci.-Revs.* **C7**, 161.

Sawamoto, M., Masuda, T., Nishii, H., and Higashimura, T. (1975), *J. Polym. Sci. Polym. Lett.* **13**, 279.

Sawamoto, M., Masuda, T., and Higashimura, T., (1976), *Makromol. Chem.* **177**, 2995.

Sawamoto, M., Masuda, T., Higashimura, T., Kobayashi, S., and Saegusa, T. (1977), *Makromol. Chem.* **178**, 389.

Schildknecht, C. E., Gross, S. T., Davidson, H. R., Lambert, J. M., and Zoss, A. O. (1948), *Ind. Eng. Chem.* **40**, 2105.

Schlag, E. W., and Sparapany, J. J. (1964), *J. Am. Chem. Soc.* **86**, 1875.

Schmitt, G. J., and Schuerch, C. (1961), *J. Polym. Sci.* **49**, 287.

Schnabel, W., and Schmidt, W. F. (1971), *Ber. Phys. Chem.* **75**, 654.

Schnabel, W., and Schmidt, W. F. (1973), *J. Polym. Sci. Polym. Symp.* **42**, 273.

Schuerch, C., Fowells, W., Yamada, Y., Bovey, F. A., Hood, E. P., and Anderson, E. W. (1964), *J. Am. Chem. Soc.* **86**, 4481.

Scott, H., Miller, G. A., and Labes, M. M. (1963), *Tetrahedron Lett.* **1**, 1073.

Sherrington, D. (1969), Ph.D. Thesis, University of Liverpool.

Shigeru, O., and Hiroshi, S. (1972), *Secki Daigaku Kogakubu Kogaku Hokobu* **14**, 1127.

Shirota, Y., Tada, K., Shimidzu, M., Kusayabashi, S., and Mikawa, H. (1970), *Chem. Commun.*, 1110.

Shirota, Y., Fujioka, H., and Mikawa, H. (1978a), *J. Polym. Sci. Polym. Lett.* **16**, 425.

Shirota, Y., and Mikawa, H. (1978b), *J. Macromol. Sci.-Rev., Macromol. Chem. Polym. Symp.*, 1875.

Sigwalt, P., Lapeyre, W., and Cheradame, H. (1973), *Int. Symp. Cationic Polym., Rouen*, Abstract C-33.

Sigwalt, P. (1974), *Makromol. Chem.* **175**, 1077.

Sigwalt, P., and Szwarc, M. (1976), discussion relative to Cesca, *J. Pol. Sci. Symp.* **56**, 159.

Sivola, A., and Harva, O., (1970), *Suomi Kemistil.* **B43**, 475.

Skell, P. S., and Hall, W. L. (1963), *J. Am. Chem. Soc.* **85**, 2851.

Snyder, C., McIver, W., and Sheffer, H. (1977), *J. Appl. Polym. Sci.* **21**, 131.

Solich, J. M., Chmelir, M., and Marek, M. (1969), *Coll. Czech. Chem. Commun.* **34**, 2611.

Sommer, F., and Breitenbach, J. W. (1969), *IUPAC Int. Symp. Macromol. Chem. Budapest* **1**, 257.

Sparapany, J. J. (1966), *J. Am. Chem. Soc.* **88**, 1357.

Spoerri, P. E., and Rosen, M. J. (1950), *J. Am. Chem. Soc.* **72**, 4918.

Stang, P. J., and Anderson, A. G. (1978), *J. Am. Chem. Soc.* **100**, 1520.

Stille, J. K., and Chung, D. C. (1975a), *Macromolecules* **8**, 83.

Stille, J. K., and Chung, D. C. (1975b), *Macromolecules* **8**, 114.

Stille, J. K., Oguni, N., Chung, D. C., Tarvin, R. F., Aoki, S., and Kamachi, M. (1975c), *J. Macromol. Sci. Chem.* **A9**, 745.

Stoermer, R., and Kootz, H. (1928), *Chem. Ber.* **61**, 2330.

Subira, F., Polton, A., and Sigwalt, P. (1973), *Int. Symp. Cationic Polym., Rouen, Fr.*

Suzuki, M., Yamamoto, Y., Irie, M., and Hayashi, K. (1976), *J. Macromol. Sci. Chem.* **A10**, 1607.

Szell, T., and Eastham, A. M. (1966), *J. Chem. Soc. B*, 30.

Tabata, Y. (1976), *J. Polym. Sci. Polym. Symp.* **56**, 409.

Tada, K., Saegusa, T., and Furukawa, J. (1967), *Makromol. Chem.* **102**, 47.

Tada, K., Shirota, Y., Kusabayashi, S., and Mikawa, H. (1970), *Chem. Commun.* 682.

Tada, K., Shirota, Y., and Mikawa, H. (1973a), *Macromol.* **6**, 9.

Tada, K., Shirota, Y. and Mikawa, H. (1973b), *J. Polym. Sci. Polym. Chem. Ed.* **11**, 2961.

Tada, K., Shirota, Y., and Mikawa, H. (1974), *Macromolecules* **7**, 549.

Taft, R. W., Purlee, E. L., Reisz, P., and DeFazio, C. A. (1955), *J. Am. Chem. Soc.* **77**, 1584.

Takashashi, I., and Yamashita, Y. (1965), *J. Polym. Sci. Polym. Lett.* **3**, 251.

Tanaka, Y., and Sato, H. (1976), *J. Polym. Sci. Polym. Lett.* **14**, 335.

Tarvin, R. F., Aoki, S., and Stille, J. K. (1972), *Macromolecules* **5**, 663.

Tazuke, S., Tjoa, T. B., and Okamura, S. (1967), *J. Polym. Sci.* **A1 5**, 1911.

Thomas, R. M., Sparks, W. J., Frolich, P. K., Otto, M., and Mueller-Cunradi, M. (1940), *J. Am. Chem. Soc.* **62**, 276.

Throssel, J. J., Sood, S. P., Szwarc, M., and Stannett, V. (1956), *J. Am. Chem. Soc.* **78**, 1122.

Tidswell, B. M., and Doughty, A. G. (1971), *Polymer* **12**, 431.

Tipper, C. F. H., and Walker, D. A. (1959), *J. Chem. Soc.* 1352.

Toman, L., Marek, M., and Jokl, J. (1974), *J. Polym. Sci. Chem. Ed.* **12**, 1897.

Toman, L., and Marek, M. (1976), *Makromol. Chem.* **177**, 3325.

Toncheva, V., Velichkova, R., and Panayotov, I. M. (1974), *Bull. Soc. Chim. Fr.*, 1033.

Trifan, D. S., and Bartlett, P. D. (1959), *J. Am. Chem. Soc.* **81**, 5573.

Turcot, L., Glasel, A., and Funt, B. L. (1974), *J. Polym. Sci. Polym. Lett.* **12**, 687.

Vairon, J. P. and Sigwalt, P. (1971), *Bull. Soc. Chim. Fr.*, 559.

Van Lohuizen, O. E., and de Vries, K. S. (1968), *J. Polym. Sci., Polym. Symp.* **16**, 3943.

Velichkova, R. S., and Panayotov, I. M. (1970), *Makromol. Chem.* **138**, 171.

Vermeil, C. (1964a), *C. R. Acad. Sci.* **259**, 369.

Vermeil, C., Mathéson, M., Leach, S., and Muller, F. (1964b), *J. Chim. Phys.* **61**, 596.

Vesely, K. (1958), *J. Polym. Sci.* **30**, 375.

Villesange, M., Sauvet, G., Vairon, J. P., and Sigwalt, P. (1980), *Polym. Bull.* **2**, 131.

Viswanathan, N. S., and Kevan, L. (1967), *J. Am. Chem. Soc.* **89**, 2482.

Viswanathan, N. S., and Kevan, L. (1968), *J. Am. Chem. Soc.* **90**, 1375.

Wablat, W., Schmidt, W. F., and Schnabel, W. (1974), *Makromol. Chem.* **175**, 2687.

Wanless, G. G., and Kennedy, J. P. (1965), *Polymer* **6**, 111.

Watson, J. M., and Rysselberge, J. V. (1978), *J. Polym. Sci., Chem. Ed.* **16**, 1173.

Wichterle, O., Marek, M., and Trekoval, J. (1959a), *IUPAC Symp. Macromol. Wiesbaden,* Sect. III A 13.

Wichterle, O., Kolinsky, M. and Marek, M. (1959b), *Coll. Czech. Chem. Commun.* **24**, 2473.

Wichterle, O., Marek, M., and Trekoval, J. (1961), *J. Polym. Sci.* **53**, 281.

Williams, F. (1964), *J. Am. Chem. Soc.* **86**, 3954.

Williams, F. (1968), in *Fundamental Processes in Radiation Chemistry*, P. Ausloos, Ed., Interscience, New York, p. 515.

Williams, F. (1976), discussion relative to Cesca, *J. Pol. Sci. Symp.* **56**, 159.

Wislicenus, J. (1878), *Ann.* **92**, 106.

Woolhouse, R. A., and Eastham, A. M. (1966), *J. Chem. Soc. B*, 33 and nine previous papers.

Worrall, R., and Charlesby, A. (1959a), *Int. J. Appl. Radiation Isotopes* **4**, 84.

Worrall, R., and Pinner, S. H. (1959b), *J. Polym. Sci.* **34**, 229.

Worsfold, D. J., and Bywater, S. (1957), *J. Am. Chem. Soc.* **79**, 4917.

Yamada, Y., Schimada, K., and Hayashi, T. (1966), *J. Polym. Sci. Polym. Lett.* **4**, 477.

Yamada, N., Schimada, K., and Takemura, T. (1967), U. S. Patent 3,326,879.

Yamada, K., Tanaka, H., and Kawazura, H. (1976), *J. Polym. Sci. Polym. Lett.* **14**, 517.

Yamamoto, R., Imanishi, Y., and Higashimura, T. (1967a), *Kobunshi Kagaku* **24**, 397.

Yamamoto, R., Imanishi, Y., and Higashimura, T. (1967b), *Kobunshi Kagaku* **24**, 405.

Yamamoto, R., Imanishi, Y., and Higashimura, T. (1967c), *Kobunshi Kagaku* **24**, 412.

Yamamoto, R., Imanishi, Y., and Higashimura, T. (1967d), *Kobunshi Kagaku* **24**, 479.

Yamamoto, R., Imanishi, Y., and Higashimura, T. (1967e), *Kobunshi Kagaku* **24**, 486.

Yamaoka, H., Takakura, K., Hayashi, K., and Okamura, S. (1966), *Polym. Lett.* **4**, 509.

Yonezawa, T., Higashimura, T., Katagiri, K., Hayashi, K., Okamura, S., and Fukui, K. (1957), *J. Polym. Sci.* **26**, 311.

Yoshida, H., and Hayashi, K. (1969), *Adv. Polym. Sci.* **6**, 401.

Zlamal, Z. (1959), in *Kinetics and Mechanism of Polymerization*, G. E. Ham, Ed., Vol. 1, Chapter 6.

Zorn, H. (1958), *Angew. Chem.* **A60**, 185.

□ 5 □

Kinetics
of Carbocationic
Polymerization

5.1 □ INTRODUCTION

The organization and final writing of this chapter may have been the most frustrating part of our mission to rejuvenate the field of carbocationic polymerization. Our frustration grew by gradually realizing that "kinetics of carbocationic polymerization" as such does not really exist. No doubt there have been some outstanding investigations of kinetic nature; however, even the total of these studies falls far short of a coherent, generalizable discipline of carbocationic polymerization *kinetics*. Critical analysis of pertinent publications leads to the conclusion that the derivation of simple kinetic expressions of general validity, such as those available for radical or anionic polymerizations, is impossible in the field of carbocations at this time.

Among the several presently insurmountable difficulties that contribute to this somewhat disheartening conclusion, the following are discussed in detail in subsequent sections:

1 □ *Multiplicity of active chain carriers* □ Although authors have often considered only one growing carbocation in their analyses, recently it has become increasingly evident that carbocationic chains are carried by a multiplicity of species (ions, ion pairs, solvated ions, aggregates) having different chemical characteristics.

2 □ *Limited validity of stationary state assumption* □ For convenience authors have often assumed the existence of steady state conditions during polymerization. This assumption is probably erroneous and steady state conditions are probably more the exception than the rule in carbocationic polymerizations.

3 □ *Neglecting the rate of transfer reactions* □ For simplicity rate expressions have been derived by neglecting chain transfer; this is often a questionable if not objectionable practice.

4 □ *Underestimation of the effect of experimental conditions on rate expressions* □ The importance of conditions on rates of carbocationic reactions, particularly solvent and counteranion, cannot be overemphasized. Even minute endogenous changes may invalidate conclusions (i.e., decreasing the temperature a few degrees).

5 □ *Assumption of second-order propagation and chain transfer steps* □ In most kinetic propositions second-order propagation and chain transfer reactions have been assumed to operate without direct confirmation. Evidence to the contrary is accumulating. Propagation could be first- or second- or even higher-order; desolvation rates are almost universally neglected.

6 □ *Fragmentary knowledge of mechanism* □ It is impossible to construct a mechanistic scheme without detailed knowledge of the underlying chemistry; nonetheless schemes have often been proposed on the basis of assumptions in the absence of precise chemical information. Similar to this difficulty is the next point.

7 □ *Complexity of carbocationic reaction paths* □ In view of their intrinsic high reactivities, carbocations may undergo a bewildering variety of reactions; only few systems involving these species have been defined in detail sufficient for kinetic analysis.

Against this background, "surveying" the kinetics of carbocationic polymerizations presents a truly hazardous undertaking. In view of the large amount of published information of kinetic nature it was decided to select specific systems the analysis of which appeared to be of interest from the fundamental point of view, and to search for areas where some cautious generalizations may be attempted. Thus this chapter is less of a survey of generalizable systems since very few are comparable, but more of a presentation of principles and difficulties that need solutions before a coherent system of carbocationic polymerization kinetics can be developed.

The following symbolism is used (brackets indicate concentrations): $[I]$ = initiator, $[CoI]$ = coinitiator, $[M]$ = monomer, $[P]$ = polymer, $[M^{\oplus}]$ = active or growing cation; initial concentrations are denoted by subscript zero, for example, $[M]_0$; concentrations at time t are indicated by subscript t, for example, $[M]_t$; k_i, k_p, k_{tr}, and k_t = rate constants for initiation, propagation, chain transfer, and termination; $k_{tr,M}$ = chain transfer to monomer; R_i, R_p, R_{tr}, and R_t = rates of initiation, propagation, chain transfer, and termination, R_o = overall rate; $\overline{DP}_n$ = number average degree of polymerization.

5.2 □ VALIDITY OF THE STEADY STATE ASSUMPTION IN CARBOCATIONIC POLYMERIZATIONS

Two types of steady or stationary states have been distinguished (Plesch, 1963): (1) when $R_i = R_t$ and (2) when R_i = instantaneous and $R_t = 0$. In practice the phase of the reaction during which the rate of change of active center concentration $d[M^{\oplus}]/dt$ is negligible as compared to its rate of formation and destruction is regarded to be stationary (Bamford et al., 1958). Thus under stationary conditions:

$$\frac{d[M^{\oplus}]}{dt} = R_i - R_t = k_i[I][M] - k_t[M^{\oplus}] = 0 \tag{5.1}$$

Whatever the rate of initiation and termination, in the presence of unimolecular termination and absence of chain transfer,

$$\overline{DP}_n = \nu = \frac{R_p}{R_i} = \frac{R_p}{R_t} = \frac{k_p[M^{\oplus}][M]}{k_t[M^{\oplus}]} = \frac{k_p[M]}{k_t} \tag{5.2}$$

where ν = kinetic chain length. In the presence of chain transfer,

$$\overline{DP}_n = \frac{R_p}{R_t + \sum R_{tr,T}} = \frac{k_p[M]}{k_t + \sum k_{tr,T}[T]} \tag{5.3}$$

where $R_{tr,T}$ and $k_{tr,T}$ are, respectively, the rate and rate constant of chain transfer to transfer agent T.

According to equations 5.2 and 5.3 under stationary conditions the number average degree of polymerization $\overline{DP}_n$ is independent of active center concentration and propagation, chain transfer, and termination are first order with respect to $[M^{\oplus}]$.

In most cases researchers could not corroborate their assumption in regard to stationary conditions. Many carbocationic polymerizations are so rapid (for example, the polymerization isobutylene by "H_2O"/$AlCl_3$ is over within a few seconds even at $-100°C$) that stationary conditions cannot be proved by experiment. Even in polymerizations that may be considered very slow, such as styrene polymerization with $ReCl_5$ at $0°C$ (Kamachi and Miyama, 1965), the stationary state is not reached after 10^3 to 10^4 sec, after 2 to 30% conversion.

Only very few studies concern the experimental determination of active center concentration during carbocationic polymerization. According to Vairon and Sigwalt (1971), stationary conditions do not prevail in the H_2O/n-BuOTiCl$_3$/cyclopentadiene/CH_2Cl_2/$-70°C$ system since the $\overline{DP}_n$ of polycyclopentadiene decreases with increasing $[CoI]_0$. In this system

$$(R_0)_t = (k_p + k_{tr,M})[M^{\oplus}]_t[M]_t \tag{5.4}$$

which leads to

$$[M^{\oplus}]_t = \frac{1}{k_p + k_{tr,M}} \frac{(R_0)_t}{[M]_t} \tag{5.5}$$

Experimental determination of $(R_0)_t$ and $[M]_t$ yields the relation of $[M^{\oplus}]_t$ with time shown in Figure 5.1. Apparently $[M^{\oplus}]$ rapidly reaches a maximum.

These observations may be explained by assuming either that the active centers are formed instantaneously and are slowly destroyed (Burton and Pepper, 1961; Hayes and Pepper, 1961) or that the active centers form slowly and, except at the very beginning of the reaction, $R_t > R_i$.

According to the first assumption the initial number of polymer chains should be approximately equal to the moles of initiator and $\overline{DP}_n$ should increase during polymerization. In contrast, it has been shown (Vairon and Sigwalt, 1971) that the initial number of polymer molecules is smaller by a factor of 10^3 than the moles of initiator used, and $\overline{DP}_n$ decreases with time. Thus the first assumption cannot be correct and the observations are compatible with the second hypothesis.

Similar conclusions in regard to the absence of stationary state have been reached by authors who examined the $Ph_3C^{\oplus}SbCl_6^{\ominus}$/isobutyl vinyl

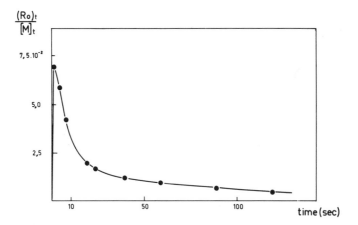

Figure 5.1 □ Active center concentration as a function of time for the H_2O/n-BuOTiCl$_3$/cyclopentadiene/CH$_2$Cl$_2$/ $-70°C$ system ([H$_2$O]$_0$ = 8.5 10^{-4} mole/l; [n-BuOTiCl$_3$]$_0$ = 10^{-3} mole/l; [M]$_0$ = 0.503 mole/l (from Vairon and Sigwalt, 1971).

ether/CH$_2$Cl$_2$/-40 to $+20°C$ (Subira et al., 1976) and Ph$_3$C$^\oplus$SbCl$_6^\ominus$/cyclopentadiene/CH$_2$Cl$_2$/$-30°C$ (Sauvet et al., 1974) systems. Figure 5.2 shows $(R_0)_t/[M]_t$ as a function of time at various temperatures (Subira et al., 1976). The shape of the curve obtained at 20°C suggests relatively rapid initiation (complete initiator consumption before complete

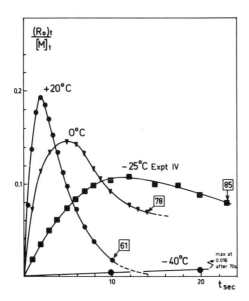

Figure 5.2 □ Variation of active center concentration with temperature for the Ph$_3$C$^\oplus$SbCl$_6^\ominus$/isobutyl vinyl ether/CH$_2$Cl$_2$ system. Numbers in squares refer to monomer conversion (Subira et al., 1976).

polymerization) and significant termination. Variations in $[M^{\oplus}]$ at -25 and $-40°C$ suggest slow rate of formation of active centers. According to complementary experiments the stability of active centers was probably low at $0°C$. In a similar vein, as shown in Figure 5.3, $[M^{\oplus}]$ steadily increases in the $Ph_3C^{\oplus}SbCl_6^{\ominus}/p$-methoxystyrene/$CH_2Cl_2$ system in the -2 to $+25°C$ range (Cotrel et al., 1976). In this instance k_t must be very small.

Since the variation of $[M^{\oplus}]$ with time is a function of k_i and k_t, one type of curve should change into another type by changing the temperature. This has indeed been observed in the H_2O or HCl/n-$BuOTiCl_3/\alpha$-methyl-styrene/CH_2Cl_2 system in the -30 to $-70°C$ range, as shown in Figure 5.4 (Villesange et al., 1977).

Other examples of nonstationary cationic polymerization have been reported (Pepper and Reilly, 1966); however, in most cases quantitative evaluation of $[M^{\oplus}]$ is absent. Sawamoto and Higashimura (1978a, b) used stop-flow spectroscopy to study the cationic polymerization of p-methoxystyrene in dichloroethane at $-30°C$. The spectra of active polymerization charges initiated by I_2, CH_3SO_3H, $BF_3 \cdot OEt_2$, and $SnCl_4$ were essentially identical and exhibited a strong absorption at 380 nm. The authors assigned this absorption to a single propagating carbocation. Figure 5.5 shows ΔOD_{380} as a function of time at various iodine concentrations. Depending on [I], ΔOD_{380} increases with time and reaches a plateau after 10 to 100 sec. Monomer consumption measured at 295 nm indicated fastest rate at maximum ΔOD_{380}. Let ϵ_{380} be the molar absorptivity of the peak at 380 nm; then

$$-\frac{d[M]_t}{dt} = k_p[M^{\oplus}][M]_t = k_p \frac{\Delta OD_{380}}{\epsilon_{380}}[M]_t \qquad (5.6)$$

Integration of equation 5.6 leads to

$$\ln\frac{[M]_{t1}}{[M]_{t2}} = \frac{k_p}{\epsilon_{380}}\int_{t1}^{t2} \Delta OD_{380}\, dt \qquad (5.7)$$

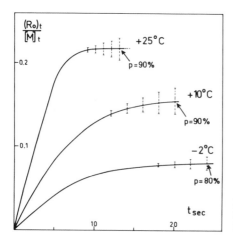

Figure 5.3 ☐ Variation of active center concentration with temperature for the $Ph_3C^{\oplus}SbCl_6^{\ominus}/p$-methoxystyrene/$CH_2Cl_2$ system (p = monomer conversion at the last significant point of each plot: $[M_0] \times 10^2 = 1\text{--}8$ and $[I_0] \times 10^5 = 2.5 - 11$). [Reprinted with permission from Cotrel et al., *Macromolecules* 9, 93 (1976). Copyright 1976 American Chemical Society.]

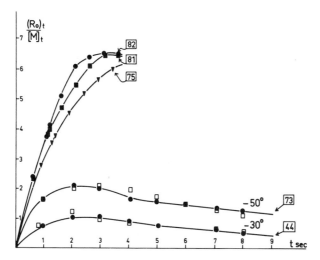

Figure 5.4 □ Variation of active center concentration in the H_2O or HCl/n-$BuOTiCl_3/\alpha$-methylstyrene/$CH_2Cl_2/-30$ to $-70°$ system; $[CoI] = 1.05\ M$. Numbers in squares represent monomer conversion (%) at the last point of each plot. At $-70°C$; $[M]_0 \times 10^2$ (mole/l) = ● 11.4; ■ 30; ▲ 22.6. At -50 and $-30°C$, $[M]_0 = 0.3$ mole/l. □ = Experimental points; ● = theoretical points obtained by simulation using the best fit for k_t, k_i, k_p. [Reprinted from Villesange et al., (1977), p. 391, by courtesy of Marcel Dekker, Inc.]

The left side of equation 5.7 is experimentally available by measuring monomer consumption at 295 nm and graphic integration of ΔOD_{380} versus time plots affords the integral. In this manner plots shown in Figure 5.6 are obtained and the slopes of the lines yield k_p. According to these data the nature of the propagating species is independent of the counteranion, at least in these systems.

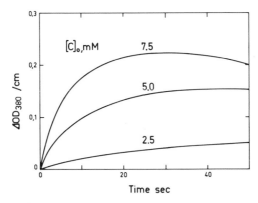

Figure 5.5 □ Change in the absorption at 380 nm for the I_2/p-methoxystyrene/1, 2-dichloroethane/30° system. ($[M]_0 = 5 \times 10^{-3}\ M$, $[I] = 7.5, 5.0, 2.5$ mM). [Reprinted with permission from Sawamoto and Higashimura, *Macromolecules* **11**, 328 (1978). Copyright 1978 American Chemical Society.]

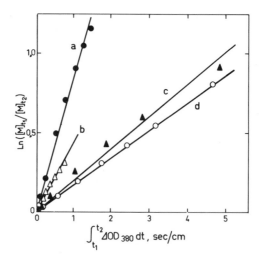

Figure 5.6 □ Plots of equation 5.7 for the initiator/p-methoxystyrene/1, 2-dichloroethane/30°C system. Initiator = a = CH$_3$SO$_3$H (5 mM); b = SnCl$_4$ (0.030 mM); c = BF$_3$ · OEt$_2$ (0.5 mM); d = I$_2$ (5.0 mM). [Reprinted with permission from Sawamoto and Higashimura, *Macromolecules* **11**, 328 (1978). Copyright 1978 American Chemical Society.]

In contrast to k_p, [M$^\oplus$]$_{max}$ values are strongly affected by the nature of the initiator. With I$_2$ and CH$_3$SO$_3$H [M$^\oplus$]$_{max}$ was much smaller than [I$_0$] (i.e., ~0.1 of [I]$_0$), whereas with BF$_3$ · OEt$_2$ and SnCl$_4$ [M$^\oplus$]$_{max}$ was 5 to 25% of [I]$_0$. These observations led Sawamoto et al. 1978a to conclude that the empirical reactivity order of initiators reflects [M$^\oplus$]$_{max}$/[I]$_0$ and not k_p.

In summary, according to a survey of published information a well authenticated steady state carbocationic polymerization system has not yet been described and the many kinetic rate constants and some rate constant ratios that have been derived by assuming the existence of steady state conditions should be critically reexamined. It appears that stationary conditions in carbocationic polymerizations are probably more the exception than the rule and the constancy of [M$^\oplus$] has not been experimentally verified by authors who assumed steady state conditions.

5.3 □ DETERMINATION OF RATES AND RATE CONSTANTS

Difficulties Relative to k_p Determination

The k_p's in radical and cationic polymerizations are affected by fundamentally different factors. In radical polymerization k_p is mainly determined by the structure of active centers and monomer, and temperature. In addition to these factors, k_p in cationic polymerizations is affected by the nature of the counteranion and solvent, that is, ionicity and solvation of the active site. Unfortunately, the effect(s) of solvent and temperature, particularly with polar systems, is very difficult if not impossible to separate (see also Section 5.2).

Plesch (1971) has analyzed the reasons that render the determination of k_p difficult and cast doubt on many published data. By definition,

$$R_p = k_p[M^{\oplus}][M] \tag{5.8}$$

To obtain k_p it is necessary to know the overall rate of polymerization:

$$R_o = k_0[C]^x[M]^y \tag{5.9}$$

where [C] is the nominal initiator or coinitiator concentration, whichever is rate determining or a function of these. If R_0 is known, the relation between the coefficients of [M] in equations 5.8 and 5.9 can be established, that is, between $k_p[M^{\oplus}]$ and $k_0[C]^x[M]^{y-1}$, and thus $[M^{\oplus}]$ can be obtained.

In order to obtain $[M^{\oplus}]$ one must know the type and the nature of active centers, the relationships that govern their relative concentrations and their dependence on [C], the ionicity of the active species, and the relative concentration of active species if more than one active center coexist. Since the chain carriers are charged species, their state of solvation by solvent, monomer, or polymer must be known and changes of solvation as a function of the extent of reaction must be assessed. No wonder that, in view of such an extensive list of demands, k_p could be determined in only a few instances.

The problem of the definition of the nature of the active species has been analyzed by several authors (Plesch, 1968, 1971; Szwarc, 1968, 1969; Ledwith and Sherrington, 1976). Except in nonpolar solvents, for example, CCl_4 and n-alkanes, free ions may always be present during cationic polymerization. The contribution of free ions to the overall rate cannot be neglected even when their concentration is very low because the reactivity of free ions is probably far higher than that of ion pairs.

The definition of the ionicity, that is, nature of the various active species, involved in carbocationic-polymerization is rendered difficult not only because of the complex equilibria between free ions and ion pairs but also because these entities may be chemically different (Plesch, 1968). This may be the case, for example, in a system in which both water and solvent RX are initiators so that two kinds of counteranions $MX_nOH^{\ominus}$ and $MX_{n+1}^{\ominus}$ may coexist. Higashimura et al. (1974, 1976, 1977, 1979; Masuda and Higashimura, 1971, Masuda et al., 1976; Sawamoto et al., 1977) have shown the coexistence of various active species in carbocationic polymerizations (see Section 4.1) and demonstrated the effect of reaction conditions on their relative concentrations. According to Plesch (1971) the relative contributions of free ions and ion pairs to propagation may be treated as follows ($P_n^{\oplus}$ = growing chain, $A^{\ominus}$ counteranion, K = equilibrium constant):

$$P_n^{\oplus}A^{\ominus} \rightleftharpoons P_n^{\oplus} + A^{\ominus} \tag{5.10}$$

$$[P_n^\oplus] = [A^\ominus] = f; \quad [P_n^\oplus A^\ominus] = p; \quad [p_n^\oplus] + [p_n^\oplus A^\ominus] = C; \quad f + p = C \quad (5.11)$$

$$K = \frac{f^2}{p} = \frac{f^2}{C - f} \quad (5.12)$$

and

$$f = K^{1/2} p^{1/2} = \tfrac{1}{2}[-K + (K^2 + 4KC)^{1/2}] \quad (5.13)$$

In principle K can be calculated according to Dennison and Ramsey (1955) or to Fuoss (1958).

Since active carbocations are inherently unstable species, K must often be estimated from data obtained with more stable, relatively large carbocations under conditions similar to those of polymerizations (Jones and Plesch, 1970; Sangster and Worsfold, 1972; Bowyer et al., 1971b; Longworth and Mason, 1966; Kalfoglou and Szwarc, 1968; Lee and Treloar, 1969; Ledwith, 1969). K varies little even in the presence of significant structural differences. Thus Bowyer et al. (1971b) obtained a narrow range of K values for various hexachloroantimonate salts despite relatively wide variations in the structure of the cations, for example, $C_7H_7^\oplus$, $\phi_3C^\oplus$, and xanthylium. However, approximations of this kind might be much less valid with cations structurally related to the growing site in styrene, isobutylene, or alkyl vinyl ether polymerizations that is, with carbocations having much less delocalized charges (Plesch et al., 1971).

Let $k_p^\oplus$, $k_p^\oplus$, $R_p^\oplus$, and $R_p^\oplus$ be, respectively, the rate constants and rates of propagation of free ions and ion pairs; then the overall rate is

$$R_0 = -\frac{d[M]}{dt} = R_p^\oplus + R_p^\oplus = k_p^\oplus f[M] + k_p^\oplus p[M] \quad (5.14)$$

and

$$Q = \frac{R_p^\oplus}{R_p^\oplus} = 2\frac{k_p^\oplus}{k_p^\oplus} \frac{1}{(1 + 4C/K)^{1/2} - 1} \quad (5.15)$$

The value of $k_p^\oplus / k_p^\oplus$ for equal contribution of free ions and ion pairs to $R_0(Q = 1)$ has been calculated. According to Plesch (1971) C/K may range from 10^2 to 10^{-5}; thus, assuming $C/K = 10^2$, one obtains $k_p^\oplus / k_p^\oplus = 10$, which is probably too low. In contrast, assuming $C/K = 10^2$ and $k_p^\oplus / k_p^\oplus < 100$, one obtains $Q \sim 10$, suggesting that ion pairs contribute less than 10% to the overall rate. These calculations have made for solvents with $\epsilon < 10$. Thus in nonpolar media, for example, hydrocarbons and CCl_4, even small concentrations of ion pairs or ion pair aggregates may significantly contribute to k_p. Conversely, in solvents with $\epsilon > 10$ ion pairs do not exist and only free ions determine k_p.

Considerable efforts have been exerted for the determination of $[M^\oplus]$. Investigators employed direct and indirect methods: (1) using a direct characteristic physical property of the active species (UV, ^{1}H- or ^{13}C-NMR, etc.) but $[M^\oplus]$ is usually too low and k_p too high for accurate measurements

(Sawamoto and Higashimura, 1978a, b; De Sorgo et al., 1973; Lorimer and Pepper, 1976); and (2) using an indirect, "short-stop" method, that is, stopping propagation by the addition of an ingredient B having extremely high specific activity toward the chain end:

$$P_n^{\oplus} + B \rightarrow P_n B^{\oplus} = \text{stable entity}$$

Depending on the nature of B and experimental conditions, the newly formed species $P_n B^{\oplus}$ is or is converted into a stable entity. Obviously, short-stopping must be quantitative, essentially a diffusion-controlled reaction, and should not change the nature of the active species. For example, the $BF_3 \cdot OEt_2$/styrene system has been short-stopped with 2-bromothiophene and the bromine content of the chain ends was analyzed by radioactivation (Higashimura et al., 1971). This short-stop method promises to be a reliable technique for the determination of k_p.

The values of k_i, k_p, $k_{tr,M}$, and k_t can be determined simultaneously by adiabatic calorimetry (Vairon and Sigwalt, 1971; Villesange et al., 1977; Subira et al., 1976; Cotrel et al., 1976; Bowyer et al., 1971a), dilatometry (Hayashi et al., 1971, Sauvet et al., 1974; Kanoh et al., 1965; Okamura et al., 1961), iodometry by UV (Okamura et al., 1961, Kanoh et al., 1965), and gravimetry (Taylor and Williams, 1969). Select examples are discussed below.

Kinetic Studies of Representative Systems

This section concerns an examination of some representative systems which yielded reliable rate constants.

Polymerization of α-Methylstyrene Coinitiated by n-BuOTiCl₃

Villesange et al. (1977) studied the polymerization of H_2O or HCl/n-BuOTiCl₃/α-methylstyrene/CH_2Cl_2/−70°C system. Under superdry conditions yields were 2 to 5%. Preliminary studies showed that even very small amounts ($5 \times 10^{-4} M$) of coinitiator led to complete monomer consumption and that initial polymerizations rates were first-order in initiator (H_2O) and monomer. Initial rates were also first-order in coinitiator but only if [I]/[CoI] > 1. It was found that

$$R_i = k_i[H_2O][CoI] \tag{5.16}$$

$$R_p = k_p[M^{\oplus}][M] \tag{5.17}$$

Since I and CoI consumption was less than 15%, in first approximation:

$$R_i = k_i[H_2O]_0[CoI]_0 \tag{5.18}$$

and assuming first-order termination:

$$\frac{d[M^{\oplus}]}{dt} = k_i[H_2O]_0[CoI]_0 - k_t[M^{\oplus}] \tag{5.19}$$

which on integration between 0 and 1 yields:

$$\frac{R_p}{[M]} = k_p[M^{\oplus}] = \frac{k_i k_p}{k_t} [H_2O]_0 [CoI]_0 (1 - e^{-k_t t}) \qquad (5.20)$$

Equation 5.20 provides information as to the change in $[M^{\oplus}]$ during polymerization (this has already been analyzed; see Figure 5.4). Also, k_t can be obtained from linear $R_p/[M]$ versus $1 - \exp(-k_t t)$ plots shown in Figure 5.7. In a series of experiments Villesange et al. (1977) obtained $k_t = 0.54 \pm 0.05$ sec^{-1} at $-70°C$.

The total concentration of polymer [P] is experimentally available from $\bar{M}_n$ and $[M]_0$ and is equal to the sum of concentrations of chains obtained by initiation $[P_i]$ plus chain transfer $[P_{tr}]$:

$$[P] = [P_i] + [P_{tr}] \qquad (5.21)$$

If $[H_2O]_0 = [CoI]_0$,

$$\frac{d[P_i]}{dt} = k_i[H_2O][CoI] = k_i[CoI]^2 \qquad (5.22)$$

Equation 5.22 yields

$$\frac{d[P_i]}{dt} = k_i([CoI]_0 - [P_i])^2 \qquad (5.23)$$

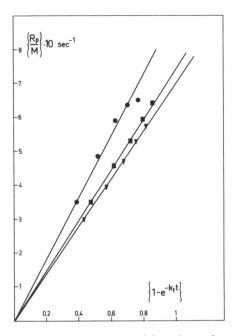

Figure 5.7 □ Determination of k_t and $k_i k_p$ for the H$_2$O/n-BuOTiCl$_3$/α-methylstyrene/CH$_2$Cl$_2$/$-70°C$ system $[CoI]_0 \times 10^3 = 1.05\ M$; $[H_2O]_0/[I]_0 = 1$; $[M]_0 \times 10^2 =$ ● 11.4 M, ■ 30 M, ▼ 22.6 M. [Reprinted from Villesange et al. (1977), p. 391, by courtesy of Marcel Dekker, Inc.]

which on integration gives:

$$p_i = \frac{k_i[CoI]_0^2 t}{1 + k_i[CoI]_0 t} \tag{5.24}$$

If chain transfer to monomer is the predominant chain transfer process:

$$\frac{d[P_{tr,M}]}{dt} = k_{tr,M}[M^\oplus][M] \tag{5.25}$$

Substituting $[M^\oplus]$ by equation 5.22 and integrating,

$$P_{tr,M} = \frac{k_i k_{tr,M}}{k_t}[CoI]_0^2[M]_0 \int_0^t (1 - e^{-k_t t})(1 - p)\, dt \tag{5.26}$$

where p is the extent of polymerization and p is the overall concentration of chains formed during the polymerization.

$$[P] = \frac{k_i[CoI]_0 t_\infty}{1 + k_i[I]_0 t_\infty} + \frac{k_i k_{tr,M}}{k_t}[CoI]_0^2[M]_0 \int_0^{t_\infty} (1 - e^{-k_t t})(1 - p)\, dt \tag{5.27}$$

In equation 5.27 $[M]_0$, $[CoI]_0$, and p are experimentally readily available and graphic integration can be carried out. As shown in Figure 5.8, $[P]$ versus $[M]_0$ yields a straight line, the intercept of which yields $[P_i]$ in the absence of transfer or $[I]$ consumed during initiation. This value is ~15% of $[I]_0$ at $-70°C$. For $t_\infty \approx 10$ sec the intercept gives $k_i = 17 \pm 6$ l/mole · sec and the slope yields $k_{tr,M} = 30 \pm 15$ l/mole · sec. From k_i and the product $k_i k_p$ (Figure 5.7) $k_p = (2.2 \pm 1.1)10^4$ l/mole · sec.

Polymerization of Isobutyl Vinyl Ether Initiated by Trityl Salts

The kinetics of isobutyl vinyl ether polymerization initiated by $Ph_3C^\oplus SbCl_6^\ominus$ in CH_2Cl_2 in the $-40°$ to $+20°C$ range has been studied by adiabatic calorimetry (Subira et al., 1976). Initiation was noticeably slower than propagation and unconsumed trityl salt remained after polymerization

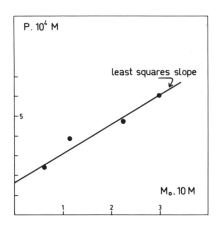

Figure 5.8 □ Total macromolecule concentration as a function of monomer concentration in the $H_2O/n\text{-BuOTiCl}_3/\alpha\text{-methylstyrene}/CH_2Cl_2/-70°C$ system. $[n\text{-BuOTiCl}_3]_0 = [H_2O]_0 = 1.5 \times 10^{-3}$ M. [Reprinted from Villesange et al. (1977), p. 391, by courtesy of Marcel Dekker, Inc.]

was complete at 0°C or below. This finding was contrary to the assumption that initiation is instantaneous, as proposed by Chung et al. (1975) and Bawn et al. (1971). At − 20°C the initiating salt had been consumed before the end of the polymerization; however, conversion was incomplete. Evidently the stability of active centers is quite low at this temperature.

Figure 5.9 shows the relationship between R_i and the product [M][I], and the slopes of the lines yield k_i and $\Delta H_i^{\ddagger}$ the enthalpy of activation of initiation.

Monomer conversion versus time curves were S-shaped, characteristic of slow initiation. This behavior is further illustrated by the curves in Figure 5.2, which show $(R_0)_t/[M]_t = (k_p + k_{tr,M})[M^{\oplus}]$ as a function of time.

At − 40°C [M$^{\oplus}$] slowly increases, indicating slow initiation and the existence of relatively stable active centers. At − 20°C the curve shows a maximum, suggesting rapid initiation and relatively rapid termination. Between these temperatures intermediate situations prevail.

The data obtained at − 25°C have been treated by plotting $\ln(R_{max}/[I_c]_{max})$ as a function of $\ln(M)_{max}$ (where R_{max} = maximum rate of polymerization and $[I_c]_{max}$ and $(M)_{max}$ are corresponding concentrations of consumed initiator and residual monomer).

This treatment leads to an overall order of 2 for monomer. Thus this system cannot be "living" because that would yield first-order monomer dependence.

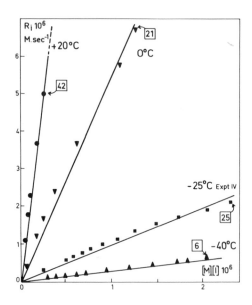

Figure 5.9 ☐   First-order plots for the $\phi_3 C^{\oplus} SbCl_6^{\ominus}$/isobutyl vinyl ether/$CH_2Cl_2$/ + 20° to − 40° system with respect to initiator and to monomer (Subira et al., 1976). Significant points for high [M] · [I] values are out of scale. Numbers in squares are monomer concentrations in M.

If unimolecular termination is assumed,

$$\frac{d[M^{\oplus}]}{dt} = R_i - R_t = k_i[M][I] - k_t[M^{\oplus}] \tag{5.28}$$

$$R_p = (k_p + k_{tr,M})[M][M^{\oplus}] \tag{5.29}$$

leads to

$$\left[\frac{d}{dt}\left(\frac{R_p}{M}\right)\right]_0 = (k_p + k_{tr,M})(R_i)_0 \tag{5.30}$$

where the subscript zero indicates values extrapolated to 0. Thus

$$-\frac{1}{\ln([M]_0/[M])}\frac{d}{dt}\ln[M] = (k_p + k_{tr,M})\frac{[I]_0 - [I]}{\ln([M]_0/[M])} - k_t \tag{5.31}$$

From equations 5.30 and 5.31 and experimental values $[I]_t$, $[M]_t$, and $(R_0)_t/[M]_t$, internal and external values for $k_p + k_{tr,M}$ and k_t can be obtained. At a given temperature k_p is independent of $[M]_0$ which is also incompatible with a living system (Bawn et al., 1971). By treating the experimental values of Subira et al. (1976) as if they were representative of a living system, one obtains k_p values similar to those of Bawn et al. (1971) and as a function of $[M]_0$. This shows that the experimental conditions (purity, etc.) were the same in both Subira and Bawn's experiments; the differences in kinetic values are due to different interpretations.

Determination of $k_{tr,M}$: Polymerization of p-Methoxystyrene Initiated by Trityl Salt

Cotrel et al. (1976) have recently published a method which may be generally useful for determination of $k_{tr,M}$. These workers investigated the $Ph_3C^{\oplus}SbCl_6^{\ominus}$/p-methoxystyrene/$CH_2Cl_2$/– 15 to 25°C system. Initiation was relatively slow. The overall order in monomer was 2 and for propagation it was 1. Assuming the existence of transfer to monomer and unimolecular termination, equations 5.28 to 5.31 have been obtained. If termination is unimolecular and propagation first-order with respect to $[M]$, a straight relationship is obtained by plotting $-(d\ln[M]/dt)/\ln([M]_0/[M])$ versus $([I]_0 - [I])/\ln([M]_0/[M])$.

This linear representation gives external and internal values of $k_p + k_{tr,M}$ and k_t. Figure 5.10 shows such a plot for the above-mentioned system. $k_{tr,M}$ is obtained from the number average degree of polymerization:

$$\overline{DP}_n = \frac{\int R_p\,dt}{\int R_i\,dt + \int R_{tr,M}\,dt} \tag{5.32}$$

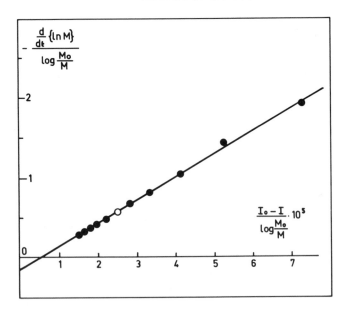

Figure 5.10 □ Determination of $(k_p + k_{tr,M})$ and of k_t values for the $Ph_3C^{\oplus}SbCl_6^{\ominus}/p$-methoxystyrene/$CH_2Cl_2$/10°C system ($[M]_0 = 4.0 \times 10^{-2}\ M$; $[I]_0 = 5.8 \times 10^{-5}\ M$) (from Sauvet et al., 1974).

where $[I]_c = [I]_0 - [I]$ is the initiator consumed, whence

$$\frac{1}{\overline{DP}_n} = \frac{[I]_c}{[M]_0} + \frac{k_{tr,M}}{k_p} \tag{5.33}$$

since conversion is always complete. Thus

$$\frac{k_p}{k_{tr,M}} = \frac{[M]_0/[I]_c \overline{DP}_n}{[M]_0/[I]_c - \overline{DP}_n} \tag{5.34}$$

Since $\overline{DP}_n$ can be determined and $[M]_0/[I]_c$ calculated, $k_p/k_{tr,M}$ is readily obtainable. Thus one obtains $k_p + k_{tr,M}$ and $k_p/k_{tr,M}$ for each temperature, and both constants can be obtained separately.

Polymerization of Isobutyl Vinyl Ether Initiated by X-Rays

The rate of bulk polymerization of isobutyl vinyl ether induced by X-ray radiation has been followed by dilatometry and conductometry at 0, 25, and 50°C (Hayashi et al., 1971). According to the authors, radiation impinges on monomer M and produces a cation radical $M^{\cdot\oplus}$ and counteranion $A^{\ominus}$ (equation 5.36). Cationation (equations 5.37 and 5.38) and chain transfer to monomer (equation 5.39) and to impurity X, most likely residual water (equation 5.40), may occur:

$$M \rightsquigarrow M^{\cdot\oplus} + e$$

$$M^{\cdot\oplus} + e \rightsquigarrow M'^{\oplus} + A^{\ominus}$$

$$M'^{\oplus} + M \xrightarrow{k_p} M_2^{\oplus} \tag{5.37}$$

$$M_n^{\oplus} + M \xrightarrow{k_p} M_{n+1}^{\oplus} \tag{5.38}$$

$$M_n^{\oplus} + M \xrightarrow{k_{tr,M}} M_n + M'^{\oplus} \tag{5.39}$$

$$M_n^{\oplus} + X \xrightarrow{k_{tr,X}} M_n + X^{\oplus} \tag{5.40}$$

where $M'^{\oplus}$ is a more stable cationic species than $M^{\cdot\oplus}$, probably a carbenium ion.

Termination may be ion recombination:

$$M_n^{\oplus} + A^{\ominus} \xrightarrow{k_t} M_n - A \text{ or } M_n + A' \tag{5.41}$$

$$X^{\oplus} + A^{\ominus} \xrightarrow{k_{t,X}} X - A \text{ or } X' + A'' \tag{5.42}$$

Reactions 5.41 and 5.42 are diffusion controlled.

The authors assume steady state concentration for every ionic species in the system. The recombination constants k_t or $k_{t,x}$ have been calculated by using classical diffusion constants. Calculation led to

$$R_p = k_p[M] \sum_{i=1}^{\infty} [M_i^{\oplus}] = \frac{k_P[M](10IG/N)}{pk_t[A^{\ominus}] + qk_{t,x}[X]} \tag{5.43}$$

where $k_{t,x}$ and k_t are relative to the case where $n = 1$, p and q are constants obtainable from diffusion constants, I is the dose rate (eV/cm$^3 \cdot$ sec), and G is the yield of free ions. The specific conductivity of the system is given by (Hayashi et al., 1967)

$$\sigma = \frac{(\epsilon/4\pi e)10IG/N}{[A^{\ominus}]} \tag{5.44}$$

It is known (Hayashi et al., 1967; Williams et al., 1967) that specific conductivity is a function of the square root of dose rate. In line with these considerations, equation 5.43 indicates that at small dose rates and/or in the presence of high concentrations of impurities, R_p is a function of I, and at high dose rates and/or in the presence of insignificant amounts of impurities R_p is a function of $I^{1/2}$. Under "superdry" conditions, that is,

[X] = 0 (Hayashi et al., 1971),

$$\frac{R_p}{\tau_{A^\ominus}} = \left(\frac{10k_p[M]G}{N}\right) I \qquad (5.45)$$

where $\tau_{A^\ominus}$ is the lifetime of $A^\ominus$.

Figure 5.11 shows R_p as a function of I. Evidently at dose rates higher than 4×10^{14} eV/cm$^3 \cdot$ sec R_p is proportional to $I^{1/2}$, which proves superdry conditions and that termination by impurities is negligible relative to termination by recombination (equation 5.41).

In agreement with the statement in equation 5.45 $R_p/\tau_{A^\ominus}$ versus I plots give straight lines, as shown in Figure 5.12. From the slopes of these straight lines k_p (in l/mole $\cdot$ sec) = 3.8×10^4, 1.2×10^5, and 6.0×10^5 at 0, 25 and 50°C, respectively.

Further, for dose rates higher than 5×10^{14} eV/cm$^3 \cdot$ sec $k_{tr,M}$ values have been calculated from $k_{tr,M} = k_p/\overline{DP}_n$ by the use of experimentally obtainable $\overline{DP}_n$ data (in l/mole $\cdot$ sec), that is, $k_{tr,M} = 130$ and 490 at 0, 25 and 50°C, respectively.

Very recently Ma et al. (1979) obtained reliable k_p data for the Ph$_2$MeC$^\oplus$SbCl$_6^\ominus$/isopropyl vinyl ether/CH$_2$Cl$_2$/ -25 to $+45$°C and γ-rays/isopropyl vinyl ether/bulk/0°C systems. The k_p values obtained for the chemically initiated system at 0°C range from 4.2 to 15.7 l/mole $\cdot$ sec depending on the initial concentrations of monomer and coinitiator; the k_p obtained for the radiation-initiated polymerization at 0°C is 3.9×10^5 l/mole $\cdot$ sec, which is

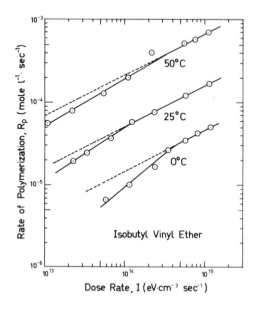

Figure 5.11 ☐   Rate of polymerization of isobutyl vinyl ether at 0, 25, and 50°C as a function of dose rate (Hayashi et al., 1971).

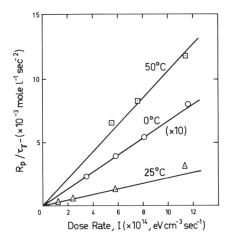

Figure 5.12 □ Variations of $R_p/\tau_{A^\ominus}$ with dose rate I for the polymerization of isobutyl vinyl ether at 0, 25, and 50°C (Hayashi et al., 1971) (values at 0°C have been multiplied by 10).

somewhat below the value previously obtained by Goineau et al. (1977) under the same conditions (9.0×10^5 l/mole · sec). The discrepancy is probably due to differences in the impurity levels.

5.4 □ THE EFFECT OF SOLVENT AND TEMPERATURE ON RATES, RATE CONSTANTS, AND ACTIVATION PARAMETERS

At present it appears impossible to separate the effects of solvent and temperature on kinetic parameters of carbocationic polymerization; therefore these effects are examined jointly in this section.

Rates and Rate Constants

Since cationic polymerizations are commonly carried out in polar media and temperature affects the dielectric constant ($\epsilon_T = \epsilon_{T'} - \alpha(T' - T')$, where $\alpha = $ constant), a change in temperature may affect the rate in a variety of interdependent ways. For example, the distance between ions in ion pairs, that is, ionicity, in moderately polar solvents increases by decreasing the temperature (Beard and Plesch, 1964b), which should affect the rate. Similarly ion aggregation, solvation, and mobility are affected by the nature of the solvent, solvent polarity, and temperature. Since temperature influences the nature of the solvent including its polarity, each system has to be treated individually.

Unfortunately, the effect of solvent on the rate of a polar reaction such as carbocationic polymerization cannot be predicted. It is even worse that not only is a quantitative understanding of the effect of solvent or solvation on rates of carbocationic polymerization lacking, but only in a few instances do we have even a qualitative appreciation of these effects on the molecular level. Though Hughes and Ingold's rules help to rationalize some effects of polar or nonpolar solvents on observed rates, ambiguity prevails, and more than one interpretation can usually be advanced for a particular effect. Although polarity and/or polarizability of solvents play an extremely important role in shaping the rate of cationic polymerizations, these terms remain ill-defined, loose, and imprecise. The literature is replete with references to polar and nonpolar solvents; however, distinction between these media is a matter of prejudice or preference of the individual investigator. Similarly, interactions between charge/dipole, charge/charge, induced dipoles, and aggregates, for example, must exist, but at which point general solvation ends and specific solvation, coordination, or complexation starts is more in the realm of intuition than science. To discern a true, kinetically meaningful solute–solvent effect, first of all the ionicity of the system must be defined. Unfortunately, the identification of active cationic species and their quantitative definition is an extremely demanding undertaking and except for a very few cases is beyond the capabilities of contemporary polymer scientists. In the absence of well-characterized ionicities only apparent solvent effects can be obtained and the data reflect not only true solute–solvent effects but also the effect of solvent on ionicity and other solvent-sensitive parameters.

The term "solvating power" has sometimes been used qualitatively to express the relative ability of solvents to interact with charged species. Although at first glance solvating power appears to be a specific quantity, it has never been defined and used to advantage in cationic polymerizations. Problems include the ill-defined nature of charged species involved and the difficulty of measuring the energy of solvation accurately.

That the dielectric constant is a rather poor, sometimes even misleading measure of "polarity" has repeatedly been pointed out (Beard and Plesch, 1964a).

From the point of view of elementary steps of carbocationic polymerization, solvation by polar medium should in general increase the rate of ionization since the process involves the generation of charged species; that is, the activation energy difference of ionization should be lowered by a polar medium that solvates the charges in the transition state: $RA \rightleftharpoons [R^{\oplus} \cdots A^{\ominus}]^{\neq} \rightleftharpoons R^{\oplus}//A^{\ominus}$. In contrast, the rates of cationation and/or propagation are expected to decrease by increasingly polar media; that is, endoenergetic desolvation would reduce the rate of spreading the charge in the transition state: $\sim C_n^{\oplus} + M \rightarrow [\sim C^{\delta\oplus} \cdots M^{\delta\oplus}]^{\neq} \rightarrow \sim C_{n+1}^{\oplus}$. The effect of medium polarity on the many kinds of chain transfer processes is difficult to generalize. As discussed in Section 4.3, there is evidence for the existence of many types of chain transfers, and a detailed analysis of the effect of solvent polarity on such a multitude of possibilities would be too speculative and unprofitable. As a common element, the charge of the growing site probably tends to spread while moving toward the transition states of the various chain transfer steps, so that increasing solvent polarity should tend to decrease the rate (desolvation, similar to the phenomenon during propagation). Termination involves the destruction of charges so that the rate of this process would decrease in the presence of solvating polar solvents. Earlier workers have often remarked that the rate of polymerizations increased exponentially with solvent ϵ. Evidently k_i and k_t would increase and decrease, respectively, with increasing ϵ and would result in increasing rates.

Sometimes the effect of polar solvent may be quite dramatic. For example, the polymerization of isobutylene or styrene cannot be initiated by t-BuCl/Me$_3$Al in nonpolar medium (n-pentane); however, addition of a polar solvent (CH$_3$Cl) to a quiescent t-BuCl/Me$_3$Al/i-C$_4$H$_8$/n-C$_5$H$_{12}$/−50°C system results in immediate polymerization. Evidently, absence of ion generation (t-BuCl + Me$_3$Al → t-Bu$^\oplus$Me$_3$AlCl$^\ominus$) in the nonpolar medium prevents cationation. To the uninitiated eye, the rapid polymerization ensuing upon CH$_3$Cl addition may appear as an unusual "co-coinitiation" phenomenon (i.e., the necessity for the addition of a second coinitiator besides Me$_3$Al in conjunction with t-BuCl) (Kennedy and Milliman, 1969).

The effect of solvent polarity on molecular weights is generally much less pronounced than that on rates. An explanation for this generalization may be the fact that in many systems molecular weights are solely affected by $R_p/(R_t + \Sigma R_{tr})$; R_i, which is highly solvation sensitive, does not affect DP.

The interpretation of kinetic measurements would be greatly simplified in the presence of only one well-defined chain carrier. Closer examination of this requirement leads to the conclusion that such a scenario may be unattainable in most carbocationic polymerizations. It is proposed that well-defined charge carriers could be expected to prevail only at either end of the "Winstein spectrum":

$$R\text{–}A \rightleftharpoons R^\oplus A^\ominus \rightleftharpoons R^\oplus/A^\ominus \rightleftharpoons R^\oplus//A^\ominus \rightleftharpoons \cdots \rightleftharpoons R^\oplus + A^\ominus$$

If covalent propagating entities (for a discussion of pseudocationic polymerizations see Section 3.1) are neglected for a moment, uniform propagating species could mean either tight ion pairs or "free" ions, since intermediate species would rapidly equilibrate by even slightly disturbing the system and would result in more than one kind of active cations.

Tight ion pairs could exist only in nonpolar solvents; however, kinetic investigation of carbocationic polymerization is rendered extremely difficult by the use of nonpolar media since reactants and/or reaction products would solvate the charged species under these conditions. Monomer solvation would certainly prevail in aliphatic hydrocarbons or CCl$_4$ since vinyl monomers are more nucleophilic than these solvents. For example, the overall order in monomer decreases one (or two) units by changing the solvent from CCl$_4$ to the somewhat more polar or solvating SnCl$_4$/styrene/solvent system (Higashimura and Okamura, 1956; Plesch, 1963).

Monomer solvation is also a possibility in vinyl ether polymerizations since vinyl ethers (although of rather low dielectric constants) are highly nucleophilic toward carbocations.

According to several authors (Fontana and Kidder, 1948; Longworth and Plesch, 1959; Biddulph et al., 1965; Higashimura, 1969; Sigwalt, 1971), k_p is strongly affected by complexation of the active site and monomer. This proposition has been quantitatively treated by Fontana and Kidder (1948) (see Section 4.2).

In general k_p increases with increasing solvent dielectric constant ϵ (Kanoh et al., 1965; Pepper and Reilly, 1966). According to Ledwith and Sherrington (1976), this phenomenon may be due to two major effects: (1) Active species may be transformed into other more active entities; for example, contact ion pairs into solvated ion pairs or free ions by increasing ϵ. (2) The k_p values are not true but apparent rate constants of propagation and are affected by the nature of the counteranion and degree of dissociation of ion pairs, for example. In such cases the rate of initiation (always visualized to be controlled by ion generation) decreases by decreasing ϵ and diminution of k_p is in fact due to diminution of rate of ionization.

The effect of nature of the solvent on k_p has not yet been studied systematically.

At the other extreme of the Winstein spectrum, kinetic orders attributed to "free" cations also have to be carefully defined because free ions as such simply do not exist in condensed phase. Free ions in solution are at best completely ionized *and* solvated to a larger or lesser extent and exist beyond the influence sphere of counteranions. But solvation is determined by a series of complicated physical and chemical factors, and when the interaction between solvent and solute becomes "strong," "complexation" commences. Thus kinetic measurements cannot be carried out in highly polar, that is, too strongly interacting, solvents. For that matter, carbocationic polymerizations may stop altogether in the presence of a solvent of high nucleophilicity, for example, oxygenated compounds. Thus kinetic measurements can be carried out only in moderately polar media. But in moderately polar media, equilibrium between more than one species prevails. This restriction necessitates the use of very high dilutions ($>10^{-4}$ M), which in turn render kinetic measurements difficult to execute. In practice, therefore, meaningful kinetic investigations could only have been carried out by using preformed initiators (i.e., stable cation salts) in high dilution (10^{-4}–10^{-5} M) in solvents of medium polarity (CH_2Cl_2). However, in line with this analysis, the data should not be generalized and the kinetics of each system must be investigated on an individual basis.

The use of mixed solvent systems, of course, is fraught with danger and should not be used for kinetic study. No one can predict the ultimate ionicity of even the simplest ion pair in a mixture of a polar/nonpolar solvent.

Since truly "free" ions exist only in the gas phase, one is forced to the absurd conclusion that in order to generate fundamental information, kinetic studies of a carbocationic polymerization should be carried out in the gas phase.

Until a more precise, quantitatively describable treatment emerges, this facet of carbocationic (or in general, ionic) polymerization chemistry and kinetics will remain extremely unsatisfactory.

Activation Parameters

Activation parameters may provide valuable information relative to the nature of active species, including their state of solvation.

Activation enthalpy differences of propagation are in general very low, more or less comparable to those obtained in radical polymerizations. In certain cases (Ledwith et al., 1975; Goineau et al., 1977) these values are close to zero, indicating that the transition state resembles the carbocation. A comparison of results obtained in carbocationic polymerizations by free ions with those of radical polymerizations for various monomers gives k_p (for free cations)$/k_p$ (radical) $\approx 10^5$. This ratio is determined not only by the activation enthalpy but also by the preexponential factor since $k_p = A_p \exp(-\Delta H^\ddagger/RT)$, where $A_p = \exp(\Delta S_p^\ddagger/R)$.

The entropy of activation difference $\Delta S_p^\ddagger$ in ionic polymerizations has been analyzed by various authors (Pepper and Reilly, 1966; Sawada, 1972; Higashimura et al., 1969; Ledwith and Sherrington, 1976). According to Pepper and Reilly (1966) the activation entropy difference during polymerization:

$$\Delta S_p^\ddagger = \Delta S_{p(immob)}^\ddagger + \Delta S_{p(solv)}^\ddagger \qquad (5.46)$$

where $\Delta S_{p(immob)}^\ddagger =$ the activation entropy difference of a previously free monomer which becomes immobilized by polymerization, and $\Delta S_{p(solv)}^\ddagger =$ the activation entropy difference between the reagent and transition states.

According to Higashimura (1969) the mobility of a monomer in a transition complex is determined by the nature of the chain end (free ion or radical, or ion pair):

rotation   vibration

$C^\oplus$---M → vibration $C^\oplus$---M--- $B^\ominus$

rotation   vibration

Free ion or radical   Ion pair

In line with these assumptions,

$$\Delta S_p^\ddagger = R \ln\left(\frac{f_T}{f_i}\right) + \frac{RT\, d\, \ln(f_T/f_i)}{dT} \qquad (5.47)$$

where f_T and f_i are partition functions of monomer in the transition and initial state, respectively. Except for the active end, the rest of the growing chain is the same in both instances. The second term in equation 5.47 is negligible and one obtains

$$\frac{A_{p(\text{free ion})}}{A_{p(\text{radical})}} = 1 \qquad (5.48)$$

and

$$\frac{A_{p(\text{free ion})}}{A_{p(\text{ion pair})}} = \frac{f_{\text{rot}}^2 f_{\text{vib}}}{f_{\text{vib}}^3} \simeq \frac{f_{\text{rot}}^2}{f_{\text{vib}}^2} \qquad (5.49)$$

which, considering conventional partition function, yields

$$\frac{A_{p(\text{free cation or radical})}}{A_{p(\text{ion pair})}} \approx 10^3 \qquad (5.50)$$

According to this expression insertion of a monomer molecule between the cation and counteranion reduces its mobility much more than its addition to a free chain end. Consequently the activation entropy difference is more negative for ion pairs than for free ions.

Such data, however, must be analyzed with caution. For example, A_p values obtained for polymerization of styrene by $HClO_4$ and free radicals are the same. This is contrary to our understanding of polymerization in the $HClO_4$/styrene system, which most likely does not contain free ions as propagating species (Pepper and Reilly, 1966).

This anomaly could at least partly be explained by considering the $\Delta S_{p(\text{solv})}^{\ddagger}$ term in equation 5.46. This quantity is positive for small ions because in the transition state the charge is spread and the degree of solvation decreases. A positive contribution of $\Delta S_{p(\text{solv})}^{\ddagger}$ tends to increase A_p, which may explain why A_p values even of ion pairs approach those of free radicals. However, $\Delta S_{p(\text{solv})}^{\ddagger}$ is smaller for ion pairs than for free ions because the difference in solvation between the initial and transition states are smaller for the former than the latter.

The contribution of $\Delta S_{p(\text{solv})}^{\ddagger}$ can be appreciated by examining values obtained in irradiation polymerization (Bowyer et al., 1971a). In these systems $A_p^{\ddagger}$ is higher than in those obtained in radical polymerization of the same monomers and therefore higher than those prevailing in free ion polymerization (since $A_{p(\text{free ion})}^{\ddagger} \simeq 10^{10}$ l/mole $\cdot$ sec, which is larger than that obtained by free radicals 1.2×10^6) (Hughes and North, 1966); the difference is probably due to solvation.

Difficulties involved in trying to unravel the effect of experimental conditions on the rate of carbocationic polymerization are illustrated by examining the $Ph_3C^{\oplus}SbCl_6^{\ominus}$/cyclopentadiene/$CH_2Cl_2$/$-20$ to $-70°C$ system. Sauvet et al. (1974) studied this polymerization and found for the enthalpy of activation of propagation and termination $\Delta H_p^{\ddagger} = -0.8 \pm 0.5$ and $\Delta H_t^{\ddagger} = 0.3 \pm 0.1$ kcal/mole, respectively. To explain the negative values the authors assumed simultaneous propagation by free ions and ion pairs whose proportion would vary with temperature. The equilibrium

$$P^{\oplus}SbCl_6^{\ominus} \overset{K_D}{\rightleftharpoons} P^{\oplus} + SbCl_6^{\ominus} \tag{5.51}$$

is shifted toward the formation of free ions by decreasing the temperature. The apparent rate of propagation is

$$k_{p(app)} = k_p^{\oplus}(1-\alpha) + k_p^{\oplus}\alpha \approx k_p^{\oplus} + k_p^{\oplus}\alpha \tag{5.52}$$

where $k_p^{\oplus} \ll k_p^{\oplus}$.

Owing to the presence of excess initiating stable cation salt, equilibrium 5.51 may be treated as a common salt equilibrium:

$$Ph_3C^{\oplus}SbCl_6^{\ominus} \overset{K_s}{\rightleftharpoons} Ph_3C^{\oplus} + SbCl_6^{\ominus} \tag{5.53}$$

so that

$$K_D = \frac{[P^{\oplus}][SbCl_6^{\ominus}]}{[P^{\oplus}SbCl_6^{\ominus}]} = \frac{\alpha}{1-\alpha}[SbCl_6^{\ominus}] \simeq \alpha[SbCl_6^{\ominus}] \tag{5.54}$$

whence

$$k_{p(app)} = k_p^{\oplus} + k_p^{\oplus}\left(\frac{K_D}{[SbCl_6^{\ominus}]}\right) \tag{5.55}$$

where

$$[SbCl_6^{\ominus}] = \frac{1}{2(K_s^2 + 4K_s[I])^{1/2} - K_s} \tag{5.56}$$

where [I] is the total initiator concentration.

The variation of K_s with temperature is known (Bowyer et al., 1971b), which permits evaluation of $k_{p(app)}[SbCl_6^{\ominus}]$ as a function of $1/T$, as shown in Figure 5.13. The linear relationship in combination with equation 5.55 yields

$$k_{p(app)} \approx k_p^{\oplus}\frac{K_D}{[SbCl_6^{\ominus}]} \tag{5.57}$$

Thus the contribution of contact ion pairs is much less than that of free ions. If we assume the equilibrium contains 10% fewer ion pairs than free ions,

$$k_p^{\oplus} < \frac{1}{10}k_p^{\oplus}\frac{K_D}{[SbCl_6^{\ominus}]} \tag{5.58}$$

With reasonable values for K_D and $[SbCl_6^{\ominus}]$, $k_p^{\oplus}$ is 500 to 5000 times smaller than $k_p^{\oplus}$.

Figure 5.13 yields a composite activation enthalpy $\Delta H_p^{\ddagger\oplus} + \Delta H_D =$

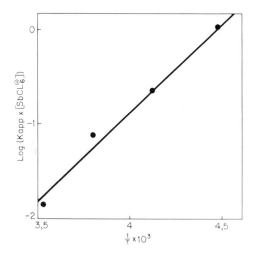

Figure 5.13 ☐ Variation of $\log k_{p(app_0)}[SbCl_6^{\ominus}])$ with $1/T$ for the $Ph_3C^{\oplus}SbCl_6^{\ominus}/$ cyclopentadiene/CH_2Cl_2/ -20 to $-70°C$ system (Sauvet et al., 1974).

-8.7 kcal/mole, in which $\Delta H_p^{\ddagger\oplus}$ is the activation enthalpy difference for propagation by free ions and ΔH_D is the dissociation enthalpy of the active chain end. Thus the values for $\Delta H_p^{\ddagger\oplus}$ may be between 0 and 2.0 kcal/mole (Bonin et al., 1965), which leads to

$$-8.7 < \Delta H_D < -10.7 \text{ kcal/mole}$$

These dissociation enthalpy values are too high for a salt comprising such large constituents as the cyclopentadienyl cation and $SbCl_6^{\ominus}$ counteranion. To account for this anomaly the authors propose a secondary solvation equilibrium:

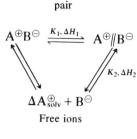

The apparent dissociation constant is thus

$$K_D = K_2 \frac{K_1}{1 + K_1} \tag{5.59}$$

Such behavior has been observed for alkali salts of fluorene in THF solutions (Smid and Hogen-Esch, 1966).

In the $Ph_3C^{\oplus}SbCl_6^{\ominus}$/cyclopentadiene/$CH_2Cl_2$/$-78°C$ system in which the common salt effect caused by the excess stable cation salt initiator decreases the concentration of free ions relative to that of contact ion pairs and solvent separated ion pairs, one can show that

$$k_{p(app)} = \frac{k_p^{\oplus}}{1+K_1} + k_{p(solv)}^{\oplus} \frac{K_D}{K_2} + k_p^{\oplus} \frac{K_D}{[SbCl_6^{\ominus}]} \tag{5.60}$$

According to the data in Figure 5.11 $\log(k_{p(app)}[SbCl_6^{\ominus}])$ is a linear function of $1/T$ so that the first two terms in equation 5.60 can be neglected; that is, the contribution of free ions dominates $k_{p(app)}$.

If $K_D \approx 10^{-5}$, the contribution of contact ion pairs can be neglected if $k_p^{\oplus}$ is 500 times smaller than $k_p^{\oplus}$. However, if we assume that the reactivities of solvent separated ion pairs and free ions are comparable, their contribution is negligible if $K_2 > 5 \times 10^{-3}$, which yields $K_1 < 5 \times 10^{-2}$ and justifies the approximation $\Delta H_D \approx \Delta H_1 + \Delta H_2$. Under these conditions the experimental value obtained for ΔH_D becomes reasonable.

It should be emphasized that $k_p = 1$ to $40\ l/mole \cdot sec$ is much lower than that obtained with the same monomer using n-$BuOTiCl_3$ coinitiator ($>10^3\ l/mole \cdot sec$). This difference substantiates the presence of a common salt effect in the above system. Evidently the unconsumed common stable cation salt suppresses dissociation.

Similarly complex interrelated influences have been analyzed in the $Ph_3C^{\oplus}SbCl_6^{\ominus}$/$p$-methoxystyrene/$CH_2Cl_2$/$+25$ to $-15°C$ system (Cotrel et al., 1976). The overall activation enthalpy is -6.0 ± 1.0 kcal/mole in this system. Two possibilities may be considered. (1) Significant shift in the dissociation equilibrium toward propagating free ions with decreasing temperatures, this shift being accompanied by strong solvation. (2) Existence of a solvation equilibrium between active centers and the monomer followed by a separate propagation step:

$$\sim P_n^{\oplus} + M \underset{k_{-1}}{\overset{k_1}{\rightleftharpoons}} \sim P_n^{\oplus} M \xrightarrow{k_p} \sim P_{n+1}^{\oplus} \tag{5.61}$$

and $K_s = k_1/k_{-1}$. If $k_1 \gg k_p$ the mass action law applies:

$$R_p = \frac{k_p K_s}{1 + K_s[M]} [P^*][M] \tag{5.62}$$

where $[P^*]$ is the overall concentration of active sites and $K_s[M] \ll 1$. Thus, by definition

$$k_{p(app)} \approx k_p K_s \tag{5.63}$$

and

$$E_{p(app)} \approx E_p + \Delta H_s \approx -6\ kcal/mole \tag{5.64}$$

Although E_p is unknown, it may be assumed that it is 0 to 4 kcal/mole, that is, the value obtained in bulk irradiation polymerization. This yields $\Delta H_s^{\ddagger} = -6$ to -10 kcal/mole, the order of magnitude of which is similar to that found by the same authors (Sauvet et al., 1975) for the equilibrium

$$CH_3\overset{\oplus}{C}Ph_2 + CH_2{=}CPh_2 \rightleftharpoons CH_3\overset{\oplus}{C}Ph_2 \cdot CH_2{=}CPh_2$$

5.5 □ RATE CONSTANT RATIOS BY MOLECULAR WEIGHT DETERMINATION

Molecular weights are important repositories of kinetic information. Molecular weights represent an absolute, readily determined quantity that reflect the sometimes quite complex sequence of elementary events leading to the polymer molecule. Analysis of molecular weights in combination with molecular weight distributions provide further extremely valuable insight into the overall mechanism of polymerizations.

The average degree of polymerization $\overline{DP}_n$ is by definition the ratio of rates of chain-building over chain-limiting events:

$$\overline{DP}_n = \frac{N_{monomer\ units}}{N_{macromolecules}} \approx \frac{R_p}{R_t + R_{tr,M}} \tag{5.65}$$

This fundamental equation leads to the well-known Mayo equation (Mayo, 1943; Mayo et al., 1951) originally derived for the study of free radical polymerizations:

$$\frac{1}{\overline{DP}_n} = \frac{k_{tr,M}}{k_p} + \frac{k_t}{k_p}\frac{1}{[M]} \tag{5.66}$$

The assumptions implicit in this equation are as follows:

$$R_p = k_p[P^{\oplus}][M] \tag{5.67}$$

$$R_t = k_t[P^{\oplus}] \tag{5.68}$$

$$R_{tr,M} = k_{tr,M}[P^{\oplus}][M] \tag{5.69}$$

Further, the rate of termination and chain transfer to monomer are independent of each other, chain transfer reactions are absent with the exception of a second-order chain transfer to monomer, and the effect of impurities is negligible. Provided the molecular weights have been obtained at low conversions, the slope and intercept of $1/DP_n$ versus $1/[M]$ plots (i.e., Mayo plots), provide k_t/k_p and $k_{tr,M}/k_p$, respectively. The ratio of rate constants obtained from Mayo plots is independent of the rate of initiation, which greatly facilitates the unraveling of mechanistic information. Examples illustrating the usefulness of the Mayo equation in generating kinetic information are critically and concisely presented by older (Plesch, 1963) and more up-to-date (Alley and Patrick, 1974) texts.

Information generated by Mayo analyses are particularly important in instances when direct rate measurements cannot be carried out, for example, owing to excessive velocity of polymerization.

In addition to the usual precautions (Plesch, 1963), care must be exercised when using the Mayo equation for studying carbocationic olefin polymerizations carried out in polar solvents. Since the polarity (dielectric constant) of the monomer and solvent may be quite different, monomer concentration cannot be changed by simply diluting with polar solvent because then not only [M] but overall polarity would also change, and such a change within an experimental series would obviously vitiate the results. To overcome this problem system polarity is maintained by the use of a compensating solvent. For example, the monomer concentration for Mayo analysis in an isobutylene/methylene chloride system may be changed by introducing an inert diluent whose polar characteristics are similar to isobutylene, for example, n-pentane or isobutane. Thus the volume change due to changing [M] is compensated by appropriate amounts of the inert diluent and the overall polarity remains constant throughout the experiment.

The applicability of the Mayo equation in carbocationic polymerizations has recently been scrutinized. Mandal and Kennedy (1978) found that Mayo analysis may produce erroneous results in polymerizations in which both free ions and ion pairs coexist. Thus for a polymerization which proceeds by free ions and ion pairs the simple Mayo equation must be extended to (Plesch, 1968a, b)

$$\frac{1}{\overline{DP}_n} = \frac{k_{tr,M}^{\oplus}[M^{\oplus}] + k_{tr,M}^{\ominus}[M^{\oplus}X^{\ominus}]}{k_p^{\oplus}[M^{\oplus}] + k_p^{\ominus}[M^{\oplus}X^{\ominus}]} + \frac{k_t[M^{\oplus}]^2 + k_t^{\ominus}[M^{\oplus}X^{\ominus}]}{(k_p^{\oplus}[M^{\oplus}] + k_p^{\ominus}[M^{\oplus}X^{\ominus}])[M]} \quad (5.70)$$

where the rate constants with $\oplus$ and $\ominus$ superscripts refer to reactions involving free ions and ion pairs, and $M^{\oplus}$ and $M^{\oplus}X^{\ominus}$ indicate propagating free ions and ion pairs, respectively. Let α and [M*] be the degree of dissociation and total active center concentration, respectively:

$$[M^{\oplus}] = \alpha[M^*] \quad \text{and} \quad [M^{\oplus}X^{\ominus}] = (1 - \alpha)[M^*] \quad (5.71)$$

and

$$K_d = \frac{[M^{\oplus}]^2}{[M^{\oplus}X^{\ominus}]} = \frac{\alpha^2}{1 - \alpha}[M^*] \quad (5.72)$$

Thus equation 5.70 may be extended in terms of the degree of dissociation:

$$\frac{1}{\overline{DP}_n} = \frac{\alpha k_{tr,M}^{\oplus} + (1 - \alpha)k_{tr,M}^{\ominus}}{\alpha k_p^{\oplus} + (1 - \alpha)k_p^{\ominus}} + \frac{(1 - \alpha)k_t^{\oplus}K_d + (1 - \alpha)k_t^{\ominus}}{(\alpha k_p^{\oplus} + (1 - \alpha)k_p^{\ominus})[M]} \quad (5.73)$$

This equation of overall validity has been examined for three cases:

1 □   If $\alpha = 0$, that is, in the presence of ion pairs only, equation 5.73 assumes the well-known form of the Mayo equation:

$$\frac{1}{\overline{DP}_n} = \frac{k_{tr,M}^{\oplus}}{k_p^{\oplus}} + \frac{k_t^{\oplus}}{k_p^{\oplus}} \cdot \frac{1}{[M]} \tag{5.74}$$

2 □   If $\alpha = 1$, that is, in the presence of free ions only, equation 5.73 becomes

$$\frac{1}{\overline{DP}_n} = \frac{k_{tr,M}^{\oplus}}{k_p^{\oplus}} + \frac{k_t^{\oplus}[M^{\oplus}]}{k_p^{\oplus}} \cdot \frac{1}{[M]} \tag{5.75}$$

3 □   If $0 < \alpha < 1$, that is, in the presence of free ions and ion pairs only, the general equation is meaningful only if α remains constant over the whole range of [M] investigated.

Thus the Mayo equation is valid only for polymerizations that proceed exclusively by ion pairs, that is, if termination is unimolecular. Since in most carbocationic polymerizations both free ions and ion pairs coexist, many published $k_{tr,M}/k_p$ and k_t/k_p values are probably composites and, unless constancy of α over the experimental range can be established, are of little quantitative significance.

     Mayo plots render invaluable service by revealing the absence of chain transfer to monomer. In conformity with the general equation 5.70, the absence of intercept in $1/\overline{DP}_n$ versus $1/[M]$ plots is proof for the absence of chain transfer to monomer. In contrast, the presence of an intercept does not necessarily indicate chain transfer to monomer during polymerization (Mandal and Kennedy, 1978).

5.6 □   THE EFFECT OF TEMPERATURE ON MOLECULAR WEIGHT

Molecular weights invariably increase with decreasing temperature in carbocationic polymerizations. Indeed this inverse relationship between molecular weights and polymerization temperature is a characteristic feature of carbocationic homopolymerizations. The large effect of temperature on product molecular weights was recognized by the earliest workers (for a review on this subject see Kennedy and Trivedi, 1978a, b), who readily produced very high molecular weight ($< 10^5$) polyisobutylenes by decreasing the polymerization temperature to $-78°C$ or below. Depending on the particular system, the temperature coefficient of molecular weight may be large (e.g., isobutylene) or less pronounced (e.g., styrene), but it is invariably present and is a most valuable parameter for molecular weight control. Temperature control in commercial butyl rubber or polyisobutylene synthesis is of great importance and a few degrees of temperature fluctuation during the polymerization step may endanger product homogeneity and thus quality.

Flory, who attempted to gain insight into the kinetics of isobutylene polymerization but could not obtain rate constants on account of the extreme velocity of this reaction, was the first to use Arrhenius plots (log $\bar{M}_v$ versus $1/T$) to express the effect of temperature on molecular weights (Flory, 1953). This author, using molecular weight data published by Thomas et al. (1940), obtained a linear relationship between log $\bar{M}_v$ and $1/T$ and from the slope derived $\Delta H_{\bar{M}_v}^{\ddagger}$, the activation enthalpy difference of molecular weight. Since these earliest investigations, the Arrhenius representation of the effect of temperature on molecular weight turned out to be a most useful aid in studying the kinetics of carbocationic polymerizations. Such studies were particularly helpful in generating rate constant ratios for systems in which extreme rates prevent direct velocity determinations, for example, isobutylene polymerizations. Thus data relative to the effect of temperature on $\bar{M}_v$ of polyisobutylene obtained under various conditions have been assembled by Kennedy and Trivedi (1978b) and further developed in Table 5.1. Specifically, Table 5.1 contains $\Delta H_{\bar{M}_v}^{\ddagger}$, $\Delta S_{\bar{M}_v}^{\ddagger}$ and $\Delta G_{\bar{M}_v}^{\ddagger}$ values, that is, activation free enthalpy, entropy, and energy differences of molecular weights, calculated from log $\bar{M}_v$ (or $\bar{M}_n$) versus $1/T$ Arrhenius lines available for a large variety of isobutylene polymerization systems. Analysis of these data led to significant mechanistic insight as follows.

Previous workers (Biddulph et al., 1965; Kennedy and Thomas, 1962; Kennedy and Squires, 1965) determined $\Delta H_{\bar{M}_v}^{\ddagger}$ values for isobutylene polymerizations and tried to correlate these with experimental conditions and mechanisms. Thus Kennedy and Thomas (1962) found $\Delta H_{\bar{M}_v}^{\ddagger} = -3.5$ kcal/mole and -0.22 kcal/mole in the range -50 to $-11°C$ and -110 to $-145°C$, respectively, for the "H_2O"/$AlCl_3$/isobutylene/propane system. This change in $\Delta H_{\bar{M}_v}^{\ddagger}$ was explained by postulating that below $-110°C$ diffusion became molecular weight controlling. Kennedy and Squires (1965) reported $\Delta H_{\bar{M}_v}^{\ddagger} = -6.6$ and -0.7 kcal/mole in the ranges -30 to $-80°C$ and -100 to $-145°C$, respectively, for "H_2O"/$AlCl_3$, "H_2O"/BF_3, and "H_2O"/$EtAlCl_2$ using MeCl. The change in $\Delta H_{\bar{M}_v}^{\ddagger}$ was attributed to the change from chain transfer to solvent at higher temperatures to chain transfer to monomer at lower temperatures. Plesch (Biddulph and Plesch, 1965) reported $\Delta H_{\bar{M}_v}^{\ddagger} = -8.2$ kcal/mole for H_2O/$TiCl_4$/isobutylene/CH_2Cl_2 in the range from $+19$ to $-70°C$, whereas the slope of the Arrhenius plot decreased below $-70°C$, indicating a decrease in $\Delta H_{\bar{M}_v}^{\ddagger}$. Plesch explained these results by postulating that the growing species changed from ion pairs to free ions below $-70°C$. Analysis of the data in Table 5.1 reveals that the $\Delta H_{\bar{M}_v}^{\ddagger}$ data fall into three groups characterized by the following values (in kcal/mole): -6.6 ± 1.0, -4.6 ± 1.0, and -1.8 ± 1.0.

Analysis of magnitudes of activation enthalpies characteristic of chain transfer, termination, and propagation led Kennedy and Trivedi (1978b) to

Table 5.1 □ $\Delta H^{\ddagger}_{M_v}$, $\Delta S^{\ddagger}_{M_v}$, and $\Delta G^{\ddagger}_{M_v}$ Values of Isobutylene Polymerization[a]

Initiator	Coinitiator	Solvent	Temperature Range (°C) t_1	t_2	$\Delta H^{\ddagger}_{M_v}$ (cal/mole)	$\Delta S^{\ddagger}_{M_v}$ (cal/mole · deg)	$-T\,\Delta S^{\ddagger}_{M_v}$ (cal/mole)	$\Delta G^{\ddagger}_{M_v}$ (cal/mole)	Reference
			First Class: $\Delta H^{\ddagger}_{M_v} = -6.6 \pm 1$ kcal/mole; $\Delta S^{\ddagger}_{M_v} < 0$ cal/mole · deg						
"H$_2$O"	EtAlCl$_2$	n-Pentane	-30	-40	-5800	-2.4	600	-5200	Kennedy and Trivedi (1978a, b)
t-BuI	Et$_2$AlI	MeCl	-30	-60	-5700	-1.2	300	-5400	Kennedy and Trivedi (1978a, b)
Radiation induced		—	29	-31	-7400	-3.9	1100	-6300	Kennedy et al. (1971)
"H$_2$O"	BF$_3$	MeCl	-30	-100	-5900	-3.4	700	-5200	Kennedy and Squires (1965)
"H$_2$O"	Et$_2$AlCl	MeCl	-30	-100	-7100	-7.7	1600	-5500	Kennedy and Squires (1965)
"H$_2$O"	AlCl$_3$	MeCl	-30	-100	-6100	-4.7	1000	-5100	Kennedy and Squires (1965)
t-BuCl	Me$_2$AlCl	MeCl	-25	-50	-7300	-5.3	1200	-6100	Kennedy and Renchegary (1974)
t-BuBr	Me$_2$AlCl	MeCl	-25	-50	-8000	-9.1	2100	-5900	Kennedy and Renchegary (1974)
"H$_2$O"	Me$_2$AlCl	MeCl	-40	-70	-8000	-12.3	2700	-5300	Kennedy and Renchegany 1974
"H$_2$O"	EtAlCl$_2$	n-Heptane	0	-55	-5800	-10	2500	-3300	Malinska-Solich et al. (1969)
"H$_2$O"	TiCl$_4$	EtCl	-30	-112	-5500	0 to -0.1	0	-5500	Biddulph and Plesch (1960)
			Second Class: $\Delta H^{\ddagger}_{M_v} = -4.6 \pm 1$ kcal/mole; $\Delta S^{\ddagger}_{M_v} = 5 \pm 5$ cal/mole · deg						
t-BuCl	Et$_2$AlCl	MeBr	-30	-45	-5100	3.0	-700	-5800	Kennedy and Trivedi (1978a, b)
t-BuBr	Et$_2$AlCl	MeCl	-30	-50	-4900	4.5	-1100	-6000	Kennedy and Trivedi (1978a, b)
t-BuCl	Et$_2$AlBr	MeBr	-30	-55	-5100	2.2	-500	-5600	Kennedy and Trivedi (1978a, b)
t-BuBr	Et$_2$AlBr	MeCl	-30	-65	-4800	3.8	-900	-5700	Kennedy and Trivedi (1978a, b)
t-BuCl	Et$_2$AlI	MeCl	-30	-60	-5000	2.7	-600	-5600	Kennedy and Trivedi (1978a, b)
t-BuBr	Et$_2$AlI	MeCl	-30	-60	-4800	3.9	-900	-5700	Kennedy and Trivedi (1978a, b)
"H$_2$O"	Et$_2$AlI	MeCl	-30	-55	-4500	5.6	-1300	-5800	Kennedy and Trivedi (1978a, b)
Me$_3$CCl	Et$_2$AlCl$_2$	MeCl	-30	-50	-4600	7.4	-1700	-6300	Cesca et al. (1975)
"H$_2$O"	MgCl$_2$	n-Heptane	-10	-78	-4700	6	-1400	-6100	Addecott et al. (1967)

			t_1	t_2			$-T\Delta S^\ddagger$		
Cl₂	Et₂AlCl	MeCl	−30	−50	−5800	2.9	−700	−6500	Cesca et al. (1975)
$h\nu$	VCl₄	n-Heptane	−20	−78	−5000	9.5	−2100	−7100	Toman and Marek (1976)
t-BuBr	Et₂AlCl	MeCl	−20	−50	−3600	9.5	−2300	−5900	Kennedy and Renchegary (1974)
Third Class: $\Delta H^\ddagger_{M_v} = -1.8 \pm 1.1$ kcal/mole; $\Delta S^\ddagger_{M_v} = 20 \pm 10$ cal/mole · deg									
t-BuCl	Et₂AlCl	MeCl	−30	−70	−1300	20.8	−4600	−5900	Kennedy and Trivedi (1978a, b)
t-BuCl	Et₂AlCl	MeBr	−50	−70	−1300	19.8	−4200	−5500	Kennedy and Trivedi (1978a, b)
t-BuBr	Et₂AlCl	MeCl	−50	−70	−1200	20.2	−4300	−5500	Kennedy and Trivedi (1978a, b)
t-BuCl	Et₂AlBr	MeCl	−50	−65	−1400	19.0	−4100	−5500	Kennedy and Trivedi (1978a, b)
CF₃COOH	TiCl₄	n-Hexane	−60	−80	−2300	17.5	−3600	−5900	Biddulph and Plesch (1960)
t-BuCl	Et₃Al	MeCl	−40	−80	−2500	10.2	−3600	−6100	Kennedy and Renchegary (1974)
t-BuCl	Et₂AlCl	MeCl	−40	−80	−2300	15.5	−3300	−5600	Kennedy and Renchegany (1974)
t-BuBr	Me₂AlCl	MeCl	−40	−80	−1300	21.5	−4600	−5900	Kennedy and Renchegary (1974)
t-BuCl	Me₂AlCl	MeCl	−40	−80	−2700	18.1	−3600	−5700	Kennedy and Renchegary (1974)
"H₂O"	MeAlCl₂	MeCl	−80	−100	−1900	18.0	−3300	−5200	Kennedy and Renchegary (1974)
Cl₂	Me₃Al	MeCl	−40	−100	−1800	14.2	−2900	−4700	Kennedy and Sivaram (1973)
t-BuCl	Me₃Al	MeCl	−40	−100	−1900	13.2	−2700	−4600	Kennedy and Sivaram (1973)
Cl₂	Et₂AlCl	MeCl	−40	−100	−2100	16.5	−3400	−5500	Kennedy and Sivaram (1973)
Br₂	Et₂AlCl	MeCl	−30	−50	−2200	16.8	−3900	−6100	Kennedy and Sivaram (1973)
t-BuCl	Et₂AlCl	MeCl	−50	−100	−2000	16.1	−3200	−5200	Kennedy and Sivaram (1973)
"H₂O"	BF₃	MeCl	−100	−146	−700	28	−4200	−4900	Kennedy and Squires (1965)
"H₂O"	Et₂AlCl	MeCl	−100	−146	−900	26	−3900	−4800	Kennedy and Squires (1965)
"H₂O"	AlCl₃	MeCl	−100	−146	−700	27	−4100	−4800	Kennedy and Squires (1965)
"H₂O"	EtAlCl₂	MeCl	−80	−102	−700	25.6	−4700	−5400	Kennedy and Trivedi (1978a, b)

[a] Temperatures in the $-T\Delta S^\ddagger$ column are average T values of the t_1, t_2 range indicated.

287

propose that molecular weight control in these three groups reflects the existence of two fundamentally different (limiting) and one intermediate mechanism. According to these authors the highest $\Delta H_{\bar{M}_v}^{\ddagger}$ ($-6.6 \pm$ 1.0 kcal/mole) is characteristic of a regime in which the molecular weight-controlling event is chain transfer to monomer, the intermediate $\Delta H_{\bar{M}_v}^{\ddagger}$ ($4.6 \pm$ 10 kcal/mole) appears when molecular weight is controlled by a combination of chain transfer to monomer and termination, and the lowest $\Delta H_{\bar{M}_v}^{\ddagger}$ values (-1.8 ± 1.0 kcal/mole) are obtained in polymerizations in which termination determines molecular weights. If $\Delta H_{\bar{M}_v}^{\ddagger}$ changes, for example, by changing the nature of coinitiator or temperature or solvent, this change reflects a shift in the molecular weight-controlling mechanism.

According to this hypothesis, $\Delta H_{\bar{M}_v}^{\ddagger} = -6.6$ and -1.8 kcal/mole are extreme values, characteristic of two fundamentally different molecular weight-governing processes, and intermediate values between these extremes indicate the existence of a transition region in which both molecular weight-controlling processes are operative. Although a transition region between these extremes must always exist, its experimental definition may be difficult or even impossible, particularly when its operational range is narrow and/or insufficient data are available.

Unfavorable experimental conditions may prevent the detection of termination or transfer region altogether. For example, the transfer for the t-BuCl/Et$_2$AlCl/isobutylene/MeCl system may lie above $-30°$C, that is, in a region where polymerization cannot be carried out owing to the boiling point of MeCl ($-24.2°$C), or the termination region for t-BuX/Et$_2$AlI/iso-butylene/MeCl may lie below $-70°$C, where initiation does not take place.

According to the data in Table 5.1 termination is the dominant molecular weight-determining event at low temperatures, whereas chain transfer to monomer is molecular weight-controlling at higher temperatures; by raising the temperature a transferless isobutylene polymerization may be gradually converted into one in which molecular weights are predominantly controlled by chain transfer. All these conclusions are in agreement with results obtained from an analysis of Mayo plots, model experiments, and consideration of findings of γ-ray-induced isobutylene polymerizations (Kennedy and Trivedi, 1978b).

Further analysis of published information leads to $\Delta S_{\bar{M}_v}^{\ddagger}$, the activation entropy difference of viscosity average molecular weight values, although the precision of these data is less than those of $\Delta H_{\bar{M}_v}^{\ddagger}$. Table 5.1 is a summary of $\Delta H_{\bar{M}_v}^{\ddagger}$, $\Delta S_{\bar{M}_v}^{\ddagger}$, and $\Delta G_{\bar{M}_v}^{\ddagger}$ data, that is, differences of activation free enthalpy of viscosity average molecular weights calculated by $\Delta G_{\bar{M}_v}^{\ddagger} = \Delta H_{\bar{M}_v}^{\ddagger} - T\Delta S_{\bar{M}_v}^{\ddagger}$ for available isobutylene polymerization systems. $\Delta G_{\bar{M}_v}^{\ddagger}$ is the molecular weight-determining thermodynamic function that combines an enthalpic ($\Delta H_{\bar{M}_v}^{\ddagger}$) and entropic ($\Delta S_{\bar{M}_v}^{\ddagger}$) term. According to the relative magnitude of $\Delta H_{\bar{M}_v}^{\ddagger}$ and $-T\Delta S_{\bar{M}_v}^{\ddagger}$ molecular weight control will be enthalpic or entropic.

As suggested by the data in Table 5.1, for systems in Class 1 $\Delta H_{\bar{M}_v}^{\ddagger} =$

-6.6 ± 1 kcal/mole and $\Delta S_{\bar{M}_v}^{\ddagger} < 0$ cal/mole $\cdot$ deg. Consequently in these polymerizations molecular weights are determined by enthalpy and the influence of entropy is negligible. According to Kennedy and Trivedi (1978b), in Class 1 systems molecular weights are controlled mainly by chain transfer to monomer. Conceivably, in these instances the number of degrees of freedom of the system for the solvated growing ion pair does not change appreciably before and after chain transfer to monomer. Schematically,

$$
\left. \begin{array}{c} \sim (P^{\oplus}X^{\ominus})_n S \\ \text{or} \\ \sim (P^{\oplus}//X^{\ominus})_n S \end{array} \right\} + M \rightarrow \; \sim P^{=} + \left\{ \begin{array}{c} (M^{\oplus}X^{\ominus})_n S \\ \text{or} \\ (M^{\oplus}//X^{\ominus})_n S \end{array} \right.
$$

where S = solvent. The entropic term is slightly negative probably because solvation of $M^{\oplus}$, that is, protonated monomer (see Chapter 4) is better than that of $\sim P^{\oplus}$, that is, growing macrocation.

In Class 2 $\Delta H_{\bar{M}_v}^{\ddagger} = 4.6 \pm 1$ kcal/mole and $0 < \Delta S_{\bar{M}_v}^{\ddagger} < 10$ cal/mole $\cdot$ deg. Evidently molecular weights are still mainly determined by enthalpy; however a small positive $\Delta S_{\bar{M}_v}^{\ddagger}$ appears, indicating a nonnegligible entropy contribution during molecular molecular weight control.

Finally, in Class 3 systems $\Delta H_{\bar{M}_v}^{\ddagger}$ is smallest (-1.8 ± 1.1 kcal/mole) and $\Delta S_{\bar{M}_v}^{\ddagger}$ is largest (10 to 30 cal/mole $\cdot$ deg); therefore molecular weights are controlled mainly by entropy. Analysis of the enthalpic term (Kennedy and Trivedi, 1978b) has led to the conclusion that in this class termination controls molecular weight. In line with this hypothesis, termination is most likely accompanied by a large decrease in solvation (i.e., release of ordered solvent molecules in the solvate shell owing to the disappearance of charged entities), and consequently to a large increase in the degrees of freedom of the system:

$$
\begin{array}{c} \sim (P^{\oplus}X^{\ominus})_n S \\ \text{or} \qquad\qquad \longrightarrow \; \sim PX + n S \\ \sim (P^{\oplus}//X^{\ominus})_n S \end{array}
$$

In the last few systems of Class 3 shown in Table 5.1 the entropic term appears to be particularly prominent, which suggests that termination is particularly important in determining molecular weights.

5.7 □ MOLECULAR WEIGHT DISTRIBUTIONS

Since the introduction of moderately priced high pressure (speed) gel permeation chromatography GPC instrumentation a few years ago molecular weight distribution, that is, $\bar{M}_w/\bar{M}_n$, data have been increasingly used for the investigation of the mechanism of ionic polymerization. Although a

Table 5.2 □ Systems in Which M_w/M_n is Less Than 2

System	$\bar{M}_w/\bar{M}_n$	Reference
CH₃COClO₄/styrene/CH₂Cl₂/0°C	1.34–1.45	Higashimura and Kishiro (1974)
HClO₄/styrene/CH₂Cl₂	1.3–1.6 for L and H[a]	Pepper (1974)
γ-Ray/p-fluorostyrene/bulk/25–27°C	1.3[b]	Hayashi and Pepper (1976)
γ-Ray/p-chlorostyrene/bulk/25–27°C	1.3[b]	Hayashi and Pepper (1976)
γ-Ray/2, 4-dimethylstyrene/bulk/10–17°C	1.5[b]	Hayashi and Pepper (1976)
γ-Ray/2, 6-trimethylstyrene/bulk/10–17°C	1.7[b]	Hayashi and Pepper (1976)
t-BuCl/Et₂AlCl/α-methylstyrene/CH₂Cl₂–MeOH(40:60)/–50°C	1.5	Kennedy and Fehérvári (1979b)
"H₂O"/Et₂AlCl/α-methylstyrene/CH₂Cl₂–MeOH(40:60)/–50°C	1.5	Kennedy and Fehérvári (1979b)
"H₂O"/Et₂AlCl/α-methylstyrene/CH₂Cl₂–MeOH(25:75)/–50°C	1.5	Kennedy and Fehérvári (1979b)
HSiMe₂CH₂CH₂C₆H₄CH₂Cl/Me₃Al/α-methylstyrene/CH₂Cl₂–n-C₆H₁₄(65:35)/–50 to –70°C	1.5–1.6	Kennedy and Chang (1979a)
Tetracyanobenzene/hν/α-methylstyrene/CH₂Cl₂/ –74°C	[b]	Yamamoto et al. (1976)
	1.4 (H)	Yamamoto et al. (1976)
	1.9 (L)	Yamamoto et al. (1976)
	1.5 (H)	Yamamoto et al. (1976)
	1.8 (L)	Yamamoto et al. (1976)
–60°C	1.3 (H)	Yamamoto et al. (1976)
	1.9 (L)	Yamamoto et al. (1976)
–50°C	1.8 (H)	Yamamoto et al. (1976)
	2.2 (L)	Yamamoto et al. (1976)
–30°C	1.7	Yamamoto et al. (1976)

"H_2O"/$TiCl_4$/p-chloro-α-methylstyrene/ CH_2Cl_2/−78°C	1.5	Lenz and Westfelt (1976)
"H_2O"/BCl_3/isobutylene/CH_3Cl or $CH_2Cl/−40$ to −70°C	1.5–1.7	Kennedy (1978c)
$HSiMe_2CH_2CH_2C_6H_4CH_2Cl/Me_3Al$/isobutylene/ CH_2Cl_2/n-C_6H_{14} (65:35)/−50 to −70°C	1.5–1.6	Kennedy and Chang (1979a)
$CH_2=CHC_6H_4CH_2Cl/H_2O/Me_3Al$/isobutylene/ $CH_3Cl/−60°C$	1.5–1.6	Kennedy and Frisch (1979c)
$CH_2=CHC_6H_4CH_2Cl/"H_2O"/Et_2AlCl$/isobutylene/ $CH_3Cl/−60°C$	1.6–1.7	Kennedy and Frisch (1979c)
Alkylphenols/$SnCl_4$/isobutylene/$C_2H_5Cl/78°C$	1.3 to 1.6[c]	Bauer et al. (1970).
I_2/n-butyl vinyl ether/$CH_2Cl_2/−60°$		
1st addition[d]	1.40	Johnson and Young (1976)
2nd addition[d]	1.54	Johnson and Young (1976)
3rd addition[d]	1.49	Johnson and Young (1976)
4th addition[d]	1.40	Johnson and Young (1976)

[a] The letters H and L refer to high and low fractions in a binodal distribution.

[b] Conversions are low (<1% in most cases).

[c] With phenol as initiator $\bar{M}_w/\bar{M}_n \simeq 2.8$.

[d] The polymerizations have been carried out by sequential additions of monomer.

large body of $\bar{M}_w/\bar{M}_n$ data is available that deals with anionic polymerizations, there is a dearth of information relative to cationic systems. Important advances in the understanding of the mechanism of cationic polymerization of styrene and styrene derivatives have been obtained mainly by Japanese investigators (Yamamoto et al., 1976; Higashimura and Kishiro, 1974) who, on the basis of GPC data, proposed the existence of multiple active species, that is, free and associated ion pairs. These matters are discussed in some depth in Section 4.2 on propagation.

Similar to molecular weights, molecular weight distributions, that is, $\bar{M}_w/\bar{M}_n$ values, are also important respositories of kinetic information. Thus the interpretation of $\bar{M}_w/\bar{M}_n = 2.0$, 1.5, or 1.0 in kinetic terms have been vigorously developed (Peebles, 1971); these values are very important guideposts for the mechanistic understanding of polymerization processes. In order to gain insight into the mechanism of carbocationic polymerizations $\bar{M}_w/\bar{M}_n$ data were assembled and analyzed. As a result of this survey it was discovered that numerous carbocationic polymerizations exist in which $\bar{M}_w/\bar{M}_n$ is significantly less than the expected 2.0. Systems in which $\bar{M}_w/\bar{M}_n < 2$, that is, narrower than the "most probable distribution," are compiled in Table 5.2. According to these data $\bar{M}_w/\bar{M}_n < 2$ for various aromatic and aliphatic olefin polymers and at least for one vinyl ether polymer obtained under a wide variety of conditions. In spite of great efforts to explain these unexpected data a satisfactory model could not be found. A recent model proposed by Morawetz (1979) includes the assumption of rapid simultaneous initiation which is probably of limited validity. Similarly, none of the numerous scenarios proposed to describe a wide variety of polymerizations (Peebles, 1971) seem to hold for most carbocationic systems: relatively slow initiation, fast propagation, and random chain breaking by chain transfer or termination. Combination of two chains, akin to the process frequently encountered in free radical polymerizations which results in $\bar{M}_w/\bar{M}_n = 1.5$ is very difficult to visualize in cationic systems. Similarly, the scenario which is the basis of "living" polymerizations (i.e., rapid simultaneous initiation with relatively slow propagation), is unlikely to occur in carbocationic polymerizations. It is tantalizing to be confronted with a rather large body of seemingly reliable data whose interpretation is totally obscure. Indeed, one is forced to the rather disturbing conclusion that at present even the overall (much less detailed) mechanism of polymerizations is unknown for a large number of systems, that is, systems with $\bar{M}_w/\bar{M}_n < 2$.

5.8 □ CONCLUSIONS: COMPILATION AND ANALYSIS OF RELIABLE KINETIC DATA

Even a cursory examination of the foregoing sections in this chapter indicates the sometimes insurmountable handicaps facing the kineticist in the field of carbocationic polymerizations and substantiates the list of difficulties outlined in the introduction.

Obviously, valid conclusions relative to the kinetics of carbocationic polymerizations can be drawn only if reliable quantitative information exists. However, the gathering of such data is extremely difficult, and only recently have investigators commenced to generate meaningful kinetic information in this field. Again referring to the difficulties listed in the introduction, among the perennial problems are the lack of quantitative understanding of the physical–chemical factors controlling ionicity of active species and the fragmentary understanding of chemical transformations that occur in the course of polymerization reactions.

In order to obtain reliable quantitative information for kinetic analysis a large amount of data has been compiled and examined in depth. Data that appear to have been generated under rigorously controlled conditions and are of general significance have been identified and are compiled in Table 5.3. Examination of the information in this Table 5.3 leads to valuable insight and conclusions relative to kinetic aspects of carbocationic polymerizations.

Apparently the most reliable data to date have been obtained with monomers leading to relatively stable propagating carbocations, for example, styrene derivatives, vinyl ethers, and cyclopentadiene, using stable cation salt initiator systems in the presence of the stable counteranion $SbCl_6^{\ominus}$ and CH_2Cl_2 solvent.

However, even data obtained by different authors under the most carefully controlled conditions seem to diverge. For example, depending on the experimental conditions, k_p's obtained with ethyl vinyl ether in the presence of $SbCl_6^{\ominus}$ in CH_2Cl_2 at 0°C are 1500, 5100, or 7100 l/mole · sec. Or, for isobutyl vinyl ether in the same system, the k_p's are 2000, 6800, 7000, or 9200. In view of such divergence among "reliable" data, it is well-nigh impossible to arrive at quantitative expressions, and even qualitative conclusions should be drawn with great caution.

Different rate constants obtained by different groups may be due to errors in experimental technique or in theory. Owing to the extreme sensitivity of carbocations toward impurities and the often high rates of polymerizations, it is difficult to operate under unobjectionable experimental conditions. Thus it is very difficult to determine whether characteristic phenomena that often appear during kinetic experiments (e.g., induction periods of various lengths, sigmoidal conversions) are in fact due to the reaction under investigation or to the presence of impurities. Among theoretical problems, the most vexing is the uncertainty as to the existence of stationary state. Stationary conditions are often assumed to prevail without direct experimental evidence; unfortunately in most instances when this assumption has been experimentally tested, stationary conditions could not be proved. The presence of steady state kinetics can be assumed only by finding the concentration of active species to be invariant at various intervals during the course of the polymerization.

Table 5.3 □ Selected Reliable Kinetic Data

System	k_i (l/mole·sec)	$\Delta H_i^\ddagger$ (kcal/mole)	k_p (l/mole·sec)	$\Delta H_p^\ddagger$ (kcal/mole)	k_t (sec^{-1})	$\Delta H_t^\ddagger$ (kcal/mole)	$k_{tr,M}$ (l/mole)	$\Delta H_{tr,M}^\ddagger$ (kcal/mole)	Reference
Styrene									
$\phi_3C^\oplus SnCl_5^\ominus/C_2H_4Cl_2/30°C$	0.0019	6.7	—	—	—	—	—	—	Higashimura et al. (1967)
$\phi_3C^\oplus SnCl_5^\ominus/C_2H_4Cl_2(80\ vol\ \%) + C_6H_6\ (10\ vol\ \%)/30°C$	0.0015	—	—	—	—	—	—	—	Higashimura et al. (1967)
$\phi_3 C^\oplus SnCl_5^\ominus/C_2 H_4 Cl_2\ (70\ vol\ \% + C_6H_6(20\ vol\ \%)/30°C$	0.00045	—	—	—	—	—	—	—	Higashimura et al. (1967)
$HClO_4/C_2H_4Cl_2/25°C$	—	—	17.0	—	—	—	—	—	Pepper et al. (1961)
$H_2SO_4/C_2H_4Cl_2/25°C$	—	—	7.6	—	—	—	—	—	Pepper and Reilly (1966)
$H_2SO_4/CCl_4/25°C$	—	—	0.0012	—	—	—	—	—	Pepper and Reilly (1966)
"H_2O"/$SnCl_4/C_3H_4Cl_2/30°C$	—	—	0.42	—	—	—	—	—	Dainton (1953)
$BF_3OEt_2/C_6H_6/30°C$	—	—	0.25	—	—	—	—	—	Higashimura et al. (1971)
$I_2/C_2 H_4 Cl_2/30°C$	—	—	0.0035	—	—	—	—	—	Kanoh et al. (1962)
α-Methylstyrene									
$I_2/C_2H_4Cl_2/30°C$	0.0048	5.0	—	—	—	—	—	—	Higashimura et al. (1967)
$I_2/C_2H_4Cl_2(86\ vol\ \%) + C_6H_6(20\ vol\ \%)$	0.0033	—	—	—	—	—	—	—	Higashimura et al. (1967)
$I_2/C_2H_4Cl_2(76\ vol\ \%) + C_6H_6(20\ vol\ \%)$	0.0023	—	—	—	—	—	—	—	Higashimura et al. (1967)

Conditions									Reference
H$_2$O/n-BuOTiCl$_3$/CH$_2$Cl$_2$/−70°C	17±6		22000±11000		0.54±0.05		30±15		Villesange et al. (1977)
H$_2$O/n-BuOTiCl$_3$/CH$_2$Cl$_2$/−50°C	60±10	4.5–5	4500±500	−6.5 to −7.5	1±0.1	0.5 to 2	—		Villesange et al. (1977)
H$_2$O/n-BuOTiCl$_3$/CH$_2$Cl$_2$/−30°C	120±20		1300±500		1±0.1		—		Villesange et al. (1977)
p-Methoxystyrene									
ϕ_3C$^{\oplus}$SbCl$_6^{\ominus}$/CH$_2$Cl$_2$/−15°C	0.03		74000		0.25	—	17		
ϕ_3C$^{\oplus}$SbCl$_6^{\ominus}$/CH$_2$Cl$_2$/−2°C	0.10	13.3±0.6	39000	−6±1	0.07	—	25		Cotrel et al. (1976)
ϕ_3C$^{\oplus}$SbCl$_6^{\ominus}$/CH$_2$Cl$_2$/10°C	0.28		(28±5)000		0.08±0.05	—	53±16	5.9±0.05	
ϕ_3C$^{\oplus}$SbCl$_6^{\ominus}$/CH$_2$Cl$_2$/25°C	0.89		16000		0.08	—	73		
I$_2$/C$_2$H$_4$Cl$_2$/30°C	0.13	—	—	—	—	—	—		Higashimura and Sawamoto (1978)
HOSCH$_3$/C$_2$H$_4$Cl$_2$/30°C	0.64	—	—	—	—	—	—		Higashimura and Sawamoto (1978)
"H$_2$O"/SnCl$_4$/C$_2$H$_4$Cl$_2$/30°C	49	—	—	—	—	—	—		Higashimura and Sawamoto (1978)
BF$_3$OEt$_2$/C$_2$H$_4$Cl$_2$/30°C	14	—	—	—	—	—	—		Higashimura and Sawamoto (1978)
BF$_3$OEt$_2$/C$_2$H$_4$Cl$_2$(90)+CCl$_4$(10)/30°C	7.5	—	—	—	—	—	—		Higashimura and Sawamoto (1978)
BF$_3$OEt$_2$/C$_2$H$_4$Cl$_2$(80)+CCl$_4$(10)/30°C	3.3	—	—	—	—	—	—		Higashimura and Sawamoto (1978)
BF$_3$OEt$_2$/C$_2$H$_4$Cl$_2$(60)+CCl$_4$(40)/30°C	0.74	—	—	—	—	—	—		Higashimura and Sawamoto (1978)

Table 5.3 □ *(Continued)*

System	k_i (l/mole·sec)	$\Delta H_i^{\ddagger}$ (kcal/mole)	k_p (l/mole·sec)	$\Delta H_p^{\ddagger}$ (kcal/mole)	k_t (sec^{-1})	$\Delta H_t^{\ddagger}$ (kcal/mole)	$k_{tr,M}$ (l/mole)	$\Delta H_{tr,M}^{\ddagger}$ (kcal/mole)	Reference
Cyclopentadiene									
H_2O/n-BuOTiCl$_3$/CH$_2$Cl$_2$/−70°C	11.1	—	2190	—	—	—	—	—	Vairon and Sigwalt (1971)
$C_7H_7^{\oplus}SbCl_6^{\ominus}$/CH$_2Cl_2$/25°C	0.018	—	—	—	—	—	—	—	Sauvet et al. (1969, 1974)
$C_7H_7^{\oplus}SbCl_6^{\ominus}$/CH$_2Cl_2$/10°C	—		0.98		0.006		—	—	Sauvet et al. (1969, 1974)
$C_7H_7^{\oplus}SbCl_6^{\ominus}$/CH$_2Cl_2$/10°C	0.0033		—				—	—	Sauvet et al. (1969, 1974)
$C_7H_7^{\oplus}SbCl_6^{\ominus}$/CH$_2Cl_2$/−10°C	0.0041	9.2 ± 1	3.78	−8 ± 0.5	0.0062	−0.3 ± 0.1	—	—	Sauvet et al. (1969, 1974)
$C_7H_7^{\oplus}SbCl_6^{\ominus}$/CH$_2Cl_2$/−30°C	0.0041		11.03		0.0083		—	—	Sauvet et al. (1969, 1974)
$C_7H_7^{\oplus}SbCl_6^{\ominus}$/CH$_2Cl_2$/−50°C	—		45.33		0.0068		—	—	Sauvet et al. (1969, 1974)
$C_7H_7^{\oplus}SbCl_6^{\ominus}$/CH$_2Cl_2$/−70°C	0.000017		—				—	—	Sauvet et al. (1969, 1974)
Methyl vinyl ether									
$C_7H_7^{\oplus}SbCl_6^{\ominus}$/CH$_2Cl_2$/0°C	—	—	140	14	—	—	—	—	Ledwith et al. (1975)
$C_7H_7^{\oplus}SbCl_6^{\ominus}$/CH$_2Cl_2$/11.8°C	—	—	400		—	—	—	—	
$\phi_3C^{\oplus}SbCl_6^{\ominus}$/CH$_2Cl_2$/−20°C	0.18		50		0.003	—	—	—	Subira et al. (to be published)
$\phi_3C^{\oplus}SbCl_6^{\ominus}$/CH$_2Cl_2$/−10°C	0.31	10.4	120	12	0.01	—	—	—	Subira et al. (to be published)
$\phi_3C^{\oplus}SbCl_6^{\ominus}$/CH$_2Cl_2$/−20°C	0.60		265		—	—	—	—	Subira et al. (to be published)

									Reference
Ethyl vinyl ether									
$C_7H_7^{\oplus}\,SbCl_6^{\ominus}/CH_2Cl_2/0°C$	—	—	1500	10		—	—	—	Ledwith et al. (1975)
$C_7H_7^{\oplus}\,SbCl_6^{\ominus}/CH_2Cl_2/11.8°C$	—	—	3400			—	—	—	Ledwith et al. (1975)
$\phi_3C^{\oplus}SbCl_6^{\ominus}/CH_2Cl_2/-40°C$	0.27		230		0	—	—	—	Subira et al. (to be published)
$\phi_3C^{\oplus}SbCl_6^{\ominus}/CH_2Cl_2/-25°C$	0.8	6.5	1030	10.2	0.01	—	—	—	Subira et al. (to be published)
$\phi_3C^{\oplus}SbCl_6^{\ominus}/CH_2Cl_2/0°C$	2.3		7100		0.2	—	—	—	Subira et al. to be published
$\phi_3C^{\oplus}SbCl_6^{\ominus}/CH_2Cl_2/-25°C$	—		660			—	—	—	Chung et al. (1975)
$\phi_3C^{\oplus}SbCl_6^{\ominus}/CH_2Cl_2/0°C$	—		5100	9.9		—	—	—	Chung et al. (1975)
$\phi_3C^{\oplus}SbCl_6^{\ominus}/CH_2Cl_2/15°C$	—		13200			—	—	—	Chung et al. (1975)
$\gamma/bulk/0°C$	—		8300	10.8	—	—	—	—	Suzuki et al. (1977)
Isopropyl vinyl ether									
$\gamma/bulk/0°C$	—		900000	1.8	—	—	—	—	Goineau et al. (1977)
$\phi_3C^{\oplus}SbCl_6^{\ominus}/CH_2Cl_2/-25°C$	6.0		6550		0.38	—	—	—	Subira et al. (to be published)
$\phi_3C^{\oplus}SbCl_6^{\ominus}/CH_2Cl_2/0°C$	15		11300	3.5	0.26	—	—	—	Subira et al. (to be published)
$\phi_3C^{\oplus}SbCl_6^{\ominus}/CH_2Cl_2/40°C$	2.5		3700		0.11	—	—	—	Subira et al. (to be published)
Isobutyl vinyl ether									
$C_7H_7^{\oplus}SbCl_6^{\ominus}/CH_2Cl_2/-25°C$	—		2000			—	—	—	Bawn et al. (1971)
$C_7H_7^{\oplus}SbCl_6^{\ominus}/CH_2Cl_2/0°C$	—		6800	6.0		—	—	—	Bawn et al. (1971)
$\phi_3C^{\oplus}SbCl_6^{\ominus}/CH_2Cl_2/-25°C$	—		1500			—	—	—	Bawn et al. (1971)
$\phi_3C^{\oplus}SbCl_6^{\ominus}/CH_2Cl_2/0°C$	—		4000			—	—	—	Bawn et al. (1971)

Table 5.3 □ *(Continued)*

System	k_i (l/mole·sec)	$\Delta H_i^{\ddagger}$ (kcal/mole)	k_p (l/mole·sec)	$\Delta H_p^{\ddagger}$ (kcal/mole)	k_1 (sec^{-1})	$\Delta H_1^{\ddagger}$ (kcal/mole)	$k_{tr,M}$ (l/mole)	$\Delta H_{tr,M}^{\ddagger}$ (kcal/mole)	Reference
Isobutyl vinyl ether									
$\phi_3C^{\oplus}SbCl_6^{\ominus}/CH_2Cl_2/-25°C$	—	—	3400	—	—	—	—	—	Chung et al. (1975)
$\phi_3C^{\oplus}SbCl_6^{\ominus}/CH_2Cl_2/0°C$	—	—	9200	7.1	—	—	—	—	Chung et al. (1975)
$\phi_3C^{\oplus}SbCl_6^{\ominus}/CH_2Cl_2/15°C$	—	—	24400	—	—	—	—	—	Chung et al. (1975)
$\phi_3C^{\oplus}SbCl_6^{\ominus}/CH_2Cl_2/-40°C$	0.3	—	930	—	0.01	—	8	—	Subira et al. (1976)
$\phi_3C^{\oplus}SbCl_6^{\ominus}/CH_2Cl_2/-25°C$	0.9	—	3250	—	0.05	—	27	—	Subira et al. (1976)
$\phi_3C^{\oplus}SbCl_6^{\ominus}/CH_2Cl_2/0°C$	5.4	9.4	7000	6.9	0.19	9.6	185	8.9	Subira et al. (1976)
$\phi_3C^{\oplus}SbCl_6^{\ominus}/CH_2Cl_2/20°C$	16	—	16000	—	0.8	—	590	—	Subira et al. (1976)
$\phi_3C^{\oplus}BF_4^{\ominus}/CH_2Cl_2/0°C$	—	—	2800	—	—	—	—	—	Bawn et al. (1971)
$I_2/C_2H_4Cl_2/30°C$	—	—	6.5	—	—	—	—	—	Kanoh and Higashimura (1966)
$I_2/CCl_4/30°C$	—	—	0.083	—	—	—	—	—	Okamura et al. (1961)
X-Rays/bulk/0°C	—	—	3800	—	—	—	50	—	Hayashi et al. (1971)
X-Rays/bulk/25°C	—	—	120000	9.6	—	—	135	—	Hayashi et al. (1971)
X-Rays/bulk/50°C	—	—	600000	—	—	—	490	—	Hayashi et al. (1971)

γ-Rays/bulk/0°C	—	—	—	—	38000	—	—	Goineau et al. (1977)
γ-Rays/bulk/15°C	—	—	—	7.6	95000	—	—	Goineau et al. (1977)
tert-Butyl vinyl ether								
$C_7H_7^{\oplus}SbCl_6^{\ominus}$/$CH_2Cl_2$/0°C	—	—	—	1	3500	—	—	Ledwith et al. (1975)
$C_7H_7^{\oplus}SbCl_6^{\ominus}$/$CH_2Cl_2$/11.8°C	—	—	—	—	3800	—	—	Goineau et al. (1977)
$C_7H_7^{\oplus}SbCl_6^{\ominus}$/Bulk/0°C	—	—	—	4.9	50000	—	—	
2-Chloroethyl vinyl ether								
$C_7H_7^{\oplus}SbCl_6^{\ominus}$/$CH_2Cl_2$/0°C	—	—	—	—	200	—	—	Ledwith et al. (1975)
$C_7H_7^{\oplus}SbCl_6^{\ominus}$/$CH_2Cl_2$/11.8°	—	—	—	7	310	—	—	
Cyclohexyl vinyl ether								
$C_7H_7^{\oplus}SbCl_6^{\ominus}$/$CH_2Cl_2$/0°C	—	—	—	—	3300	—	—	Ledwith et al. (1975)
$C_7H_7^{\oplus}SbCl_6^{\ominus}$/$CH_2Cl_2$/11.8°C	—	—	—	9	6700	—	—	Ledwith et al. (1975)

An insurmountable problem facing the kineticist in evaluating carbocationic polymerization data is his ignorance as to the exact nature of the active species. The rate constants shown in Table 5.3 are in most cases "apparent," not absolute, since most experiments have been carried out in the liquid phase where the influence of solvent and/or counteranion cannot be neglected. The ionicities of active species have been sufficiently characterized in many carbanionic systems; however, the relative contributions of the various species, for example, free ions, solvent-separated ions, and contact ions, to the measured rate or rate constant remain largely obscure in carbocationic polymerizations.

Analysis of the effect of experimental conditions on k_p is a most difficult undertaking; most published information provides at best an order of magnitude for k_p. Frequently k_p values reported as such are not pure constants of propagation and include nonnegligible contributions by initiation.

According to the data in Table 5.3, chemical initiation gives at least one order of magnitude lower k_p than that of irradiation. Apparently, k_p values obtained in γ-ray initiated polymerizations of vinyl ethers are much higher (one to three orders of magnitude) than those obtained by chemical initiation in moderately polar solvents, for example, CH_2Cl_2 and $C_2H_4Cl_2$. Free ions are most likely the active species in the former systems and, considering the nature of the media (the dielectric constant of isobutyl vinyl ether is below 4), they are probably little solvated. The difference between k_p's is larger with styrene than with isobutyl vinyl ether, where the difference is quite small. This may be due to differences in solvation: the growing isobutyl vinyl ether cations due to charge delocalization to the oxygen may be less solvated than the growing polystyryl cations having less delocalized charge.

Valid comparison of k_p's obtained in irradiation- and chemical-initiated polymerizations is difficult, however, because the latter have usually been obtained by the use of chlorinated solvents, which are sensitive to radiolysis.

In view of the appreciable dissociation constants of stable cation salts, initiation in these systems occurs probably by solvated (by CH_2Cl_2) free ions. That the active species obtained by stable cation salts are quite similar (i.e., solvated free ions) is indicated by the similar k_p values obtained in the presence of various counteranions for example, $SbCl_6^{\ominus}$ and $BF_4^{\ominus}$.

In contrast, the situation is quite different with initiating systems such as $HClO_4$, H_2SO_4, "H_2O"/$SnCl_4$, $BF_3 \cdot OEt_2$, and I_2 (counteranion $I^{\ominus}$ or $I_3^{\ominus}$). Although some of the values are probably too low (incorrect data), a considerable difference in k_p values is apparent. In these instances the propagating carbocations are probably more or less dissociated ion pairs and the contribution by free ions to the equilibrium cannot be neglected. This is, for example, the case with $k_p = 0.25$ l/mole $\cdot$ sec for styrene polymerization in benzene solvent (Higashimura et al., 1971).

Free radical initiated polymerization rates invariably increase by raising the temperature. However, in carbocationic systems, as indicated by the data in Table 5.3, overall activation enthalpies may be negative or positive, and raising the temperature may increase or decrease R_0. Differences of activation enthalpy of initiation of carbocationic polymerizations $\Delta H_i^{\ddagger}$ are smaller (5 to 13 kcal/mole) than those obtained in radical polymerizations (20 to 30 kcal/mole) and are reasonably close to activation enthalpy differences of ionization of the type $AB \rightleftharpoons A^{\oplus}B^{\ominus}$. Increasing solvent polarity further reduces $\Delta H_i^{\ddagger}$.

The effect of solvation could be studied by polymerizing a particular monomer by various initiation methods, that is, by determining $k_p^{\oplus}$'s for the same monomer polymerized by various initiating methods, that is, irradiation, electric field, or chemical, at different temperatures and thus obtaining strongly solvation-dependent enthalpies and entropies of activation.

Solvation by monomer (Fontana and Kidder, 1948; Sigwalt, 1971; Higashimura, 1969) and complexation by monomer seem to affect numerous systems (Pepper, 1975; Kucera, 1973; Chmelir, 1973; Plesch, 1974). These phenomena should be further investigated and defined.

The absence of generalizable kinetics in the field of carbocationic polymerizations is a real, though not insurmountable handicap. Indeed, we consider future research in carbocationic kinetics much facilitated by the recognition that generalization from one system to another, even to a similar one, is not possible. According to the analysis in this chapter every system must be individually scrutinized and application of kinetic schemes by similarity between systems should be made with utmost care. The realization that generalizations cannot be made should go a long way toward clarifying the opinion *"Rudis indigestaque moles,"** made by an occasional visitor to the land of carbocationic polymerizations. Researchers must be highly selective in choosing their particular target and should undertake demanding kinetic studies only if they are convinced that their cost will be worth the effort. In a sense kinetic and analytical-instrumental analyses fulfill the same mission: both provide badly needed information as to the detailed mechanism of polymerizations.

□ REFERENCES

Addecott, K. S. B., Mayor, L., and Turton, C. N. (1967), *Eur. Polym. J.* 3, 601.

Alley, P. E. M., and Patrick (1974), *Kinetics and Mechanisms of Polymerization Reactions*, Wiley, New York.

Bamford, C. H., Barb, W. G., Jenkins, A. D. and Onyon, P. F. (1958), *Kinetics of Vinyl Polymerization by Radical Mechanism*, Butterworth, London.

Bauer, R. F., Laflair, R. T. and Russell, K. E. (1970), *Can. J. Chem.* **48**, 1251.

*"A crude and confused mass" (F. S. Dainton in *Cationic Polymerization and Related Complexes*, P. H. Plesch, Heffer, Cambridge, 1953, p. 148).

Bawn, C. E. H., Fitzsimmons, C., Ledwith, A., Penfold, J., Sherrington, D. C., and Weightman, J. A. (1971), *Polymer* **12**, 119.

Beard, T. H. and Plesch, P. H. (1964a), *J. Chem. Soc.*, 3682.

Beard, T. H., and Plesch, P. H. (1964b), *J. Chem. Soc.*, 4879.

Biddulph, R. H., and Plesch, P. H. (1960), *J. Chem. Soc.*, 3913.

Biddulph, R. H., Plesch, P. H., and Rutherford, P. P. (1965), *J. Chem. Soc.*, 275.

Bonin, M. A., Busler, W. R., and Williams, F. (1965), *J. Am. Chem. Soc.* **87**, 199.

Bowyer, P. M., Ledwith, A., and Sherrington, D. C. (1971a), *Polymer* **12**, 509.

Bowyer, P. M., Ledwith, A., and Sherrington, D. C. (1971b), *J. Chem. Soc. B.*, 1511.

Brown, C. P., and Mathieson, A. R. (1957), *J. Chem. Soc.*, 3612.

Burton, E., and Pepper, D. C. (1961), *Proc. Roy. Soc. A* **263**, 58.

Cesca, S., Priola, A., Bruzzone, M., Ferraris, G., and Giusti, P. (1975), *Makromol. Chem.* **176**, 2339.

Chmelir, M. (1973), *Int. Symp. Cationic Polym. Rouen*, Commun. no. 7.

Chung, Y. J., Rooney, J. P., Squire, D. R., and Stannett, V. (1975), *Polymer* **16**, 527.

Cotrel, R., Sauvet, G., Vairon, J. P., and Sigwalt, P. (1976), *Macromolecules* **9**, 931.

Dainton, F. S. (1953), in *Cationic Polymerization and Related Complexes*, P. H. Plesch, Ed., Heffer, Cambridge,. 148.

Dennison, J. T., and Ramsey, J. B. (1955), *J. Am. Chem. Soc.* **77**, 2615.

De Sorgo, M., Pepper, D. C., and Szwarc, M. (1973), *Chem. Commun.*, 419.

Flory, P. J. (1953), *Principles of Polymer Chemistry*, Cornell University Press, New York.

Fontana, C. M., and Kidder, G. A. (1948), *J. Am. Chem. Soc.* **70**, 3745.

Fuoss, R. (1958), *J. Am. Chem. Soc.* **80**, 5059.

Goineau, A. M., Kohler, J., and Stannett, V. (1977), *J. Macromol. Sci. Chem.* **A11**, 99.

Hayashi, K., Yamazawa, Y., Takagaki, T., Williams, F., Hayashi, K., and Okamura, S. (1967), *Trans. Faraday Soc.* **63**, 1489.

Hayashi, K., Hayashi, H., and Okamura, S. (1971), *J. Polym. Sci. Al*, **9**, 2305.

Hayashi, K., and Pepper, D. C. (1976), *Polym. J.* **8**, 1.

Hayes, M. J., and Pepper, D. C. (1961), *Proc. Roy. Soc. A* **263**, 63.

Heublein, G. (1975), *Zum Ablauf ionischer Polymerisation sreaktionen*, Akademie Verlag, Berlin.

Higashimura, T., and Okamura, S. (1956), *Chem. High Polym. Jap.* **13**, 338.

Higashimura, T., Fukushima, T., and Okamura, S. (1967), *J. Macromol. Sci. Chem.* **A1**, 683.

Higashimura, T. (1969), in T. Tsuruta and K. F. O'Driscoll, Eds., *Structure and Mechanism in Vinyl Polymerization*, Dekker, New York.

Higashimura, T., Kusano, H., Masuda, T., and Okamura, S. (1971), *Polym. Lett.* **9**, 463.

Higashimura, T., and Kishiro, O. (1974), *J. Polym. Sci., Polym. Chem.* **12**, 967.

Higashimura, T., Kishiro, O., and Takeda, T. (1976), *J. Polym. Sci. Polym. Chem.* **14**, 1089.

Higashimura, T., Sawamoto, M., and Masuda, T. (1977), *Polym. Colloq., Kyoto*, 26.

Higashimura, T., and Sawamoto, M. (1978), *Polym. Bull.* **1**, 11.

Higashimura, T. (1979), *Polym. Prep. (ACS)*, **20**(1) 161.

Hughes, J., and North, A. M. (1966), *Trans. Faraday Soc.* **62**, 1866.

Imanishi, Y., Higashimura, T., and Okamura, S. (1960), *Kobunshi Kagaku* **17**, 357.

Jenkinson, D. H., and Pepper, D. C. (1961), *Proc. Roy. Soc. (London)* **A263**, 82.

Johnson, A. F., and Young, R. N. (1976), *J. Polym. Sci. Symp.* **56**, 211.

Jones, F. R., and Plesch, P. H. (1970), *Chem. Commun.*, 1018.

Kalfoglou, N., and Szwarc, M. (1968), *J. Phys. Chem.* **72**, 2233.

Kamachi, M., and Miyama, H. (1965), *J. Polym. Sci. A* **3**, 1337.

Kanoh, N., Higashimura, T., and Okamura, S. (1962), *Macromol. Chem.* **56**, 65.

Kanoh, N., Ikeda, K., Gotoh, A., Higashimura, T., and Okamura, S. (1965), *Makromol. Chem.* **86**, 200.

Kanoh, N. and Higashimura, T. (1966), *Kobunshi Kagaku* **23**, 114.

Kennedy, J. P., and Thomas, R. M. (1961), *Int. Symp. Macromol. Chem., Montreal.*

Kennedy, J. P., and Thomas, R. M. (1962), *Adv. Chem. Ser.* **34**, 111.

Kennedy, J. P., and Squires, R. G. (1965), *Polymer* **6**, 579.

Kennedy, J. P., and Milliman, G. E. (1969), *Adv. Chem. Ser.* **91**, 287.

Kennedy, J. P., Shinkawa, A., and Williams, F. (1971), *J. Polym. Sci. Al* **9**, 1551.

Kennedy, J. P., and Sivaram (1973), *J. Macromol. Sci. Chem.* **A7**, 969.

Kennedy, J. P., and Rengachary, S. (1974), *Adv. Polym. Sci.* **14**, 1.

Kennedy, J. P., and Trivedi (1978a), *Adv. Polym. Sci.* **28**, 83.

Kennedy, J. P., and Trivedi (1978b), *Adv. Polym. Sci.* **28**, 113.

Kennedy, J. P. (1978c), unpublished data.

Kennedy, J. P., and Chang, S. C. (1979a), unpublished data.

Kennedy, J. P., and Fehervari, A. (1979b), unpublished data.

Kennedy, J. P., and Frisch, K., Jr. (1979c), unpublished data.

Kucera, M. (1973), *Int. Symp. Cationic Polym., Rouen,* Commun. no. 21.

Ledwith, A. (1969), *Adv. Chem. Ser.* **91**, 317.

Ledwith, A., Lockett, E., and Sherrington, D. C. (1975), *Polymer* **16**, 31.

Ledwith, A., and Sherrington, D. C. (1976), in *Comprehensive Chemical Kinetics,* C. H. Bamford and C. F. H. Tipper, Eds., Elsevier, Vol. 15, p. 67.

Lee, W. Y., and Treloar, F. E. (1969), *J. Phys. Chem.* **73**, 2458.

Lenz, W. L., and Westfelt, L. C. (1976), *J. Polym. Sci., Polym. Chem.* **12**, 2147.

Longworth, W. R. and Plesch, P. H. (1959), *Proc. Int. Symp. Macromol. Chem. Wiesbaden,* Paper III A11.

Longworth, W. R., and Mason, C. P. (1966), *J. Chem. Soc. A,* 1164.

Lorimer, J. P., and Pepper, D. C. (1976), *Proc. R. Soc. London, Ser. A* **351**, 551.

Ma, C. C., Kubota, H., Rooney, J. M., Squire, D. R., and Stannett, V. (1979), *Polymer* **20**, 317.

Malinska-Solich, J., Chmelir, M., and Marek, M. (1969), *Coll. Czech. Chem. Commun.* **34**, 2611.

Mandal, B. M., and Kennedy, J. P. (1978), *J. Polym. Sci. Polym. Chem. Ed.* **16**, 833.

Masuda, T., and Higashimura, T. (1971), *Polym. Lett.* **9**, 783.

Masuda, T., Sawamoto, M., and Higashimura, T. (1976), *Makromol. Chem.* **177**, 2981.

Mayo, F. A. (1943), *J. Am. Chem. Soc.* **65**, 2324.

Mayo, F. A., Gregg, R. A., and Matheson, M. S. (1951), *J. Am. Chem. Soc.* **73**, 1691.

Morawetz, H. (1979), *Macromolecules* **12**, 532.

Okamura, S., Kanoh, N., and Higashimura, T. (1961), *Makromol. Chem.* **47**, 19, 35.

Peebles, L. H., *Molecular Weight Distributions in Polymers,* Wiley-Interscience, New York, 1971.

Pepper, D. C., Burton, R. E., Hayes, M. J., Albert, A., and Jenkinson, D. H. (1961), *Proc. R. Soc. Ser. A* **263**, 58, 63, 75, 82.

Pepper, D. C., and Reilly, P. J. (1966), *Proc. R. Soc. Ser. A* **291**, 41.

Pepper, D. C. (1974), *Makromol. Chem.* **175**, 1077.

Pepper, D. C. (1975), *J. Polym. Sci., Polym. Symp.* **50**, 51.

Plesch, P. H. (1963), *The Chemistry of Cationic Polymerization*, Macmillan, New York.

Plesch, P. H. (1968), *Progress in High Polymers*, J. C. Robb and W. H. Peaker, Eds., Heywood Books, London, Vol. **2**, p. 137.

Plesch, P. H. (1971), *Adv. Poly. Sci.* **8**, 137.

Plesch, P. H. (1974), *Makromol. Chem.* **175**, 1065.

Sangster, J. M., and Worsfold, D. J. (1972), *Polym. Prepr. Am. Chem. Soc., Div. Polym. Chem.* **13**, 72.

Sauvet, G., Vairon, J. P., and Sigwalt, P. (1969), *J. Polym. Sci.* **A1, 7**, 983.

Sauvet, G., Vairon, J. P., and Sigwalt, P. (1970), *Bull. Soc. Chim.*, 4031.

Sauvet, G., Vairon, J. P., and Sigwalt, P. (1974), *Eur. Polym. J.* **10**, 501.

Sauvet, G., Vairon, J. P., and Sigwalt, P. (1975), *J. Polym. Sci., Polym. Symp.* **52**, 173.

Sauvet, G., Vairon, J. P., and Sigwalt, P. (1978), *J. Polym. Sci., Polym. Symp.* **16**, 3047.

Sawada, H. (1972), *Thermodynamics of Polymerization*, Dekker New York, p. 183.

Sawamoto, M., Masuda, T., Higashimura, T., Kobayashi, S., and Saegusa, T. (1977), *Makromol. Chem.* **178**, 389.

Sawamoto, M., and Higashimura, T. (1978a) *Macromolecules* **11**, 328.

Sawamoto, M., and Higashimura, T. (1978b), *Macromolecules* **11**, 501.

Sigwalt, P. (1971), *23rd Int. Cong. Pure Appl. Chem., Boston.*

Smid, J., and Hogen-Esch, T. E. (1966), *J. Am. Chem. Soc.* **88**, 318.

Subira, F., Sauvet, G., Vairon, J. P., and Sigwalt, P. (1976), *J. Polym. Sci. Symp.*, **56**, 221.

Subira, F., Sauvet, G., Vairon, J. P., Sigwalt, P. To be published.

Suzuki, Y., Chudgar, A., Rooney, J. M., and Stannett, V. (1977), *J. Macromol. Sci. Chem.* **A11**, 115.

Szwᵣrc, M. (1968), *Carbanions, Living Polymers and Electron Transfer Processes*, Wiley-Interscience, New York.

Szwarc M. (1969), *Addition and Condensation Polymerization Processes, Adv. Chem. Ser.* **91**, 236.

Taylor, R. B., and Williams, F. (1969), *J. Am. Chem. Soc.* **91**, 3728.

Thomas, R. M., Sparks, W. J., Frolich, P. K., Otto, M., and Mueller-Cunradi, M. (1940), *J. Am. Chem. Soc.* **62**, 276.

Toman, L., and Marek, M. (1976), *Makromol. Chem.* **177**, 3325.

Ueno, K., Hayashi, K., and Okamura, S. (1968), *J. Macromol. Sci. Chem.* **A2**, 209.

Vairon, J. P., and Sigwalt, P. (1971), *Bull. Soc. Chim.*, 569.

Villesange, M., Sauvet, G., Vairon J. P., and Sigwalt, P. (1977), *J. Macromol. Sci. Chem.* **A11**, 391.

Williams, F., Hayashi, K., Ueno, K., Hayashi, K., and Okamura, S. (1967), *Trans. Faraday Soc.* **19**, 1501.

Yamamoto, Y., Irie, M., and Hayashi, K. (1976), *Polym. J.* **8**, 437.

□ 6 □

Copolymerization and Reactivity

6.1 ☐ INTRODUCTION

During the past 25 years numerous studies have been published relative to the problem of reactivity in radical polymerization; however, the study of reactivity in ionic, particularly cationic, copolymerization lagged owing to experimental/theoretical difficulties. In contrast to radical copolymerizations, reactivity ratios in cationic polymerizations are strongly affected by reaction conditions and it is much more difficult to develop a general theory of reactivity in cationic than in radical systems. This chapter mainly concerns a survey of determination methods of reactivity ratios in cationic copolymerization, experimental and theoretical studies of reactivity, and a discussion of the influence of experimental and structural factors on reactivity.

6.2 ☐ DEFINITIONS AND FUNDAMENTALS

Numerous cationic initiating systems are able to copolymerize mixtures of two or more monomers; the copolymers contain monomer units in proportions determined by their respective concentrations in the charge and reactivity toward the growing carbenium ions. Surveys on cationic copolymerization concerning early developments have been published (Kennedy, 1964; Cundall, 1963; Ham, 1964).

In 1944 Mayo and Lewis (1944) and Alfrey and Goldfinger (1944) independently proposed an equation that, independent of the nature of the active species, quantitatively describes the effect of the terminal units on growing chain reactivity. The four propagation steps in a carbocationic copolymerization are as follows:

$$\text{\textasciitilde}M_1^{\oplus} + M_1 \rightarrow \text{\textasciitilde}M_1M_1^{\oplus} \qquad k_{11} \tag{6.1}$$

$$\text{\textasciitilde}M_1^{\oplus} + M_2 \rightarrow \text{\textasciitilde}M_1M_2^{\oplus} \qquad k_{12} \tag{6.2}$$

$$\text{\textasciitilde}M_2^{\oplus} + M_1 \rightarrow \text{\textasciitilde}M_2M_1^{\oplus} \qquad k_{21} \tag{6.3}$$

$$\text{\textasciitilde}M_2^{\oplus} + M_2 \rightarrow \text{\textasciitilde}M_2M_2^{\oplus} \qquad k_{22} \tag{6.4}$$

where M_1, M_2, $M_1^{\oplus}$, and $M_2^{\oplus}$ are the two monomers and corresponding carbenium ions, k is a rate constant, and $m_1 = d[M_1]$ and $m_2 = d[M_2]$ are the change in monomer concentrations during dt. It can be demonstrated (Mayo and Lewis, 1944; Alfrey and Goldfinger, 1944) that

$$\frac{m_1}{m_2} = \frac{r_1[M_1]/[M_2] + 1}{r_2[M_2]/[M_1] + 1} = \frac{[M_1]}{[M_2]} \frac{r_1[M_1] + [M_2]}{r_2[M_2] + [M_1]} \tag{6.5}$$

where $r_1 = k_{11}/k_{12}$ and $r_2 = k_{22}/k_{21}$ are the reactivity ratios. The copolymer composition equation 6.5 implies that the chains are long, so that monomer consumption is due only to propagation. The equation is valid for low conversions, two monomers, and two propagating species.

The knowledge of individual homopolymerization rates is largely meaningless in predicting the characteristics of copolymer composition; no such extrapolation should be made (Kennedy, 1964).

Reactivity ratio values are acceptable only when obtained in experiments in which the change in the initial monomer concentrations may be considered negligible. Since the composition of a copolymer always differs from the feed—except in the case of azeotropic copolymerization—this requirement can be met only in continuous systems or approached in batch operations at low conversions. However, it is possible to obtain a complete description. of a copolymerization by integrating the copolymerization equation. The product obtained at higher conversions reflects the drift in charge composition with progressing conversion (Kennedy, 1964).

The integrated form of the copolymerization equation was derived by Mayo and Lewis (1944):

$$\log \frac{[M_1]}{[M_1]_0} = \frac{r_1}{1-r_1} \log \frac{[M_1]_0[M_2]}{[M_1][M_2]_0} -$$

$$\frac{1-r_1r_2}{(1-r_1)(1-r_2)} \log \frac{(r_2-1)\frac{[M_2]}{[M_1]} - r_1 + 1}{(r_2-1)\frac{[M_2]_0}{[M_1]_0} - r_1 + 1} \qquad (6.6)$$

or by using a parametric form:

$$r_2 = \frac{\log \frac{[M_2]_0}{[M_2]} - \frac{1}{p} \log \frac{1-p\frac{[M_1]}{[M_2]}}{1-p\frac{[M_1]_0}{[M_2]_0}}}{\log \frac{[M_1]_0}{[M_1]} + \log \frac{1-p\frac{[M_1]}{[M_2]}}{1-p\frac{[M_1]_0}{[M_2]_0}}} \qquad (6.7)$$

and

$$r_1 = p(r_2-1) + 1$$

where $[M_1]_0$ and $[M_2]_0$ and $[M_1]$ and $[M_2]$ are the mole fractions of monomers 1 and 2 in the charge at $t = 0$ and after copolymerization, respectively.

6.3 □ DETERMINATION OF REACTIVITY RATIOS

The determination of reactivity ratios is particularly difficult in cationic polymerizations. The various methods based on the copolymerization equation have been discussed in detail (Ham, 1964; Kennedy, 1964; Tidwell and Mortimer, 1970; Chan and Meyer, 1968; Kelen and Tüdős, 1974, 1975). A new method has been developed by Kelen and Tüdős both for low (Kelen and Tüdős, 1974, 1975; Kennedy et al., 1975c) and high conversion (Tüdős et al., 1975, 1976; Kelen et al., 1977).

This section includes a brief discussion of the options available for the determination of reactivity ratios before the Kelen–Tüdős method became available, a description of the Kelen–Tüdős (KT) method and Kennedy's application to cationic copolymerization, and a critical survey of reactivity ratios (Kennedy et al., 1975c; Kelen et al., 1977).

Differential Methods

These methods are based on the differential form of the copolymer composition equation and are valid only at low monomer conversions. Equation 6.5 contains four parameters (monomer molar concentration in the charge, $[M_1]$ and $[M_2]$, and molar amount of monomer in copolymer, m_1 and m_2) and two "unknowns," r_1 and r_2. $[M_1]$ and $[M_2]$ are arbitrarily chosen and m_1 and m_2 are determined by experiment.

Various methods have been proposed to obtain the "best" r_1, r_2 pair (Alfrey et al., 1951).

1 □ In the *curve-fitting method* several monomer mixtures (preferentially eight or more) over the entire mole fraction range are copolymerized and the compositions of the products obtained at low conversion are analyzed. Results are plotted according to Alfrey et al. (1951). The best r_1 and r_2 values are obtained by trial and error fitting of the theoretical curve with experimental points. As a guide, $1/r_1$ is equal to the initial slope of the composition curve at 100% $[M_2]$ (Ham, 1954a).

2 □ For the *intersection method* (Mayo and Lewis, 1944), equation 6.5 is rewritten:

$$r_2 = \frac{[M_1]}{[M_2]}\left[\frac{m_2}{m_1}\left(1 + \frac{[M_1]}{[M_2]}r_1\right) - 1\right] \tag{6.9}$$

or

$$r_2 = ar_1 + b$$

where a and b are parameters computable from $[M_1]$, $[M_2]$, m_1, and m_2. Plotting r_2 versus r_1 produces a straight line. Each experiment produces a straight line the intersection of which provides r_1 and r_2. The intersections define an area the size of which is characteristic of the error involved in the determination of r_1 and r_2.

3 □ According to the *Fineman and Ross method* (1950), equation 6.5 is rewritten as

$$\frac{x(y-1)}{y} = r_1\frac{x^2}{y} - r_2 \tag{6.10}$$

where $x = [M_1]/[M_2]$ and $y = m_1/m_2$. A plot of x^2/y gives a straight line whose slope is r_1 and intercept r_2.

4 ☐ The *approximation method* mentioned by Tidwell and Mortimer (1970) exploits the fact that at very low M_2 mole fraction in the charge, copolymer composition almost entirely depends on r_1:

$$r_1 \simeq \frac{[M_2]}{m_2} \tag{6.11}$$

Thus a single experiment provides an approximate value for r_2. Extremely sensitive analytical procedures are required to determine m_2. If $r < 0.1$ or $r > 10$ the computed r will be seriously biased.

5 ☐ The *nonlinear least-squares method* (Tidwell and Mortimer, 1965) is an extension of the curve-fitting method. It differs from the latter in that the values of r_1 and r_2 satisfy the criterion that, for the selected values of these ratios, the sum of the squares of the differences between the observed and computed polymer composition is minimized. In this sense the values obtained are unique for a given set of data. A FORTRAN program for the determination of r_1 and r_2 by differential methods is available (Harwood et al., 1968).

Integral Method

Relations 6.6, 6.7, and 6.8 can be used to obtain r_1 and r_2 by the "integral method" (Tidwell and Mortimer, 1965; Chan and Meyer, 1968; Tortai and Maréchal, 1971; Harwood et al., 1968). Recently a "trapezoidal" integral method has been suggested for reactivity ratio determination (Heublein, 1975a; Heublein et al., 1976a); however has not yet been used by other workers. The application of integral methods is difficult and prone to error.

Discussion of Reactivity Ratio Determination Methods

According to Tidwell and Mortimer (1970) a good method for the determination of reactivity ratios should (1) give an unbiased estimate of the parameters, (2) use all (or nearly all) the information resident in the data with regard to the parameters to be estimated, thus providing precise estimates, (3) supply a measure of errors, and (4) be reasonably easy to use; finally, (5) the calculated parameters should not be affected by arbitrary factors (i.e., monomer indexing). It is particularly difficult to obtain accurate reactivity ratios in cationic copolymerizations. Differential methods cannot be used if conversions are high and monomer concentration in the feed is not constant. Cationic polymerizations are generally very fast and it may be difficult to stop the reaction at low conversions. The extent of error at high conversions depends on the initial monomer concentration and relative reaction rates; it may be as large as 100%.

Copolymer composition may change with molecular weight. For instance, Borg and Maréchal (1977) have fractionated by preparative GPC copolymers of 4-chlorostyrene and styrene and found significant ($\sim 20\%$) compositional variations by IR spectroscopy.

For precise information it is advisable to obtain approximate r_1 and r_2 values in preliminary runs and to follow up by detailed investigation.

The Kelen–Tüdős Method

Relation 6.5 has been linearized by Fineman and Ross (see above):

$$G = r_1 F - r_2 \tag{6.12}$$

$$\frac{G}{F} = -r_2 \frac{1}{F} + r_1 \tag{6.13}$$

where $G = x(y-1)/y$ and $F = x^2/y$, x and y being, respectively, $[M_1]/[M_2]$ and m_1/m_2. According to Tidwell and Mortimer (1970) the data are unequally weighted by relations 6.12 and 6.13. The data obtained under extreme conditions (low $[M_2]$ in equation 6.12 or $[M_1]$ in equation 6.13) greatly influence the slope of the line calculated by the usual linear least-squares method. Consequently, r_1 and r_2 depend on arbitrary factors such as which monomer is indexed as M_1. Considerable deviations between the r values obtained by using equations 6.12 or 6.13 may be observed.

Another disadvantage of the Fineman and Ross (FR) method is that if the charge concentration is changing systematically during copolymerization, the r values appear. along the ordinate at increasing intervals.

Kelen and Tüdős (1974, 1975; Tüdős et al., 1975, 1976) showed that the disadvantages of the FR method can be abolished by using the following graphically evaluable linear equation:

$$\frac{G}{\alpha + F} = \left(r_1 + \frac{r_2}{\alpha}\right) \frac{F}{\alpha + F} - \frac{r_2}{\alpha} \tag{6.14}$$

where α is an arbitrary positive constant.

By introducing

$$\eta = \frac{G}{\alpha + F} \quad \text{and} \quad \xi = \frac{F}{\alpha + F} \tag{6.15}$$

relation 6.14 can be written

$$\eta = \left(r_1 + \frac{r_2}{\alpha}\right) \xi - \frac{r_2}{\alpha} \tag{6.16}$$

or

$$\eta = r_1 \xi - \frac{r_2}{\alpha}(1 - \xi) \tag{6.17}$$

with

$$0 < \xi < 1 \tag{6.18}$$

Plotting η against ξ, a straight line is obtained which on extrapolation to $\xi = 0$ and $\xi = 1$ gives $-r_2/\alpha$ and r_1 (as intercepts), respectively. The experimental data can be distributed symmetrically along the line when α is obtained by $\alpha = \sqrt{F_m \cdot F_M}$, where F_m and F_M are the lowest and highest F values, respectively.

By means of equation 6.16 or 6.17 the data can be evaluated by both the graphic and least-squares methods. In the latter case,

$$r_1 = \frac{\sum \eta\xi\left(n - \sum \xi\right) - \sum \eta\left(\sum \xi - \sum \xi^2\right)}{n \sum \xi^2 - \left(\sum \xi\right)^2} \tag{6.19}$$

$$r_2 = \alpha \frac{\sum \eta\xi \sum - \sum \eta \sum \xi^2}{\eta \sum \xi^2 - \left(\sum \xi\right)^2} \tag{6.20}$$

where n is the number of experimental data, and the summation refers to all data:

$$\sum = \sum_{i=1}^{n} \xi i \tag{6.21}$$

The r_1 and r_2 values calculated by this procedure give the best straight line obtained by the graphic method. The Kelen–Tüdős (KT) equation is invariant to reindexing of monomers. Linearity of KT plots rigorously proves the validity of the assumptions implicit in the copolymer composition equation and can be used as a diagnostic tool in copolymerization mechanism studies (Kennedy et al., 1975c). The extended KT method (Tüdős et al., 1975, 1976; Kelen et al., 1977) yields highly reliable reactivity ratios for practically all copolymerization systems up to relatively high (>60%) conversions. In this method, instead of using $G = x(y - 1)/y$ and $F = x^2/y$ as with low conversions, one uses for high conversions in equation 6.15, $G = (y - 1)z$ and $F = y/z^2$, where

$$z = \frac{\log ([M_1]/[M_1]_0)}{\log ([M_2]/[M_2]_0)} \tag{6.22}$$

The error of approximation is generally <0.5%, with r values ranging from 0.01 to 100.

Kennedy et al. (1975c) and Kelen et al. (1977) recalculated several hundred published r_1 and r_2 values and compared them with those obtained by other methods (e.g., FR, intersection). According to the reliability and availability of published data four classes have been distinguished (Kennedy et al., 1975c; Kelen et al., 1977):

•Class I □   The data yield a well-defined straight line in the η versus ξ plot. The calculated r_1, r_2 values are satisfactory for quantitative studies.

The simplifying assumptions implicit in the copolymer composition equation used to derive the reactivity ratios are valid for these systems; that is, the two-parameter model adequately describes the experimental results.

• Class I(!) □   The experimental points exhibit a noticeable error spread so that the numerical data are somewhat less reliable; some caution should be exercised when employing these reactivity ratios for quantitative purposes.

• Class II □   The η versus ξ plot exhibits a curvature. In these instances the conventional copolymer composition equation 6.5 does not adequately describe the system and the r_1, r_2 values calculated by the original authors are erroneous and misleading. The simple two-parameter model is insufficient to characterize the system, and more than two parameters would be necessary to express the copolymerization results quantitatively.

• Class III □   The η versus ξ plots exhibit completely unacceptable data spread. The reactivity ratios derived by the original authors are meaningless and the results cannot be used even for qualitative discussion. In some instances even the existence of true copolymers is questioned.

A Comprehensive Compilation of Reactivity Ratios

Table 6.1 is a comprehensive updated compilation of reactivity ratios evaluated and classified by the KT method. The monomers have been listed alphabetically. Columns 1 and 2 show the monomers, column 3 specifies initiator/coinitiator/solvent/temperature, columns 4 and 5 are published r values, column 6 contains references, columns 7 and 8 are recalculated values by the KT method, column 9 is the classification as defined above, and column 10 shows comments as follows: (*a*) cannot be evaluated; the paper does not contain necessary data. (*b*) Not evaluated because original paper gives composition data only in diagram form and does not give conversion data. (*c*) Composition data obtained from diagrams and calculated by assuming very low conversions. (*d*) Serious data spread. (*e*) Only three data points. (*f*) Only four data points. (*g*) Contradictory data. (*h*) 226 and 227, and 229 and 230 differ only with regard to initiator concentration; 225 differs in the manner of mixing from the formers. (*i*) Narrow compositional range. (*j*) Original does not give conversion data; calculated by assuming very low conversions. (*k*) The r value of the original reference is an average value of the original data. (*l*) No reaction conditions given. (*m*) Only two data points. (*n*) Unacceptable data spread. (*o*) Data obtained from diagram; conversion very high. (*p*) One of the parameters assumed to be 0. (*r*) Scatter may indicate a systematic discrepancy. (*s*) One conspicuously erroneous data point ignored. (*t*) Copolymer formation has not been established.

The following abbreviations have been used in the table: Bz = benzene, Tol = toluene, EtCl = ethyl chloride, $MeCl_2$ = methylene chloride, $EtCl_2$ = ethylene dichloride, NoMe = nitromethane, NoEt nitroethane, NoBz =

Table 6.1 □ Compilation of Reactivity Ratios Determined and Classified by the KT Method

No.	Monomer 1.	Monomer 2.	Initiating System or Coinitiatior/Solvent /Temp (°C)	Published Values		Ref.	Values calculated by KT Method			Comments
				r_1	r_2		r_1	r_2	Class	
1	Acenaphthylene	2-Vinylfluorene	TiCl$_4$/MeCl$_2$/−78	0.63	2.07	Cohen and Maréchal (1975)	0.7	2.0	I	**
2		n-Butyl vinyl ether	BF$_3$/Bz or Tol/0	0.24 ± 0.04	4.2 ± 0.8	Imoto and Saotome (1958)	—	—	—	a
3		n-Butyl vinyl ether	BF$_3$/Bz or Tol/30	0.38 ± 0.04	1.3 ± 0.3	Imoto and Saotome (1958)	—	—	—	a
4		n-Butyl vinyl ether	BF$_3$/Bz or Tol/−78	0.04 ± 0.02	≃20	Imoto and Saotome (1958)	—	—	—	a
5		n-Butyl vinyl ether	BF$_3$/Bz or Tol/−20	0.14 ± 0.03	6.0 ± 1.0	Imoto and Saotome (1958)	—	—	—	a
6	Benzofuran	Benzothiophene	BF$_3$/Bz/−78	3.75 ± 0.25	0.5 ± 0.15	Zaffran and Maréchal (1970)	3.5	0.5	I	a
7	1,3-Butadiene	1-Butene	AlEtCl$_2$–H$_2$O/EtCl/≃−40	1.81	0.15	Bogomoenyi et al. (1964)	—	—	—	a
8		1-Butene	AlEtCl$_2$–H$_2$O/Tol/−78	0.19	0.26	Bogomoenyi et al. (1964)	—	—	—	a
9		2-Butene	AlEtCl$_2$–H$_2$O/EtCl/−78	0.45	0.15	Bogomoenyi et al. (1964)	—	—	—	a
10		2-Butene	AlEtCl$_2$–H$_2$O/Tol/−78	0.17	0.51	Bogomoenyi et al. (1964)	—	—	—	a
11		2,3-Dimethyl-1,3-butadiene	AliBu$_3$–TiCl$_4$/Bz/30	2.16 ± 0.20	0.392 ± 0.036	Bresler et al. (1963)	—	—	—	a
12		2,3-Dimethyl-1,3-butadiene	AliBu$_2$Cl–CoCl$_2$/Bz/0	2.71 ± 0.5	1.13 ± 0.05	Bresler et al. (1963)	—	—	—	a
13		2,3-Dimethyl-1,3-butadiene	AlEtCl$_2$/Bz/12	0.053 ± 0.005	4.24 ± 0.411	Bresler et al. (1963)	—	—	—	a
14	1-Butene	2-Butene	BF$_3$/none/−3 to −6	3.1 ± 0.07	0.15 ± 0.05	Meier (1950)	—	—	—	a
15		2-Butene	BF$_3$/MeCl/−20	3.1 ± 0.07	0.15 ± 0.05	Meier (1950)	—	—	—	a
16	Isobutyl-Vinyl Ether	cis-Isobutylvinyl ethyl ether	BF$_3$/MeCl$_2$/−78	0.20 ± 0.10	0.032 ± 0.038	Okuyama et al. (1968)	—	—	—	a
17		trans-Isobutylvinyl ethyl ether	BF$_3$/MeCl$_2$/−78	2.48 ± 0.10	0.33 ± 0.03	Okuyama et al. (1968)	—	—	—	a
18		cis-n-Butylvinyl ethyl ether	BF$_3$/MeCl$_2$/−78	0.17 ± 0.05	0.46 ± 0.05	Okuyama et al. (1968)	—	—	—	a
19		trans-n-Butylvinyl ethyl ether	BF$_3$/MeCl$_2$/−78	2.10 ± 0.21	0.59 ± 0.09	Okuyama et al. (1968)	—	—	—	a
20		trans-tert-Butylvinyl ethyl ether	BF$_3$/MeCl$_2$/−78	~100	~0.01	Okuyama et al. (1968)	—	—	—	i
21		Chloroethyl vinyl ether	I$_2$/Et$_2$O/25	1.90 ± 0.05	0.70 ± 0.05	Eley and Saunders (1954)	1.83	0.52	I	i
22		Chloroethyl vinyl ether	I$_2$/Et$_2$O/30	2.03 ± 0.55	0.34 ± 0.14	Eley and Saunders (1954)	1.77	0.29	I(?)	d
23		Chloroethyl vinyl ether	SnCl$_4$+TCA/Tol/0 to −78	2.17	0.58	Masuda and Higashimura (1971a)	—	—	—	b
24		Chloroethyl vinyl ether	I$_2$/MeCl$_2$/0	4.25 ± 0.41	0.30 ± 0.05	Yamamoto and Higashimura (1976)	—	—	—	b
25		Chloroethyl vinyl ether	I$_2$/MeCl$_2$–CCl$_4$ (1:1)/0	5.14 ± 0.28	0.27 ± 0.28	Yamamoto and Higashimura (1976)	—	—	—	b
26		Chloroethyl vinyl ether	I$_2$+(n-Bu)$_4$NI/MeCl$_2$–CCl$_4$ (1:1)/0	4.59 ± 0.39	0.23 ± 0.03	Yamamoto and Higashimura (1976)	—	—	—	b
27		Chloroethyl vinyl ether	I$_2$+(n-Bu)$_4$NI/CCl$_4$/0	4.46 ± 0.37	0.35 ± 0.05	Yamamoto and Higashimura (1976)	—	—	—	b
28		ββ-Dimethyl vinyl ethyl ether	BF$_3$/MeCl$_2$/−78	1.48 ± 0.03	0.00 ± 0.01	Okuyama et al. (1968)	—	—	—	a
29		cis-β-Ethylvinyl i-butyl ether	BF$_3$/Tol/−78	0.37 ± 0.03	0.70 ± 0.05	Higashimura et al. (1973)	—	—	—	b
30		cis-β-Ethylvinyl i-butyl ether	BF$_3$/NOEt/−78	0.45 ± 0.03	0.82 ± 0.05	Higashimura et al. (1973)	—	—	—	b
31		cis-β-Ethylvinyl i-butyl ether	BF$_3$/Tol/0	0.58 ± 0.07	0.56 ± 0.07	Higashimura et al. (1973)	—	—	—	b
32		cis-β-Ethylvinyl ethyl ether	BF$_3$/MeCl$_2$/−78	0.26 ± 0.04	0.81 ± 0.03	Okuyama et al. (1968)	—	—	—	a

Table 6.1 □ (Continued)

No.	Monomer 2.	Initiating System or Coinitiator/Solvent /Temp (°C)	Published Values r_1	Published Values r_2	Ref.	KT Method r_1	KT Method r_2	Class	Comments
	Isobutyl Vinyl Ether								
33	trans-β-Ethylvinyl ethyl ether	BF₃/MeCl₂/−78	0.68 ± 0.03	1.54 ± 0.14	Okuyama et al. (1968)	—	—	—	a
34	cis-β-Ethylvinyl i-propyl ether	BF₃/MeCl₂/−78	0.32 ± 0.05	0.93 ± 0.06	Okuyama et al. (1968)	—	—	—	a
35	trans-β-Ethylvinyl i-propyl ether	BF₃/MeCl₂/−78	0.50 ± 0.06	0.90 ± 0.05	Okuyama et al. (1968)	—	—	—	a
36	cis-β-Methylvinyl i-butyl ether	AlEt₃–SnCl₄/Tol/−78	0.13 ± 0.04	1.78 ± 0.08	Okuyama et al. (1968)	—	—	—	a
37	cis-β-Methylvinyl i-butyl ether	Al(SO₄)₃–H₂SO₄/Tol/0	1.56 ± 0.16	0.29 ± 0.05	Okuyama et al. (1968)	—	—	—	a
38	cis-β-Ethylvinyl i-butyl ether	BF₃/MeCl₂/−78	0.29 ± 0.05	2.20 ± 0.14	Okuyama et al. (1968)	—	—	—	a
39	cis-β-Methylvinyl i-butyl ether	BF₃/MeCl₂/0	0.29 ± 0.03	0.85 ± 0.12	Okuyama et al. (1968)	—	—	—	a
40	cis-β-Methylvinyl i-butyl ether	BF₃/NOEt/−78	0.67 ± 0.06	0.78 ± 0.05	Okuyama et al. (1968)	—	—	—	b
41	cis-β-Methylvinyl isobutyl ether	BF₃/NOEt/−78	0.58 ± 0.03	1.03 ± 0.05	Higashimura et al. (1973)	—	—	—	b
42	cis-β-Methylvinyl isobutyl ether	BF₃/Tol/−78	0.18 ± 0.04	2.17 ± 0.09	Okuyama et al. (1968)	—	—	—	a
43	cis-β-Methylvinyl isobutyl ether	BF₃/Tol/−78	0.21 ± 0.05	2.38 ± 0.26	Higashimura et al. (1973)	—	—	—	b
44	cis-β-Methylvinyl isobutyl ether	BF₃/Tol/0	0.30 ± 0.04	0.76 ± 0.08	Higashimura et al. (1973)	—	—	—	b
45	cis-β-Methylvinyl isobutyl ether	SnCl₄/NOEt/0	0.44 ± 0.03	0.81 ± 0.03	Okuyama et al. (1968)	—	—	—	a
46	cis-β-Methylvinyl isobutyl ether	SnCl₄/Tol/−78	0.70 ± 0.16	1.47 ± 0.31	Higashimura et al. (1973)	—	—	—	b
47	trans-β-Methylvinyl isobutyl ether	AlEt₃–SnCl₄/Tol/−78	1.67 ± 0.02	0.65 ± 0.01	Okuyama et al. (1968)	—	—	—	a
48	trans-β-Methylvinyl isobutyl ether	Al₂(SO₄)₃–H₂SO₄/Tol/−78	31.4 ± 2.3	0.10 ± 0.04	Okuyama et al. (1968)	—	—	—	a
49	trans-β-Methylvinyl isobutyl ether	BF₃/MeCl₂/−78	1.04 ± 0.05	0.90 ± 0.03	Okuyama et al. (1968)	—	—	—	a
50	trans-β-Methylvinyl isobutyl ether	BF₃/MeCl₂/0	1.05 ± 0.03	0.71 ± 0.02	Okuyama et al. (1968)	—	—	—	a
51	trans-β-Methylvinyl isobutyl ether	BF₃/NOEt/−78	1.36 ± 0.10	1.01 ± 0.08	Higashimura et al. (1973)	—	—	—	b
52	trans-β-Methylvinyl isobutyl ether	BF₃/Tol/−78	1.72 ± 0.11	0.60 ± 0.04	Higashimura et al. (1973)	—	—	—	a
53	trans-β-Methylvinyl isobutyl ether	BF₃/Tol/−78	2.64 ± 0.16	0.35 ± 0.04	Okuyama et al. (1968)	—	—	—	a
54	trans-β-Methylvinyl isobutyl ether	SnCl₄/NOEt/0	0.96 ± 0.14	0.79 ± 0.06	Okuyama et al. (1968)	—	—	—	a
56	trans-β-Methylvinyl isobutyl ether	SnCl₄–TCA/Tol/−78	1.51 ± 0.15	0.47 ± 0.06	Higashimura et al. (1973)	—	—	—	b
57	cis-β-Methylvinyl ethyl ether	BF₃/MeCl₂/−78	0.23 ± 0.07	2.25 ± 0.10	Okuyama et al. (1968)	—	—	—	a
58	trans-β-Methylvinyl ethyl ether	BF₃/MeCl₂/−78	0.56 ± 0.03	1.44 ± 0.02	Okuyama et al. (1968)	—	—	—	a
59	trans-β-Methylvinyl isopropyl ether	BF₃/MeCl₂/−78	0.18 ± 0.08	2.37 ± 0.19	Okuyama et al. (1968)	—	—	—	a
60	Isopropyl vinyl ether	BF₃/MeCl₂/−78	0.22 ± 0.03	2.70 ± 0.11	Okuyama et al. (1968)	—	—	—	p
61	cis-β-Propylvinyl ethyl ether	BF₃/MeCl₂/−78	1.71 ± 0.20	0.09 ± 0.04	Okuyama et al. (1968)	—	—	—	a
62	trans-β-Propylvinyl ethyl ether	BF₃/MeCl₂/−78	14.44 ± 0.21	0.09 ± .01	Okuyama et al. (1968)	—	—	—	a
	n-Butyl vinyl ether								
63	cis-β-methyl n-butyl ether	BF₃/MeCl₂/−78	0.5 ± 0.2	4.0 ± 1.0	Mizote et al. (1967)	—	—	—	a
64	trans-β-Methylvinyl n-butyl ether	BF₃/MeCl₂/−78	0.8 ± 0.3	2.3 ± 1.0	Mizote et al. (1967)	—	—	—	a
	t-Butyl vinyl ether								
65	cis-Propenyl tert-butyl ether	BF₃·OEt₂/Tol/−78	2.2 ± 0.4	0.28 ± 0.08	Higashimura (1968b)	—	—	—	b

No.	Monomer	Conditions			Reference				
66	cis-Isobutylvinyl ethyl ether	BF₃/MeCl₂/−78	0.46±0.15	0.74±0.07	Okuyama et al. (1968)	—	—	—	a
67	cis-β-n-Butylvinyl ethyl ether	BF₃/MeCl₂/−78	1.16±0.05	0.88±0.03	Okuyama et al. (1968)	—	—	—	a
68	2-Chloroethyl vinyl ether	SnCl₄−TCA/MeCl₂/−78	0.95±0.10	1.85±0.19	Hasegawa and Asami (1978)	—	—	—	b
69	1,3-Dimethylbutadiene	BF₃·OEt₂/MeCl₂/−78	1.94±0.27	1.14±0.26	Hasegawa and Asami (1978)	1.64	0.51	I (?)	b
70	α,4-Dimethylstyrene	SnCl₄/Bz/30	1.7	0.525	Marvel and Dunphy (1960)	1.65	0.52	I (?)	k
71	α,4-Dimethylstyrene	SnCl₄/Bz/30	1.73±0.24	0.54±0.24	Dunphy and Marvel (1960)	1.61	1.03	I (?)	r
72	α,4-Dimethylstyrene	SnCl₄/HOBz/30	1.7	0.64	Marvel and Dunphy (1960)	0.55	1.15	I (?)	r
73	α-Methyl-4-Methoxystyrene	SnCl₄/Bz/30	0.42±0.20	1.09±0.27	Dunphy and Marvel (1960)	0.54	1.19	I (?)	r
74	α-Methyl-4-Methoxystyrene	SnCl₄/Bz/30	0.42	1.1	Marvel and Dunphy (1960)	0.87	1.62	I	—
75	α-Methyl-4-Methoxystyrene	SnCl₄/NOBz/25	0.73	1.3	Marvel and Dunphy (1960)	4.75	0.29	I	—
76	α-Methyl-4-Methoxystyrene	SnCl₄/Bz/30	5.03±0.66	0.33±0.05	Dunphy and Marvel (1960)	4.83	0.28	I	b
77	α-Methoxystyrene	SnCl₄/Bz/30	5.0	0.33	Marvel and Dunphy (1960)	—	—	I	a
78	α-Methoxystyrene	SnCl₄+TCA/Tol/−78	3.46±0.25	0.46±0.07	Masuda and Higashimura (1971a)	3.45	0.43	I (?)	k
79	α-Methoxystyrene	SnCl₄+TCA/Tol/−23	5.00	0.42	Masuda and Higashimura (1971a)	3.29	1.11	I	c
80	α-Methoxystyrene	SnCl₄/NOBz/30	2.95	0.38	Marvel and Dunphy (1960)	3.40	5.02	III	c
81	α-Methoxystyrene	SnCl₄+TCA/MeCl₂/−78	3.31±0.17	1.12±0.06	Masuda and Higashimura (1971a)	—	—	I	c
82	4-Methoxystyrene	BF₃/MeCl₂/−78	3.08±0.26	4.55±0.35	Masuda and Higashimura (1971a)	2.56	3.63	I (?)	c
83	4-Methoxystyrene	BF₃/MeCl₂/0	1.63	9.25	Masuda and Higashimura (1971a)	2.61	1.16	I	c
84	4-Methoxystyrene	BF₃/Tol/−78	2.81±0.30	4.37±0.30	Masuda and Higashimura (1971a)	4.57	0.74	—	b
85	4-Methoxystyrene	BF₃/Tol/−36	5.06	2.36	Masuda and Higashimura (1971a)	—	—	—	a
86	4-Methoxystyrene	BF₃/Tol/0	9.32	1.55	Masuda and Higashimura (1971a)	—	—	—	b
87	4-Methoxystyrene	SnCl₄+TCA/MeCl₂/−78	1.56±0.04	7.80±0.20	Masuda and Higashimura (1971a)	—	—	—	b
88	4-Methoxystyrene	SnCl₄+TCA/MeCl₂/0	1.24	12.1	Masuda and Higashimura (1971a)	—	—	—	b
89	4-Methoxystyrene	SnCl₄−TCA/Tol/−78	1.73±0.19	6.93±0.50	Masuda and Higashimura (1970)	—	—	—	b
90	4-Methoxystyrene	I₂/MeCl₂−CCl₄(2:1)/0	1.36±0.29	8.78±1.54	Yamamoto and Higashimura (1976)	—	—	—	b
91	4-Methoxystyrene	I₂/CCl₄/0	1.99±0.26	3.92±0.63	Yamamoto and Higashimura (1976)	—	—	—	b
92	4-Methoxystyrene	I₂+(n-Bu)₄NI/MeCl₂/0	0.99±0.15	9.04±1.08	Yamamoto and Higashimura (1976)	—	—	—	b
93	4-Methoxystyrene	I₂+(n-Bu)₄NI/MeCl₂−CCl₄ (2:1)/0	1.87±0.18	8.73±0.67	Yamamoto and Higashimura (1976)	—	—	—	b
94	4-Methoxystyrene	I₂+(n-Bu)₄NI/MeCl₂−CCl₄ (2:1)/0	1.88±0.33	6.02±0.90	Yamamoto and Higashimura (1976)	—	—	—	b
95	4-Methoxystyrene	I₂+(n-Bu)₄NI/MeCl₂−CCl₄ (2:1)/0	2.00±0.23	6.50±0.65	Yamamoto and Higashimura (1976)	—	—	—	b
96	4-Methoxystyrene	I₂+(n-Bu)₄NI/MeCl₂/0	1.72±0.02	6.65±0.08	Yamamoto and Higashimura (1976)	—	—	—	b
97	4-Methoxystyrene	I₂/MeCl₂/0	7.62±0.35	0.66±0.05	Yamamoto and Higashimura (1976)	—	—	—	b
98	α-Methylstyrene	BF₃/MeCl₂/−23	6.02	0.42	Masuda and Higashimura (1971a)	—	—	—	a
99	α-Methylstyrene	SnCl₄+TCA/MeCl₂/−23	2.50	0.76	Masuda and Higashimura (1971a)	—	—	—	b
100	α-Methylstyrene	BF₃·OEt₂/Tol/−78	5.72±0.70	0.31±0.05	Masuda and Higashimura (1970)	—	—	—	b
101	α-Methylstyrene	BF₃·OEt₂/MeCl₂/−78	2.05±0.22	0.68±0.14	Masuda and Higashimura (1970)	—	—	—	b
102	α-Methylstyrene	BF₃·OEt₂/Tol/−78	3.46±0.25	0.48±0.07	Masuda and Higashimura (1970)	—	—	—	b
103	α-Methylstyrene	BF₃·OEt₂/MeCl₂/−78	1.02±0.10	1.00±0.10	Masuda and Higashimura (1970)	—	—	—	b

Table 6.1 □ (Continued)

No.	Monomer 1.	Monomer 2.	Initiating System or Coinitiation/Solvent /Temp (°C)	Published Values		Ref.	Values calculated by KT Method		Class	Comments
				r_1	r_2		r_1	r_2		
104	2-Chloroethyl vinyl ether	4-Methylstyrene	I_2/MeCl$_2$–CCl$_4$(2:1)/0	13.5 ± 1.66	0.22 ± 0.10	Yamamoto and Higashimura (1976)	—	—	—	b
105		4-Methylstyrene	I_2/CCl$_4$/0	22.9 ± 0.96	0.00 ± 0.06	Yamamoto and Higashimura (1976)	—	—	—	
106		4-Methylstyrene	$I_2 + (n\text{-Bu})_4$NI/CH$_2$Cl$_2$–CCl$_4$ (2:1)/0	15.5 ± 1.49	0.03 ± 0.06	Yamamoto and Higashimura (1976)	—	—	—	b
107		4-Methylstyrene	$I_2 + (n\text{-Bu})_4$NI/CH$_2$Cl$_2$–CCl$_4$ (2:1)/0	25.9 ± 3.33	0.00 ± 0.11	Yamamoto and Higashimura (1976)	—	—	—	b
108		4-Methylstyrene	BF$_3$·OEt$_2$/MeCl$_2$/−78	8.80 ± 0.45	0.40 ± 0.03	Masuda and Higashimura (1970)	—	—	—	b
109		4-Methylstyrene	SnCl$_4$–TCA/Tol/−78	10.1 ± 1.8	0.50 ± 0.07	Masuda and Higashimura (1970)	—	—	—	b
110		4-Methylstyrene	SnCl$_4$–TCA/MeCl$_2$/−78	2.31 ± 0.17	1.12 ± 0.06	Masuda and Higashimura (1970)	—	—	—	b
111		4-Methylstyrene	SnCl$_4$+TCA/MeCl$_2$/−78	3.31 ± 0.17	1.12 ± 0.06	Masuda and Higashimura (1971a)	3.29	1.11	I	c
112		4-Methylstyrene	SnCl$_4$+TCA/MeCl$_2$/−23	8.00	0.88	Masuda and Higashimura (1971a)	—	—	—	a
113	2-Chlorostyrene	4-Methoxypropenylbenzene (anethole)	SnCl$_4$/CCl$_4$/0	0.03 ± 0.005	18 ± 3	Alfrey et al. (1949)	0	14.8	I (!)	p
114	4-Chlorostyrene	4-Bromostyrene	SnCl$_4$/CCl$_4$ + NOBz/0	1.0 ± 0.1	1.0 ± 0.1	Overberger et al. (1952)	1.18	1.34	I	—
115		1,3-Cyclooctadiene	SnCl$_4$/MeCl$_2$/−20	0.30 ± 0.05	0.84 ± 0.05	Mondal and Young (1971)	—	—	—	a, k
116		1,3-Cyclooctadiene	SnCl$_4$/MeCl$_2$/0	0.46 ± 0.05	0.73 ± 0.05	Mondal and Young (1971)	—	—	—	a, k
117		1,3-Cyclooctadiene	TiCl$_4$/MeCl$_2$/−20	0.52 ± 0.05	0.52 ± 0.05	Mondal and Young (1971)	—	—	—	a
118		1,3-Cyclooctadiene	TiCl$_4$/MeCl$_2$/0	0.56 ± 0.05	0.42 ± 0.05	Mondal and Young (1971)	—	—	—	a
119		β-Ethylstyrene	SnCl$_4$(CCl$_4$–NOBz(1:1)/0	0.88 ± 0.30	0	Overberger et al. (1958)	1.12	0.08	I	a
120		4-Ethylstyrene	SnCl$_4$(CCl$_4$–NOBz(1:1)/0	0.29 ± 0.04	4.1 ± 0.5	Overberger et al. (1958)	—	—	—	—
121		Isobutylene	AlBr$_3$/n-Hexane/0	1.02	1.01	Overberger and Kamath (1959)	—	—	—	a
122		Isobutylene	AlBr$_3$/NOBz/0	0.15	14.7	Overberger and Kamath (1959)	—	—	—	a
123		Isobutylene	AlBr$_3$/NOMe/0	0.7	22.5	Overberger and Kamath (1959)	—	—	—	a
124		Isobutylene	SnCl$_4$/NOBz/0	1.2	8.6	Overberger and Kamath (1959)	—	—	—	a
125		Isoprene	AlCl$_3$/EtCl/−78	0.23	0.50	Jones et al. (1961a)	—	—	—	—
126		3-Methoxystyrene	SnCl$_4$/CCl$_4$–NOBz(1:1)/0	0.38 ± 0.05	2.6 ± 0.4	Overberger et al. (1952)	0.43	2.58	I (!)	—
127		α-Methylstyrene	SnCl$_4$/CCl$_4$/−78	0.12 ± 0.03	28 ± 2	Smets and De Haes (1950)	0.24	72.25	I (!)	d, f
128		α-Methylstyrene	SnCl$_4$/CCl$_4$/0	0.35 ± 0.05	15.5 ± 1.5	Overberger et al. (1951)	0.10	9.55	I (!)	d
129		cis-β-Methylstyrene	SnCl$_4$/CCl$_4$–NOBz(1:1)/0	1.0 ± 0.1	0.32 ± 0.02	Overberger et al. (1958)	1.08	0.38	I	—
130		trans-β-Methylstyrene	SnCl$_4$/CCl$_4$–NOBz(1:1)/0	0.74 ± 0.06	0.32 ± 0.04	Overberger et al. (1958)	0.68	0.27	I	—
131		4-Methylstyrene	TiCl$_4$/MeCl$_2$/0	0.10 ± 0.05	10.0 ± 0.5	Visse and Maréchal (1974)	0.1	10.5	I	**
132		4-Methylstyrene	SnCl$_4$/CCl$_4$–NOBz/0	0.22 ± 0.05	4.5 ± 0.7	Overberger et al. (1952)	0.35	5.99	I (!)	c, n
133		β-Propylstyrene	SnCl$_4$/CCl$_4$–NOBz/0	—	—	Overberger et al. (1958)	—	—	III	a
134		Vinyl benzyl chloride	AlCl$_3$/EtCl/−65 to −30	1	1	Jones et al. (1961b)	—	—	—	a

No.	Monomer	Comonomer	Conditions	r_1	r_2	Reference				
135	cis-β-Chlorovinyl ethyl ether	trans-β-Chlorovinyl ethyl ether	BF₃/MeCl₂/−78	1.39±0.04	0.28±0.04	Okuyama et al. (1969a)	1.40	0.25	I	b
136	Cyclopentadiene	Methylcyclopentadiene	SnCl₄–TCA/MeCl₂/−78	0.42±0.23	14.9±5.6	Kohjiya et al. (1968)	—	—	—	b
137		Methylcyclopentadiene	SnCl₄–TCA/Tol/−78	0.36±0.26	8.5±3.5	Kohjiya et al. (1968)	—	—	—	b
138		1,3-Dimethylcyclopentadiene	BF₃OEt₂/Tol/−78	0.30±0.1	6.85±1.10	Aso and Ohara (1969)	—	—	—	a
139		β-Cyclohexylstyrene	TiCl₄/MeCl₂/−78	25.3±3.5	—	Heublein and Barth (1974)	—	—	—	a
140		β-Cyclohexylstyrene	TiCl₄/Tol/−78	97±6	—	Heublein and Barth (1974)	—	—	—	a
141		β-Cyclohexylstyrene	TiCl₄/MeCl₂/10	47±4	—	Heublein and Barth (1974)	—	—	—	a
142		β-Cyclohexylstyrene	TiCl₄/Tol/10	123±23	—	Heublein and Barth (1974)	—	—	—	a
143		β-Cyclohexylstyrene	TiCl₄/NOMe/10	118±16	—	Heublein and Barth (1974)	—	—	—	a
144		β-Cyclohexylstyrene	TiCl₄/NOBz/10	89±12	—	Heublein and Barth (1974)	—	—	—	a
145		β-Cyclohexylstyrene	TiCl₄/MeCl₂/10	23.5±3.5	—	Heublein and Barth (1974)	—	—	—	a
146		β-Cyclohexylstyrene	TiCl₄+DMF(1:1)/MeCl₃/10	15.8±2.5	—	Heublein and Barth (1974)	—	—	—	a
147		β-Cyclohexylstyrene	TiCl₄/Tol/10	97±6	—	Heublein and Barth (1974)	—	—	—	a
148		β-Cyclohexylstyrene	TiCl₄+DMF(1:1)/Tol/10	73±4	—	Heublein and Barth (1974)	—	—	—	a
149		β-Cyclohexylstyrene	TiCl₄/MeCl₂/10	25.3±3.5	—	Heublein and Barth (1974)	—	—	—	a
150		β-Cyclohexylstyrene	TiCl₄+CA(1:2)/MeCl₂/10	25.7±3	—	Heublein and Barth (1974)	—	—	—	a
151		β-Cyclohexylstyrene	TiCl₄/Tol/10	97±6	—	Heublein and Barth (1974)	—	—	—	a
152		β-Cyclohexylstyrene	TiCl₄+CA(1:2)/Tol/10	70.5±4	—	Heublein and Barth (1974)	—	—	—	a
153		α-Methylstyrene	SnCl₄/CCl₄/0	31±19	0.035±0.015	Overberger et al. (1952)	—	—	III	n
154	4-(Dimethylamino)styrene	4,7-Dimethylindene	TiCl₄/MeCl₂/0	0.60±0.04	0.60±0.04	Maréchal and Evrard (1969a)	0.63	0.66	I	—
155	4,6-Dimethylindene	4,5,6,7-Tetramethylindene	TiCl₄/MeCl₂/0	0.53±0.08	2.2±0.3	Maréchal and Evrard (1969a)	0.79	2.83	I	—
156	4,7-Dimethylindene	4,5,6,7-Tetramethylindene	TiCl₄/MeCl₂/0	0.30±0.08	2.0±0.2	Maréchal and Evrard (1969a)	0.20	2.00	I	—
157	Ethyl vinyl ether	2-tert-Butoxybutadiene	BF₃·OEt₂/Tol/−78	1.24±0.11	4.70±0.38	Masuda and Higashimura (1973a)	—	—	—	b
158		Isobutyl vinyl ether	BF₃/MeCl₂/−78	1.30±0.02	0.92±0.02	Okuyama et al. (1968)	—	—	—	a
159		2-(2-Chloroethoxy)butadiene	BF₃OEt₂/Tol/−78	0.65±0.11	1.96±0.26	Masuda and Higashimura (1973a)	—	—	—	b
160		1-Ethoxybutadiene	BF₂OEt₂/Tol/−78	1.15±0.06	2.62±0.11	Otsuki et al. (1976)	—	—	—	b
161		2-Ethoxybutadiene	BF₃OEt₂/Tol/−78	0.49±0.07	1.92±0.17	Masuda and Higashimura (1973a)	—	—	—	b
162		α-Ethylvinyl ethyl ether	BF₃·1Et₂/MeCl₂/−78	0.11±0.05	13.8±1.6	Masuda and Higashimura (1973b)	—	—	—	b
163		1-Isopropoxybutadiene	BF₃·OEt₂/MeCl₂/−78	0.61±0.009	7.02±0.85	Otsuki et al. (1976)	—	—	—	b
164		2-Isopropoxybutadiene	BF₃·OEt₂/MeCl₂/−78	0.28±0.04	3.38±0.40	Masuda and Higashimura (1973a)	—	—	—	b
165		1-Methoxybutadiene	BF₃·OEt₂/Tol/−78	2.12±0.16	1.38±0.14	Otsuki et al. (1976)	—	—	—	b
166		2-Methoxybutadiene	BF₃·OEt₂/Tol/−78	0.79±0.06	1.28±0.10	Masuda and Higashimura (1973b)	—	—	—	b
167		α-Methylvinyl ethyl ether	BF₃·OEt₂/MeCl₂/−78	0.13±0.04	—	Masuda and Higashimura (1973b)	—	—	—	b
168		cis-β-Methylvinyl ethyl ether	BF₃/NOBz/−78	0.34±0.03	1.37±0.09	Higashimura et al. (1973)	—	—	—	b
169		cis-β-Methylvinyl ethyl ether	BF₃/Tol/−78	0.12±0.05	1.58±0.13	Higashimura et al. (1973)	—	—	—	b
170	α-Phenylvinyl ethyl ether		BF₃/Tol/−78	$r_1 = r_1' = 0.5$	$r_2' \approx 20$	Masuda and Higashimura (1973a)	—	—	mate effect	b
171	cis-β-Ethylvinyl ethyl ether	trans-Propenyl ethyl ether	BF₃·OEt₂/Tol/−78	0.94±0.10	0.94±0.10	Higashimura (1968b)	—	—	—	b
172		trans-β-Ethylvinyl ethyl ether	BF₃/MeCl₂/−78	1.23±0.06	0.92±0.04	Okuyama et al. (1968)	—	—	—	a

Table 6.1 □ (Continued)

No.	Monomer 1.	Monomer 2.	Initiating System or Coinitiator/Solvent/Temp (°C)	Published Values		Ref.	Values calculated by KT Method		Class	Comments
				r_1	r_2		r_1	r_2		
173	cis-β-Ethylvinyl isopropyl ether	trans-β-Isopropylvinyl ethyl ether	BF₃/MeCl₂/-78	1.18±0.05	0.49±0.02	Okuyama et al. (1968)	—	—	—	a
174	Indene	Benzofuran	TiCl₄/MeCl₂/-78	1.85±0.25	0.15±0.01	Zaffran and Maréchal (1970)	1.90	0.20	I	**
175		Benzoguran	BF₃·OEt₂/Tol/30	2.08±0.05	0.06±0.02	Mizote et al. (1966b)	—	—	—	b
176		Benzofuran	BF₂·OEt₂/EtCl₂/30	5.5±0.6	0.07±0.04	Mizote et al. (1966b)	—	—	—	b
177		Benzofuran	SnCl₄TCA/EtCl₂/30	5.4±0.5	0.09±0.05	Mizote et al. (1966b)	—	—	—	b
178		4,7-Dimethylbenzofuran	SnCl₄TCA/EtCl₂/30	0.7±0.1	1.3±0.1	Mizote et al. (1966b)	0.70	1.30	I	**
179		Benzothiophene	SnCl₄TCA/EtCl₂/30	2.4±0.1	0.08±0.01	Mizote et al. (1966b)	2.40	0.10	I	**
180		Bis(1-indenyl)-1,2-ethane (meso isomer)	TiCl₄/MaCl₂/0	0.4±0.1	0.3±0.1	Maréchal et al. (1969e)	0.40	0.30	I	**
181		Bis(1-indenyl)-1,2-ethane (racemic isomer)	TiCl₄/MeCl₂/0	0.60±0.05	0.40±0.05	Maréchal et al. (1969e)	0.60	0.35	I	**
182		1,4-Bis(1-indenyl)butane	TiCl₄/MeCl₂/0	0.5±0.1	0.5±0.1	Maréchal et al. (1969e)	0.53	0.48	I	**
183		1,4-Bis(1-indenyl)-2-butene	TiCl₄/MeCl₂/0	0.20±0.05	0.75±0.05	Maréchal et al. (1969e)	0.20	0.70	I	**
184		Isobutene	TiCl₄/MeCl₂/-72	2.2±0.2	1.1±0.2	Maréchal et al. (1968a)	2.20	1.10	I	—
185		1-Methylindene	TiCl₄/MeCl₂/-72	3.18±0.05	0.14±0.01	Sigwalt and Maréchal (1966)	3.27	0.11	I	—
186		2-Methylindene	TiCl₄/MeCl₂/-72	0.99±0.03	0.052±0.004	Sigwalt and Maréchal (1966)	—	—	—	—
187		3-Methylindene	TiCl₄/MeCl₂/-72	0.37±0.02	2.80±0.25	Sigwalt and Maréchal (1966)	—	—	—	Penultimate effect
188		5-Methylindene	TiCl₄/MeCl₂/0	1.15±0.05	0.85±0.05	Caillaud et al. (1970)	1.13	0.59	I	—
189		6-Methylindene	TiCl₄/MeCl₂/-72	0.45±0.05	2.50±0.25	Caillaud et al. (1970)	0.24	3.00	I (!)	—
190		7-Methylindene	TiCl₄/MeCl₂/-72	0.80±0.01	1.15±0.05	Caillaud et al. (1970)	0.54	1.21	I (!)	d
191		4,6-Dimethylindene	TiCl₄/MeCl₂/-72	0.40±0.08	2.1±0.1	Maréchal and Evrard (1969a)	0.40	2.30	I	—
192		4,7-Dimethylindene	BF₃E₂O/MeCl₂/-72	0.40±0.05	3.7±0.3	Maréchal and Evrard (1969a)	0.37	3.78	I	r
193		4,7-Dimethylindene	TiCl₄/MeCl₂/-72	0.15±0.03	2.5±0.2	Maréchal and Evrard (1969a)	0.12	2.44	I (!)	—
194		5,6-Dimethylindene	TiCl₄/MeCl₂/-72	0.50±0.01	4.2±0.4	Maréchal and Evrard (1969a)	0.60	4.97	I	a
195		5,7-Dimethylindene	TiCl₄/MeCl₂/-30	0.10±0.05	3.80±0.04	Anton et al. (1970)	0.07	3.98	I	—
196		6,7-Dimethylindene	TiCl₄/MeCl₂/-30	0.18±0.05	5.1±0.3	Anton and Maréchal (1971a)	0.11	5.04	I	a
197		4,6,7-Trimethylindene	TiCl₄/MeCl₂/-72	0.15±0.05	3.4±0.1	Anton et al. (1970)	—	2.43	I (!)	d,p
198		4,5,6,7-Tetramethylindene	TiCl₄/MeCl₂/0	0.20±0.04	7.5±0.5	Maréchal and Evrard (1969a)	0	7.4	I (!)	d,p
199		4,5,6,7-Tetramethylindene	H₂SO₄/MeCl₂/0	0.75±0.10	16.0±0.5	Maréchal and Evrard (1969a)	0.76	17.59	I	—
200		4-Methoxropenylbenzene (anethole)	TiCl₄/MeCl₂/0	0.30±0.15	1.60±0.30	Maréchal (1968b)	—	—	—	—
201	Isobutylene	1,3-Butadiene	AlEtCl₂/MeCl/-100	43	0	Kennedy and Canter (1967a)	—	—	—	a
202		1,3-Butadiene	AlCl₃/MeCl/-103	115±15	0.01±0.01	Thomas and Sparks (1944)	—	—	—	—

No.	Monomer	Conditions	Separate polymerization					Reference		Notes
203	Chloroprene	BF$_3$/EtCl/-103	0.60 ± 0.15	4.5 ± 0.5	—	—	—	Anosov and Korotkov (1960)	—	b
204	Cyclopentadiene	BF$_3\cdot$OEt$_2$/Tol/-78	0.73 ± 0.17	1.86 ± 0.20	—	—	—	Imanishi et al. (1966)	—	b
205	Cyclopentadiene	BF$_3\cdot$OEt$_2$/MeCl$_2$/-78	0.21 ± 0.02	6.3 ± 0.6	—	—	—	Imanishi et al. (1966)	—	b
206	Cyclopentadiene	SnCl$_4$-TCA/Tol/-78	0.80 ± 0.15	1.55 ± 0.20	—	—	—	Imanishi et al. (1966)	—	b
207	Cyclopentadiene	SnCl$_4$-TCA/MeCl$_2$/-78	0.77 ± 0.20	1.05 ± 0.25	—	—	—	Imanishi et al. (1966)	—	b
208	Cyclohexadiene	SnCl$_4$-TCA/Tol/-78	0.98 ± 0.20	1.07 ± 0.45	—	—	—	Imanishi et al. (1966)	—	b
209	Cyclohexadiene	SnCl$_4$-TCA/MeCl$_2$/-78	2.5 ± 0.5	0.4 ± 0.1	—	—	—	Imanishi et al. (1966)	—	b
210	Isoprene	AlCl$_3$/MeCl/-103	0.80	1.28	—	—	—	Thomas and Sparks (1944)	—	a
211	Isoprene	AlEtCl$_2$/hexane 100%; MeCl0)/-80	1.17	1.08	—	—	—	Thaler and Buckley (1976)	—	a
212	Isoprene	AlEtCl$_2$/hexane 88%; MeCl 12%/-80	1.90	1.05	—	—	—	Thaler and Buckley (1976)	—	a
213	Isoprene	AlEtCl$_2$/hexane 50%; MeCl 50%/-80	2.15	1.03	—	—	—	Thaler and Buckley (1976)	—	a
214	Isoprene	AlEtCl$_2$/hexane 12%; MeCl 88%/-80	2.5 ± 0.5	—	—	—	—	Thaler and Buckley (1976)	—	a
215	Isoprene	AlEtCl$_2$ + Cl$_2$/MeCl/-35	0.4	1	—	—	—	Cesca et al. (1975)	—	g
216	Isopropenylbenzyl chloride	BF$_3$/EtCl/-105 to -96	0.27 ± 0.08	1.41 ± 0.25	—	—	—	Jones et al. (1961a)	—	a
217	α-Methylstyrene	TiCl$_4$/MeCl$_2$/-78	1.2	5.5	—	—	—	Okamura et al. (1967)	—	a
218	α-Methylstyrene	TiCl$_4$/Tol/-78	1.1 ± 0.2	2.2 ± 0.2	—	—	—	Okamura et al. (1967)	—	a
219	α-Methylstyrene	TiCl$_4$/MeCl$_2$/-72	5.0	—	—	—	—	Maréchal et al. (1968a)	—	a
220	cis-1,3-Pentadiene	AlEt$_2$Cl + Me$_3$CCl/EtCl/-55	2.3	—	—	—	—	Priola et al. (1975)	—	—
221	trans-1,3-PEntadiene	AlEt$_2$Cl + Me$_3$CCl/EtCl/-55	0.3	3.4	—	—	—	Priola et al. (1975)	—	—
222	β-Pinene	AlEtCl$_2$/EtCl/-50	1.60	0.17	—	0.20	2.96	Kennedy and Chou (1975a)	I (!)	a, h
223	Styrene	SnCl$_4$/EtCl/0	3.5	0.33	—	1.50	0.17	Lyudvig et al. (1959a, b)	I (!)	a, h
224	Styrene	γ Rays/EtCl/-78	—	1.99 ± 0.24	—	—	—	Abkin et al. (1961)	—	a, h
225	Styrene	AlCl$_3$/MeCl/-92	9.02 ± 0.77	1.21 ± 0.06	—	—	—	Rehner et al. (1953)	—	a, h
226	Styrene	AlCl$_3$/MeCl/-30	2.51 ± 0.05	0.76 ± 0.13	—	—	—	Rehner et al. (1953)	—	a, b
227	Styrene	AlCl$_3$/MeCl/-30	2.36 ± 0.06	1.4 ± 0.2	—	—	—	Rehner et al. (1953)	—	a, b
228	Styrene	EtAlCl$_2$/MeCl$_2$/-100	2.8 ± 0.2	0.42 ± 0.02	—	—	—	Kennedy and Chou (1975b)	II	a
229	Styrene	AlCl$_3$/MeCl/-90	1.66 ± 0.02	0.24 ± 0.02	—	—	—	Rehner et al. (1953)	—	—
230	Styrene	AlCl$_3$/MeCl/-90	1.79 ± 0.02	0.6 ± 0.3	—	—	—	Rehner et al. (1953)	—	—
231	Styrene	AlCl$_3$/MeCl/-103	3 ± 1	1.20 ± 0.10	—	—	—	Tegge (1953)	—	a
232	Styrene	TiCl$_4$/Tol/-78	1.78 ± 0.10	2.41 ± 0.12	—	—	—	Imanishi et al. (1965)	—	—
233	Styrene	TiCl$_4$/n-Hexane/-78	0.37 ± 0.07	1.20 ± 0.11	—	—	—	Okamura et al. (1961)	II	c
234	Styrene	TiCl$_4$/n-Hexane/-20	0.54 ± 0.24	5.50 ± 0.55	—	—	—	Imanishi et al. (1965)	—	a
235	Styrene	TiCl$_4$/n-Hex-MeCl$_2$ (75:25)/-20	2.63 ± 0.52	—	—	—	—	Imanishi et al. (1965)	—	a

Table 6.1 □ *(Continued)*

No.	Monomer 1.	Monomer 2.	Initiating System or Coinitiation/Solvent /Temp (°C)	Published Values		Ref.	Values calculated by KT Method		Class	Comments
				r_1	r_2		r_1	r_2		
236	Isobutylene	Styrene	$TiCl_4$/n-Hex-$MeCl_2$ (50:50)/−20	3.25 ± 0.25	2.75 ± 0.25	Imanishi et al (1965)	—	—	II	c
237		Styrene	$TiCl_4$/n-Hex-$MeCl_2$ (25:75)/−20	4.11 ± 0.19	1.70 ± 0.07	Imanishi et al (1965)	—	—	II	—
238		Styrene	$TiCl_4$/$MeCl_2$/−20	4.48 ± 0.28	1.08 ± 0.07	Imanishi et al (1965)	2.70	0.70	I (!)	c
239		Styrene	$SnCl_4$/n-Hex-$MeCl_2$ (25:75)/−20	3.75 ± 0.45	1.92 ± 0.41	Imanishi et al (1965)	—	—	—	a
240		Styrene	$TiCl_4$/CS_2/−20	1.51 ± 0.15	2.44 ± 0.15	Okamura et al (1961)	—	—	—	a
241		Styrene	$TiCl_4$/decalin/−20	1.53 ± 0.03	2.61 ± 0.4	Okamura et al (1961)	—	—	—	a
242		Styrene	$TiCl_4$/Me-Cyclohexane/−20	0.75 ± 0.15	1.30 ± 0.15	Okamura et al (1961)	—	—	—	a
243		Styrene	$SnCl_4$/SO_2/−78	3.1	1.1	Iino and Tokura (1964)	—	—	—	b
244		Styrene	$SnCl_4$/SO_2−NOEt/−78	2.2	1.1	Iino and Tokura (1964)	—	—	—	b
245		Styrene	$SnCl_4$/SO_2−$EtCl_2$/−78	1.5	1.0	Iino and Tokura (1964)	—	—	—	b
246		Styrene	$SnCl_4$/SO_2−NOBz/−78	0.8	0.5	Iino and Tokura (1964)	—	—	—	b
247		Styrene	$SnCl_4$/SO_2−Bz/−78	1.9	0.6	Iino and Tokura (1960)	—	—	—	b
248	Isoprene	Propylene	$AlCl_3$/EtCl/−78	0.23	0.50	Immergut et al (1960)	—	—	—	a
249	3-Methoxystyrene	α-Methylstyrene	$SnCl_4$/NOBz−CCl_4/0	0.3 ± 0.1	5 ± 1	Overberger et al (1952)	0.02	3.94	I (!)	d
250	4-Methoxystyrene	trans-4-Methoxypropenylbenzene (anethole)	$BF_3 \cdot OEt_2$/Tol/−78	3.17 ± 0.05	0.00 ± 0.01	Higashimura et al (1972)	—	—	—	b
251		trans-4-Methoxypropenylbenzene (anethole)	$BF_3 \cdot OEt_2$/Tol/0	2.11 ± 0.24	0.00 ± 0.07	Higashimura et al (1972)	—	—	—	b
252		trans-4-Methoxypropenylbenzene (anethole)	$SnCl_4$−Tol/−78	1.71 ± 0.09	0.01 ± 0.02	Higashimura et al (1972)	—	—	—	b
253		trans-4-Methoxypropenylbenzene (anethole)	$BF_3 \cdot OEt_2$/NOEt/−78	1.40 ± 0.010	0.00 ± 0.03	Higashimura et al (1972)	—	—	—	b
254		trans-4-Methoxypropenylbenzene (anethole)	$BF_3 \cdot OEt_2$/$EtCl_2$/30	1.2 ± 0.2	0.04 ± 0.02	Mizote et al (1965)	—	—	—	b
255		cis-4-Methoxypropenylbenzene	$BF_3 \cdot OEt_2$/Tol/30	9.03 ± 0.65	0.00 ± 0.02	Higashimura et al (1972)	—	—	—	b
256		cis-4-Methoxypropenylbenzene	$BF_3 \cdot OEt_2$/Tol/0	7.73 ± 0.57	0.00 ± 0.03	Higashimura et al (1972)	—	—	—	b
257		cis-4-Methoxypropenylbenzene	$SnCl_4$−TCA/Tol/−78	3.60 ± 0.23	0.00 ± 0.02	Higashimura et al (1972)	—	—	—	b
258		cis-4-Methoxypropenylbenzene	$BF_3 \cdot OEt_2$/NOEt/−78	4.42 ± 0.27	0.00 ± 0.02	Higashimura et al (1972)	—	—	—	b
259		2-Methoxystyrene	$BF_3 \cdot OEt_2$/Tol/30	2.9 ± 0.7	0.35 ± 0.03	Imanishi et al (1963a)	—	—	—	b
260		2-Methoxystyrene	$BF_3 \cdot OEt_2$/Tol/−20	3.9 ± 1	0.35 ± 0.09	Imanishi et al (1963a)	—	—	—	b

No.	M_1	M_2	System	r_1	r_2	Reference				Note
261		2-Methoxystyrene	$BF_3\cdot OEt_2$/Tol/−78	6.4±0.4	0.45±0.05	Imanishi et al. (1963a)	—	—	—	b
262		4-Methylstyrene	I_2/$MeCl_2$/0	11.0±0.96	0.56±0.30	Yamamoto and Higashimura (1976)	—	—	—	b
263		4-Methylstyrene	$I_2+(n\text{-}Bu)_4NI$/$MeCl_2$/0	26.3±6.50	0.41±0.25	Yamamoto and Higashimura (1976)	—	—	—	b
264		4-Methylstyrene	I_2/$MeCl_2\text{–}CCl_4$(2:1)/0	26.5±6.49	0.28±0.06	Yamamoto and Higashimura (1976)	—	—	—	b
265		4-Methylstyrene	I_2/CCl_4/0	28.7±7.40	0.22±0.05	Yamamoto and Higashimura (1976)	—	—	—	b
266		4-Methylstyrene	I_2/CCl_4/30	10	0.1	Kanoh et al. (1965)	—	—	—	b
267		4-Methylstyrene	I_2/$EtCl_2$/30	4.3	0.3	Kanoh et al. (1965)	—	—	—	a
268		4-Methylstyrene	$TiCl_4$/NOBz/0	1.54±0.10	0.52±0.06	Tobolsky and Boudreau (1961)	—	—	—	a
269		4-Methylstyrene	$TiCl_4$/CCl_4/0	3.7±0.4	0.03±0.07	Tobolsky and Boudreau (1961)	—	—	—	a
270		4-Methylstyrene	$TiCl_4$/NOBz–CCl_4(1:1)/0	1.9±0.3	0.14±0.14	Tobolsky and Boudreau (1961)	—	—	—	a
271		cis-Phenyl propenyl ether	$BF_3\cdot OEt_2$/Tol/−78	13.6±1.3	0.52±0.16	Yamamoto and Higashimura (1975)	—	—	—	a
272		cis-Phenyl propenyl ether	$BF_3\cdot OEt_2$/$MeCl_2$/−78	12.8±1.1	0.68±0.16	Yamamoto and Higashimura (1975)	—	—	—	a
273		Phenyl vinyl ether	$BF_3\cdot OEt_2$/Tol/−78	12.2±1.3	0.27±0.09	Yamamoto and Higashimura (1975)	—	—	—	a
274		Phenyl vinyl ether	$BF_3\cdot OEt_2$/$MeCl_2$/−78	11.1±1.1	0.15±0.11	Yamamoto and Higashimura (1975)	—	—	—	b
275	cis-4-Methoxy-α-methylstyrene	trans-4-Methoxy-α-methylstyrene	$BF_3\cdot OEt_2$/Tol/0	2.96±0.05	0.72±0.02	Higashimura et al. (1972)	—	—	—	b
276		trans-4-Methoxy-α-methylstyrene	$BF_3\cdot OEt_2$/NOEt/0	1.02±0.08	0.29±0.04	Higashimura et al. (1972)	—	—	—	b
276a	1-Methylcyclo pentadiene	2-Methylcyclopentadiene	$BF_3\cdot OEt_2$/Tol/−78	0.01±0.05	1.1±0.2	Aso and Chasa (1967)	—	—	—	b
277	Methyl methacrylate	Methyl acrylate	$SnCl_4$/Bz/20	≪1	>1	Landler (1952)	—	—	—	a
278	α-Methylstyrene	Bis(1-indenyl)	$TiCl_4$/$MeCl_2$/−72	3.60±0.20	0.15±0.15	Maréchal et al. (1966)	—	—	II	
279		1,3-Dimethylbutadiene	$SnCl_4\text{–}TCA$/$MeCl_2$/−78	0.42±0.03	1.08±0.04	Hasegawa and Asami (1978)	—	—	—	b
280		Isoprene	$SnCl_4$/EtCl/0	7.1	1.6	Lipatova et al. (1966)	—	—	III	a
281		4-Methylstyrene	$SnCl_4$/CCl_4/0	0.3±0.1	15±5	Overberger et al. (1952)	3.1	0.3	I	a
282		trans-1-Phenyl-1,3-butadiene	$SnCl_4\text{–}TCA$/$MeCl_2$/−78	0.29±0.04	3.12±0.18	Asami et al. (1976d)	—	—	—	
283		cis-Phenyl propenyl ether	$BF_3\cdot OEt_2$/Tol/−78	2.16±0.16	0.75±0.09	Yamamoto and Higashimura (1975)	—	—	—	a
284		cis-Phenyl propenyl ether	$BF_3\cdot OEt_2$/$MeCl_2$/−78	5.14±0.32	0.22±0.07	Yamamoto and Higashimura (1975)	—	—	—	a
285		Phenyl vinyl ether	$BF_3\cdot OEt_2$/Tol/−78	2.99±0.10	0.60±0.04	Yamamoto and Higashimura (1975)	—	—	—	a
286	β-Methylstyrene	Phenyl vinyl ether	$BF_3\cdot OEt_2$/$MeCl_2$/−78	1.98±0.07	0.57±0.04	Masuda and Higashimura (1973a)	—	—	—	a
287		2-Phenyl-1,3-butadiene	$SnCl_4\text{–}TCA$/$MeCl_2$/0	0.05±0.05	1.67±0.15	Hasegawa and Asami (1976)	—	—	—	b
288		2-Phenyl-1,3-butadiene	$SnCl_4\text{–}TCA$/$MeCl_2$/−78	0.42±0.07	1.07±0.11	Hasegawa and Asami (1976)	—	—	—	b
289	4-Methylstyrene	β,4-Dimethylstyrene	BF_3/$EtCl_2$/30	1.3±0.3	0.04±0.04	Mizote et al. (1965)	—	—	—	b
290		4-Methoxypropenylbenzene (acetole)	$BF_3\cdot OEt_2$/$EtCl_2$/30	1.2±0.3	0.04±0.04	Mizote et al. (1965)	—	—	—	b
291		Phenyl vinyl ether	$BF_3\cdot OEt_2$/Tol/−78	0.61±0.14	8.96±0.72	Yamamoto and Higashimura (1975)	—	—	—	a
292		Phenyl vinyl ether	$BF_3\cdot OEt_2$/$MeCl_2$/−78	0.89±0.02	1.45±0.03	Yamamoto and Higashimura (1975)	—	—	—	a
293	cis-β-Methylvinyl isobutyl ether	trans-β-Methylvinyl isobutyl ether	$AlEt_3\text{–}SnCl_4$/Tol/−78	2.88±0.33	0.27±0.08	Okuyama et al. (1968)	—	—	—	a
294		trans-β-Methylvinyl isobutyl ether	$Al_2(SO_4)_3\text{–}H_2SO_4$/Tol/0	5.94±0.50	0.17±0.06	Okuyama et al. (1968)	—	—	—	a
295		trans-β-Methylvinyl isobutyl ether	BF_3/$MeCl_2$/−78	2.21±0.02	0.46±0.01	Okuyama et al. (1968)	—	—	—	a
296		trans-β-Methylvinyl isobutyl ether	BF_3/$MeCl_2$/0	1.85±0.09	0.33±0.03	Okuyama et al. (1968)	—	—	—	a
297		trans-β-Methylvinyl isobutyl ether	BF_3/NOEt/−78	1.37±0.19	0.78±0.15	Higashimura et al. (1973)	—	—	—	b
298		trans-β-Methylvinyl isobutyl ether	BF_3/Tol/−78	1.92±0.15	0.13±0.05	Higashimura et al. (1973)	—	—	—	b

Table 6.1 □ (Continued)

No.	Monomer 1.	Monomer 2.	Initiating System or Coinitiator/Solvent/Temp (°C)	Published Values			Values calculated by KT Method		Class	Comments
				r_1	r_2	Ref.	r_1	r_2		
299	cis-β-Methylvinyl isobutyl ether	trans-β-Methylvinyl isobutyl ether	BF₃/Tol/−78	4.45±0.53	0.25±0.09	Okuyama et al. (1968)	—	—	—	a
300		trans-β-Methylvinyl isobutyl ether	SnCl₄/NOEt/0	1.16±0.03	0.59±0.02	Okuyama et al. (1968)	—	—	—	a
301		trans-β-Methylvinyl isobutyl ether	BF₃/NOEt/−78	2.56±0.27	1.30±0.18	Higashimura et al. (1973)	—	—	—	b
302		trans-β-Methylvinyl isobutyl ether	BF₃/Tol/0	1.36±0.06	0.93±0.04	Higashimura et al. (1973)	—	—	—	b
303		trans-β-Methylvinyl isobutyl ether	BF₃/Tol/0	1.27±0.02	1.07±0.02	Higashimura et al. (1973)	—	—	—	b
304	cis-β-Methylvinyl isopropyl ether	trans-β-Methylvinyl ethyl ether	BF₃/MeCl₂/−78	1.37±0.29	0.72±0.16	Okuyama (1968)	—	—	—	a
305		trans-β-Methylvinyl isopropyl ether	BF₃/MeCl₂/−78	2.04±0.05	0.48±0.02	Okuyama (1968)	—	—	—	a
306	Octadecyl vinyl ether	2-Chloroethyl vinyl ether	SnCl₄/Bz/30	2.67±0.06	0.21±0.06	Dunphy and Marvel (1960)	2.70	0.24	I	c
307	1-Phenyl-1,3-pentadiene	(3-Chlorophenyl)-1,3-butadiene	SnCl₄/CCl₄/0	3.23	0.24	Hiroyasu and Isao (1975)	—	—	—	b
308		(4-Chlorophenyl)-1,3-butadiene	SnCl₄/CCl₄/0	1.58	0.24	Hiroyasu and Isao (1975)	—	—	—	b
309		(4-Methylphenyl)-1,3-butadiene	SnCl₄/CCl₄/0	0.32	2.79	Hitoyasu and Isao (1975)	—	—	—	b
310		(4-Nitrophenyl)-1,3-butadiene	SnCl₄/NOBz/0	7.89	0.35	Hiroyasi and Isao (1975)	—	—	—	b
311	1-Phenyl-1,3-butadiene	1,3-Pentadiene	SnCl₄-TCA/Tol/−78	0.60±0.07	5.61±0.35	Masuda et al. (1974a)			—	b
312		2-Phenyl-1,3-butadiene	SnCl₄-TCA/MeCl₂/0	1.43±0.04	0.50±0.02	Hasegawa and Asami (1976)			—	b
313		2-Phenyl-1,3-butadiene	SnCl₄-TCA/MeCl₂/−78	2.09±0.13	0.42±0.08	Hasegawa and Asami (1976)			—	b
314		Acenaphthylene	BF₃-OEt₂/MeCl₂/30	0.3±0.1	4.4±0.3	Saotome and Imoto (1958)	—		III	g
315		Acenaphthylene	TiCl₄/MeCl₂/0	0.25±0.05	17.5±1.5	Belliard and Maréchal (1972)			III	o
316		Acenaphthylene	TiCl₄/MeCl₂/−78	0.17	11.3	Cohen and Maréchal (1975)	0.20	12	I	**
317		Acenaphthylene	TiCl₄/MeCl₂/−45	0.16	15.2	Cohen and Maréchal (1975)	0.15	15	I	**
318		Acenaphthylene**	TiCl₄/MeCl₂/−20	0.14	16.8	Cohen and Maréchal (1975)	0.15	17	I	**
319		Acenaphthylene	TiCl₄/MeCl₂/0	0.14	21.8	Cohen and Maréchal (1975)	0.15	22	I	**
320		Acenaphthylene	TiCl₄/MeCl₂/18	0.13	24.6	Cohen and Maréchal (1975)	0.2	25	I	**
321		Acenaphthylene	TiCl₄/NOEt/−20	0.13	14.1	Cohen and Maréchal (1975)	0.2	14	I	**
322		Acenaphthylene	−SnCl₄/MeCl₂/−78	0.17	14.2	Cohen and Maréchal (1975)	0.2	14	I	**
323		Benzofuran	BF₃·OEt₂/EtCl₂/30	0.80±0.1	0.95±0.1	Mizote et al. (1966b)	—		I	—
324		Benzofuran	TiCl₄/MeCl₂/−75	0.55	0.53	Garreau and Maréchal (1977)	0.55	0.53	I	—
325		Benzofuran	TiCl₄/MeCl₂/−50	0.62	0.55	Garreau and Maréchal (1977)	0.59	0.55	I	—
326		Benzofuran	TiCl₄/MeCl₂/−30	0.80	0.71	Garreau and Maréchal (1977)	0.80	0.71	I	—
327		Benzofuran	TiCl₄/MeCl₂/0	0.96	0.85	Garreau and Maréchal (1977)	0.95	0.85	I	—
328		Benzofuran	TiCl₄/MeCl₂/24	1.08	1.00	Garreau and Maréchal (1977)	1.08	1.00	I	—
329		Bis(1-indenyl)	TiCl₄/MeCl₂/−72	1.95±0.10	0.17±0.05	Maréchal et al. (1966)	—	—	II	—
330		6-Bromoindene	TiCl₄/MeCl₂/−30	4.2±0.04	0.21±0.04	Quére and Maréchal (1971)	—	—	—	a

No.	Monomer	Catalyst/Solvent/Temp			Reference				
331	4-Bromostyrene	BF₃·OEt₂/EtCl₂/10	1.8 ± 0.4	0.3 ± 0.2	Tokura et al. (1963)	—	—	—	b
332	4-Bromostyrene	BF₃·OEt₂/CCl₄/10	1.8 ± 0.6	0.3 ± 0.2	Tokura et al. (1963)	—	—	—	b
333	4-Bromostyrene	BF₃·OEt₂/Bz/10	1.4 ± 0.4	0.5 ± 0.1	Tokura et al. (1963)	—	—	—	b
334	4-Bromostyrene	BF₃·OEt₂/NOBz/10	1.8 ± 0.3	0.3 ± 0.1	Tokura et al. (1963)	—	—	—	b
335	4-Bromostyrene	BF₃·OEt₂/SO₂/−15	0.8 ± 0.3	0.4 ± 0.1	Tokura et al. (1963)	—	—	—	b
336	4-Bromostyrene	BF₃·OEt₂/SO₂−NOBz/−15	0.7 ± 0.3	0.3 ± 0.1	Tokura et al. (1963)	—	—	—	b
337	4-Bromostyrene	BF₃·OEt₂/SO₂−Bz/−15	0.9 ± 0.4	0.4 ± 0.2	Tokura et al. (1963)	—	—	—	b
338	4-Bromostyrene	BF₃·OEt₂/SO₂−CCl₄/−15	1.0 ± 0.4	0.5 ± 0.2	Tokura et al. (1963)	—	—	—	a
339	4-tert-Butylstyrene	SnCl₄/EtCl₂/0	0.88 ± 0.09	0.94 ± 0.11	Furukawa et al. (1974b)	—	—	—	a
340	4-tert-Butylstyrene	SnCl₄/EtCl₂/57	0.58 ± 0.05	0.50 ± 0.07	Furukawa et al. (1974b)	—	—	—	a
341	4-tert-Butylstyrene	AlBr₃/EtCl₂/57	1.13 ± 0.13	1.08 ± 0.14	Furukawa et al. (1974b)	—	—	—	a
342	4-tert-Butylstyrene	AlBr₃/EtCl₂/57	0.86 ± 0.07	1.11 ± 0.09	Furukawa et al. (1974b)	—	—	—	a
343	4-tert-Butylstyrene	BBr₃/EtCl₂/−9	0.74 ± 0.07	0.61 ± 0.07	Furukawa et al. (1974b)	—	—	—	a
344	4-tert-Butylstyrene	BBr₃/EtCl₂/57	0.62 ± 0.15	1.13 ± 0.22	Furukawa et al. (1974b)	—	—	—	a
345	4-tert-Butylstyrene	SnCl₄/NOBz/−10	0.85 ± 0.02	0.35 ± 0.02	Furukawa et al. (1974b)	—	—	—	a
346	4-tert-Butylstyrene	SnCl₄/NOBz/70	0.54 ± 0.02	0.64 ± 0.04	Furukawa et al. (1974b)	—	—	—	a
347	4-tert-Butylstyrene	AlBr₃/NOBz/−10	0.68 ± 0.03	0.31 ± 0.04	Furukawa et al. (1974b)	—	—	—	a
348	4-tert-Butylstyrene	AlBr₃/NOBz/70	0.64 ± 0.04	0.91 ± 0.08	Furukawa et al. (1974b)	—	—	—	a
349	4-tert-Butylstyrene	BBr₃/NOBz/−10	0.78 ± 0.02	0.32 ± 0.03	Furukawa et al. (1974b)	—	—	—	a
350	4-tert-Butylstyrene	BBr₃/NOBz/70	0.66 ± 0.03	0.90 ± 0.05	Furukawa et al. (1974b)	—	—	—	a
351	5-Chloroindene	TiCl₄/MeCl₂/−40	1.35	0.93	Olivier and Maréchal (1974a)	1.4	0.9	—	**
352	6-Chloroindene	TiCl₄/MeCl₂/−40	0.59	0.04	Olivier and Maréchal (1974b)	0.6	0.05	—	**
353	7-Chloroindene	TiCl₄/MeCl₂/−40	1.42	0.93	Olivier and Maréchal (1974b)	1.4	0.9	—	**
354	2-Chlorostyrene	AlCl₃/CCl₄/0	1.6 to 3.8	0.54 ± 0.35	Florin (1951a, b)	—	—	—	—
355	2-Chlorostyrene	AlBr₃/CCl₄/0	}		Florin (1951a, b)	—	—	—	—
356	2-Chlorostyrene	SnCl₄/CCl₄/0	}		Florin (1951a, b)	—	—	—	—
357	2-Chlorostyrene	TiCl₄/CCl₄/0	1.6 to 3.8	0.54 to 0.35	Florin (1951a, b)	—	—	—	—
358	2-Chlorostyrne	BF₃/CCl₄/0			Florin (1951a, b)	—	—	—	—
359	2-Chlorostyrene	H₂SO₄/NOBz/0	3.2 to 3.8	0.25 to 0.30	Florin (1951a, b)	—	—	—	—
360	2 Chlorostyrene	AlCl₃/CCl₄/23	1.82	0.68	Florin (1951a, b)	—	—	—	—
361	2-Chlorostyrene	AlCl₃/NOBz/23	3.1	0.38	Florin (1951a, b)	—	—	—	—
362	2-Chlorostyrene	TiCl₄/MeCl₂/25	4.3 ± 0.3	0.8 ± 0.1	Borg and Maréchal (1977)	4.30	0.80	—	**
363	2-Chlorostyrene	TiCl₄/MeCl₂/0	4.1 ± 0.1	0.66 ± 0.06	Borg and Maréchal (1977)	4.10	0.60	—	**
364	2-Chlorostyrene	TiCl₄/MeCl₂/−30	3.5 ± 0.1	0.66 ± 0.06	Borg and Maréchal (1977)	3.50	0.70	—	**
365	2-Chlorostyrene	TiCl₄/MeCl₂/−78	2.31 ± 0.07	0.71 ± 0.03	Borg and Maréchal (1977)	2.30	0.60	—	**
366	2-Chlorostyrene	Ph₃⊕SbF₆⊖MeCl₂/25	1.1 ± 0.1	0.55 ± 0.07	Borg and Maréchal (1977)	1.10	0.30	—	**
367	2-Chlorostyrene	SnCl₄−TCA/Tol/30	1.6 ± 0.2	0.5 ± 0.1	Imanishi et al. (1963a)	—	—	—	—
368	2-Chlorostyrene	SnCl₄/CCl₄−NOBz(1:1)/0	3.3 ± 0.4	0.3 ± 0.05	Overberger et al. (1952)	2.91	0.28	—	**
369	3-Chlorostyrene	TiCl₄/MeCl₂/25	3.7 ± 0.2	0.3 ± 0.03	Borg and Maréchal (1977)	3.7	0.3	—	**

Table 6.1 □ (Continued)

No.	Monomer 1.	Monomer 2.	Initiating System or Coinitiatior/Solvent /Temp (°C)	Published Values r_1	Published Values r_2	Ref.	Values calculated by KT Method r_1	Values calculated by KT Method r_2	Class	Comments
370	1-Phenyl-1,3-butadiene	3-Chlorostyrene	$TiCl_4/MeCl_2/0$	3.2±0.1	2.38±0.04	Borg and Maréchal (1977)	3.2	2.3	I	**
371		3-Chlorostyrene	$TiCl_4/MeCl_2/-30$	2.5±0.1	0.37±0.03	Borg and Maréchal (1977)	2.5	0.4	I	**
372		3-Chlorostyrene	$TiCl_4/MeCl_2/-70$	2.0±0.1	0.35±0.06	Borg and Maréchal (1977)	2.0	0.4	I	**
373		3-Chlorostyrene	$Ph_3C^{\oplus}SbF_6^{\ominus}/MeCl_2/25$	3.5±0.1	0.50±0.04	Borg and Maréchal (1977)	3.5	0.5	I	**
374		4-Chlorostyrene	$TiCl_4/MeCl_2/25$	3.9±0.9	0.4±0.2	Borg and Maréchal (1977)	3.9	0.4	I	**
375		4-Chlorostyrene	$TiCl_4/MeCl_2/0$	3.8±0.3	0.4±0.9	Borg and Maréchal (1977)	3.8	0.4	I	**
376		4-Chlorostyrene	$TiCl_4/MeCl_2/-30$	3.5±0.4	0.40±0.06	Borg and Maréchal (1977)	3.5	0.4	I	**
377		4-Chlorostyrene	$TiCl_4/MeCl_2/-75$	3.33±0.06	0.54±0.02	Borg and Maréchal (1974)	3.3	0.5	I	**
378		4-Chlorostyrene	$TiCl_4/MeCl_2/5$	3.7±0.07	0.24±0.08	Borg and Maréchal (1974)	3.8	0.15	I	**
379		4-Chlorostyrene	$TiCl_4/CCl_4/5$	1.9±0.2	0.51±0.09	Borg and Maréchal (1974)	3.7	0.5	I	**
380		4-Chlorostyrene	$TiCl_4/NOBz/5$	2.0±0.2	0.43±0.09	Borg and Maréchal (1974)	2.0	0.6	I	**
381		4-Chlorostyrene	$Ph_3C^{\oplus}SbF_6^{\ominus}/MeCl_2/25$	4.0±0.3	0.64±0.07	Borg and Maréchal (1974)	4.0	0.6	I	**
382		4-Chlorostyrene	$SnCl_4/CCl_4/32$	2.7±0.3	0.35±0.05	Alfrey and Wechsler (1948)	1.97	0.28	I	—
383		4-Chlorostyrene	$SnCl_4/CCl_4/0$	2.5±0.4	0.30±0.03	Overberger et al. (1951)	2.31	0.27	I	—
384		4-Chlorostyrene	$SnCl_4/NOBz–CCl_4(1:1)/0$	2.1±0.2	0.35±0.02	Overberger et al. (1951)	1.69	0.21	I (!)	a
385		4-Chlorostyrene	$SnCl_4/CCl_4/-20$	2.5±0.4	0.3±0.03	Overberger et al. (1951)	2.71	0.42	I (!)	a
386		4-Chlorostyrene	$SnCl_4/NOBz/0$	2.2±0.2	0.45±0.02	Overberger et al. (1954)	2.15	0.44	I	—
387		4-Chlorostyrene	$TiCl_4/NOBz/0$	2.2±0.2	0.45±0.02	Overberger et al. (1954)	2.05	0.47	I	—
388		4-Chlorostyrene	$AlBr_3/NOBz/0$	2.3±0.4	0.36±0.05	Overberger et al. (1954)	2.26	0.37	I	—
389		4-Chlorostyrene	$TiCl_4/NOBz–CCl_4(1:1)/0$	2.2±0.2	0.45±0.02	Overberger et al. (1954)	2.49	0.47	I (!)	—
390		4-Chlorostyrene	$AlBr_3/NOBz–CCl_4(1:)/0$	2.0±0.2	0.34±0.05	Overberger et al. (1954)	1.90	0.34	I	—
391		4-Chlorostyrene	$SbCl_5/NOBz–CCl_4(1:1)/0$	1.7±0.2	0.55±0.05	Overberger et al. (1954)	—	—	I	a
392		4-Chlorostyrene	$FeCl_3/NOBz–CCl_4(1:1)/0$	2.0±0.1	0.43±0.03	Overberger et al. (1954)	1.87	0.36	I	—
393		4-Chlorostyrene	$HClO_4/EtCl_2/25$	2±0.20	0.43±0.05	Brown and Pepper (1965)	—	—	—	b
394		4-Chlorostyrene	$SnCl_4–TCA/Bz–Tol/-23$	2.72	0.48	Masuda and Higashimura (1971a)	—	—	—	b
395		4-Chlorostyrene	$SnCl_4–TCA/Bz–Tol/30$	2.82	0.43	Masuda and Higashimura (1971a)	—	—	—	b
396		4-Chlorostyrene	$TiCl_4/CCl_4–NOBz/0$	2.2±0.2	0.45±0.02	Overberger et al. (1954)	2.49	0.43	I (!)	—
397		4-Chlorostyrene	$TiCl_4–TCA//CCl_4–NOBz/0$	2.0±0.4	0.50±0.05	Overberger et al. (1954)	2.10	0.53	I	—
398		4-Chlorostyrene	$TiCl_4/NOBz/5$	1.9±0.2	0.43±0.09	Overberger et al. (1954)	—	—	—	a
399		4-Chlorostyrene	$SnCl_4/Tol/30$	2.5±0.4	0.30±0.03	Furukawa et al. (1973)	—	—	—	—
400		4-Chlorostyrene	$SnBr_4/Tol/30$	1.93±0.09	0.28±0.04	Furukawa et al. (1973)	—	—	—	—
401		4-Chlorostyrene	$BF_3 \cdot OEt_2/Tol/30$	1.55±0.04	0.55±0.05	Furukawa et al. (1973)	—	—	—	—
402		4-Chlorostyrene	$AlBr_3/Tol/30$	1.59±0.07	0.79±0.05	Furukawa et al. (1973)	—	—	—	—

No.	Monomer	System			Reference				
403	4-Chlorostyrene	BCl₃/Tol/30	0.62 ± 0.07	1.44 ± 0.11	Furukawa et al. (1973)	—	—	—	—
404	4-Chlorostyrene	BBr₃/Tol/30	0.29 ± 0.06	1.41 ± 0.11	Furukawa et al. (1973)	—	—	—	b
405	1-(3-Chlorophenyl)-1,3-butadiene	BBr₃-OEt₂/MeCl₂/0	2.07 ± 0.32	2.30 ± 0.45	Hasegawa and Asami (1976)	—	—	—	b
406	1-(3-Chlorophenyl)-1,3-butadiene	SnCl₄-TCA/Tol/-78	1.53 ± 0.08	3.56 ± 0.15	Masuda et al. (1974b)	—	—	—	b
407	1-(4-Chlorophenyl)-1,3-butadiene	BF₃-OEt₂/MeCl₂/0	0.98 ± 0.10	2.87 ± 0.24	Hasegawa and Asami (1976)	—	—	—	b
408	1-(4-Chlorophenyl)-1,3-butadiene	SnCl₄-TCA/Tol/-78	2.10 ± 0.15	0.77 ± 0.05	Masuda et al. (1974b)	—	—	—	b
409	2-(4-Chlorophenyl)-1,3-butadiene	SnCl₄-TCA/Tol/-78	4.22 ± 0.36	0.33 ± 0.06	Masuda et al. (1974a)	—	—	—	b
410	2-(4-Chlorophenyl)-1,3-butadiene	SnCl₄-TCA/MeCl₂/-78	2.06 ± 0.14	0.45 ± 0.07	Hasegawa and Asami (1976)	—	—	—	i
411	Chloroprene	AlBr₃/Bz/0	6.9 ± 0.5	0.04 ± 0.01	Overberger and Kamath (1963)	7.51	0.07	I	i
412	Chloroprene	AlBr₃/NOBz/0	16.0 ± 0.05	0.065 ± 0.01	Overberger and Kamath (1963)	15.87	0.05	I	e,i
413	Chloroprene	BF₃-OEt₂/n-hexane/0	12.8 ± 0.3	0.06 ± 0.01	Overberger and Kamath (1963)	15.11	0.09	I	—
414	Chloroprene	BF₃-OEt₂/NOBz/0	33.0 ± 0.5	0.15 ± 0.05	Overberger and Kamath (1963)	37.47	0.10	I	i
415	Chloroprene	BF₃-OEt₂/cyclohexane/-18	15.6	0.24	Foster (1950)	14.04	0.20	I (!)	a,k
416	1,3-Cyclooctadiene	SnCl₄/MeCl₂/-20	1.64 ± 0.05	0.485 ± 0.05	Mondal and Young (1971)	—	—	—	a,k
417	1,3-Cyclooctadiene	SnCl₄/MeCl₂/0	1.63 ± 0.05	0.495 ± 0.05	Mondal and Young (1971)	—	—	—	a
418	1,3-Cyclooctadiene	TiCl₄/MeCl₂/-20	1.66 ± 0.05	0.56 ± 0.05	Mondal and Young (1971)	—	—	—	a
419	1,3-Cyclooctadiene	TiCl₄/MeCl₂/0	1.90 ± 0.05	0.32 ± 0.05	Mondal and Young (1971)	—	—	—	d,f
420	2,5-Dichlorostyrene	AlCl₃/EtCl/0	14.8 ± 2	0.34 ± 0.2	Florin (1949)	15.97	0.49	I (!)	m
421	3,4-Dichlorostyrene	AlCl₃/CCl₄/23	2.8 ± 0.2	0.45 ± 0.10	Florin (1951a)	—	—	—	m
422	3,4-Dichlorostyrene	AlCl₃/NOBz/23	3.5 ± 0.5	0.0 ± 0.2	Florin (1951b)	—	—	—	—
423	3,4-Dichlorostyrene	AlBr₃/CCl₄/23	6.8 ± 0.8	0.0 ± 0.2	Florin (1951a, b)	2.93	0.39	I	m
424	3,4-Dichlorostyrene	TiCl₄/CCl₄/23	7.2 ± 0.5	0.38 ± 0.2	Florin (1951a, b)	—	—	—	m
425	3,4-Dichlorostyrene	BF₃-OEt₂/CCl₄/23	5.9 ± 0.2	0.27 ± 0.07	Florin (1951a, b)	—	—	—	m
426	3,4-Dichlorostyrene	ZnCl₂, solid/Et₂O/30	4.2 ± 0.2	0.10 ± 0.05	Florin (1951a, b)	—	—	—	e
427	3,4-Dichlorostyrene	H₂SO₄/NOBz/0	3.0 ± 0.5	0.20 ± 0.15	Florin (1951a, b)	3.07	0.25	I (!)	b
428	1,2-Dihydronaphthalene	BF₃-OEt₂/EtCl₂/30	1.0 ± 0.3	0.4 ± 0.2	Mizote et al. (1966a)	—	—	—	**
429	2,5-Dimethoxystyrene	TiCl₄/MeCl₂/-72	0.3 ± 0.1	0.75 ± 0.1	Mayen and Maréchal (1972)	0.3	0.8	I	b
430	1,2-Dimethylbutadiene	SnCl₄-TCA/MeCl₂/-78	0.61 ± 0.03	1.66 ± 0.08	Hasegawa and Asami (1978)	—	—	—	b
431	1,3-Dimethylbutadiene	SnCl₄-TCA/MeCl₂/-78	0.55 ± 0.11	11.96 ± 0.09	Hasegawa and Asami (1978)	—	—	—	b
432	1,3-Dimethylbutadiene	SnCl₄-TCA/MeCl₂/-78	0.51 ± 0.06	21.74 ± 1.04	Hasegawa and Asami (1978)	—	—	—	b
433	1,3-Dimethylbutadiene	BF₃-OEt₂/MeCl₂/0	0.39 ± 0.38	19.19 ± 6.04	Hasegawa and Asami (1978)	—	—	—	b
434	2,3-Dimethylbutadiene	SnCl₄-TCA/MeCl₂/-78	0.77 ± 0.05	1.03 ± 0.06	Hasegawa and Asami (1978)	—	—	—	b
435	β,4-Dimethylstyrene	BF₃/EtCl₂/30	0.75 ± 0.1	0.36 ± 0.05	Mizote et al. (1965)	—	—	—	—
436	2,4-Dimethylstyrene	BF₃/EtCl₂/-30	0.60 ± 0.05	1.35 ± 0.04	Anton and Maréchal (1971b)	0.50	1.34	I	b
437	Diphenylethylene	BF₃-OEt₂/EtCl₂/-78	0.88 ± 0.08	0.03 ± 0.03	Masuda and Higashimura (1973b)	—	—	—	—
438	4-Ethoxypropenylbenzene	BF₃-OEt₂/EtCl₂/0	0.5 ± 0.2	4.9 ± 0.07	DaSilva and Maréchal (1974)	0.5	5	I	b
439	α-Ethylstyrene	SnCl₄-TCA/MeCl₂/-78	0.53 ± 0.17	3.2 ± 0.5	Masuda and Higashimura (1973b)	—	—	—	—
440	5-Fluoroindene	TiCl₄/MeCl₂/-30	0.97	0.85	Olivier and Maréchal (1974)	1.0	0.9	I	**
441	6-Fluoroindene	TiCl₄/MeCl₂/-30	0.56	1.8	Olivier and Maréchal (1974)	0.6	1.8	I	**

Table 6.1 □ (Continued)

No.	Monomer 1.	Monomer 2.	Initiating System or Coinitiatior/Solvent /Temp (°C)	Published Values		Ref.	Values calculated by KT Method		Class	Comments
				r_1	r_2		r_1	r_2		
442	1-Phenyl-1,3-butadiene	7-Fluoroindene	TiCl$_4$/MeCl$_2$/−40	1.33	0.42	Olivier and Maréchal (1974)	1.4	0.4	I	**
443		2-Fluorostyrene	TiCl$_4$/MeCl$_2$/25	3.16	0.59	Laval and Maréchal (1977)	3.2	0.6	I	***
444		2-Fluorostyrene	TiCl$_4$/MeCl$_2$/0	2.88	0.49	Laval and Maréchal (1977)	2.9	0.5	I	***
445		2-Fluorostyrene	TiCl$_4$/MeCl$_2$/−40	2.14	0.55	Laval and Maréchal (1977)	2.2	0.6	I	***
446		2-Fluorostyrene	TiCl$_4$/MeCl$_2$/−78	1.69	0.50	Laval and Maréchal (1977)	1.7	0.5	I	***
447		2-Fluorostyrene	Ph$_3$C$^\oplus$SF$_6^\ominus$/MeCl$_2$/25	5.34	0.97	Laval and Maréchal (1977)	5.3	1.0	I	***
448		3-Fluorostyrene	TiCl$_4$/MeCl$_2$/25	6.76	0.34	Laval and Maréchal (1977)	6.8	0.3	I	***
449		3-Fluorostyrene	TiCl$_4$/MeCl$_2$/0	5.68	0.44	Laval and Maréchal (1977)	5.7	0.4	I	***
450		3-Fluorostyrene	TiCl$_4$/MeCl$_2$/−23	4.65	0.56	Laval and Maréchal (1977)	4.7	0.6	I	***
451		3-Fluorostyrene	TiCl$_4$/MeCl$_2$/−70	3.93	0.66	Laval and Maréchal (1977)	3.9	0.7	I	***
452		3-Fluorostyrene	Ph$_3$C$^\oplus$SbF$_6^\ominus$/MeCl$_2$/25	6.10	0.73	Laval and Maréchal (1977)	6.1	0.7	I	***
453		4-Fluorostyrene	TiCl$_4$/MeCl$_2$/25	2.40	0.47	Laval and Maréchal (1977)	2.4	0.5	I	***
454		4-Fluorostyrene	TiCl$_4$/MeCl$_2$/0	2.11	0.45	Laval and Maréchal (1977)	2.1	0.5	I	***
455		4-Fluorostyrene	TiCl$_4$/MeCl$_2$/−25	2.10	0.39	Laval and Maréchal (1977)	2.1	0.4	I	***
456		4-Fluorostyrene	TiCl$_4$/MeCl$_2$/−78	1.51	0.25	Laval and Maréchal (1977)	1.7	0.3	I	***
457		4-Fluorostyrene	TiCl$_4$/MeCl$_2$/0	1.68	0.55	Laval and Maréchal (1977)	1.7	0.6	I	***
458		4-Fluorostyrene	TiCl$_4$/NOBz/0	1.78	0.75	Laval and Maréchal (1977)	1.8	0.8	I	***
459		4Fluorostyrene	TiCl$_4$/CCl$_4$/0	1.40	0.58	Laval and Maréchal (1977)	1.4	0.6	I	***
460		4-Fluorostyrene	Ph$_3$C$^\oplus$SbF$_6^\ominus$/MeCl$_2$/25	1.07	0.38	Laval and Maréchal (1977)	1.1	0.4	I	**
461		2,4-Hexadiene	SnCl$_4$-TCA/MeCl$_2$/−78	1.18±0.06	1.55±0.07	Hasegawa and Asami (1978)	—	—	—	b
462		2,4-Hexadiene	SnCl$_4$-TCA/Tol/−78	0.68±0.04	0.67±0.03	Hasegawa and Asami (1978)	—	—	I (‡)	b
463		Indene	TiCl$_4$/MeCl$_2$/−72	0.97±0.07	2.16±0.15	Sigwalt and Maréchal (1966)	0.69	1.18	I	d
464		Indene	TiCl$_4$/MeCl$_2$/−30	0.90±0.01	5.05±0.03	Anton and Maréchal (1971b)	0.63	4.78	I	
465		Indene	BF$_3$·OEt$_2$/Tol/30	0.6±0.1	3.7±0.2	Mizote et al. (1966b)	—	—	—	a
466		Indene	BF$_3$·OEt$_2$/EtCl$_2$/30	0.6±0.1	3.7±0.2	Mizote et al. (1966b)	—	—	—	a
467		Indene	BF$_3$·OEt$_2$/NOBz/30	0.33±0.1	2.6±0.5	Mizote et al. (1966b)	—	—	—	a
468		Indene	I$_2$/EtCl$_2$/30	0.28±0.05	4.3±0.3	Mizote et al. (1966b)	—	—	—	a
469		Indene	SnCl$_4$-TCA/Tol/30	0.29±0.05	3.0±0.2	Mizote et al. (1966b)	—	—	—	a
470		Indene	AlEtCl$_2$/Tol/30	0.74±0.06	2.0±0.15	Mizote et al. (1966b)	—	—	—	a
471		Indene	AlEtCl$_2$/EtCl$_2$/30	0.85±0.03	1.32±0.05	Mizote et al. (1966b)	—	—	—	a
472		Indene	BF$_3$·OEt$_2$/EtCl$_2$/7	0.73±0.35	3.20±0.72	Mizote et al. (1966b)	—	—	—	a
473		Indene	BF$_3$·OEt$_2$/EtCl$_2$-TCE/7	1.57±0.36	3.28±0.52	Mizote et al. (1966b)	—	—	—	a
474		Isobutyl vinyl ether	AlCl$_3$/MeCl$_2$/−78	0.2	30±10	Heublein and Heublein (1975b)	—	—	III	e, n
475		Isobutyl vinyl ether	AlCl$_3$ + TCA/MeCl$_2$/−78	0.43	10±3	Heublein and Heublein (1975b)	—	—	III	e, n

No.	Monomer	Conditions	r_1	r_2	Reference				
476	Isobutyl vinyl ether	AlCl₃ + TCE/MeCl₂/−78	0.7	5 ± 0.4	Heublein and Heublein (1975b)	—	—	III	e, n
477	Isobutyl vinyl ether	AlCl₃/CHCl₃/−50	0.4	100 ± 20	Heublein and Heublein (1975b)	—	—	III	e, n
478	Isobutyl vinyl ether	AlCl₃ + TCA/CHCl₃/−50	0.28	65 ± 11	Heublein and Heublein (1975b)	—	—	—	—
479	Isobutyl vinyl ether	AlEt₃ + TCE/CHCl₃/−50	0.13	42 ± 8	Heublein and Heublein (1975b)	—	—	I	j
480	Isoprene	AlEt₃−t-BuCl/Bz/room	0.46	0.50	Lipscomb and Matthews (1971)	0.45	0.52	—	d
481	Isoprene	SnCl₄/EtCl/0	0.8	0.1	Lipatova et al. (1956)	—	—	—	—
482	Isoprene	SnCl₄−TCA/MeCl₂/−78	1.13 ± 0.02	0.90 ± 0.02	Hasegawa and Asami (1978)	0.15	16.49	I (!)	d
483	4-Methoxyindene	TiCl₄/MeCl₂/−78	0.15 ± 0.05	26 ± 1	Tortai and Maréchal (1971)	0.36	5.64	I	n
484	5-Methoxyindene	TiCl₄/MeCl₂/−50	0.15 ± 0.05	6.3 ± 0.5	Tortai and Maréchal (1971)	0.63	7.12	I (!)	n
485	5-Methoxyindene	TiCl₄/MeCl₂/−30	0.36 ± 0.05	5.8 ± 0.5	Tortai and Maréchal (1971)	—	—	III	**
486	6-Methoxyindene	TiCl₄/MeCl₂/−78	0.1 ± 0.05	4.1 ± 0.5	Tortai and Maréchal (1971)	—	—	II	b
487	4-Methoxy-2-Phenyl Propenylbenzene	TiCl₄/MeCl₂/−72	0.7 ± 0.4	2.0 ± 0.9	Barre and Maréchal (1974)	—	—	—	b
488	2-(4-Methoxyphenyl)-1,3-Butadiene	BF₃·OEt₂/MeCl₂/0	0.06 ± 0.05	40.09 ± 4.39	Hasegawa and Asami (1976)	—	—	—	—
489	2-(4-Methoxyphenyl)-1,3-Butadiene	SnCl₄−TCA/MeCl₂/−78	0.05 ± 0.01	42.3 ± 0.4	Masuda et al. (1974b)	—	—	—	b
490	2-Methoxypropenylbenzene	BF₃·OEt₂/NOEt/0	0.46 ± 0.05	0.67 ± 0.05	Da Silva and Maréchal (1974)	0.5	0.7	I	b
491	cis-4-Methoxypropenylbenzene	BF₃·OEt₂/NOEt/0	0.42 ± 0.12	8.24 ± 0.86	Higashimura et al. (1972)	—	—	—	b
492	trans-4-Methoxypropenylbenzene (anethole)	BF₃·OEt₂/EtCl₂/30	0.45 ± 0.2	2.8 ± 0.2	Mizote et al. (1965)	—	—	—	b
493	trans-4-Methoxypropenylbenzene (anethole)	BF₃·OEt₂/EtCl₂/−72	0.60 ± 0.05	4.8 ± 0.1	Da Silva and Maréchal (1974)	0.6	4.8	I	—
494	trans-4-Methoxypropenylbenzene (anethole)	BF₃·OEt₂/NOEt/−72	0.7 ± 0.1	3.9 ± 0.2	Da Silva and Maréchal (1974)	0.7	4.0	I	—
495	trans-4-Methoxypropenylbenzene (anethole)	BF₃·OEt₂/NOEt/0	0.13 ± 0.4	1.56 ± 0.11	Higashimura et al. (1972)	0.2	1.7	I	—
496	2-Methoxystyrene	SnCl₃−TCA/Tol/30	0.20 ± 0.2	3.9 ± 0.7	Imanishi et al. (1963a)	—	—	—	b
497	3-Methoxystyrene	SnCl₄/CCl₄/0	0.90 ± 0.15	1.1 ± 0.15	Overberger et al. (1952)	0.80	0.99	I	—
498	4-Methoxystyrene	SnCl₄/CCl₄−NOBz(1:1)/0	−0.04 ± 0.04	19 ± 3	Tobolsky and Boudreau (1961)	—	—	—	a
499	4-Methoxystyrene	SnCl₄/CCl₄−NOBz(1:1)/0	−0.02 ± 0.07	29 ± 5	Tobolsky and Boudreau (1961)	—	—	—	a
500	4-Methoxystyrene	AlCl₃/CCl₄−NOBz/0	0.34 ± 0.05	11 ± 1	Tobolsky and Boudreau (1961)	—	—	—	a
501	4-Methoxystyrene	TiCl₄/CCl₄−NOBz/0	0.12 ± 0.07	14 ± 3	Tobolsky and Boudreau (1961)	—	—	—	a
502	4-Methoxystyrene	TiCl₄/CCl₄−NOBz/0	0.38 ± 0.04	11.5 ± 0.7	Tobolsky and Boudreau (1961)	—	—	—	a
503	4-Methoxystyrene	TiCl₄/CCl₄/0	0.05 ± 0.04	46 ± 0	Tobolsky and Boudreau (1961)	—	—	—	a
504	4-Methoxystyrene	TiCl₄/CCl₄/0	0.00 ± 0.03	31 ± 6	Tobolsky and Boudreau (1961)	—	—	—	a
505	4-Methoxystyrene	TiCl₄/CCl₄−NOBz (95:5)/0	−0.12 ± 0.12	35 ± 5	Tobolsky and Boudreau (1961)	—	—	—	a
506	4-Methoxystyrene	TiCl₄/NOBz/0	0.48 ± 0.08	5.6 ± 0.8	Tobolsky and Boudreau (1961)	—	—	—	a
507	4-Methoxystyrene	TiCl₄/NOBz/0	0.22	7.6	Tobolsky and Boudreau (1961)	—	—	—	a
508	4-Methoxystyrene	TiCl₄/Tol/0	0.01	72	Tobolsky and Boudreau (1961)	—	—	—	a

Table 6.1 □ (Continued)

No.	Monomer 1.	Monomer 2.	Initiating System or Coinitiator/Solvent /Temp (°C)	Ref.	Published Values r_1	Published Values r_2	KT r_1	KT r_2	Class	Comments
509	1-Phenyl-1,3-butadiene	4-Methoxystyrene	TiCl$_4$/Tol/0	Tobolsky and Boudreau (1961)	−0.33 ± 0.03	12 ± 2	—	—	—	a
510		4-Methoxystyrene	TiCl$_4$/CCl$_4$/0	Overberger et al. (1952)	0.01 approx.	100 approx.	—	—	—	a
511		4-Methoxystyrene	SnCl$_4$/CCl$_4$/0	Kamath and Haas (1957)	0.025	400	—	—	—	a, i
512		1-Methylacenaphthylene	TiCl$_4$/MeCl$_2$/0	Belliard and Maréchal (1972)	0.11 ± 0.05	0.4 ± 0.5	0.27	9.7	III	g
513		3-Methylacenaphthylene	TiCl$_4$/MeCl$_2$/0	Belliard and Maréchal (1972)	0.23 ± 0.05	9.30 ± 0.5	—	—	I (!)	**
514		5-Methylacenaphthylene	TiCl$_4$/MeCl$_2$/0	Belliard and Maréchal (1972)	0.15 ± 0.05	11.7 ± 1	—	—	—	g
515		1-Methylindene	TiCl$_4$/MeCl$_2$/−72	Sigwalt and Maréchal (1966)	1.15 ± 0.1	0.06 ± 0.01	1.21	0.06	I	e
516		2-Methylindene	TiCl$_4$/MeCl$_2$/−72	Sigwalt and Maréchal (1966)	2.84 ± 0.2	0.038 ± 0.015	3.00	0.04	I	e
517		Methyl acrylate	SnCl$_4$/Bz/25	Landler (1952)	2.2 ± 0.2	0.4 ± 0.2	—	—	—	a
518		Methyl acrylate	SnCl$_4$/EtCl$_3$/30	Higashimura and Ikamura (1960)	$r_1 \gg r_2$		—	—	—	a
519		Methyl methacrylate	SnBr$_4$/NOBz/25	Landler (1952)	10.5 ± 0.2	0.1 ± 0.5	—	—	—	b
520		Methyl methacrylate	SnCl$_4$/EtCl$_3$/30	Higashimura and Ikamura (1960)	$r_1 \gg r_2$		—	—	—	b
521	Styrene	1-(4-Methylphenyl)-1,3-butadiene	BF$_3$·OEt$_2$/MeCl$_2$/−78	Hasegawa and Asami (1976)	0.18 ± 0.11	12.38 ± 2.26	—	—	—	b
522		1-(4-Methylphenyl)-1,3-butadiene	SnCl$_4$-TCA/Tol/−78	Masuda et al. (1974b)	0.30 ± 0.05	32.8 ± 2.5	—	—	—	b
523		2-(4-Methylphenyl)-1,3-butadiene	BF$_3$·OEt$_2$/MeCl$_2$/−78	Hasegawa and Asami (1976)	0.30 ± 0.05	4.02 ± 0.25	—	—	—	b
524		2-(4-Methylphenyl)-1,3-butadiene	SnCl$_4$-TCA/MeCl$_2$/−78	Masuda et al. (1974b)	0.62 ± 0.25	2.89 ± 0.78	—	—	—	b
525		α-Methylstyrene	SnCl$_4$/EtCl/0	Lyudvig et al. (1959a, b)	0.05	2.90	—	—	—	b
526		α-Methylstyrene	TiCl$_4$/Tol/0	Tobolsky and Boudreau (1961)	—	—	0.22	1.34	I (!)	c
527		α-Methylstyrene	TiCl$_4$/NOBz/0	Tobolsky and Boudreau (1961)	< 0.1	> 20	0.20	9.02	I (!)	c
528		α-Methylstyrene	BF$_3$·OEt$_2$/SO$_2$/−40	Iino and Tokura (1965)	0.2 ± 0.5	12 ± 2	—	—	—	b
529		α-Methylstyrene	BF$_3$·OEt$_2$/MeCl$_2$/−20	Iino and Tokura (1965)	0.24 ± 0.05	1.12 ± 0.09	—	—	—	b
530		α-Methylstyrene	TiCl$_4$-TCA/MeCl$_2$/−78	Okamura et al. (1967)	0.49 ± 0.12	1.19 ± 0.73	—	—	—	a
531		α-Methylstyrene	TiCl$_4$-TCA/Tol/−78	Okamura et al. (1967)	0.05	2.90	0.15	3.91	I (!)	b
532		α-Methylstyrene	SnCl$_4$/EtCl/0	Lyudvig et al. (1958, 1959a, b)	0.65 ± 0.22	5.86 ± 0.32	—	—	—	b
533		α-Methylstyrene	BF$_3$·OEt$_2$/EtCl$_2$/7	Panayotov et al. (1974)	0.28 ± 0.08	3.78 ± 0.43	—	—	—	b
534		α-Methylstyrene	BF$_3$·OEt$_2$/EtCl$_2$ + TCE/7	Panayotov et al. (1974)	0.25 ± 0.08	20.6 ± 3.5	—	—	—	b
535		α-Methylstyrene	BF$_3$·OEt$_2$/MeCl$_2$/−78	Masuda and Higashimura (1973b)	1.8 ± 0.2	0.07 ± 0.02	—	—	—	a
536		β-Methylstyrene	BF$_3$·OEt$_2$/EtCl$_2$/30	Mizote et al. (1965, 1966a)	1.88	0.115	—	—	—	b
537		β-Methylstyrene	SnCl$_4$/MeCl$_2$/0	Panayotov and Heublein (1977)	1.82	0.12	—	—	—	b
538		β-Methylstyrene	SnCl$_4$-TCE(1:0.025)/MeCl$_2$/0	Panayotov and Heublein (1977)	1.60	0.14	—	—	—	b
539		β-Methylstyrene	SnCl$_4$-TCE(1:0.20)/MeCl$_2$/0	Panayotov and Heublein (1977)	1.55	0.16	—	—	—	b
540		β-Methylstyrene	SnCl$_4$-TCE(1:0.50)/MeCl$_2$/0	Panayotov and Heublein (1977)	3.16	0.05	—	—	—	b
541		cis-β-Methylstyrene	BF$_3$·OEt$_2$/MeCl$_2$/0	Heublein and Schütz (1976b)			—	—	—	b

542	cis-β-Methylstyrene	TiCl₄/CCl₄/0	2.72	0.10	Heublein and Schütz (1976b)	—	—	—	b
543	cis-β-Methylstyrene	TiCl₄/Tol/10	2.60	0.10	Heublein and Schütz (1976b)	—	—	—	a
544	cis-β-Methylstyrene	TiCl₄/MeCl₂/0	1.90	0.10	Heublein and Schütz (1976b)	—	—	—	a
545	cis-β-Methylstyrene	TiCl₄/EtCl₂/0	1.83	0.10	Heublein and Schütz (1976b)	—	—	—	a
546	cis-β-Methylstyrene	TiCl₄/NOEt/0	1.53	0.12	Heublein and Schütz (1976b)	—	—	—	a
547	cis-β-Methylstyrene	TiCl₄/MeCl₂/0	2.72	0.08	Heublein and Schütz (1976b)	—	—	—	a
548	trans-β-Methylstyrene	BF₃-OEt₂/MeCl₂/0	1.80	0.08	Heublein and Schütz (1976b)	—	—	—	a
549	trans-β-Methylstyrene	TiCl₄/CCl₄/0	1.90	0.14	Heublein and Schütz (1976b)	—	—	—	a
550	trans-β-Methylstyrene	TiCl₄/Tol/0	1.85	0.14	Heublein and Schütz (1976b)	—	—	—	a
551	trans-β-Methylstyrene	TiCl₄/MeCl₂/0	1.52	0.14	Heublein and Schütz (1976b)	—	—	—	a
552	trans-β-Methylstyrene	TiCl₄/EtCl₂/0	1.50	0.15	Heublein and Schütz (1976b)	—	—	—	a
553	trans-β-Methylstyrene	TiCl₄/NOEt/0	1.35	0.16	Heublein and Schütz (1976b)	—	—	—	a
554	trans-β-Methylstyrene	SnCl₄/MeCl₂/0	1.80	0.10	Heublein and Schütz (1976b)	—	—	—	a
555	2-Methylstyrene	SnCl₄/MeCl₂/-72	0.8	0.8	Visse (1974)	≈0.8	≈0.8	—	*
556	3-Methylstyrene	TiCl₄/MeCl₂/-15	2.3 ± 0.1	1.30 ± 0.1	Visse and Maréchal (1974a)	2.25	1.28	I (¹)	*
557	3-Methylstyrene	TiCl₄/MeCl₂/-55	1.55 ± 0.1	1.2 ± 0.3	Visse and Maréchal (1974b)	1.06	0.9	I	*
558	3-Methylstyrene	TiCl₄/MeCl₂/-78	1.3 ± 0.1	0.85 ± 0.1	Visse and Maréchal (1974b)	1.34	0.85	I (¹)	*
559	4-Methylstyrene	TiCl₄/MeCl₂/25	0.36 ± 0.10	3.15 ± 0.2	Visse and Maréchal (1974b)	0.35	3.1	I	*
560	4-Methylstyrene	TiCl₄/MeCl₂/0	0.48 ± 0.06	2.9 ± 0.3	Visse and Maréchal (1974b)	0.66	2.9	I	*
561	4-Methylstyrene	TiCl₄/MeCl₂/-30	0.65 ± 0.10	2.5 ± 0.3	Visse and Maréchal (1974b)	0.65	2.5	—	*
562	4-Methylstyrene	TiCl₄/MeCl₂/-78	0.9 ± 0.06	2.0 ± 0.3	Visse and Maréchal (1974b)	—	—	II	a
563	4-Methylstyrene	SnCl₄/CCl₄/-78	0.33 ± 0.03	1.74 ± 0.03	Furukawa et al. (1973)	—	—	—	a
564	4-Methylstyrene	SnBr₄/EtCl₂/-30	0.59 ± 0.10	2.39 ± 0.37	Furukawa et al. (1973, 1974b)	—	—	—	a
565	4-Methylstyrene	SnBr₄/EtCl₂/-60	0.38 ± 0.20	3.50 ± 1.09	Furukawa et al. (1973, 1974b)	—	—	—	a
566	4-Methylstyrene	AlBr₃/EtCl₂/-30	0.59 ± 0.03	1.09 ± 0.08	Furukawa et al. (1974)	—	—	—	a
567	4-Methylstyrene	AlBr₃/EtCl₂/60	0.77 ± 0.12	2.83 ± 0.41	Furukawa et al. (1974)	—	—	—	a
568	4-Methylstyrene	BCl₃/EtCl₂/-30	0.53 ± 0.01	1.42 ± 0.03	Furukawa et al. (1974)	—	—	—	a
569	4-Methylstyrene	BCl₃/EtCl₂/60	0.51 ± 0.15	4.15 ± 0.80	Furukawa et al. (1974)	—	—	—	a
570	4-Methylstyrene	BBr₃/EtCl₂/-30	0.64 ± 0.05	2.20 ± 0.15	Furukawa et al. (1974)	—	—	—	a
571	4-Methylstyrene	BBr₃/EtCl₂/60	0.51 ± 0.10	3.27 ± 0.48	Furukawa et al. (1974)	—	—	—	a
572	4-Methylstyrene	SnBr₄/NOBz/-5	0.51 ± 0.05	1.96 ± 0.17	Furukawa et al. (1974)	—	—	—	a
573	4-Methylstyrene	SnBr₄/NOBz/80	0.47 ± 0.05	3.12 ± 0.22	Furukawa et al. (1974)	—	—	—	a
574	4-Methylstyrene	AlBr₃/NOBz/-5	0.51 ± 0.03	1.50 ± 0.09	Furukawa et al. (1974)	—	—	—	a
575	4-Methylstyrene	AlBr₃/NOBz/80	0.63 ± 0.10	2.50 ± 0.35	Furukawa et al. (1974)	—	—	—	a
576	4-Methylstyrene	BCl₃/NOBz/-5	0.74 ± 0.10	2.03 ± 0.29	Furukawa et al. (1974)	—	—	—	a
577	4-Methylstyrene	BCl₃/NOBz/80	0.82 ± 0.12	4.25 ± 0.49	Furukaw aet al. (1974)	—	—	—	a
578	4-Methylstyrene	BBr₃/NOBz/-5	0.54 ± 0.03	1.73 ± 0.09	Furukawa et al. (1974)	—	—	—	a
579	4-Methylstyrene	BBr₃/NOBz/80	0.68 ± 0.13	3.36 ± 0.52	Furukawa et al. (1974)	—	—	—	a

Table 6.1 □ *(Continued)*

No.	Monomer 1.	Monomer 2.	Initiating System or Coinitiator/Solvent /Temp (°C)	Published Values r_1	Published Values r_2	Values calculated by KT Method r_1	Values calculated by KT Method r_2	Ref.	Class	Comments
580	Styrene	4-Methylstyrene	TiCl$_4$/CCl$_4$/0	0.32±0.10	1.08±0.14	—	—	Tobolsky and Boudreau (1961)	—	a
581		4-Methylstyrene	SnCl$_4$/CCl$_4$–NOBz/0	0.48±0.07	1.18±0.14	—	—	Tobolsky and Boudreau (1961)	—	a
582		4-Methylstyrene	TiCl$_4$/NOBz/0	0.68±0.04	1.10±0.05	—	—	Tobolsky and Boudreau (1961)	—	a
583		4-Methylstyrene	TiCl$_4$/Tol/0	0.54±0.04	3.6±0.1	—	—	Tobolsky and Boudreau (1961)	—	a
584		4-Methylstyrene	TiCl$_4$/Tol/–78	0.55±0.06	1.18±0.06	—	—	Tobolsky and Boudreau (1961)	—	a
585		3-Nitrostyrene	SnCl$_4$/CCl$_4$–NOBz(1:1)/0	20±4	0.03±0.03	—	—	Overberger et al. (1952)	II	b
586		1,3-Pentadiene	SnCl$_4$–TCA/MeCl$_2$/–78	1.03±0.02	0.90±0.03	—	—	Hasegawa and Asami (1978)	—	b
587		1,3-Pentadiene	SnCl$_4$–TCA/Tol/–78	0.94±0.04	0.29±0.04	—	—	Hasegawa and Asami (1978)	—	b
588		trans-1-Phenyl-1,3-butadiene	BF$_3$·OEt$_2$/NOBz/0	0.34±0.13	7.35±3.05	0.35	7.4	Asami et al. (1976)	I	**
589		trans-1-Phenyl-1,3-butadiene	BF$_3$·OEt$_2$/NOBz/–40	0.46±0.13	5.55±1.25	0.5	5.6	Asami et al. (1976)	I	**
590		trans-1-Phenyl-1,3-butadiene	BF$_3$·OEt$_2$/NOBz/–78	0.60±0.09	3.03±0.47	0.6	3.1	Asami et al. (1976)	I	**
591		trans-1-Phenyl-1,3-butadiene	BF$_3$·OEt$_2$/MeCl$_2$/0	0.60±0.13	5.30±0.47	0.6	5.30	Asami et al. (1976)	I	**
592		trans-1-Phenyl-1,3-butadiene	BF$_3$·OEt$_2$/Tol/30	0.58±0.09	3.70±0.53	0.58	3.7	Asami et al. (1976)	I	**
593		trans-1-Phenyl-1,3-butadiene	BF$_3$·OEt$_2$/Tol/0	0.66±0.013	3.30±0.42	0.7	3.3	Asami et al. (1976)	I	**
594		trans-1-Phenyl-1,3-butadiene	BF$_3$·OEt$_2$/Tol/–40	0.70±0.10	2.98±0.40	0.7	3.0	Asami et al. (1976)	I	**
595		trans-1-Phenyl-1,3-butadiene	BF$_3$·OEt$_2$/Tol/–78	0.74±0.08	2.60±0.60	0.75	2.6	Asami et al. (1976)	I	**
596		trans-1-Phenyl-1,3-butadiene	SnCl$_4$–TCA/MeCl$_2$/–78	0.37±0.06	4.55±0.22	0.4	4.6	Asami et al. (1976)	I	**
597		trans-1-Phenyl-1,3-butadiene	SnCl$_4$–TCA/MeCl$_2$/–78	0.60±0.09	3.58±0.34	0.6	3.7	Asami et al. (1976)	I	**
598		trans-1-Phenyl-1,3-butadiene	SnCl$_4$–TCA/MeCl$_2$/–78	0.56±0.06	4.90±0.27	—	—	Masuda et al. (1974a)	—	b
599		trans-1-Phenyl-1,3-butadiene	SnCl$_4$–TCA/Tol/–78	0.80±0.14	4.59±0.45	—	—	Masuda et al. (1974a)	—	b
600		trans-1-Phenyl-1,3-butadiene	SnCl$_4$–TCA/Tol/0	0.33±0.07	4.91±0.35	—	—	Masuda et al. (1974a)	—	b
601		2-Phenyl-1,3-butadiene	BF$_3$·OEt$_2$/MeCl$_2$/0	0.92±0.03	1.62±0.04	—	—	Hasegawa and Asami (1976)	—	b
602		2-Phenyl-1,3-butadiene	SnCl$_4$–TCA/MeCl$_2$/0	1.07±0.04	1.70±0.09	—	—	Hasegawa and Asami (1976)	—	b
603		2-Phenyl-1,3-butadiene	SnCl$_4$–TCA/MeCl$_2$/–78	1.70±0.04	0.98±0.04	—	—	Hasegawa and Asami (1976)	—	b
604		2-Phenyl-1,3-butadiene	SnCl$_4$–TCA/MeCl$_2$/–78	1.60±0.09	0.86±0.06	—	—	Masuda et al. (1974a)	—	b
605		Isopropoxy-4-propenylbenzene	BF$_3$·OEt$_2$/EtCl$_2$/–25	0.32±0.05	10.2±0.5	0.30	9.8	Da Silva and Maréchal (1974)	I	—
606		Isopropoxy-4-propenylbenzene	BF$_3$·OEt$_2$/EtCl$_2$/0	0.7±0.2	6.9±0.7	0.70	7.0	Da Silva and Maréchal (1974)	I	—
607		Isopropoxy-4-propenylbenzene	BF$_3$·OEt$_2$/EtCl$_2$/13	0.97±0.09	9.3±0.3	1.0	9.3	Da Silva and Maréchal (1974)	I	—
608		2,4,6-Trimethylstyrene	TiCl$_4$/MeCl$_2$/0	1.8±0.2	0.36±0.05	1.58	0.27	Zwegers and Maréchal (1972)	I	*
609		2,4,5-Trimethylstyrene	TiCl$_4$/MeCl$_2$/0	1.3±0.2	6.5±0.5	1.95	10.05	Zwegers and Maréchal (1972)	I	*
610		2,3,4-Trimethylstyrene	TiCl$_4$/MeCl$_2$/–72	1.0±0.1	2.2±0.2	1.09	2.22	Zwegers and Maréchal (1972)	I	*
611		3,4,5-Trimethylstyrene	TiCl$_4$/MeCl$_2$/–72	1.7±0.5	3.8±0.5	1.55	3.42	Zwegers and Maréchal (1972)	I	*
612		Vinyl acetate	SnBr$_4$/Bz/25	8.25±0.05	0.015±0.015	—	—	Landler (1952)	—	—
613		Vinyl acetate	SnCl$_4$/EtCl$_2$/30	2.65±0.35	0.35±0.15	—	—	Landler (1952)	—	—

No.	Monomer 1	Monomer 2	Conditions	r_1	r_2	Reference	r_1	r_2	Class	Rating
614		Vinyl acetate	SnCl$_4$/EtCl$_2$/30	2.65±0.35	0.30±0.15	Landler (1952)	—	—	—	—
615		Vinyl acetate	SnCl$_2$/NOBz/30	6.1±0.8	0.18±0.08	Landler (1952)	—	—	—	—
616		Vinyl acetate	BF$_3$·OEt$_2$/NOBz/30	6.1±0.8	0.18±0.08	Landler (1952)	—	—	—	—
617		Vinyl-1-anthracene	TiCl$_4$/MeCl$_2$/−72	0.20±0.09	42±9	Laguerre and Maréchal (1974)	0.20	45	I	**
618		Vinyl-2-anthracene	TiCl$_4$/MeCl$_2$/19	0.40±0.05	38±6	laguerre and Maréchal (1974)	0.45	43	I	**
619		Vinyl-9-anthracene	TiCl$_4$/MeCl$_2$/−72	0.10±0.05	115±20	Laguerre and Maréchal (1974)	0.20	>100	I (!)	***
620		4-Vinylbiphenyl	TiCl$_4$/MeCl$_2$/−20	1	0.9	Cohen and Maréchal (1975)	—	1	I	***
621		4-Vinylbiphenyl	TiCl$_4$/MeCl$_2$/25	0.8	1.0	Cohen and Maréchal (1975)	—	—	I	***
622		4-Vinylbiphenyl	TiCl$_4$/NOBz/25	0.6	1.1	Cohen and Maréchal (1975)	0.7	—	I	***
623		4-Vinylbiphenyl	TiCl$_4$/NOBz/25	0.5	3.6	Cohen and Maréchal (1975)	0.5	—	I	***
624		4-Vinylbiphenyl	TiCl$_4$/Tol/−78	0.2	1	Cohen and Maréchal (1975)	0.5	3.8	I	***
625		4-Vinylbiphenyl	SnCl$_4$/MeCl$_2$/−78	1.4	Very large	Cohen and Maréchal (1975)	1.5	1	I	***
626		2-Vinylfluorene	SnCl$_4$/MeCl$_2$/−78	Very small	3.08	Cohen and Maréchal (1975)	—	—	—	—
627		1-Vinylnaphthalene	TiCl$_4$/MeCl$_2$/−72	0.42	2.4±0.4	Blin et al. (1978)	0.40	3.10	I	**
628		2-Vinylnaphthalene	TiCl$_4$/MeCl$_2$/−72	0.85±0.2	2.29	Blin et al. (1978)	0.85	2.40	I	***
629		4-Bromo-1-vinylnaphthalene	TiCl$_4$/MeCl$_2$/−75	1.11	2.17	Cho et al. (1980)	1.10	2.3	I	***
630		4-Chloro-1-vinylnaphthalene	TiCl$_4$/MeCl$_2$/−75	1.01	2.51	Cho et al. (1980)	1.0	2.2	I	***
631		4-Fluoro-1-vinylnaphthalene	TiCl$_4$/MeCl$_2$/−75	1.14	8.1±1.5	Cho et al. (1980)	1.1	2.5	I	***
632		2-Vinylphenanthrene	TiCl$_4$/CCl$_4$/0	0.95±0.3	9.25±2	Blin et al. (1978)	1.0	8.1	I	***
633		2-Vinylphenanthrene	TiCl$_4$/MeCl$_2$/0	0.3±0.15	5.5±0.3	Blin et al. (1978)	0.3	9.3	I	***
634		2-Vinylphenanthrene	TiCl$_4$/NOBz/0	1.2±0.1	2.95±0.4	Blin et al. (1978)	1.2	5.5	I	***
635		2-Vinylphenanthrene	TiCl$_4$/MeCl$_2$/−72	0.35±0.1	4.7±0.7	Blin et al. (1978)	0.35	3.0	I	***
636		2-Vinylphenanthrene	TiCl$_4$/MeCl$_2$/−40	0.4±0.15	9.25±2	Blin et al. (1978)	0.4	4.7	I	***
637		2-Vinylphenanthrene	TiCl$_4$/MeCl$_2$/0	0.3±0.15	5.8±0.8	Blin et al. (1978)	0.3	9.3	I	***
638		2-Vinylphenanthrene	TiCl$_4$/MeCl$_2$/−72	0.5±0.15	6.7±1	Blin et al. (1978)	0.5	5.8	I	***
639		2-Vinylphenanthrene	TiCl$_4$/MeCl$_2$/−50	0.35±0.2	7.6±1	Blin et al. (1978)	0.4	6.7	I	***
640		2-Vinylphenanthrene	TiCl$_4$/MeCl$_2$/−30	0.3±0.2	8.8±1	Blin et al. (1978)	0.3	7.6	I	***
641		2-Vinylphenanthrene	BF$_3$/EtCl/−115 to −60	0.2±0.15	0.7±0.1	Blin et al. (1978)	0.2	8.8	I	***
642	Vinylbenzyl chloride	Isobutylene		4.5±1	2.6	Jones et al. (1961a)	—	—	III	d
643	4-Vinylbiphenyl	2-Vinylfluorene	TiCl$_4$/MeCl$_2$/−78	0.21		Cohen and Maréchal (1975)	0.3	2.5	I	**

nitrobenzene, TCE = tetracyanoethylene, TCA = trichloracetic acid, CA = chloroanil. * KT determination by Garreau (private communication); ** by Maréchal; all other data by Kelen et al. (1977).

6.4 □ PENULTIMATE EFFECT

During the preceding discussion reactivity ratios have been calculated by relation 6.5 considering only the effect of the ultimate monomer unit. Merz et al. (1946) suggested that in some cases the penultimate unit might also influence the rate of addition of monomers to a growing chain. The four propagating steps discussed in Section 6.2 have been substituted by eight reactions:

$$\text{wwM}_2\text{M}_1^{\oplus} + \text{M}_1 \rightarrow \text{M}_1\text{M}_1\text{M}_1^{\oplus} \qquad k_{111} \qquad (6.23)$$

$$\text{wwM}_1\text{M}_1^{\oplus} + \text{M}_2 \rightarrow \text{M}_1\text{M}_1\text{M}_2^{\oplus} \qquad k_{112} \qquad (6.24)$$

$$\text{wwM}_2\text{M}_2^{\oplus} + \text{M}_2 \rightarrow \text{M}_2\text{M}_2\text{M}_2^{\oplus} \qquad k_{222} \qquad (6.25)$$

$$\text{wwM}_2\text{M}_2^{\oplus} + \text{M}_1 \rightarrow \text{M}_2\text{M}_2\text{M}_1^{\oplus} \qquad k_{221} \qquad (6.26)$$

$$\text{wwM}_2\text{M}_1^{\oplus} + \text{M}_1 \rightarrow \text{M}_2\text{M}_1\text{M}_1^{\oplus} \qquad k_{211} \qquad (6.27)$$

$$\text{wwM}_2\text{M}_1^{\oplus} + \text{M}_2 \rightarrow \text{M}_2\text{M}_1\text{M}_2^{\oplus} \qquad k_{212} \qquad (6.28)$$

$$\text{wwM}_1\text{M}_2^{\oplus} + \text{M}_2 \rightarrow \text{M}_1\text{M}_2\text{M}_2^{\oplus} \qquad k_{122} \qquad (6.29)$$

$$\text{wwM}_1\text{M}_2^{\oplus} + \text{M}_1 \rightarrow \text{M}_1\text{M}_2\text{M}_1^{\oplus} \qquad k_{121} \qquad (6.30)$$

A calculation similar to that used to derive relation 6.5 leads to

$$y = \frac{1 + \dfrac{r_1'x(r_1x + 1)}{r_1'x + 1}}{1 + \dfrac{r_2'}{x}\left[\dfrac{r_2 + x}{r_2' + x}\right]} \qquad (6.31)$$

where $y = m_1/m_2$, $x = [\text{M}_1]/[\text{M}_2]$, and $m_1 \simeq d[\text{M}_1]$. This treatment gives rise to four reactivity ratios:

$$r_1 = \frac{k_{111}}{k_{112}} \qquad r_1' = \frac{k_{211}}{k_{212}}$$

$$r_2 = \frac{k_{222}}{k_{221}} \qquad r_2' = \frac{k_{122}}{k_{121}} \qquad (6.32)$$

Although several examples of the penultimate effect have been published in the field of radical polymerizations it seems that the first penultimate effect in vinyl cationic polymerization was recognized by Sigwalt and Maréchal (1966). They discussed briefly the research that led to the postulation of the penultimate effect in the 3-methylindene/indene copolymerization system.

Whereas the reactivity ratios obtained for 1-methyl- and 2-methylindenes with indene are in good agreement with expected values, those for the copolymerization of 3-methylindene and indene are not. Thus the reactivity ratios for 3-methylindene, M, and indene, I, from equation 6.5, that is, neglecting penultimate effect, are

$$r_1 = \frac{k_{MM}}{k_{MI}} = 0.72 \pm 0.15 \quad \text{and} \quad r_2 = \frac{k_{II}}{k_{IM}} = 0.25 \pm 0.05$$

That $r_2 < 1$ is unexpected since the inductive effect by the methyl group should increase the reactivity of 3-methylindene with respect to indene. The shape of the m_1 versus $[M_1]$ and m_2 versus $[M_2]$ plots and $r_2 < 1$ indicate that 3-methylindene is more reactive than indene. In contrast, the value $r_1 = 0.72$ is inconsistent with these results; moreover, the accuracy of the data is not as good for 3-methyl- as for 2- or 1-methylindene. This discrepancy was taken to mean that owing to the penultimate effect equation 6.5 does not describe the experimental observations. Therefore, the data should be analyzed by equation 6.31, not by 6.5.

The four reactivity ratios according to equation 6.31 are

$$r_1 = \frac{k_{MMM}}{k_{MMI}}, \; r_1' = \frac{k_{IMM}}{k_{IMI}}, \; r_2 = \frac{k_{III}}{k_{IIM}}, \; r_2' = \frac{k_{MII}}{k_{MIM}} \quad (6.33)$$

r_2 and r_2' show the relative reactivity of indene and 3-methylindene toward the indenyl carbenium ion. In the case of r_2' the penultimate unit to the indenyl ion is 3-methylindene. If a steric effect is involved, $r_2 < r_2'$; but this steric effect is probably very weak because the indene unit acts as a "buffer" between the two methylindene units. Consequently, the ratio of the addition rates is probably unchanged and $r_2 \approx r_2'$. On the other hand, r_1 and r_1' are probably rather different because they involve 3-methylindene sequences.

The homopolymerization of 3-methylindene leads to chains involving no more than 2 units and $k_{MMM} \approx 0$, so that $r_1 = 0$. On the other hand, the formation of a 3-methylindene-rich copolymer shows that r_1' should be higher than 1. When $r_1 = 0$ and $r_2' = r_2$, relation 6.31 changes to 6.34

$$r_2' = \frac{[M]}{[I]} \left(\frac{i}{m} - 1 \right) + \frac{[M]i}{[I]m} \frac{r_1'[M]}{r_1'[M] + I} \quad (6.34)$$

where i, m, $[I]$, and $[M]$ are indene and 3-methylindene mole fractions, respectively, in the copolymer and the feed.

A hyperbole $r_2' = f(r_1')$ connects the experimental points. The intersection area of the curves provides r_1' and r_2' and a measure of the accuracy of the data. The values obtained for the 3-methylindene–indene pair are

$$r_1' = 2.80 \pm 0.25 \quad \text{and} \quad r_2' = 0.375 \pm 0.015$$

where r_1' is the reactivity of 3-methylindene relative to indene and toward the 3-methylindene carbenium ion, and r_2' is the relative reactivity of indene relative to 3-methylindene toward the indene carbenium ion. For 3-methylindene equation 6.31 satisfactorily describes the results because k_{MMM} was allowed to be zero.

Infrared analysis of indene–3-methylindene copolymers shows that each indene unit is associated with about two 3-methylindene monomer units.

An examination of reactivity ratios,

$$\frac{k_{IMM}}{k_{IMI}} = 2.8 \quad \text{and} \quad \frac{k_{MII}}{k_{MIM}} = 0.37$$

shows that 3-methylindene is more reactive than indene. This is expected considering the inductive effect due to the methyl group. Steric hindrance due to the methyl substituents prevents the formation of sequences consisting of more than two 3-methylindene units.

6.5 ☐ PREDICTION OF IONIC COPOLYMERIZATION REACTIVITY RATIOS

O'Driscoll's work (1964) relative to the prediction of ionic copolymerization reactivity ratios, though restricted to the case where $r_{ij}r_{ji} = 1$, may sometimes be useful. O'Driscoll attempted to derive a Q, e relation for cationic systems where $r_{12}r'_{21} = 1$. However, this relation is only seldomly obeyed. O'Driscoll and Kuntz (1963) showed that

$$r_{ij}r_{ji} = 1 \tag{6.35}$$

if

$$F^{\ddagger}_{ij} - F^{\ddagger}_{ii} = F^{\ddagger}_{jj} - F^{\ddagger}_{ji} \tag{6.36}$$

where $F^{\ddagger}_{ij}$ refers to the free energy of the activated complex formed when monomer j adds to a chain ended by monomer unit i and so on for ii, jj, and ji. From probability calculations,

$$\frac{r_{ik}}{r_{jk}} = r_{ij} \tag{6.37}$$

Experimental verification of equation 6.37 requires ionic copolymerization data for three sets of monomers obtained under the same conditions. Two such sets (Kanoh et al., 1963) have been examined by O'Driscoll (1964). The results are reported in Table 6.2. Predictions are made for r_{ik} and r_{ki} on the basis of relations 6.37 and 6.38:

$$r_{ik}r_{jk}r_{kj} = r_{ij}r_{ki}r_{jk} \tag{6.38}$$

assuming the other four reactivity ratios are known and the reactivity ratio product for the monomer pair $M_i \cdot M_j$ is unity. In view of the usual low

Table 6.2 □ Prediction of Cationic Reactivity Ratios [a]

Monomers		i	j	k	r_{ij}	r_{ji}	Product	r_{ik} Predicted	r_{ik} Observed	r_{ki} Predicted	r_{ki} Observed
Styrene	M_1	1	2	3	0.20	4.2	0.84	1.3	2.5	0.80	0.45
p-Methylstyrene	M_2	3	1	2	0.45	2.5	1.1	0.09	0.19	10.5	6.5
p-Chlorostyrene	M_3	2	3	1	6.5	0.19	1.2	2.9	4.2	0.47	0.20
2-Chloroethyl vinyl ether	M_1	1	1	3	45	5	>1	—	—	—	—
p-Methylstyrene	M_2	3	1	2	11	2	>1	—	—	—	—
p-Methoxystyrene	M_3	2	3	1	0.30	4.3	1.3	3.3	5	8.6	45

[a] After O'Driscoll and Kuntz (1963) and Ham (1963).

precision of experimental reactivity ratio, data agreement between the predicted and observed values is reasonable.

Relation 6.38 has been derived by Ham (1963) for terpolymerizations obeying steady state conditions.

6.6 ☐ SEQUENCE DISTRIBUTION ANALYSIS

Sequence distributions of various isobutylene copolymers have been described (Kennedy and Canter; 1967a, b; Kennedy and Chou, 1976a, b). In the case of poly(isobutylene-*co*-β-pinene) the following groups have been distinguished:

"Fully crowded" *gem* dimethyl group: A

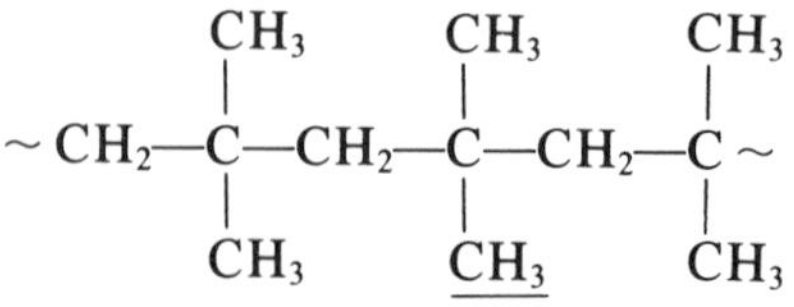

"Uncrowded" *gem* dimethyl group: B

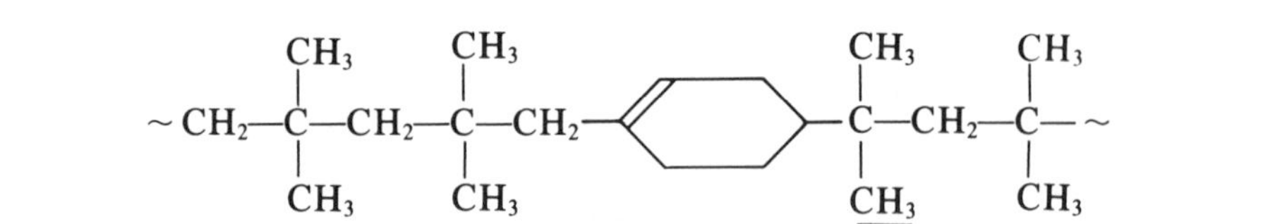

"Half-crowded" *gem* dimethyl group: C

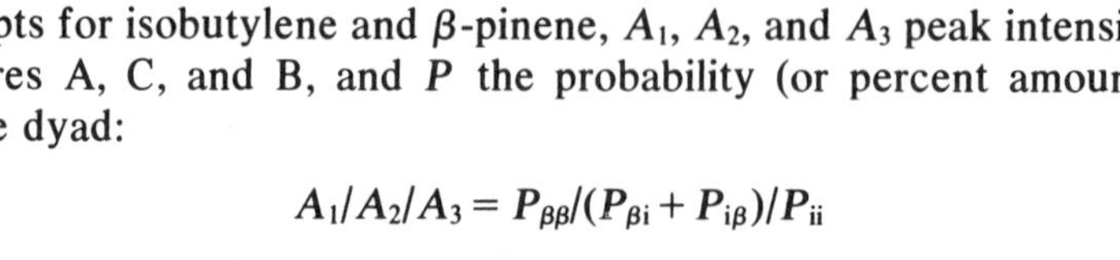

These groups have been identified by ^{1}H-NMR spectra. Let i and β be subscripts for isobutylene and β-pinene, A_1, A_2, and A_3 peak intensities for structures A, C, and B, and P the probability (or percent amount) of a possible dyad:

$$A_1/A_2/A_3 = P_{\beta\beta}/(P_{\beta i} + P_{i\beta})/P_{ii} \qquad (6.39)$$

Sequence distributions can be expressed by the run number R, that is, the average number of heterolinkages per 100 consecutive monomer link-

(Harwood and Ritchey, 1964):

$$P_{\beta\beta}/(P_{\beta i} + P_{i\beta})/P_{ii} = (100 - i\% - R/2)/R(i\% - R/2) \tag{6.40}$$

where $i\%$ is the concentration of isobutylene units in the copolymer. The run number of copolymers can also be calculated from reactivity ratios (Harwood and Ritchey, 1964):

$$R = \frac{200}{2 + r_1 \dfrac{[M_1]}{[M_2]} + r_2 \dfrac{[M_2]}{[M_1]}} \tag{6.41}$$

Agreement between calculated and experimental data shown in Figure 6.1 indicate the validity of the two-parameter model and consequently the random nature of the copolymer.

Similar analyses have been carried out with poly(isobutylene-*co*-styrene) and poly(isobutylene-*co*-isoprene) (Kennedy and Chou, 1976b). For poly(isobutylene-*co*-β-pinene) the run number has been determined both from dyad analysis and from reactivity ratios. The plot for poly(isobutylene-*co*-styrene) is shown in Figure 6.2.

Experimental points (^{1}H-NMR analysis) are consistently lower than the theoretical line for random copolymers, indicating that the reactivity ratio product is larger than unity and that the copolymer is most likely "blocky." The discrepancy between experimental and theoretical values calculated from reactivity ratios indicates that the two-parameter model is not valid for the cationic copolymerization of isobutylene and styrene.

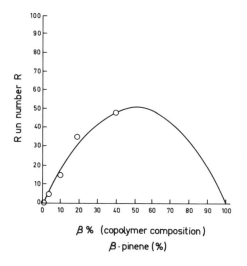

Figure 6.1 □ Run number of poly(isobutylene-*co*-β-pinene) vs. copolymer composition (from Kennedy and Chou, 1976a).

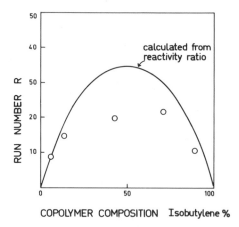

Figure 6.2 □   Run number of poly(isobutylene-*co*-styrene) vs. copolymer composition (from Kennedy and Chou, 1976b).

6.7 □   EXPERIMENTAL STUDY OF REACTIVITY

Methods used to determine the reactivity of monomers in cationic polymerization are discussed and criticized in this section.

Use of Rate Constants

The knowledge of rate constants, particularly the propagation rate constant k_p, provides an estimate of monomer reactivity toward its own cation. However, comparison of k_p's of different monomers is not always reliable for the evaluation of relative reactivities because in addition to experimental difficulties (see Chapter 5) k_p is determined not only by the nature of the monomer but also by the growing cation and experimental conditions. Since k_p is determined by at least two sets of structural factors, it is often quite difficult, if not impossible, to unravel respective contributions to overall reactivity.

Transfer constants k_{tr} have also been used to estimate reactivity of growing ions. For example, on the basis of k_{tr} for benzene and toluene in the polymerization of *o*-chlorostyrene, *p*-chlorostyrene, and styrene, Imanishi et al. (1963a, b) suggest the following order for the reactivity of growing ions:

$$o\text{-Cl-styrene}^{\oplus} > p\text{-Cl-styrene}^{\oplus} > \text{styrene}^{\oplus}$$

Use of Reactivity Ratios

Monomer reactivity ratios reflect relative monomer reactivities toward the same active species. Let R be a reference monomer (e.g., styrene) and

$M_1, \ldots, M_i$ be a set of monomers. Let r_1 and r_2 be the reactivity ratios for R and any of the monomers M_i. Then

$$r_1 = \frac{k_{ii}}{k_{ir}} \tag{6.42}$$

and

$$r_2 = \frac{k_{rr}}{k_{ri}} \tag{6.43}$$

where the subscripts i and r refer to M_i and R, respectively.

The value of $1/r_2 = k_{ri}/k_{rr}$ is a measure of the preference of the attack of $R^{\oplus}$ on M_i or its own monomer R. If k_{rr} is taken to be unity, $1/r_2$ is a measure of the reactivity of M_i toward cation $R^{\oplus}$. Consequently, $1/r_2$ values for a set of monomers may be used to express their reactivity toward the same cation.

Absolute values of $1/r_2$ depend on the nature of the reference monomer. In the absence of steric effects monomer reactivities obtained by different methods give the same reactivity values (Tortai and Maréchal, 1971; Anton and Maréchal, 1971a, b; Zwegers and Maréchal, 1972; Caillaud et al., 1970; Maréchal and Evrard, 1969a; Maréchal, 1969b–d, 1970, 1973, 1976; Belliard and Maréchal, 1972; Cohen et al., 1973; Cohen and Maréchal 1975; Bunel et al., 1975; Blin et al., 1978; Nguyen and Maréchal, 1978; Olivier and Maréchal, 1978; Visse and Maréchal, 1974; Laval and Maréchal, 1977).

Reactivity by ¹³C-NMR

Following Karplus and Pople's (1963) proposition that a ¹³C-NMR chemical shift reflects π electron density on carbon atoms, attempts have been made to correlate ¹³C-NMR chemical shifts with monomer reactivity (Masuda et al., 1973c). Thus Higashimura et al. (1969a) correlated ¹³C-NMR chemical shifts with the reactivity of β-substituted styrenes and vinyl ethers. A β-methyl group was found to decrease the reactivity of styrene and increase the reactivity of vinyl ethers. The authors explained this difference by considering the change of the π electron density caused by the β-alkyl groups and differences in the mechanisms. Table 6.3 shows ¹³C-NMR chemical shifts of β-carbons for several vinyl or alkenyl ethers.

Examination of the data in Table 6.3 shows that the chemical shift of β-carbon in n-butyl vinyl ether and isopropyl vinyl ether decreases upon methyl substitution. The decrease caused by an ethyl group is larger than that by a methyl group. The changes in chemical shifts suggest that the π electron density on the β-carbon of an olefinic double bond decreases on introduction of an alkyl group on the β-carbon, and this change in chemical shift in vinyl ethers is similar to the general trend observed in substituted hydrocarbons. Examination of the data in the last column shows that

Table 6.3 □ ¹³C-NMR Chemical Shift of β-Carbon*[a]*

AO Population of (β...

Monomer	δ_c (ppm) (from CS₂)	Model compounds	
CH₂=CHO-n-C₄H₉	107.8	CH₂=CHOCH₃	
cis-CH—CH=CHO-n-C₄H₉	93.5	cis-CH₃—CH=CHOCH₃	
cis-C₂H₅—CH′=CHO-n-C₄H₉	84.5	cis-C₂H₅—CH=CHOCH₃	
CH₂=CHO-i-C₃H₇	102.2	CH₂=CHOCH₂CH₃	
cis-CH₃—CH=CHO-i-C₃H₇	93.0	—	
CH₂=CHOCH₂CH₂Cl	104.6	—	—

*[a] After Higashimura et al. (1969a).
[b] Atomic orbital population in the π orbital (π electron density).

shielding on the β-carbon increases with increasing π electron density and that the ¹³C nucleus undergoes an upfield shift.

Evidently increase of reactivity due to β-alkyl substitution is not due to an increase of π electron density on the β-carbon. Also chemical shift does not reflect reactivity of vinyl ethers. Reactivity of vinyl ethers decreases in the following order:

$$CH_2=CH—O—i-C_3H_7 > CH_2=CH—O—n-C_4H_9$$
$$> CH_2=CH—O—CH_2CH_2Cl$$

whereas the downfield shift of β carbon resonance is

$$CH_2=CH—O—n-C_4H_9 > CH_2=CH—O—CH_2CH_2Cl$$
$$> CH_2=CH—O—i-C_3H_7$$

Since an upfield shift indicates increased π electron density on the carbon atom, the reactivity of vinyl and alkenyl ethers does not parallel π electron density on the β-carbon. Thus the carbocation attack on the β-carbon is not the rate-determining step. The same conclusion was reached indirectly from chemical evidence (Okuyama et al., 1967; Ledwith and Woods, 1966).

Hatada et al. (1970) noted that with increasingly electron-donating alkyl groups in vinyl ethers the chemical shift of the α-carbon of the vinyl group shifts to higher field and that of the β-carbon to lower field. These observations were explained by assuming that the contribution of the $\overset{\ominus}{CH_2}—CH=\overset{\oplus}{O}-R$ of resonance structure decreases with increasing electron-repelling power of the alkyl group. Good correlation between the chemical shifts of the vinyl groups and monomer reactivity was obtained.

6.8 □ THEORETICAL STUDY OF REACTIVITY

Methods and Their Evolution

Hückel's Method. A Criticism

Hückel's method, the most commonly used method in quantum chemistry of polymerization, may provide valuable information if it is used with great caution. Maréchal (1973) used Hückel's method to describe conjugated planar molecules and alternating aromatic hydrocarbons and for qualitative or semiquantitative studies. For example, using Yonezawa's relation (Yonezawa et al., 1957; O'Driscoll and Yonezawa, 1966b) for the determination of stabilization energies $(\Delta E)_r^s$ of methylindenes, the values of $(\Delta E)_r^s$ were not assumed to have absolute significance. This assumption would be meaningless because two or three parameters must be chosen to treat indene and the methyl group. Since the values of these parameters remained constant, classification of methylindenes with respect to $(\Delta E)_r^s$ was quite valuable. Experimental results confirmed the hypothesis. With not completely conjugated systems or methyl groups or heteroatoms parameters must be introduced. The problem of selecting numerical values for these parameters has not yet been satisfactorily resolved. Very often a parameter chosen for a particular property (e.g., chemical reactivity, electronegativity) was found to be unsuitable for quantitative interpretation of another property. If the same parameters are kept for the same kind of substituent, classification of monomers with respect to $(\Delta E)_r^s$ is useful; however, the values should not be viewed to have absolute significance.

Several authors (Tortai and Maréchal, 1971; Anton and Maréchal, 1971a, b; Zwegers and Maréchal, 1972; Caillaud et al. 1970; Maréchal, 1969b–d, 1970, 1973, 1976; Maréchal and Evrard, 1969a; Belliard and Maréchal, 1972; Cohen et al., 1973; Cohen and Maréchal, 1975; Bunel et al., 1975; Blin et al., 1978; Nguyen and Maréchal, 1978; Borg and Maréchal, 1977; Garreau and Maréchal, 1977; Da Silva and Maréchal, 1974; Olivier and Maréchal, 1974; Visse and Maréchal, 1974; Laval and Maréchal, 1977) compared the values obtained for similar structures. For example, a comparison between methylindenes, methoxyindenes, or methoxystyrenes was valuable but there was little significance in comparing methoxyindenes with methoxystyrenes or even indene with styrene. Since extrapolation from one system to another is impossible, the apparent good fit of a particular Hückel's parameter to one set of compounds does not justify its use with compounds of another type.

With regard to cations, the basic assumption that interelectronic and internuclear repulsions cancel each other is patently incorrect. Therefore in the preceding references care was taken always to use the same cation.

Another deficiency of Hückel's method is that it neglects molecular geometry.

Pople's Method

In Pople's method (Pople et al., 1965a; Pople and Segal, 1965b, 1966; Pople and Gordon, 1967; Pople and Beveridge, 1970) molecular orbitals are expressed as linear combinations of atomic orbitals. The coefficients in the linear combinations are chosen to minimize the total energy and to obtain LCAO SCF orbitals. (SCF = self-consistent field method). The first relations were developed by Hall (1951) and Roothaan (1951) for small molecules. The use of this method for larger systems, however, was limited by computational difficulties. It is important to extend the calculations beyond π electrons and to use all electronic valences. This improvement allows a complete treatment of the σ and π electrons in planar molecules, especially in which the σ and π systems are not separable.

Pople imagined a consistent field method which took account of σ and π valence electrons but the differential overlap was neglected (CNDO) for valence orbitals. Therefore, extension to large systems has been considered. The first version of CNDO (CNDO I) (Pople and Segal, 1966) was improved (CNDO II) (Pople and Segal, 1970) and extended to open shells of electrons.

Use of Calculations

Hückel's and Pople's methods give energy levels and eigenfunctions (C_{ij}); these values lead to quantities that give important information about monomers. Since classical values of total charge, bond order, or free valence are insufficient for the solution of many problems, frontier electron density and superdelocalizability are often used (Fukui et al., 1952, 1954, 1957, 1961). These values are particularly useful to gain insight into secondary reactions, for example, transfer or electrophilic attack on the phenyl ring by the growing chain. Garreau and Maréchal (1973) and Claus and Maréchal (1976) found good correlation between branching of methylindene polymers and superdelocalizability on phenyl carbons.

In most cases the site at which the cation attacks the monomer is known. Examination of superdelocalizability may be useful in the presence of a complexing group in the monomer when the latter and the π bond system may react simultaneously (Tortai and Maréchal, 1971; Maréchal, 1973).

Another index of reactivity is the stabilization energy $(\Delta E)_r^s$, that is, the change of energy involved in the formation of an intermediary complex between an atom s of the cation on the growing chain and an atom r of the monomer (Yonezawa et al., 1957, 1967, 1968; O'Driscoll and Yonezawa 1966b).

Maréchal (1973) applied these calculations to study various phenomena in cationic polymerization of vinyl compounds. For several series of vinyl monomers (e.g., methyl- and methoxystyrenes, methylacenaphthylenes) $1/r_2$ values were determined and the monomers classified by this parameter.

The electronic characteristics of these monomers were determined by quantum chemical methods and a second classification was developed with regard to superdelocalizabilities, localization, and stabilization energies. In most cases these classifications were in satisfactory agreement.

The values obtained with dimethyl and trimethylstyrenes were somewhat puzzling. The experimental reactivity values obtained with regard to $1/r_2$ were T246 ≃ T345 < T234 < D24, whereas theoretically the following order was expected: T345 < T245 = T234 ≃ D24 < T246 (T and D are tri- and dimethylstyrenes; the numbers after these letters indicate the positions of the substituents). Polymerization is always possible for the trimethyl-styrenes and the growing chain is not hindered by steric effect. Thus 2,4,6-trimethylstyrene polymers with average molecular weights of several millions have been obtained (Zwegers and Maréchal, 1972).

Examination of molecular models shows that the presence of one (and certainly two) ortho methyl groups strains the vinyl group out of the plane of the phenyl ring. This has been confirmed by determination of NMR coupling constants. The values shown in Table 6.4 suggest that the vinyl group is not conjugated with the phenyl ring. This conclusion is also in agreement with the classification obtained for 2,6-disubstituted styrenes whose coupling constants are close to the value obtained for ethylene.

The experimental sequence of reactivity of methylacenaphthylenes toward the styryl cation is also in agreement with theoretical reactivity (Belliard and Maréchal, 1972; Cohen et al., 1973). However, a discrepancy exists between experimental and theoretical reactivities for the acenaphthy-lene–styrene system. Experimentally, acenaphthylene is four times more

Table 6.4 □ Coupling constants J_{AB} for some Methylstyrenes[a]

Monomer	J_{AB}[b] (Hz)
Styrene	0.9
Me (3 or 4)	1.0
Me (2)	1.6
DiMe (2, 4)	1.5
DiMe (2, 5)	1.5
TriMe(2,4,5)	1.5
TriMe(2,3,4)	1.8
DiMe (2,6)	2.1
TriMe(2,4,6)	2.2
Ethylene	2.5

[a] After Zwegers and Maréchal (1972) and Maréchal (1973).
[b] For C$\underline{H}_2$ protons.

reactive than styrene, although the stabilization energy of acenaphthylene is smaller than that of styrene because Hückel's method neglects the ring strain due to the five-membered ring. It is the only discrepancy between the two systems since they are both wholly conjugated and there is no choice of parameter. Pullman (Pullman et al., 1951; Bergman et al., 1951) pointed out that Hückel's calculations for acenaphthylene give a value obviously too high for the polar moment compared with the experimental value.

The stabilization energy involved in the attack of acenaphthylene and styrene by the styryl cation by Pople's method (CNDO II) is 6.756 for styrene and 6.993 for acenaphthylene. Thus the discrepancy between theoretical and experimental results disappears by Pople's method, which takes into account the geometry (ring strain) of the molecule.

Quantum calculations were also useful in explaining aspects of the polymerization of methoxyindenes (Tortai and Maréchal, 1971). The polymerization of 5-methoxyindene with BF_3 and $TiCl_4$ does not occur below $-60°C$, but above this temperature the yield is always 100%. NMR studies have shown that below $-50°C$ the Lewis acid complexes with the methoxy group of the monomer:

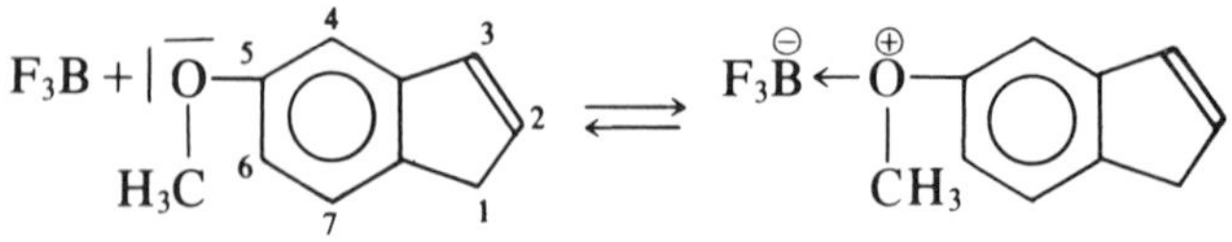

This complexation was not observed with the other methoxyindenes under the same experimental conditions. This difference can be explained by examining the data in Table 6.5, where the superdelocalizabilities on carbon-2 and oxygen for various methoxyindenes are reported. Let Δ_{O-2} be the superdelocalizability difference between oxygen and carbon-2. The smaller Δ_{O-2}, the easier the complex formation. Examination of the data in Table 6.5 shows that complex formation can occur only with 5-methoxy-indene and perhaps with 7-methoxyindene.

Table 6.5 ☐ Superdelocalizability of Oxygen and Carbon-2 for 4-Methoxy-, 5-Methoxy-, 6-Methoxy-, and 7-Methoxyindenes

Atom	4-MeO	5-MeO	6-MeO	7-MeO
Oxygen	1.181	1.169	1.180	1.156
Carbon-2	1.360	1.250	1.357	1.256
Δ_{O-2}[b]	0.179	0.081	0.177	0.100

[a] After Tortai and Maréchal (1971).
[b] Δ_{O-2} difference between superdelocalizabilities on oxygen and carbon-2.

Reactivities of Vinyl Ethers and β-Substituted Vinyl Ethers . Comparison with Unsaturated Hydrocarbons

Quantum characteristics of vinyl ethers and β-substituted vinyl ethers have been examined by Higashimura et al. (1969b). Electron distributions of monomers and related ions have been calculated by the extended Hückel method (Hoffman, 1963), because the simple Hückel method leaves some ambiguities in the parameterization for alkyl groups and in the study on the reactivity of positional isomers. Validity of the calculated π electron distribution value has been confirmed by ^{13}C-NMR chemical shifts of the β-carbon (Higashimura et al., 1969a, b).

The electron distribution in vinyl ethers and vinyl ether derivatives was computed for the eclipsed conformation, that is, the conformation in which a hydrogen atom of a methyl group eclipses the olefinic double bond. The presence of rotational isomers around the C–O bond was observed by infrared for methyl vinyl ether. Owen and Shepard (1964) reported that methyl vinyl ether existed mainly in the s-cis form (**I**) in the gas phase, and the s-trans form (**II**) was stabilized in polar solvents at room temperature.

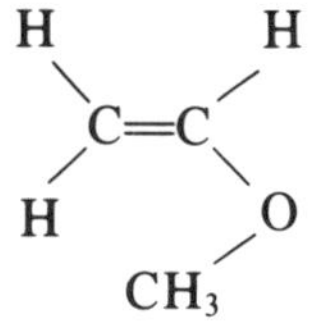

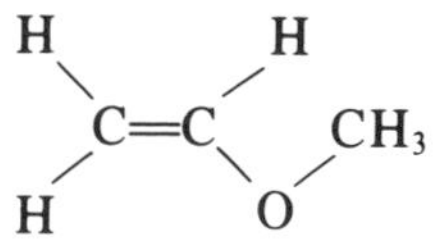

The s-cis form seems to be unstable because of steric interaction between an alkoxy group and a β-hydrogen, particularly with bulky alkyl groups. Thus to simplify the calculation only the s-trans form has been considered for vinyl ether derivatives, including methyl vinyl ether.

Higashimura et al. (1969b) studied propylene and styrene derivatives in comparison with vinyl ether derivatives. For electron distribution computation in olefins the conformation in which the hydrogen atom of a methyl group eclipses the double bond was used (Hoffman, 1963). The atomic population M_x on the olefinic carbons and the ether oxygen was calculated. M_x is the electron density:

$$M_x = \sum_r^x N_r \tag{6.44}$$

where N_r is the atomic orbital population given by equation 6.45, and C_r^j and C_s^j are the rth and sth atomic orbital coefficients in the jth molecular orbital:

$$N_r = 2\sum_j^{occ} \sum C_r^j C_s^j S_{rs} \tag{6.45}$$

where S_{rs} is the overlap integral. By introducing an electron-donating group on the β-carbon of unsaturated hydrocarbons or vinyl ether derivatives, M_x on the β-carbon decreases and M_x on the α carbon increases. Since carbenium ions attack the atom with the largest electron density, monomer reactivity may be estimated from atomic population values (M_x). However, reactivities estimated from π orbital atomic orbital population should be more suitable for prediction monomer reactivity than that from atomic population (Higashimura et al., 1969b).

Relative reactivities of olefins in cationic polymerization or oligomerization tend to decrease as $CH_2=C(CH_3)_2 \gg CH_2=CHCH_3 > cis\text{-}CH_3CH=CHCH_3 > \simeq trans\text{-}CH_3CH=CHCH_3$ and $CH_2=CHC_6H_5 > CH_3CH=CHC_6H_5$. Either atomic population or π orbital atomic orbital population coincide with relative reactivity order of monomer. Although the atomic population and β-carbon atomic orbital population follow the order of reactivities of styrene and β-methylstyrene, the atomic population on the α-carbon is larger than that of the β-carbon in β-methylstyrene. According to these values electrophilic attack should occur on the β-carbon. Experimentally, however, electrophilic attack occurs on the α-carbon, which can be explained by the frontier electron density theory (Fukui et al., 1952, 1957, 1961). The relative reactivity of vinyl ether derivatives does not correlate with atomic population, atomic orbital population, or frontier electron density.

The total electron energies of monomers or carbenium ion is

$$E = 2 \sum_{i}^{occ} E_i \tag{6.46}$$

where E_i is the energy of the ith molecular orbital. The difference between the total electron energy of a carbenium ion and that of monomer ($E_C^{\oplus} - E_M$) corresponds to the heat of a reaction. If the reaction mechanism is the same with each monomer, the difference $E_C^{\oplus} - E_M$ may serve as an index of the activation energy. Results of calculations were in agreement with experimental results for unsaturated hydrocarbons but not for ether derivatives.

Superdelocalizabilities have been calculated for vinyl ether derivatives and unsaturated hydrocarbons. Again, this reactivity index was found to be in agreement with experimental results for unsaturated hydrocarbons but could not explain reactivities of vinyl ether derivatives.

Delocalization may be visualized by assuming A, B, (Higashimura et al., 1969b), and C (Ledwith and Woods, 1966) transition states:

A   B   C

The π electron stabilization energy (ΔE) necessary for delocalization (complex formation) from the occupied orbital of the rth, sth, and so on atoms to the vacant orbital of the carbenium ion is approximated by (Fukui, 1961)

$$\Delta E = \sum_j^{occ} \left[\frac{2(C_r^j + \cdots)^2}{-E_j} \right]_\gamma \qquad (6.47)$$

where for simplicity γ, the resonance integral between the electron acceptor and donor, is assumed to be constant for any type of interactions.

By calculating the ΔE values relative to A, B, and C, Higashimura et al. (1969b) showed unambiguously that none of these states can explain the reactivity of vinyl ether derivatives. In the same way the calculation of ΔE for olefins suggests that the transition state for the rate-determining propagation step resembles model A. Thus model D was proposed to explain the increased reactivity obtained on introduction of a β-methoxy group and the higher reactivity of ethyl than methyl vinyl ether.

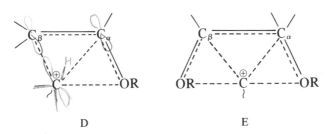

D   E

Model E was suggested as the transition state for cis-1,2-dimethoxyethylene having two ether oxygen atoms on the same side of the olefinic double bond. Structure E has a larger delocalization energy than D. $\Delta E(C_\alpha + C_\beta + O)$ values (Higashimura et al., 1969b) are in good agreement with relative reactivity of vinyl ether derivatives. Lower than calculated reactivities of trans-methyl butenyl ether and trans-1,2-dimethoxyethylene may be due to steric hindrance by the β-substituent. Thus models D or E may resemble the transition state of propagation in the polymerization of vinyl ether derivatives. The fact that steric hindrance due to the β-substituent is of minor importance in the cationic polymerization of vinyl ether derivatives is in agreement with this conclusion.

Similar experimental and theoretical work has been carried out by Laval and Maréchal (1977) and Borg and Maréchal (1977) with fluoro- and chlorostyrenes, respectively (Maréchal, 1976).

Q,e Scheme in Cationic Polymerization

According to Alfrey and Price (1947) the rate constant of propagation in radical polymerization is

$$k_p = P_R Q_M \exp(-e_R e_M) \qquad (6.48)$$

where P_R is the reactivity of the polymer radical, Q_M the general reactivity term characteristic of the monomer, and e the polarity term characteristic of the radical and the monomer. Kawabata et al. (1962a, b) assumed $e = 0$ for styrene, suggested the notation e_{RM} instead of $e_R e_M$ for the polarity term, and divided the conjugation–stabilization energy between monomer and attacking radical into four terms:

$$E_{RM} = E_{R_0 M_0} + E_R + E_M + E'_{RM} \tag{6.49}$$

where $E_{R_0 M_0}$ is the stabilization energy between polyethylene radical and ethylene monomer (the standard), E_R and E_M are increments in stabilization energy due to the substituent of the radical and mono, respectively, and E'_{RM} is the surplus increment in the stabilization energy due to the substituents of the radical and the monomer. Later, Kawabata et al. (1963) suggested

$$P_R \simeq \exp\left(\frac{E_R}{RT}\right) \tag{6.50}$$

$$\ln Q_1 - \ln Q_2 = \frac{E_{M_1} - E_{M_2}}{RT} \tag{6.51}$$

$$e_{RM} = \frac{-E'_{RM}}{RT} \tag{6.52}$$

Attempts to establish suitable relations for ionic copolymerization have been made (Overberger et al., 1952; Landler, 1952; Ham, 1954a). It was assumed that the product of monomer reactivity ratios $r_1 r_2$ was close to unity; however, this assumption is known to have limited validity. Kawabata et al. (1963) extended the Q,e scheme to ionic polymerization by considering the terms P, Q, and e to correspond to the components of the conjugation–stabilization energy between monomer and attacking ion. The activation energy in ionic polymerization was described by

$$E = C + \Delta E_\sigma + \Delta E_\pi \tag{6.53}$$

where C is a constant and ΔE_σ and ΔE_π are activation energies arising from σ and π electrons, respectively; C and ΔE_σ were assumed to be the same for ion monomer pairs.

ΔE_π is the stabilization energy between monomer and the attacking polymer ion (calculated as the total π electronic energy difference between the pair of interacting mesomeric systems and the addition product formed by their union) (Fukui et al., 1954; Yonezawa et al., 1957). In cationic polymerization $E_\pi = -E_{CM}$, where E_{CM} is the stabilization energy between the monomer and the polymer cation. In the manner adopted for radical polymerization, the stabilization energy was divided into four terms:

$$E_{CM} = E_{C_0 M_0} + E_C + E_M + E'_{CM} \tag{6.54}$$

where $E_{C_0M_0}$ (the standard) is the stabilization energy between polyethylene cation and ethylene monomer, E_C and E_M are the increments in the stabilization energies of the systems of substituted polyethylene cation–ethylene monomer and polyethylene cation–substituted ethylene monomer with respect to $E_{C_0M_0}$, and E'_{CM} is the surplus increment in the stabilization energy of the system of substituted polyethylene cation and substituted ethylene monomer. E_C, E_M, and E'_{CM} values are given by Kawabata et al. (1963).

Equations 6.50 to 6.52 derived for radical polymerization were extended to cationic polymerization as follows:

$$k_{CM} = P_C Q_M \exp(-e_{CM}) \tag{6.55}$$

$$\ln Q_{M_1} - \ln Q_{M_2} = \frac{E_{M_1} - E_{M_2}}{RT} \tag{6.56}$$

$$P_C \simeq \exp\left(\frac{E_C}{RT}\right) \tag{6.57}$$

$$e_{CM} = \frac{-E'_{CM}}{RT} \tag{6:58}$$

According to Kawabata et al. (1963) E'_{CM} is often negligible in cationic polymerization for most pairs of cations and monomers, and 6.55 reduces to

$$k_{CM} \simeq P_C Q_M \tag{6.59}$$

In this case $r_1 r_2 \simeq 1$. Thus in cationic polymerization $r_1 r_2 \simeq 1$ each time the surplus energy term is small.

However, in spite of these meritorious attempts it is difficult to develop a general Q,e scheme for cationic polymerizations because practically every counteranion/solvent/temperature combination requires a separate Q,e value.

6.9 □ EFFECT OF EXPERIMENTAL CONDITIONS ON REACTIVITY

The Effect of Temperature

The temperature dependence of r_1 and r_2 is given by

$$\ln r_1 = \ln\left(\frac{A_{11}}{A_{12}}\right) - \frac{\Delta H^\ddagger_{11} - \Delta H^\ddagger_{12}}{RT} \tag{6.60}$$

and

$$\ln r_2 = \ln\left(\frac{A_{22}}{A_{21}}\right) - \frac{\Delta H^\ddagger_{22} - \Delta H^\ddagger_{21}}{RT} \tag{6.61}$$

where A's are preexponential factors and $\Delta H^\ddagger$'s activation enthalpies. Thus

$$\ln\left(\frac{A_{11}}{A_{12}}\right) = \frac{\Delta S^\ddagger_{11} - \Delta S^\ddagger_{12}}{R} \tag{6.62}$$

Table 6.6 □ Activation Enthalpy and Entropy in Cationic Polymerization[a]

System[b]	$\Delta H^{\ddagger}_{11} - \Delta H^{\ddagger}_{12}$ (kcal/mole)	$\Delta H^{\ddagger}_{22} - \Delta H^{\ddagger}_{21}$ (kcal/mole)	$\Delta S^{\ddagger}_{11} - \Delta S^{\ddagger}_{12}$ (cal/mole · deg)	$\Delta S^{\ddagger}_{22} - \Delta S^{\ddagger}_{21}$ (cal/mole · deg)	Reference
"H₂O⁺"/TiCl₄/acenaphthylene–St/CH₂Cl₂[c]	1.57	-0.57	6.76	-4.25	Cohen et al. (1973)
"H₂O⁺"/TiCl₄/acenaphthylene–St/CH₂Cl₂[c]	1.01	-0.36	9.74	-5.14	Cohen and Maréchal (1975)
"H₂O⁺"/TiCl₄/2-chlorostyrene–St/CH₂Cl₂	0	0.77	-0.69	5.57	Borg and Maréchal (1977)
"H₂O⁺"/TiCl₄/3-chlorostyrene–St/CH₂Cl₂	0	0.81	-1.99	5.29	Borg and Maréchal (1977)
"H₂O⁺"/TiCl₄/4-chlorostyrene–St/CH₂Cl₂	0.43	0.19	-3.37	3.35	Borg and Maréchal (1977)
TCA/SnCl₄/4-chlorostyrene–St/Tol–Benz (4:6)	0.33	0.0	2.7	2.4	Masuda and Higashimura (1971a, b)
"H₂O⁺"/TiCl₄/2-fluorostyrene–St/CH₂Cl₂	0.19	0.81	5.02	0.39	Laval and Maréchal (1977)
"H₂O⁺"/TiCl₄/3-fluorostyrene–St/CH₂Cl₂	0.74	-0.93	6.25	-5.20	Laval and Maréchal (1977)
"H₂O⁺"/TiCl₄/4-fluorostyrene–St/CH₂Cl₂	1.00	0.53	3.54	1.93	Laval and Maréchal (1977)
"H₂O⁺"/TiCl₄/2-methylstyrene–St/CH₂Cl₂	0.8	0.0	2.7	0.0	Visse (1974a)
"H₂O⁺"/TiCl₄/3-methylstyrene–St/CH₂Cl₂	0.48	-0.99	3.83	-5.15	Visse (1974b)
"H₂O⁺"/SnBr₄/4-methylstyrene–St/ClCH₂CH₂Cl	0.67	1.01	3.10	5.15	Visse and Maréchal (1974)
"H₂O⁺"/TiCl₄/4-methylstyrene–St/ClCH₂CH₂Cl	-0.68	-0.76	-4.53	-4.25	Furukawa et al. (1974b)
"H₂O⁺"/AlBr₃/4-methylstyrene–St/ClCH₂CH₂Cl	-1.7	0.47	-7.27	0.68	Furukawa et al. (1974b)
"H₂O⁺"/BCl₃/4-methylstyrene–St/ClCH₂CH₂Cl	-1.91	-0.07	-8.56	-1.55	Furukawa et al. (1974b)
"H₂O⁺"/BBr₃/4-methylstyrene–St/ClCH₂CH₂Cl	-0.71	-0.4	-4.49	-2.54	Furukawa et al. (1974b)
"H₂O⁺"/SnBr₄/4-methylstyrene–St/NoBz	1.02	-0.21	-5.15	-2.11	Furukawa et al. (1974b)
"H₂O⁺"/AlBr₃/4-methylstyrene–St/NoBz	-1.13	0.43	-5.02	0.28	Furukawa et al. (1974b)
"H₂O⁺"/BCl₃/4-methylstyrene–St/NoBz	-1.62	0.23	-7.46	0.25	Furukawa et al. (1974b)
"H₂O⁺"/BBr₃/4-methylstyrene–St/NoBz	-1.47	0.51	-6.57	0.68	Furukawa et al. (1974b)
"H₂O⁺"/SnCl₄/4-methylstyrene–St/ClCH₂CH₂Cl	-0.8	-1.3	-2.7	-5.1	Furukawa et al. (1974b)
"H₂O⁺"/AlBr₃/4-methylstyrene–St/ClCH₂CH₂Cl	-0.1	-0.9	-0.5	-2.9	Furukawa et al. (1974b)
"H₂O⁺"/BBr₃/4-methylstyrene–St/ClCH₂CH₂Cl	-1.6	-0.5	-5.1	-2.4	Furukawa et al. (1974b)
"H₂O⁺"/SnCl₄/4-methylstyrene–St/NoBz	-2.7	-2.0	-8.5	-8.1	Furukawa et al. (1974b)
"H₂O⁺"/AlBr₃/4-methylstyrene–St/NoBz	-4.7	-0.3	-16.2	-1.9	Furukawa et al. (1974b)
"H₂O⁺"/BBr₃/4-methylstyrene–St/NoBz	-4.6	-0.8	-15.7	-3.4	Furukawa et al. (1974b)
"H₂O⁺"/BBr₃/4-isopropoxypropenylbenzene–St/NoBz	0	4.14	4.18	0.92	Da Silva and Maréchal (1974)
"H₂O⁺"/TiCl₄/1-vinylnaphthalene–St/CH₂Cl₂	0.94	1.54	6.98	6.10	Cho et al. (1980)
"H₂O⁺"/TiCl₄/2-vinylnaphthalene–St/CH₂Cl₂	0.73	1.1	5.6	4.9	Blin et al. (1978)
"H₂O⁺"/TiCl₄/4-fluoro-1-vinylnaphthalene–St/CH₂Cl₂	0.64	0.50	5.04	2.76	Cho et al. (1980)

					Reference
"H$_2$O"/TiCl$_4$/4-chloro-1-vinylnaphthalene–St/CH$_2$Cl$_2$	0.31	0.40	3.08	2.03	Cho et al. (1980)
"H$_2$O"/TiCl$_4$/4-bromo-1-vinylnaphthalene–St/CH$_2$Cl$_2$	0.19	0.33	2.62	1.76	Cho et al. (1980)
"H$_2$O"/TiCl$_4$/2-vinylphenanthrene–St/CH$_2$Cl$_2$	1.7	-0.2	10.5	-3	Cho et al. (1980)
"H$_2$O"/TiCl$_4$/3-vinylphenanthrene–St/CH$_2$Cl$_2$	0.63	-1.3	6.6	-8	Cho et al. (1980)
"H$_2$O"/TiCl$_4$/4-vinylbiphenyl–St/CH$_2$Cl$_2$	0	-0.56	-0.10	-2.76	Cohen and Maréchal (1975)
"H$_2$O"/TiCl$_4$/benzofuran–St/CH$_2$Cl$_2$	0.8	0.7	2.8	2.5	Garreau and Maréchal (1977)
TCA/SnCl$_4$/1-phenylbutadiene–St/CH$_2$Cl$_2$	-0.33	-0.66	-4.2	-4.4	Hasegawa and Asami (1976)
TCA/SnCl$_4$/2-phenylbutadiene–St/CH$_2$Cl$_2$	-0.75	-0.63	-3.8	-2.2	Hasegawa and Asami (1976)
TCA/SnCl$_4$/1-phenylbutadiene–2-phenylbutadiene/CH$_2$Cl$_2$	-0.52	-0.24	-1.2	0.5	Hasegawa and Asami (1976)
TCA/SnCl$_4$/2-phenylbutadiene–β-methylstyrene/CH$_2$Cl$_2$	-0.61	-2.89	-3.2	-16.5	Hasegawa and Asami (1976)
"H$_2$O"/EtAlCl$_2$/isobutylene–β-pinene/EtCl	-1.6	1.6	-8.5	8.5	Kennedy et al. (1976a)
"H$_2$O"/EtAlCl$_2$/isobutylene–butadiene/MeCl	$\Delta H_{12}^{1,4\ddagger} - \Delta H_{12}^{1,2\ddagger} = 0.34^d$			$\Delta S_{12}^{1,4\ddagger} - \Delta S_{12}^{1,2\ddagger} = 4.2^d$	Kennedy and Canter (1967b)
"H$_2$O"/BF$_3$·Et$_2$O/chloroethyl vinyl ether-4-methoxystyrene/Tol	1.9	1.6	11	4.5	Masuda and Higashimura (1971a)
"H$_2$O"/BF$_3$·Et$_2$O/chloroethyl vinyl ether-4-methoxystyrene/CH$_2$Cl$_2$	-0.97	-1.1	-2.5	-8.3	Masuda and Higashimura (1971a)
TCA/SnCl$_4$/chloroethyl vinyl ether-4-methoxystyrene/CH$_2$Cl$_2$	-0.35	-0.66	-0.78	-7.3	Masuda and Higashimura (1971a)
TCA/SnCl$_4$/chloroethyl vinyl ether-4-methylstyrene/Tol	0.09	0.34	9.0	3.1	Masuda and Higashimura (1971a)
TCA/SnCl$_4$/chloroethyl vinyl ether-4-methylstyrene/CH$_2$Cl$_2$	1.4	0.38	9.1	1.6	Masuda and Higashimura (1971a)
TCA/SnCl$_4$/chloroethyl vinyl ether-α-methylstyrene/Tol	0.77	0.19	6.3	2.5	Masuda and Higashimura (1971a)
"H$_2$O"/BF$_3$·Et$_2$O/chloroethyl vinyl ether-α-methylstyrene/CH$_2$Cl$_2$	2.1	1.0	13.0	5.9	Masuda and Higashimura (1971a)
TCA/SnCl$_4$/chloroethyl vinyl ether-α-methylstyrene/CH$_2$Cl$_2$	1.9	0.53	9.5	2.6	Masuda and Higashimura (1971a)
"H$_2$O"/TiCl$_4$/acenaphthylene-α-methylstyrene/CH$_2$Cl$_2$	1.01	-0.36	4.06	-6.25	Nguyen and Maréchal (1978)
"H$_2$O"/BF$_3$·Et$_2$O/acenaphthylene-n-butyl vinyl ether/NoBz	2.6	-2.6	6.5	-6.5	Imoto and Saotome (1958)
TCA/SnCl$_4$/chloroethyl vinyl ether-isobutyl vinyl ether/CH$_2$Cl$_2$	0.0	0.0	1.1	1.5	Masuda and Higashimura (1971a)

[a] Subscripts 1 and 2 refer to first and second monomers, respectively.

[b] St = styrene; TAC = trichloroacetic acid; Bz = benzene; NoBz = nitrobenzene; Tol = toluene.

[c] [M] and [A] are different in the two determinations.

[d] $\Delta H_{12}^{1,4\ddagger}$ or $\Delta H_{12}^{1,2\ddagger}$ indicates that butadiene (subscript 2) is incorporated as a 1,4 or 1,2 units.

where $\Delta S^{\ddagger}$ is the entropy of activation. It is affected by solvation and steric and electronic effects. Plotting $\ln r_1$ or $\ln r_2$ or $\log(r_1/r_2)$ against $1/T$ gives $\Delta H^{\ddagger}$ and $\Delta S^{\ddagger}$. Unfortunately only few reliable results are available; these are shown in Table 6.6.

In a few cases it has been possible to obtain activation parameters without determining reactivity ratios at various temperatures. Kennedy and Canter (1967a, b) studied the effect of temperature on diene incorporation into isobutylene-diene copolymers and determined activation parameters from the microstructure of the copolymers. The values are reported in Table 6.7.

According to these values the activation entropy for incorporation of butadiene as the 1,2 structure is higher than that for isobutylene incorporation, and the activation enthalpy for 1,4 entry of butadiene is still higher. Furthermore, from the entropy values it can be inferred that the transition state for 1,2 entry of butadiene is more ordered than that for 1,4 entry of diene. Also, both transition states are more ordered than that for isobutylene incorporation.

According to O'Driscoll (1969), entropies of polymerization are approximately equal and therefore entropies of activation could also be assumed to be equal. Thus $A_{11} \simeq A_{12}$ and

$$\ln r_1 = \frac{\Delta H_{11}^{\ddagger} - \Delta H_{12}^{\ddagger}}{RT} \tag{6.63}$$

or

$$\frac{d \ln r_1}{d(1/T)} = \frac{\Delta H_{11}^{\ddagger} - \Delta H_{12}^{\ddagger}}{R} \tag{6.64}$$

Combination 6.63 and 6.64 leads to

$$\frac{d \ln r_1}{d(1/t)} = T \ln r_1 \tag{6.65}$$

and

$$\frac{d \ln r_2}{d(1/T)} = T \ln r_2 \tag{6.66}$$

The term $-RT \ln r$ is the apparent activation energy for r. From equation 6.66 it appears that only large values of r (or of $1/r$) have noticeable temperature dependence. According to equation 6.65 if $r > 1$ the reactivity ratio decreases with increasing temperature, and vice versa. Moreover, r always approaches unity with increasing temperature, so that the tendency toward random copolymerization increases with increasing temperatures.

The conclusion that if $r > 1$ the reactivity ratio decreases with increasing temperature and vice versa is in agreement with Imoto and Saotome's (1958) observations, but contradicts most values obtained in cationic polymerization; this conclusion is valid only in radical polymerization. In fact, relation 6.65 was obtained because $\Delta S_{11}^{\ddagger} - \Delta S_{12}^{\ddagger} = X$ was ignored.

Table 6.7 ☐ Activation Constants for Butadiene Incorporation Relative to That for Isobutylene Incorporation in Isobutylene (1) Butadiene (2) Copolymers (AlEtCl$_2$ Coinitiator)

Activation Entropy		Activation Enthalpy	
Parameter	Value (cal/mole)	Parameter	Value (kcal/mole)
$\Delta S_{1,2}^{1,4\ddagger a}$	-1.7	$\Delta H_{1,2}^{1,4\ddagger}$	0.88
$\Delta S_{1,2}^{1,2\ddagger}$	-5.9	$\Delta H_{1,2}^{1,2\ddagger}$	0.54
$\Delta S_{1,2}^{1,4\ddagger} - \Delta S_{1,2}^{1,2\ddagger}$	4.2^b	$\Delta H_{1,2}^{1,4\ddagger} - \Delta H_{1,2}^{1,2\ddagger}$	0.34^b
$\Delta S_{1,2}^{1,4\ddagger} - \Delta S_{1,2}^{1,2\ddagger}$	5.0^b	$\Delta H_{1,2}^{1,4\ddagger} - \Delta H_{1,2}^{1,2\ddagger}$	0.50^b

[a]Superior indexes refer to the structure: 1,4 means that butadiene is incorporated as 1,4 structure in the copolymer. Inferior indexes refer to the numbering of monomers.
[b]Two sets of values obtained for the same parameter under the same experimental conditions depending on the manner of graphical determination (Kennedy and Canter, 1967b).

Taken into account X, relation 6.65 becomes

$$\ln r_1 = \frac{X}{R} - \frac{\Delta H_{11}^{\ddagger} - \Delta H_{12}^{\ddagger}}{RT} \tag{6.67}$$

and

$$\frac{d \ln r_1}{d(1/T)} = -\frac{\Delta H_{11}^{\ddagger} - \Delta H_{12}^{\ddagger}}{R} = T\left(\ln r_1 - \frac{X}{R}\right) \tag{6.68}$$

In the case of the acenaphthylene–styrene system (Cohen et al., 1973, 1975), $\ln r_1 - (X/R)$ is negative and is consequently consistent with the direction of variation of r.

According to Kennedy (1964) the elementary steps

$$\sim M_1^{\oplus} + M_1 \tag{6.69}$$

and

$$\sim M_1^{\oplus} + M_2 \tag{6.70}$$

have dissimilar activation enthalpies and entropies. Thus if the rate of step 6.69 increases faster than 6.70 with increasing temperature the reactivity ratio increases. If the opposite is true, the reactivity ratio decreases with increasing temperatures:

$$\sim M_1^{\oplus} + M_1 \xrightarrow{k_{11}} \downarrow \tag{6.71}$$

$$\sim M_1^{\oplus} + M_2 \xrightarrow{k_{12}} \uparrow \tag{6.72}$$

$$\sim M_2^{\oplus} + M_2 \xrightarrow{k_{22}} \uparrow \tag{6.73}$$

$$\sim M_2^{\oplus} + M_1 \xrightarrow{k_{21}} \downarrow \tag{6.74}$$

Equations 6.72 and 6.74 can be eliminated since it is highly unlikely that reaction rates would increase with decreasing temperatures. This means that r_1 declines because $\Delta H_{11}^{\ddagger} > \Delta H_{12}^{\ddagger}$ and r_2 because $\Delta H_{21}^{\ddagger} > \Delta H_{22}^{\ddagger}$. If steric factors are neglected,

$$r_1 = \exp\left(-\frac{\Delta H_{11}^{\ddagger} - \Delta H_{12}^{\ddagger}}{RT}\right) \tag{6.75}$$

Thus if $\Delta H_{11}^{\ddagger} > \Delta H_{12}^{\ddagger}$, r_1 increases with increasing temperatures or vice versa; if $\Delta H_{11}^{\ddagger} < \Delta H_{12}^{\ddagger}$, r_1 decreases with increasing temperatures or vice versa. The same argument is valid for r_2. These conclusions are in agreement with the results of Imoto and Saotome (1958), Cohen and Maréchal (1975), Overberger et al. (1958), and Smets and De Haes (1950).

In free radical copolymerization the product of reactivity ratios tends to move toward unity with increasing temperatures (Lewis, 1948) because the differences arising from dissimilar activation energies or entropies are eliminated and the copolymer composition changes to a more random one.

The case in which one of the reactivity ratios (r_1) decreases, whereas the other (r_2) increases with decreasing temperatures, may be analyzed as follows: the decline of r_1 may be due to decreasing k_{11} or increasing k_{12}; similarly, the rise of r_2 may stem from increasing k_{22} or decreasing k_{21}. However, with few exceptions (Borg and Maréchal, 1974, 1977; Imoto and Saotome, 1958; Visse and Maréchal, 1974), in most cationic polymerizations $r_1 r_2$ increases with increasing temperature. This unexpected effect of temperature on $r_1 r_2$ in cationic polymerization has not yet been explained. Kennedy and Chou (1976a) observed that for the system EtAlCl$_2$/isobutylene–β-pinene/EtCl, $r_1 r_2$ increases with decreasing temperature and tends toward unity. These authors speculate that lowering the temperature promotes solvation and thus the formation of freer, less discriminating carbocations.

It appears that in most cases entropy factors determine which of the two monomers is attacked by the growing cation. For instance, from activation enthalpy and entropy differences, Laval and Maréchal (1977) obtained the following relation between reactivity ratios and temperature for various fluorostyrenes (F) and styrene (S):

$$r_{2F, S} = 0.8 \exp\left(\frac{-100}{T}\right) \tag{6.76}$$

$$r_{S, 2F} = 9.0 \exp\left(\frac{-350}{T}\right) \tag{6.77}$$

$$r_{3F, S} = 0.07 \exp\left(\frac{450}{T}\right) \tag{6.78}$$

$$r_{S,3F} = 56.6 \exp\left(\frac{-500}{T}\right) \tag{6.79}$$

$$r_{4F,S} = 2.6 \exp\left(\frac{-500}{T}\right) \tag{6.80}$$

$$r_{S,4F} = \exp\left(\frac{-250}{T}\right) \tag{6.81}$$

Let $|2F, S|$, $|3F, S|$, ... be the variations of the exponential terms relative to $r_{2F,S}$, $r_{3F,S}$, ... in the 195 to 300°K range. Their values are $|2F, S| = 0.12$; $|S, 2F| = 0.15$; $|3F, S| = 5.57$; $|S, 3F| = 0.11$; $|4F, S| = 0.11$; and $|S, 4F| = 0.16$. Thus except for the 3-fluorostyryl cation the reactivity in this case is determined by the difference in entropy of activation.

The scope of this important observation has been analyzed by Bunel (1981). The variation of $\ln r_2$ with $1/T$ for the 4-methylstyrene(M_1)–styrene(M_2) and 1-vinylnaphthalene(M_1)–styrene(M_2) systems are shown in Figures 6.3 and 6.4, respectively. Let T_0 be the temperature where $\ln r_2 = 0$. According to the figures,

$$-\frac{\Delta S_{22}^{\ddagger} - \Delta S_{21}^{\ddagger}}{R} \times T_0 = -\frac{\Delta H_{22}^{\ddagger} - \Delta H_{21}^{\ddagger}}{R} \tag{6.82}$$

that is,

$$(\Delta S_{22}^{\ddagger} - \Delta S_{21}^{\ddagger})T_0 = \Delta H_{22}^{\ddagger} - \Delta H_{21}^{\ddagger} \tag{6.83}$$

From equation 6.83 the free activation enthalpy

$$\Delta G_{22}^{\ddagger} - \Delta G_{21}^{\ddagger} = (\Delta S_{22}^{\ddagger} - \Delta S_{21}^{\ddagger})(T_0 - T) \tag{6.84}$$

$$\Delta G_{22}^{\ddagger} - \Delta G_{21}^{\ddagger} = (\Delta H_{22}^{\ddagger} - \Delta H_{21}^{\ddagger})\left(1 - \frac{T}{T_0}\right) \tag{6.85}$$

Let AB be the temperature range in which r_1, r_2 are determined; in Figure 6.3 both $T_0 - T$ and $1 - (T/T_0)$ are negative. In range AB, monomer 1 is more reactive than monomer 2 ($\log r_2 < 0$) and $\Delta G_{22}^{\ddagger} - \Delta G_{21}^{\ddagger}$ is positive; thus from equations 6.84 and 6.85,

$$\Delta S_{21}^{\ddagger} > \Delta S_{22}^{\ddagger} \tag{6.86}$$

and

$$\Delta H_{21}^{\ddagger} > \Delta H_{22}^{\ddagger} \tag{6.87}$$

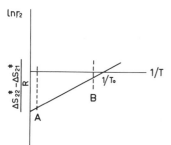

Figure 6.3 □ Variation of $\ln r_2$ with $1/T$ for the 4-methylstyrene (M_1)–styrene (M_2) system (from Visse and Maréchal, 1974).

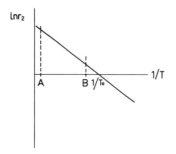

Figure 6.4 □ Variation of ln r_2 with $1/T$ for the 1-vinylnaphthalene (M₁)–styrene (M₂) system (from Blin et al., 1978).

According to these equations relative reactivity of the two monomers is controlled by entropy since ln r_2 has the sign of the entropic term. For the case shown in Figure 6.4, $T_0 - T$ and $1 - (T/T_0)$ are negative, and since $\Delta G^{\ddagger}_{22} - \Delta G^{\ddagger}_{21} < 0$,

$$\Delta S^{\ddagger}_{21} < \Delta S^{\ddagger}_{22} \tag{6.88}$$

$$\Delta H^{\ddagger}_{21} < \Delta H^{\ddagger}_{22} \tag{6.89}$$

Since $\log r_2 > 0$ (monomer 2 more reactive than monomer 1), it appears that selection is controlled by enthalpy.

The AB ranges in Figures 6.3 and 6.4 have been chosen by considering T_0 and the nature of the solvent. In the case of 4-methylstyrene (Figure 6.3) $T_0 \simeq 193°K$ (Visse and Maréchal, 1974), and for 1-vinylnaphthalene (Figure 6.4) $T_0 \simeq 263°K$ (Blin et al., 1978); above this temperature chain transfer becomes important and reasonably accurate r values cannot be obtained.

In Section 6.8 monomer reactivities obtained by $1/r_2$ (experimental) and by stabilization energy (quantum chemistry) were compared, and it was shown that except for a few cases agreement between the results was satisfactory. This agreement between reactivities obtained by experimental and quantum chemical methods is unexpected and was anticipated to occur only under well-defined conditions. The lined areas in Figure 6.5 show cases and ranges over which correlation between experimental data and theoretical calculations can be expected.

Plots A and B reflect cases in which M₁ is more reactive than M₂ and C and D when monomer M₁ is less reactive than M₂. Evidently when selection is controlled by entropy (in most cases) correlation between $1/r_2$ and stabilization energy (a purely enthalpic term) should not be expected.

An explanation of this apparent contradiction may be sought in solvation–desolvation effects (Furukawa, 1974a). The active species is solvated by solvent or monomer and in the transition complex several solvating molecules (e.g., CH_2Cl_2) may be expelled. The tighter the transition complex, the larger the number of desolvated molecules and the larger the entropy gain of the system. High stabilization energy (purely enthalpic term) indicates light transition complex, that is, high entropy gain.

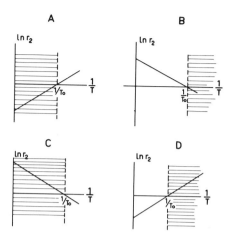

Figure 6.5 ☐ Plots of ln r_2 against $1/T$. The lined areas show where correlation between theoretical calculations and experimental reactivity ($1/r_2$) can be expected (Bunel, 1981).

Thus even when selection is controlled by the entropy of the system there is agreement between experimental reactivity and stabilization energy.

The Effect of the Nature of Solvent

Ionic polymerizations are most sensitive to the nature of the medium. However, owing to the enormous complexity of interactions between solvents and ion pairs, free ions, and aggregates, for example, the discussion of solvent effect in cationic polymerizations is a most difficult undertaking (see also Chapter 5).

Reactivity ratios are strongly affected by the nature of the solvent. Thus according to Masuda and Higashimura (1971a) the reactivity ratios of the 4-chlorostyrene–styrene pair do not change by changing the solvent from carbon tetrachloride to nitrobenzene. In contrast, Borg and Maréchal (1977) found differences by using CCl_4 and CH_2Cl_2 on the one hand and $C_6H_5NO_2$ on the other hand for the same system. Similarly, Cohen and Maréchal (1975) observed no difference for the acenaphthylene–styrene system with solvents as different as $C_6H_5NO_2$, CH_2Cl_2, and $C_6H_5CH_3$. However, Overberger and Kamath (1959) observed drastic changes in r_1 and r_2 values by changing the solvent as reported in Table 6.8.

The r_1 values increase greatly with increasing ϵ but the trend of r_2 values is somewhat erratic (Kennedy, 1964). Overberger and Kamath (1959) explain their data by suggesting preferential solvation of the propagating ion pair by the more polar monomer in nonpolar media; thus the growing cation in n-hexane would be preferentially solvated by the relatively polar p-chlorostyrene monomer, leading to an increase of p-chlorostyrene concentration in the vicinity of the active center and consequently preferred

Table 6.8 □ Effect of the Nature of the Solvent on Reactivity Ratios of 4-Chlorostyrene(1)-Isobutylene(2) at 0°C[a]. Total Monomer Concentration 20 mol %

Solvent	Dielectric Constant ϵ	Coinitiator[b]	r_1	r_2
n-C_6H_{14}	1.9	0.5 mol % $AlBr_3$	1.02	1.01
$C_6H_5NO_2$	29.7	0.1 mol % $AlBr_3$	0.15	14.7
CH_3NO_2	37.4	0.5 mol % $AlBr_3$	0.7	22.5
$C_6H_5NO_2$	29.7	0.3 mol % $SnCl_4$	1.2	8.6
n-C_6H_{14}	1.9	$SnCl_4$	No polymerization	

[a] After Overberger and Kamath (1959).
[b] Mole percent calculated on the total mixture.

incorporation and an increase of r_2. On the other hand, in nitrobenzene the ion pair is probably solvated by the polar diluent and the more reactive isobutylene monomer may dominate propagation. The dissimilarity of r values obtained with $AlBr_3$ in the two nitro solvents indicates some specific solvation effects (Kennedy, 1964; Overberger and Kamath, 1959).

The absence of copolymerization in n-hexane is probably due to the combined effect of slow initiation and ion pair association in this nonpolar solvent. However, copolymerization in the presence of $AlBr_3$ in n-hexane indicates probably faster initiation with $AlBr_3$ than with $SnCl_4$. Similar analyses have been provided for other systems (Overberger and Kamath, 1963). Imanishi et al. (1970) analyzed the influence of solvent polarity (i.e., CH_2Cl_2, n-C_6H_{14} or mixtures thereof) on the 2-methyl-2-butene–isobutylene and 2-methyl-2-butene–styrene systems. According to these authors 2-methyl-2-butene (MB2), a β-methyl derivative of isobutylene (IB), is far less reactive than IB in CH_2Cl_2. The reactivity difference between MB2 and IB becomes less pronounced with decreasing solvent polarity. In n-hexane the reactivity of MB2 was found to be equal to or larger than that of IB. Going from CH_2Cl_2 to n-hexane, the nature of the propagating species may change from dissociated ion pair to tight ion pair. In CH_2Cl_2 steric obstruction exerted by the β-methyl group in the transition state of propagation may reduce the reactivity of MB2. Alternatively in n-hexane the propagating cation may interact with either the α- or β-carbon of the double bond:

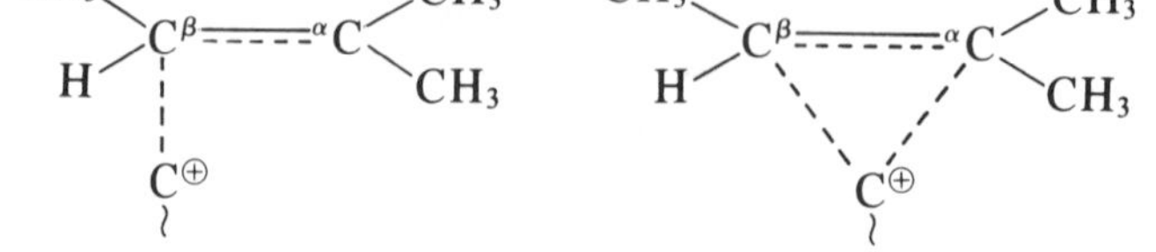

In n-hexane the β-methyl group lowers the potential energy of the transition state and thus activates MB2.

In the copolymerization of MB2 and styrene (St) in n-hexane, St was found to be more reactive than MB2. However, the relative reactivity of MB2 increased with increasing solvent polarity. In polar solvents both r_{MB2} and r_{St} were found to be larger than unity. The same phenomena have been observed in the copolymerization of isobutylene and styrene; according to Imanishi et al. (1970) this appears to be a general characteristic of aliphatic–aromatic olefin copolymerization.

The effects of copolymerization conditions have also been determined for 2-chloroethyl vinyl ether (CEVE)-α-methyl-p-methyl- or p-methoxy- styrene systems (Masuda et al., 1970). CEVE is 1.7 to 5.0 times more reactive than α-methylstyrene using $BF_3 \cdot OEt_2$ in toluene or methylene chloride or $CCl_3COOH/SnCl_4$ in toluene. However, these monomers have the same reactivity using $CCl_3COOH/SnCl_4$ in CH_2Cl_2. Masuda et al. (1970) noted that the amount of styrene derivative in the copolymer increases irrespective of the relative reactivity of monomers using $CCl_3COOH/SnCl_4$ instead of $BF_3 \cdot OEt_2$. In the presence of $CCl_3COOH/SnCl_4$ relative reactivities in the CEVE-α-methylstyrene and CEVE-p-methylstyrene systems approach unity, whereas with $BF_3 \cdot OEt_2$ they tend to be greater than unity. Evidently $CCl_3COOH/SnCl_4$ is more favorable for the polymerization of styrene derivatives than $BF_3 \cdot OEt_2$. Moreover, in the copolymerization of CEVE with nonpolar monomers such as α-methylstyrene and p-methylstyrene the CEVE content in the copolymer increases in nonpolar solvent (n-hexane). This solvent effect may be explained by assuming a decrease in cationation selectivity by increasing cation reactivity (Tobolsky and Boudreau, 1961) or, alternatively, selective solvation by polar monomer in nonpolar medium (Overberger and Kamath, 1959, 1963).

Imanishi et al. (1965) studied the effect of solvent on cationic copolymerization of isobutylene and styrene. Monomer reactivity ratios were greatly affected by solvent but only slightly by the initiating system. Isobutylene has been found to be more reactive than styrene in polymerizations in CH_2Cl_2 but not in n-hexane. In mixed solvents the growing cation exhibited a tendency to add the same kind of monomer, leading to r_1 and r_2 larger than unity. The reactivity of isobutylene tended to increase with solvent dielectric constant. Slightly more styrene was incorporated into the copolymer by $SnCl_4$ than by $TiCl_4$. Reactivity ratios obtained using various solvent mixtures are reported in Table 6.9.

The data for pure n-hexane and CH_2Cl_2 are explained by selective solvation of the growing ion pair by one of the two monomers. To explain the results obtained with mixed solvents it was proposed that the isobutyl and styryl cations are affected differently by the solvent. Specifically, the following matters were considered. The four elementary reactions in copolymerization were divided into eight steps (Taft et al.,

1955):

$$IB^{\oplus} + IB \underset{}{\overset{K_{1a}}{\rightleftharpoons}} \pi \text{ complex of } IB \underset{}{\overset{K_{1b}}{\rightleftharpoons}} IB^{\oplus} \qquad k_{11} \qquad (6.90)$$

$$IB^{\oplus} + St \underset{}{\overset{K_{2a}}{\rightleftharpoons}} \pi \text{ complex of } St \underset{}{\overset{K_{2b}}{\rightleftharpoons}} St^{\oplus} \qquad k_{12} \qquad (6.91)$$

$$St^{\oplus} + IB \underset{}{\overset{K_{3a}}{\rightleftharpoons}} \pi \text{ complex of } IB \underset{}{\overset{K_{3b}}{\rightleftharpoons}} IB^{\oplus} \qquad k_{21} \qquad (6.92)$$

$$St^{\oplus} + St \underset{}{\overset{K_{4a}}{\rightleftharpoons}} \pi \text{ complex of } St \underset{}{\overset{K_{4b}}{\rightleftharpoons}} St^{\oplus} \qquad k_{22} \qquad (6.93)$$

where $IB^{\oplus}$ and $St^{\oplus}$ are the growing isobutyl and styryl cations, and IB and St are isobutylene and styrene, respectively. Since $IB^{\oplus}$ is less stable and consequently more reactive than $St^{\oplus}$, $K_{1a} > K_{3a}$ and $K_{2a} > K_{4a}$. Since IB is more nucleophilic than St (greater rate of π complex formation) but forms a carbenium ion with more difficulty (Yonezawa et al., 1957; Taft, 1952), $K_{1a} > K_{2a}$, $K_{3a} > K_{4a}$, $K_{2b} > K_{1b}$, and $K_{4b} > K_{3b}$.

In nonpolar solvent (n-hexane) where the ion pairs are practically unsolvated, reactivity is mainly determined by carbenium ion stability. This means that K_b is determining, whence $k_{11} < k_{12}$ and $k_{21} < k_{22}$. This leads to $r_1 > 1$ and $r_2 < 1$, which is in agreement with the observations. Moreover, selective solvation of the ion pair by styrene is important in nonpolar solvent (Overberger and Kamath, 1959). This effect tends to render $r_1(<1)$ still smaller and $r_2(>1)$ still larger than unity.

A decrease in solvent dielectric constant decreases the initial rate of isobutylene polymerization more than that of styrene. Consequently stabilization by solvation is more important with the relatively unstable $IB^{\oplus}$ than

Table 6.9 □ **Influence of Solvent and Coinitiator on r_1 and r_2 for the Isobutylene (M_1)–Styrene (M_2) System at $-78°C$**

Solvent Composition		Dielectric Constant at $-78°$	Coinitiator	Reactivity Ratios	
n-C$_5$H$_{14}$ (%)	CH$_2$Cl$_2$ (%)			r_1	r_2
100	0	2.046	TiCl$_4$	0.37 ± 0.07	2.41 ± 0.12
75	25	3.95	TiCl$_4$	2.63 ± 0.52	5.50 ± 0.55
50	50	6.53	TiCl$_4$	3.25 ± 0.25	2.75 ± 0.25
25	75	10.65	TiCl$_4$	4.11 ± 0.19	1.70 ± 0.07
25	75	10.65	SnCl$_4$	3.75 ± 0.45	1.92 ± 0.41
0	100	14.89	TiCl$_4$	4.48 ± 0.28	1.08 ± 0.07

[a] After Imanishi et al. (1965).

with the stable $St^{\oplus}$; that is, the relatively unstable $IB^{\oplus}$ is formed less easily than $St^{\oplus}$ in nonpolar medium.

In a polar solvent, for example, CH_2Cl_2, both $IB^{\oplus}$ and $St^{\oplus}$ are solvated so that the stability difference of growing ion pairs is unimportant in determining reaction rate, and K_a becomes rate determining. This results in $k_{11} > k_{12}$ and $k_{21} > k_{22}$, so in methylene chloride r_1 and $r_2 < 1$. However, in practice $r_1 > 1$ and $r_2 \simeq 1$. This discrepancy may be explained by assuming selective solvation; that is, the growing ion pair is preferentially solvated by styrene, even in polar medium.

In a solvent of intermediate polarity, the reaction between the olefinic double bond and a reactive cation such as $IB^{\oplus}$ is mainly affected by K_{1a} to K_{4a} in equations 6.90 to 6.93. On the other hand, the reaction of a stable cation such as $St^{\oplus}$ is affected mainly by K_{1b} to K_{4b}. Hence $k_{11} > k_{12}$ and $k_{22} > k_{21}$, leading to r_1 and r_2 larger than unity.

Information relative to the transition state has been obtained by studying the effect of solvent on reactivity. Laval and Maréchal (1977) determined r_1 (fluorostyrene) and r_2 (styrene) with respect to the nature of the solvent. In the case of 4-fluorostyrene the plot of $\log r_2$ against $1/\epsilon$ (ϵ = dielectric constant) gives a straight line, as shown in Figure 6.6.

The rate constant k for two species A and B reacting in a medium of dielectric constant ϵ can be calculated by using a model of two conducting spheres of charges $Z_A e$ and $Z_B e$ of radii r_A and r_B brought to a distance $r^{\ddagger}$ in an isotropic continuous dielectric medium without distortion of the original charge:

$$\ln k = \ln k_0 - \frac{Z_A Z_B e^2}{(R/N) T \epsilon r^{\ddagger}} \tag{6.94}$$

where k_0 is the rate constant for the reaction that occurs in a medium of $\epsilon = \infty$ in the absence of electrostatic forces. If A and B are, respectively,

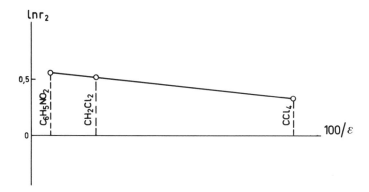

Figure 6.6 ☐ $\ln r_2$ versus $1/\epsilon$ plot for the copolymerization of 4-fluorostyrene (M_1) with styrene (M_2). $[TiCl_4] = 0.01$ M; $[M] = 0.2$ M/1; 0°C (Laval and Maréchal, 1977).

the styryl cation $St^{\oplus}$ and a substituted styrene M (and St is styrene),

$$r_2 = \frac{k_{St^{\oplus}St}}{k_{St^{\oplus}M}}$$

and

$$\ln r_2 = \ln(r_2)_0 - \frac{Z_{St^{\oplus}}(Z_{St} - Z_M)e^2}{(R/N)T\epsilon r^{\ddagger}} \tag{6.95}$$

The model predicts a linear $\ln r_2$ versus $1/\epsilon$ plot, which is in agreement with the experimental data shown in Figure 6.6.

Similar results have been obtained for the 4-methylstyrene–styrene system. From the slope $r^{\ddagger}$ can be calculated, that is, 3.0 Å for both compounds, which is the same order of magnitude as those found in the literature for many ionic reactions of this kind (Amis, 1966).

The Effect of the Nature of the Coinitiator and Counteranion

It is difficult to survey the influence of the nature of the coinitiator and solvent on reactivity since many authors have changed these experimental parameters simultaneously. The effect of the counteranion is very difficult to examine since its nature and the extent of its association with the carbocation, that is, system ionicity, are generally unknown.

Important discrepancies between results obtained by various authors have been observed. Thus according to Tobolsky and Boudreau (1961) the nature of the coinitiator determines reactivity ratios in the styrene/p-methoxystyrene copolymerization system whereas Overberger et al. (1954) found that r_1 and r_2 for styrene and p-chlorostyrene are independent of the nature of the coinitiator (TiCl$_4$ and SnCl$_4$) in nitrobenzene–carbon tetrachloride mixtures. These results may be rationalized by assuming complete dissociation and the presence of free carbenium ions in nitrobenzene (Kennedy, 1964).

The effects of solvent and counteranion on reactivity are inseparable. In nonpolar solvents the proximity of the counterion affects the behavior of the active center and the results obtained are meaningful only for the particular coinitiator employed. Theoretically a highly polar solvent should minimize the effect of counterion by giving rise to free ions; in fact, however, the active center interacts with polar solvents and it is difficult to define where solvation ends and complex formation starts (Kennedy, 1964). In electron-donating solvents, cation–solvent complexes may form and carbenium ion reactivity may decrease. If the solvent is more nucleophilic than one of the comonomers it could completely inhibit copolymerization. Reactivity ratios were found to be generally independent of coinitiator concentration (Borg and Maréchal, 1977; Landler, 1952; Overberger et al., 1954; Overberger and Kamath, 1959).

The influence of the nature of the solvent and coinitiator on $\Delta H^{\ddagger}$ and $\Delta S^{\ddagger}$

has recently been examined (Furukawa, 1974b; Maréchal, 1976). The existence of an isokinetic relationship that is, $\Delta H^{\ddagger} = \beta \Delta S^{\ddagger}$, where β is the isokinetic temperature, led Maréchal (1976) to suggest that the active species was the same in various copolymerizations. Figure 6.7 shows $\ln r_2$ versus $1/T$ plots for three fluoro- and two chlorostyrenes. The straight lines intersect at the same point $H_a(\ln r_2 = 0)$. The 3- and 4-methylstyrenes and the 2,4,6-trimethylstyrene intercept at another point (Me). The plot for 4-chlorostyrene seems to be an exception. The intercept between the straight lines $\ln r_2 = f(1/T)$ proves the existence of an isokinetic relation as shown in the $\Delta H^{\ddagger}$ versus $\Delta S^{\ddagger}$ plot in Figure 6.8.

The existence of an isokinetic temperature is possible only with identical reactions, that is, in the present case, with essentially identical active species. Furthermore, it proves that monomer reactivities estimated from $1/r_2$ relative to the same active species (styryl cation and the counteranion in the same state of association) are reasonably comparable. In contrast, an enthalpy–entropy linear relationship cannot be found by plotting $\ln r_1$ versus $1/T$; evidently in this case each reaction reflects a particular substituted styryl cation active center.

By plotting $\Delta H^{\ddagger}$ against $\Delta S^{\ddagger}$ for the copolymerization of styrenes with substituted styrenes, Furukawa et al. (1974b) obtained straight lines (Figure 6.9). However, the plots of $\Delta S^{\ddagger}$ against Lewis acid strength for the

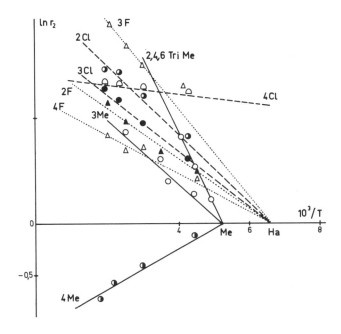

Figure 6.7 □ Variations of $\ln r_2$ (styrene: M_2) with $10^3/T$ ($T = °K$) (Maréchal, 1976) CH_2Cl_2; [TiCl₄] = 0.001 M; [M] = 0.1 M. F = fluoro; Cl = chloro; Me = methyl.

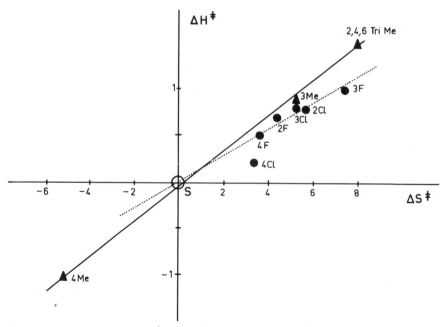

Figure 6.8 ☐ Variations of $\Delta H^\ddagger$ with $\Delta S^\ddagger$ (styrene reference) for various halogen and methyl styrenes (Maréchal, 1976). Symbols as in Figure 6.7.

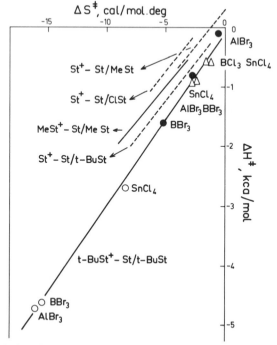

Figure 6.9 ☐ Relationship between activation entropy difference and activation enthalpy difference for the copolymerization of styrene and substituted styrenes. Solvent: ○, nitrobenzene; △, nitroethane; ● ethylene dichloride; St = styrene (from Furukawa et al., 1974b).

copolymerization of styrene with 4-methylstyrene and 4-*tert*-butylstyrene, respectively, exhibited complicated trends.

To explain these complicated results, particularly the greater sensitivity of selectivity toward strong acids than weak ones, pronounced in polar solvent, Furukawa et al. (1974b) suggested the existence of a dissociated and a nondissociated activated state. The growing cation would be solvated in both states. $\Delta H^{\ddagger}$ is small and selection is entropy-controlled; that is, the process is desolvation-controlled. The solvated growing species is desolvated in the transition state; it requires desolvation energy but acquires the entropy of liberated solvent. Solvation and desolvation of the growing species are different for the dissociated and the nondissociated states. In the dissociated state the solvent may be in the vicinity of the growing cation, but in the nondissociated state the charge is localized on the ion pair and the solvent is in the vicinity of the ion pair. Figure 6.10 serves to illustrate desolvation during propagation for a *p*-substituted styrene derivative.

In the nondissociated form interaction between the counteranion and substituent R of the monomer may be more important than between the monomer substituent R and the polymer substituent R'. On the other hand, in the dissociated form the R–R' interaction influences the removal of solvent located in the vicinity of substituent R'. The interactions increase as

$$R'\text{–}R > Cat\text{–}R > H\text{–}R \tag{6.96}$$

According to Furukawa et al. (1974b) $\Delta H^{\ddagger}$ and $\Delta S^{\ddagger}$ of desolvation consist of two terms, $a(1-x)$ arising from the nondissociated form and bx from the dissociated form:

$$\Delta H^{\ddagger} = h_s[a + (b-a)x] \tag{6.97}$$

$$\Delta S^{\ddagger} = s_s[a'(1-x) + b'x] \tag{6.98}$$

The degree of dissociation x parallels the acid strength of the Lewis acid; h_s depends on the solubility parameter and bulkiness of the solvent; a and b are related to the degree of conjugation of styryl and substituted styryl cations and the nature of the monomers.

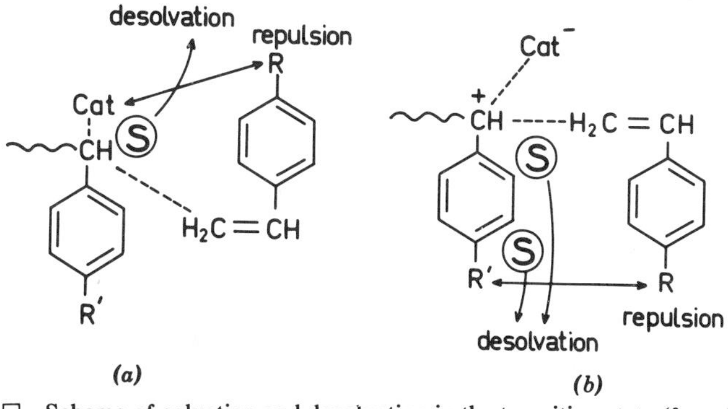

(a)   (b)

Figure 6.10 □ Scheme of solvation and desolvation in the transition state (from Furukawa et al., 1974b). Cat = counteranion. (a) Associated state; (b) dissociated state.

The Effective of Additives

The effect of additives is similar to that of counteranion, and it is often difficult to separate the various contributions. Panayotov et al. (1969, 1974), Toncheva et al. (1974), Heublein and Barth (1974), and Heublein (1975a, b) studied the effect of π electron acceptors on reactivity (see also Section 4.2). According to Panayotov (1969), in the absence of a nonelectron-donating solvent a complex is formed between the monomer M and the electron acceptor EA:

$$M + EA \rightarrow M \cdot EA \qquad (6.99)$$

On addition of initiating system $In \cdot X$ the $M \cdot EA$ complex is destroyed:

$$In \cdot X \rightleftarrows In^{\oplus} X^{\ominus} \xrightarrow{\ M \cdot EA\ } In - M^{\oplus} (X \cdot EA)^{\ominus} \qquad (6.100)$$

The influence of tetracyanoethylene TCNE on the reactivity ratios for the copolymerization of styrene (M_1) with α-methylstyrene and indene (M_2) is shown in Table 6.10.

In the case of the styrene–indene system TCNE increases r_1 without affecting r_2, suggesting a tendency toward block copolymer formation. In contrast, with the styrene–α-methylstyrene pair, TCNE addition decreases both parameters; that is, the trend is towards ideal copolymerization: $r_1 r_2 \approx 1$.

Let a_i, $k_{ij}^{\oplus}$, $k_{ij}^{\oplus}$ be the degree of dissociation of the ion pair, the rate constant for free ions, and the rate constant for ion pair, respectively. Thus

$$r = \frac{k_{11}}{k_{12}} = \frac{a_1 k_{11}^{\oplus} + (1 - a_1) k_{11}^{\oplus}}{a_1 k_{12}^{\oplus} + (1 - a_1) k_{12}^{\oplus}} \qquad (6.101)$$

$$r_2 = \frac{k_{22}}{k_{21}} = \frac{a_2 k_{22}^{\oplus} + (1 - a_2) k_{22}^{\oplus}}{a_2 k_{21}^{\oplus} + (1 - a_2) k_{21}^{\oplus}} \qquad (6.102)$$

Table 6.10 □ **Influence of TCNE on r_1 and r_2 for the Copolymerization of Styrene (M_1) with α-Methylstyrene and Indene (M^2)^{a,b}**

M_2	TCNE (mole/l)	r_1	r_2	$r_1 r_2$
α-Methylstyrene	0	0.65 ± 0.22	5.86 ± 0.32	3.8
	8×10^{-3}	0.28 ± 0.08	3.78 ± 0.43	1.06
Indene	0	0.73 ± 0.35	3.20 ± 0.72	2.33
	8×10^{-3}	1.57 ± 0.36	3.28 ± 0.52	5.15

^{a}[BF$_3 \cdot$OEt$_2$] = 0.008 M; [TCNE] = 0.008 M; [M] = 0.8 M; dichloro-ethane; 7°C.
^{b}After Panayotov et al. (1977).

In the presence of TCNE, a_1 and a_2 increase owing to complex formation (Panayotov et al., 1969, 1974; Toncheva et al., 1974) and the r values change as follows:

$$r_1 \text{ increases at } \frac{k_{11}^{\oplus}}{k_{12}^{\oplus}} > \frac{k_{11}^{\ominus}}{k_{12}^{\ominus}} > 1 \tag{6.103}$$

$$r_2 \text{ remains constant at } \frac{k_{22}^{\ominus}}{k_{21}^{\ominus}} = \frac{k_{22}^{\oplus}}{k_{21}^{\ominus}} \tag{6.104}$$

$$r_1 \text{ decreases at } \frac{k_{11}^{\oplus}}{k_{21}^{\ominus}} < \frac{k_{12}^{\oplus}}{k_{12}^{\ominus}} < 1 \tag{6.105}$$

$$r_2 \text{ decreases at } \frac{k_{22}^{\ominus}}{k_{21}^{\oplus}} < \frac{k_{22}^{\ominus}}{k_{21}^{\ominus}} < 1 \tag{6.106}$$

In the case of the styrene–indene pair expressions 6.103 and 6.104 are satisfied. In the case of the styrene–α-methylstyrene pair the experimental data correspond to relations 6.105 and 6.106. However, the four inequalities show that M_2 is more reactive than M_1 regardless of the type of the active species. Inequalities 6.104 and 106 show that the selectivity of the growing center derived of the more reactive monomer M_2 decreases more with an increase in free ions concentration than the selectivity of the styryl cation (M_1).

Higashimura and Yamamoto (1977; Yamamoto and Higashimura, 1974, 1976) studied the salt effect on the copolymerization of styrene derivatives with various vinyl ethers. Previously they had shown (see also Chapter 4) that two kinds of propagating species exist in the cationic polymerization of styrene and its derivatives initiated by iodine (Higashimura et al., 1976), acetyl perchlorate (Higashimura and Kishiro, 1974; Higashimura et al., 1975, 1976; Masuda and Higashimura, 1971b), perchloric acid (Higashimura and Kishiro, 1974; Pepper, 1974), and other strong protic acids. These conclusions are based on bimodal molecular weight distributions (Higashimura and Kishiro, 1974; Higashimura et al., 1976; Masuda and Higashimura, 1971b; Pepper, 1974) and the steric structure (Higashimura et al., 1975) of the polymers obtained. Since reliable rate constants are very difficult to obtain, the authors chose to study the effect of the nature of the propagating species on monomer reactivity ratios (Yamamoto and Higashimura, 1976). Their data show that in the copolymerization of chloroethyl vinyl ether (CEVE) with p-methoxystyrene (pMOS) initiated by iodine, the CEVE content in the copolymer increases either by decreasing solvent polarity or by addition of a common ion salt. The relative reactivities of monomers having the same structure, that is, styrene derivatives on the one hand and vinyl ethers on the other hand, are rather insensitive toward changing solvent polarity or addition of common ion salt. Decreasing solvent polarity or addition of common ion salt seems to have the same effect on the dissociation of the styryl and vinyl ether propagating ion pair.

This investigation (Yamamoto and Higashimura, 1976) is particularly important since it shows (probably for the first time) that in the same solvent propagating species possessing various degrees of dissociation may exhibit different relative reactivities, that is, selectivities, toward a pair of monomers.

Analogous results have been obtained (Higashimura and Yamamoto, 1977) with 4-methylstyrene, pMS, and CEVE using iodine or acetyl perchlorate initiators and tetra(n-butyl)ammonium perchlorate initiators and tetra(n-butyl)ammonium perchlorate added salt.

In copolymerization of styrene derivatives coinitiated by $SnCl_4$ or $TiCl_4$, the reactivity ratios are little affected by solvent polarity or the nature of coinitiator. According to Higashimura and Yamamoto (1977), counterions produced from metal halides are of relatively low nucleophilicity and interact only weakly with propagating carbocations. Owing to this weak interaction between the propagating carbocation and counterion, reaction conditions have only little effect on monomer reactivity ratios. In contrast, counterions from iodine or $AcClO_4$ are strongly interacting and affect reactivity ratios.

Quantum Study of the Effects of Solvent and Coinitiator on Reactivity

O'Driscoll and Yonezawa (1966b) have interpreted solvent effects in cationic copolymerizations by means of quantum chemistry. The calculations were carried out in terms of the Fontana–Kidder (1948) mechanism (see Chapter 4); that is, ion pair solvation affects the relative importance of complexation equilibrium K and bond formation k, which in turn affects the apparent propagation rate constant k_p:

$$P_n^{\oplus}A^{\ominus} \underset{K}{\overset{M}{\rightleftharpoons}} (P_nM)^{\oplus}A^{\ominus} \overset{k}{\longrightarrow} P_{n+1}^{\oplus}A^{\ominus}$$

$$k_p = \frac{kK}{K[M] + 1}$$

(6.107)

O'Driscoll et al. (1965a, b) showed that complex formation between a vinyl monomer and a cation is analogous to complexation between silver ion and a π system, which can be described by

$$\ln K = (\text{const})' - \frac{E^{\pi}}{RT}$$

(6.108)

where E^{π} can be calculated by a Hückel treatment (Fukui et al., 1961; Andrews and Keefer, 1950). The rate constant k is given by

$$\ln k = (\text{const})'' - \frac{E_{rs}}{RT}$$

(6.109)

where E_{rs} is the resonance stabilization energy of the transition state due to incipient bond formation between carbon atom s of the chain end and carbon r of the monomer. This quantity is calculable by quantum chemistry (see Section 6.8).

The influence of the counteranion on the molecular orbital of the chain end has been taken into account by using a corrected form for the coulomb integral of atom r:

$$\alpha_r = \alpha + \delta\beta \tag{6.110}$$

where α and β are the normal coulomb and resonance integrals and δ is a measure of the influence of the electric field

$$\delta = -0.2 \frac{Q_r e^2}{R_r} \tag{6.111}$$

Q_r is the charge on the rth atom (obtained from Hückel calculation) and R_r the distance between the center of the counterion and the rth atom. It is assumed (O'Driscoll et al., 1966a) that the center of the counterion is on a line perpendicular to the plane of the terminal carbon atom (C_1) and the phenyl ring of the chain end passing through C_1:

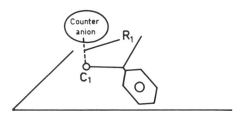

R_1 is assumed to vary continuously from 4 Å to infinity and is referred to relative values of δ; δ of unity corresponds to $R_1 = 4$ Å, whereas δ of zero corresponds to infinite R_1.

O'Driscoll et al. (1966a) described the activation energy $E^{\ddagger}$ for the apparent propagation rate constant k_p by a weighted sum of the calculated values E^{π} and E_{rs}:

$$E^{\ddagger} = (\text{const}) - (E_{rs} + aE^{\pi}) \tag{6.112}$$

Previous studies (O'Driscoll et al., 1965a, b) have shown that the electrostatic interaction may be ignored, seeking only relative rate constants for a common chain end; a would approach zero for large K and approach unity for small K. A reactivity ratio r_1 can be expressed as

$$\ln r_1 = \ln\left(\frac{k_{pii}}{k_{pij}}\right) = \frac{-\Delta S^{\ddagger}}{R} + \frac{\Delta E^{\ddagger}}{RT} \tag{6.113}$$

or

$$\ln r_1 = \frac{-\Delta S^{\ddagger}}{R} - \frac{\Delta E_{rs} + a\,\Delta E^{\pi}}{RT} \tag{6.114}$$

The (const)' term of equation 6.108 is considered to be the same for the two monomers adding to a given carbenium ion and consequently does not appear in equations 6.113 or 6.114.

It is important to know the effect of solvent and coinitiator on δ and a. Table 6.11 shows copolymer composition data obtained from equimolar styrene–p-methylstyrene charges with different solvents and coinitiators. Depending on their influence on k_p, solvents were classified as strong, medium, or weak (with a respective index of 3, 2, 1). Similarly, the order of coinitiator strength $A > B > C > D > E$ was classified to follow decreasing values of k_p. According to the results k_p decreases by following the data in Table 6.11 but copolymer composition does not follow a similar pattern. The styrene content of the copolymer decreases with increasing solvent strength (increasing k_p) at constant high coinitiator strength, or with decreasing coinitiator strength (decreasing k_p) at constant low solvent strength. In the same way, the effect of the coinitiator at high solvent strength is negligible and the effect of the solvent at any but high initiator strength is slight.

O'Driscoll et al. (1966a) analyze the results by the following chain of thought: ion pair separation probably increases with solvent strength so the relative δ should decrease with increasing solvent strength. In a weak solvent (low ϵ) increasing coinitiator strength should not affect δ since charge separation is thermodynamically unfavorable; in stronger solvent increasing the coinitiator strength should result in increased charge separation and in a corresponding decrease in δ. The quantity a which represents the extent of π complexation between monomer and chain end for determining k_p should decrease with increasing coinitiator strength in a given solvent regardless of solvent strength, since K would increase with increasing coinitiator strength. Concerning the influence of the solvent at a given coinitiator strength, at

Table 6.11 □ **Styrene Content (Wt %) of Copolymers Obtained by Using Equimolar Mixtures of Styrene/p-Methylstyrene^a**

| | | | Solvent | |
| | | Toluene | Ethylene Dichloride | Nitrobenzene |
Coinitiator		(1)	(2)	(3)
SbCl$_5$	A	46	25	28
AlX$_3$	B	34	34	28
TiCl$_4$, SnCl$_4$	C	28	27	27
BF$_3 \cdot$OEt$_2$				
SbCl$_5$				
CCl$_3$COOH	D		27	30
I$_2$	E		17	

^{a}After O'Driscoll et al. (1966a).

higher coinitiator strengths strong or medium strong solvents solvate dissociated chain ends so that the monomer must compete with the solvent for the chain end. Thus a must increase with increasing solvent strength. On the other hand, with weaker coinitiators an increase in solvent strength would increase K of poorly ionized chain ends and therefore a would decrease. The effects of polymerization conditions on δ and a are summarized in Table 6.12 and graphically presented in Figure 6.11. The absolute placement of solvent–coinitiator combinations is arbitrary, but their relative placement is dictated by the previous analysis and by the statements in Table 6.12.

The quantities E_{rs} and E^π have been calculated as a function of relative δ values for five monomer pairs: styrene–p-methylstyrene, styrene–p-chlorostyrene, styrene–p-methoxystyrene, p-chlorostyrene–p-methylstyrene, and p-methylstyrene–p-methoxystyrene. Figure 6.12 shows the

Table 6.12 □ Postulated Effects of Polymerization Conditions on δ and a[a]

Experimental Conditions			
Fixed	Increasing Strength	Change in δ	Change in a
Strong solvent	Initiator	Decrease	Decrease
Weak solvent	Initiator	None	Decrease
Strong initiator	Solvent	Decrease	Increase
Weak initiator	Solvent	Decrease	Decrease

[a] After O'Driscoll et al. (1966a).

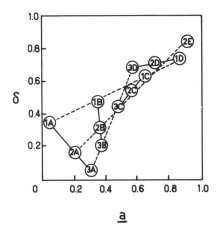

Figure 6.11 □ Variation of a and δ with changes in solvent and initiator (symbols as in Table 6.11). Solvent strength $1 < 2 < 3$; initiator strength $A < B < C < D < E$. [Reprinted from O'Driscoll et al. (1966a), p. 17, by courtesy of Marcel Dekker, Inc.]

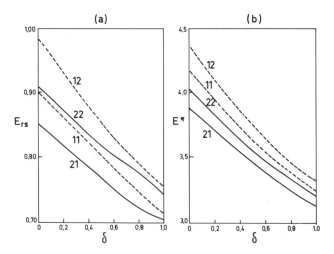

Figure 6.12 □ Effect of relative δ on (a) E_{rs} and (b) E^{π} for the copolymerization of styrene (M₁) and p-methylstyrene (M₂). The numbers 1, 2 refer to chain end 1 adding monomer 2. [Reprinted from O'Driscoll et al. (1966a), p. 17, by courtesy of Marcel Dekker, Inc.]

variation of E_{rs} and E^{π} for the styrene–p-styrene–p-methylstyrene pair. The pattern is representative of the other pairs as well.

O'Driscoll et al. (1966a) plotted the sum $\Delta E_{rs} + aE^{\pi}$ against a and δ and obtained interesting "maps." A typical map for the polystyryl carbenium ion reacting with styrene and p-methylstyrene is given in Figure 6.13.

By superimposing Figure 6.11 on Figure 6.13, one obtains the sum $\Delta E_{rs} + a\,\Delta E^{\pi}$ for various conditions of solvent and coinitiator. Figure 6.14 shows a plot of experimental reactivity ratios versus $\Delta E_{rs} + a\,\Delta E^{\pi}$ for the five monomer pairs mentioned above.

These results may be analyzed as follows: with the exception of the three points for r_2 in the styrene–4-methoxystyrene pair, the data in Figure 6.14 can be correlated by three lines having similar slopes. Line I correlates

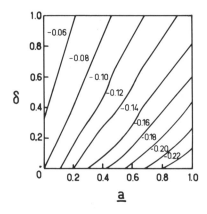

Figure 6.13 □ Contour map showing the variation of the sum ($\Delta E_{rs} + aE^{\pi}$) as a function of a and δ for the addition of styrene (or p-methylstyrene) to a polystyryl carbenium ion. The numbers give the value of the sum in $-(\Delta\beta)^2/\beta$ units (O'Driscoll et al., 1966a).

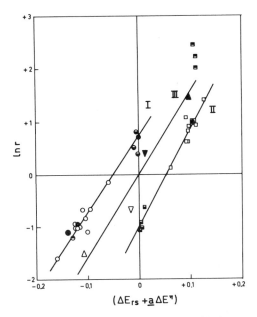

Figure 6.14 □   Correlation of experimental reactivity ratios with the sum $\Delta E_{rs} + aE^{\pi}$. Circles are for unsubstituted polystyryl carbenium ion reacting with styrene and p-methylstyrene (O), p-chlorostyrene (◐), or p-methoxystyrene (◑). Squares are for the given poly(substituted styrene) reacting with styrene and p-methylstyrene (□), p-chlorostyrene (▣), or p-methoxystyrene ▨. Triangles are for p-methylstyrene, p-methoxystyrene, and p-chlorostyrene. [Reprinted from O'Driscoll et al. (1966a), p. 17, by courtesy of Marcel Dekker, Inc.]

reactivity ratios for styrene, p-methylstyrene, p-chlorostyrene, and p-methoxystyrene reacting with the unsubstituted polystyryl carbenium ion; line II correlates the reactivity ratios for substituted styrenes and styrene reacting with a particular poly(substituted styryl) carbenium ion, and line III correlates reactivity ratios when both monomers are substituted. Lines I and II have equal intercepts on the abscissa but with opposite signs, whereas line III passes through the origin. From equation 6.114 the intercept is the difference between the entropy of activation $S^{\ddagger}$ for the addition of an unsubstituted styrene and p-substituted styrene to a chain end. The difference favors the unsubstituted monomer. From the slope of the lines $\Delta \beta / \beta$ is obtained and assuming $\beta = 18$ kcal/mole, $\Delta \beta = 0.70$, 0.78, and 0.73 for lines I, II, and III, respectively. The corresponding intercepts give $\Delta S^{\ddagger} = 1.5$, 1.9, and 0.0 cal/mole · deg, which possibly reflect the rearrangement of the solvation shell in the transition state. The monomer with bulky substituent requires a lesser degree of ordering of the solvent surrounding the ion pair than unsubstituted monomer; for two substituted monomers the difference is negligible.

These calculations work well with nine of the ten reactivity ratios considered, but fail with the highly electron-donating p-methoxy-substituted polystyryl carbenium ion. The same exception arises by using the

Hammett equation. The correlation with reactivity ratios is also poor for the isobutylene–styrene system.

The Effect of Electric Field on Reactivity

Sakurada et al. (1968) have investigated for many years the effect of electric field on cationic polymerization and copolymerization; however, recent findings (Ise, 1979) relative to the influence of impurities on the phenomena may necessitate a reevaluation of some of their theories.

6.10 □ INFLUENCE OF STRUCTURAL FACTORS ON REACTIVITY

Influence of Electronic Factors

Although electronic factors greatly affect both monomer and carbocation reactivity, the separation of these effects is enormously complex and a general treatment of quantitative validity has not yet been developed. The experimental and theoretical methods described in this chapter provide insight into the complexity of this matter.

Except for a very few instances, for example, monosubstituted styrenes, qualitative inductive or resonance concepts are insufficient to predict reactivity accurately. This is particularly true if the species under consideration contains different kinds of substituents.

Reliable reactivity information can be obtained only by suitable experiments. The point of departure is always to determine reactivity ratios with respect to a reference, usually styrene. The correct choice of the reference compound, however, is not a trivial matter: the reactivity of the reference should not be too far from that of the target monomer(s). Thus Coudane and Maréchal (to be published) investigated the reactivity of various vinylanthracenes with styrene as reference and observed very large reactivity differences; that is, styrene = 1, whereas 9-vinylanthracene = 100 to 130. In view of these huge reactivity differences the data are valid merely as indicators of orders of magnitudes. In cases such as this more meaningful data can be obtained by changing the reference to more reactive monomer that is, α-methylstyrene or indene, even though such a change may render comparison with many available styrene-based studies more cumbersome.

Further, valid relative reactivity data can be obtained only with data generated under a variety of conditions. Reactivity ratios obtained under a specific set of conditions (counteranion, solvent, temperature) reflect only reactivities prevailing under that specific set of conditions and should not be viewed to possess general validity.

Reliable reactivity information can be obtained if $\Delta H^{\ddagger}$ and $\Delta S^{\ddagger}$ data exist, and correlation between experiment and theory should be derived only on the basis of this information.

Theoretical methods together with their shortcomings were discussed in Section 6.8, steric effects are examined below.

Hammett's Postulate and Reactivity

Walling et al. (1948) found a linear correlation between the reactivity ratios and Hammett's σ values in free radical styrene/substituted styrenes copolymerizations, and Overberger et al. (1952) extended the concept to cationic polymerizations. Their results are reported in Figure 6.15, which is obtained by plotting $1/r_1$ against σ for α-methylstyrene (M_1) and various substituted styrenes (M_2).

Except for methoxy- and p-dimethylaminostyrene, relative monomer reactivities follow Hammett's σ series. Overberger et al. (1952) compared these results with those reported by Swain and Langsdorf (1951) for displacement reactions. According to the latter authors, Hammett's ρ is not constant in displacement reactions but changes with σ. Thus the $\log k/k_0$ versus σ plot for the reaction of p-substituted benzyl chlorides with trimethylamine is not linear. Overberger et al. (1952) interpreted these facts in terms of a termolecular addition of the ion pair $R^{\oplus}X^{\ominus}$ to the olefin, similar to that suggested for concerted displacement:

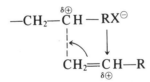

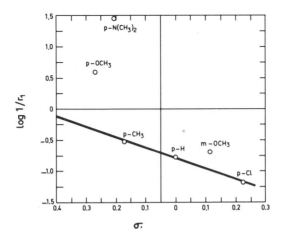

Figure 6.15 ☐ Plot of log relative reactivity $1/r_1$ of α-methylstyryl cation vs. Hammett σ value for various substituted styrenes. [Reprinted with permission from Overberger et al., *J. Am. Chem. Soc.* **74**, 4848 (1952), Copyright 1952: American Chemical Society.

Groups that stabilize partial positive charges on the α carbon in the transition state, that is, p-methoxy and p-dimethylamino, would increase ρ.

Hammett's equation has been used to elucidate the nature of the active species. Masuda et al. (1974a) plotted log $1/r_1$ versus Brown's σ^+ constant for the copolymerization of styrene (M_1) and various 1-phenylbutadienes (M_2) and obtained -1.20 for the reaction constant ρ^+. According to these workers the propagating end attacks C_4 (not C_2 or C_3) of 1-phenylbutadiene and the propagation mechanism is analogous to that of styrene. This conclusion accounts for the main 3,4 structure.

The same method has been applied (Masuda et al., 1974b) to 2-phenylbutadiene (M_2). The plot of log $1/r$ versus σ^+ gave $\rho^+ = -2.04$, which is virtually identical to -2.30, that is, for the cationic polymerization of styrene, and greater than -1.20, the value for 1-phenylbutadiene. The similarity of ρ^+ for styrene and 2-phenylbutadiene indicates that the propagating cations are similar in these systems. The difference between ρ^+ for 1- and 2-phenylbutadiene may be due to the differences in the transition states: 2-phenylbutadiene produces a benzylic, 1-phenylbutadiene a phenylallylic, carbocation.

Hasegawa and Asami (1976) have plotted log($1/r_1$) against Brown's σ^+ for the copolymerization of various substituted phenylbutadienes (M_2) with styrene (M_1) ($BF_3 \cdot OEt_2$–CH_2Cl_2 system) and obtained the lines shown in Figure 6.16. According to these data ρ^+ for 1- and 2-phenylbutadienes toward styryl cation are -1.5 and -1.67, respectively. Comparison of these values with those obtained for chlorination, bromination, and addition of sulfenyl chloride, together with other pertinent information (Okuyama et al., 1970), led Hasegawa and Asami (1976) to conclude that similarly to styrene, the transition state for propagation of 1- and 2-phenylbutadiene is a cyclic one.

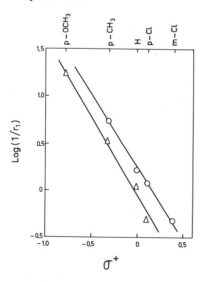

Figure 6.16 □ Plots of log ($1/r_1$) against σ^+ for the copolymerization of styrene with 1- and 2-phenylbutadienes: ○, ring-substituted 1-PB; △ ring-substituted 2-PB (Hasegawa and Asami, 1976).

Influence of Steric Factors

Steric factors greatly affect reactivity of disubstituted olefins and vinyl ethers (Okuyama et al., 1967, 1968, 1969a, b). In general, cis alkenyl ethers are several times more reactive than the corresponding trans isomers. The effect of geometric isomerism on reactivity is greater than that of the variation in alkyl group attached to the β-carbon atom. The difference between the reactivities of cis and trans isomers seems to be larger in the presence of bulky β-substituents. The relatively high reactivity of cis-β-tert-butyl compound may probably be due to relief of steric strain on forming the carbenium ion.

According to Okuyama et al. (1967), the energy difference between the transition states for a given pair of geometric isomers is small and the reactivity difference between the isomers is due to the difference in isomer stabilities. If the overall heat content of carbenium ion intermediates formed from the respective isomers is assumed to be the same, the trans isomer is more stable than the cis isomer by 1 to 3 kcal/mole.

The results of hydrolysis and copolymerization studies are in agreement with regard to reactivity; thus *cis*-propenyl isobutyl ether appears to be about four times more reactive than the *trans*-PIBE isomer (BF₃·OEt₂; −78°C). Neither isomer undergoes geometric isomerization during polymerization, and side reactions are absent during copolymerization.

Copolymerization studies of PIBE with vinyl isobutyl ether (VIBE) also indicate that *cis*-PIBE is more reactive than the trans isomer. Regardless of the structure of the attacking chain end the relative reactivity of these ethers is *cis*-PIBE > VIBE ≃ *trans*-PIBE (Okuyama et al., 1968). Isomerization equilibria also showed *cis*-PIBE to be thermochemically less stable than *trans*-PIBE, and the reactivity differences reflected relative stabilities in the ground state. Different polymerizabilities seem to be pronounced in less polar solvents. The effect of temperature on relative monomer reactivity is very small except when a growing end derived from a cis monomer is involved.

The behavior of *cis*-PIBE is remarkable. For example, copolymerization of VIBE (M_1) with *cis*-PIBE (M_2) gives $r_1 r_2$ less than unity, whereas VIBE (M_1)–*trans*-PIBE (M_2) copolymerizations give $r_1 r_2$ close to unity. Temperature affects r_2, but r_1 much less, in *cis*-PIBE–VIBE copolymerizations. The product $r_1 r_2$ is less than unity at 0°C, but close to unity at −78°C in the copolymerization of cis and trans isomers. Finally, a remarkable alternating tendency is observed in the copolymerization of the cis ether with vinyl ether (Okuyama et al., 1968). From these observations Okuyama et al. (1968) infer that the structures of the chain ends derived from cis and trans monomers are different and that the chain end derived from the cis monomer exerts a special steric effect.

Similar studies involving geometric isomers have been carried out with various other alkenyl alkyl ethers (Okuyama et al., 1968). In agreement with the results with PIBE's, none of the cis or trans ethers equilibrate during polymerization. In all instances cis alkenyl alkyl ethers were found to be more reactive than trans isomers except in copolymerization of 4-methyl-1-pentenyl ether isomers, where both r_{cis} and r_{trans} were <1 and the cis isomer appeared to be less reactive than the trans. However, copolymerizations of *cis*- and *trans*-4-methyl-1-pentenyl ether with VIBE show that the cis isomer is more reactive than the trans isomer towards the VIBE end. The cis monomer was found to be reluctant to homopolymerize ($r_1 \ll 1$ and $k_{cis,\,cis}$ very small).

Substituents exert a complex effect on relative reactivities of cis and trans monomers in cis–trans copolymerizations. Relative cis–trans reactivities judged from $1/r_1$ values (reactivities relative to VIBE end) seem to reflect the bulkiness of β-alkyl groups.

It is difficult to separate electronic and steric effects in β-substituted ethyl vinyl ether ($C_2H_5OCH=CHR$) polymerizations. With $R=CH_3$ or C_2H_5 monomer reactivities increase 4 and 1.5 times, respectively, for both geometric isomers. If R is a bulky primary alkyl group, the cis ether is more reactive than the trans ether but the latter is less reactive than the parent ethyl vinyl ether. Monomer reactivity is reduced remarkably with secondary and tertiary β-alkyl groups. In sum, β-methyl and β-ethyl groups seem to exert little steric effect on the polymerization of ethyl vinyl ether.

Dimethyl substitution on the β-carbon reduces the reactivity considerably. Isobutenyl ethyl ether is reluctant to homopolymerize even though its reactivity toward the VIBE end is moderately lower than that of vinyl ethyl ether. This is understandable in terms of steric hindrance to cationation of isobutenyl ethyl ether.

Okuyama et al. (1968) also noted interesting features relative to the reactivity of $R_1CH=CHOR_2$. When R_1 and R_2 are relatively small $r_{cis}r_{trans} \approx 1$, but this product becomes noticeably less than unity with branched alkyl substituents. Evidently with bulky R_1 and R_2 groups the chain end carbenium ions formed from cis and trans ethers must be different, and this difference influences selectivity.

Several hypotheses have been proposed to explain steric effects on the mechanism of double bond opening (Okuyama et al., 1968). When *cis*- and *trans*-propenyl isobutyl ether (PIBE)(M_2) and vinyl isobutyl ether (M_1) are copolymerized, $r_1r_2 < 1$ for cis and $r_1r_2 \approx 1$ for trans isomers. This indicates that the structures of the cis and trans PIBE chain ends are different and suggests restricted rotation of the termini. The alternating tendency in copolymerization of *cis*-PIBE with VIBE may be due to special geometry during propagation. Steric effects of growing chains derived from cis monomers are apparently determined by the bulk of β-substituents.

Steric interaction of the terminal β-substituent with coinitiator and monomer may be larger for the cis than for the trans cation during cis opening, since the coinitiator may interact with the alkoxy oxygen at the polymer end. During trans opening the terminal β-substituent seems to be too far from the coinitiator to affect stereochemistry. Thus the presence of steric effects suggests cis opening during propagation. These results concerning the relative behavior of cis and trans isomers and the steric influence of β-substituents have been confirmed by Higashimura et al. (1968a, b).

Both acetal addition and cationic polymerization have been studied with propenyl and isobutenyl ethyl ethers (PEE and IBEE) (Higashimura et al., 1968b). Relative reactivities were *cis*-PEE > *trans*-PEE ≫ IBEE. Again, β-methyl substitution exerts little steric effect toward addition of vinyl ethers to alkoxy carbenium ion; in contrast, dimethyl substitution strongly retards interaction with β-monomethyl and β,β-dimethyl-substituted carbenium ions.

Steric effects due to β-substituents have also been examined with α, β disubstituted olefins (Overberger et al., 1958; Brackman and Plesch, 1958; Mizote et al., 1966c). Overberger et al. (1958) showed that β-alkyl groups exhibit marked steric hindrance in copolymerization of β-alkylstyrenes and p-chlorostyrene $(C_6H_5NO_2-CCl_4(1:1)/SnCl_4/0°C)$. Mizote et al. (1965) determined reactivity ratios of p-substituted styrenes (M_1) and p-substituted β-methylstyrene (M_2) and found r_2 to be much smaller than unity and very close to each other. This indicates that the contribution of the steric effect of the β-methyl group on monomer reactivity can be discussed as an independent parameter.

According to Branch and Calvin (1941), the influence of monomer structure on the activation free energy of propagation can be expressed as a sum of polar, resonance, and steric factors:

$$-\Delta F^{\ddagger} = P + R + S = RT \log\left(\frac{k}{k_0}\right) \tag{6.115}$$

with

$$\log\left(\frac{k}{k_0}\right) = F_P + F_R + F_S \tag{6.116}$$

where F_P, F_R, and F_S are, respectively, the polar, resonance, and steric factors. Applying this relation to copolymerization gives

$$\log\left(\frac{1}{r_1}\right) = \log\left(\frac{k_{12}}{k_{11}}\right) = F_P + F_R + F_S \tag{6.117}$$

$$\log r_2 = \log\left(\frac{k_{22}}{k_{21}}\right) = F_P' + F_R' + F_S' \tag{6.118}$$

In copolymerization of styrene and methylstyrene both propagating cations $M_1^{\oplus}$ and $M_2^{\oplus}$ have the same electronic characteristics:

$$F_P + F_R = F_P' + F_R' \tag{6.119}$$

Since steric repulsion $H_\beta \cdots CH_{3\beta}$ (H_β in the cation approaching $CH_{3\beta}$ in the monomer, and H_β in the monomer approaching $CH_{3\beta}$ in the cation) is negligible compared to steric repulsion between β-methyl groups ($CH_{3\beta} \cdots CH_{3\beta}$),

$$F'_S \gg F_S \tag{6.120}$$

Steric repulsion between β-methyl groups is estimated by

$$F'_S = \log r_2 - \log\left(\frac{1}{r_1}\right) \tag{6.121}$$

F'_S is -1.0 to -0.13 for para substituents H, CH_3, and OCH_3. Thus the rate constant of homopolymerization decreases by a factor of $1/10$ to $1/20$ relative to the corresponding styrenes, owing to steric repulsion by the β-methyl substituent.

Although a steric effect of $F'_S = -1.3$ due to the β methyl substituent may be expected in homopropagation of anethole (4-methoxypropenyl-benzene) it does not appear in copolymerization of styrene and anethole. Conceivably, polar and resonance factors are more important than steric factors in this process.

trans-β-Methylstyrene is slightly more reactive than the cis isomer (Overberger et al., 1958); however, in almost all other β-methyl-substituted styrene derivatives, the cis isomer is more reactive than the trans isomer (Mizote et al., 1968). Heublein and Schütz (1976b) studied the copolymerization of *trans*- and *cis*-β-methylstyrene with styrene and found the trans isomer to be more reactive; however, copolymerization of the cis/trans isomer pair indicated the cis isomer to be more reactive. This difference was attributed to steric difference in propagation.

6.11 □ AN APPLICATION OF REACTIVITY ANALYSIS: AZEOTROPIC COPOLYMERIZATION

In azeotropic copolymerization the charge composition and copolymer composition are identical. Azeotropic copolymer systems are rare but of great scientific and practical interest: under azeotropic conditions copolymer composition does not change with conversion, and laborious compensation of changing monomer concentration during copolymerization becomes unnecessary (Ham, 1964). The first ionic (cationic or anionic) azeotropic copolymerization, that is, the cationic copolymerization of isobutylene/β-pinene, has recently been discovered (Kennedy and Chou, 1975b) and its fundamentals analyzed (Kennedy and Chou, 1976a).

The isobutylene/β-pinene pair represents an unusual copolymerization system. Inspection of the structures of these monomers and propagation

Table 6.13 □ The Effect of Temperature on the Reactivity Ratios of Isobutylene (r_i) and β-pinene (r_β)

Temperature (°C)	r_i	r_β	$r_i r_\beta$
− 50	0.27 ± 0.06	3.0 ± 1.5	0.81 ± 1.5
− 78	0.52	1.9	0.99
−100	0.77 ± 0.16	1.5 ± 0.5	1.15 ± 0.7
−110	1.0	0.95	0.95
−120	1.0 ± 0.1	1.0 ± 0.1	1.0 ± 0.1
−130	1.0 ± 0.1	1.0 ± 0.1	1.0 ± 0.1

[a]Kennedy and Chou (1976a).

carbocations reveals profound similarities between these species:

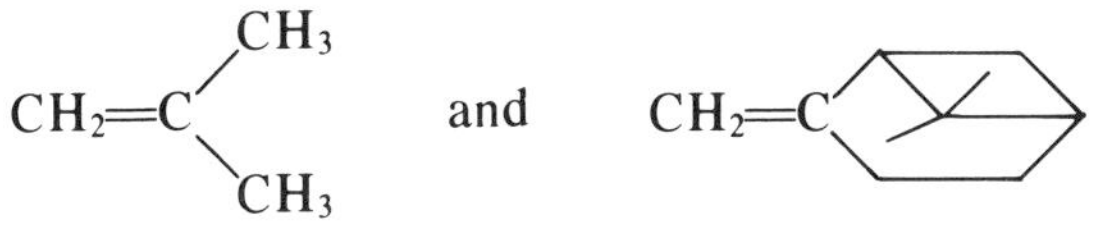

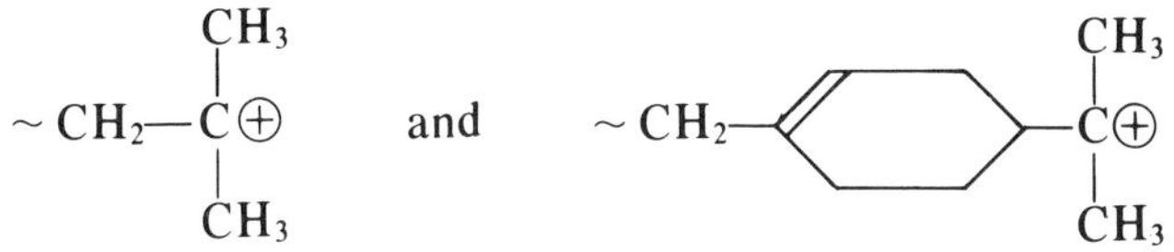

In view of the structural analogy it is not very surprising that these monomers readily copolymerize (e.g., "H_2O"/EtAlCl$_2$/EtCl/ − 50 to −130°C).

In an effort to increase copolymer molecular weights (see Section 6.12) experiments were carried out at progressively lower temperatures (from −78 to −130°C) and the reactivity ratios were determined. Table 6.13 shows representative data.

Evidently the reactivity of β-pinene is much higher than that of isobutylene at higher temperatures; however, the difference in reactivities decreases with decreasing temperatures until at ∼ −110°C (and below) it completely disappears; that is, azeotropic conditions are reached, $r_1 = r_2 = 1$. It is very difficult to explain these findings, and for speculations on possible reasons the original paper should be consulted (Kennedy and Chou, 1976a).

6.12 □ MOLECULAR WEIGHT DEPRESSION IN COPOLYMERIZATION

Analysis of a large number of (virtually all available) reliable copolymerization data (Table 6.14) indicates without exception substantial mole-

Table 6.14 □ Cationic Copolymerization Systems in which Molecular Weight Depression has been Observed

System	References
"H₂O"/EtAlCl₂/isobutylene/β-pinene/EtCl/−50 to −130°C[a]	Kennedy and Chou (1976a)
"H₂O"/TiCl₄/siobutylene/styrene/MeCl₂, Tol/−78°C	Okamura et al. (1967, 1961)
"H₂O"/TiCl₄/isobutylene/α-Methylstyrene/MeCl₂, Tol/−78°C	Okamura et al. (1967, 1971)
"H₂O"/BF₃/isobutylene/isopropene/ethylene, EtCl/−100°C	Anosov and Korotkov (1960)
"H₂O"/BF₃/isobutylene/butadiene/ethylene, EtCl/−100°C	Anosov and Korotkov (1960)
"H₂O"/BF₃/isobutylene/2,3-dimethylbutadiene/ethylene, EtCl/−100°C	Anosov and Korotkov (1960)
"H₂O"/TiCl₄/isobutylene/2-methyl-2-butene/MeCl₂, hexane/−78°C	Anosov and Korotkov (1960)
"H₂O"/BF₃·OEt₂/isobutylene/cyclopentadiene/MeCl₂, Tol/−78°C	Imanishi et al. (1966)
"H₂O"/BF₃·OEt₂/isobutylene/cyclohexadiene/MeCl₂, Tol/−78°C	Imanishi et al. (1966)
"H₂O"/AlCl₃, BF₃·OEt₂/isobutylene/acenaphthylene/EtBr, CS₂, CHCl₃/−60 to −105°C	Jones (1951)
TCA/TiCl₄/styrene/α-methylstyrene/MeCl₂, Tol/−30 to −78°C	Okamura et al. (1961, 1967)
"H₂O"/BF₃·OEt₂/styrene/acenaphthylene/Bz, Tol/−30 to −78°C	Saotome and Imoto (1958)
"H₂O"/TiCl₄/styrene/acenaphthylene/MeCl₂/−72°C	Belliard and Maréchal (1972)
"H₂O"/TiCl₄/styrene/1-methylacenaphthylene/MeCl₂/−72°C	Belliard and Maréchal (1972)
"H₂O"/TiCl₄/styrene/3-methylacenaphthylene/MeCl₂/−78°C	Belliard and Maréchal (1972)
"H₂O"/TiCl₄/styrene/2-methyl-2-butene/MeCl₂, hexane/−78°C	Imanishi et al. (1970)
"H₂O"/BF₃·OEt₂/styrene/chloroprene/cyclohexane/−78°C	Foster (1950)
"H₂O"/TiCl₄/styrene/1-methylindene/MeCl₂/−72°C	Maréchal et al. (1964)
"H₂O"/TiCl₄/styrene/5-methylindene/MeCl₂/−72°C	Tortai and Maréchal (1971)
"H₂O"/TiCl₄/styrene/1-vinylnaphthalene/MeCl₂/−72°C	Blin et al. (1978)
"H₂O"/TiCl₄/styrene/2-vinylnaphthalene/MeCl₂/−72°C	Blin et al. (1978)
"H₂O"/TiCl₄/styrene/1-chloro-1-vinylnaphthalene/MeCl₂/−72°C	Cho et al. (1980)
"H₂O"/TiCl₄/styrene/1-fluoro-1-vinylnaphthalene/MeCl₂/−72°C	Cho et al. (1980)
"H₂O"/TiCl₄/styrene/1-bromo-1-vinylnaphthalene/MeCl₂/−72°C	Cho et al. (1980)

"H$_2$O"/TiCl$_4$/styrene/1-vinylphenanthrene/MeCl$_2$/−72°C	Cho et al. (1980)
"H$_2$O"/TiCl$_4$/indene/benzofuran/MeCl$_2$/−72°C	Sigwalt (1961)
"H$_2$O"/TiCl$_4$/indene/5,7-dimethylindene/MeCl$_2$/−30°C	Anton et al. (1970)
"H$_2$O"/TiCl$_4$/indene/4,6,7-trimethylindene/MeCl$_2$/−72°C	Anton et al. (1970)
"H$_2$O"/TiCl$_4$/indene/7-methylindene/MeCl$_2$/−72°C	Caillaud et al. (1970)
"H$_2$O"/TiCl$_4$/indene/4,6-dimethylindene/MeCl$_2$/−72°C	Maréchal and Evrard (1969a)
"H$_2$O"/TiCl$_4$/indene/5,6-dimethylindene/MeCl$_2$/−72°C	Maréchal and Evrard (1969a)
"H$_2$O"/TiCl$_4$/indene/4,7-dimethylindene/MeCl$_2$/−72°C[a]	Maréchal and Evrard (1969a)
"H$_2$O"/TiCl$_4$/indene/4,5,6,7-tetramethylindene/MeCl$_2$/−72°C	Maréchal and Evrard (1969a)
"H$_2$O"/TiCl$_4$/indene/1-methylindene/MeCl$_2$/−72°C	Maréchal et al. (1964)
"H$_2$O"/TiCl$_4$/indene/2-methylindene/MeCl$_2$/−72°C	Maréchal et al. (1964)
"H$_2$O"/TiCl$_4$/indene/3-methylindene/MeCl$_2$/−72°C	Maréchal et al. (1964)
"H$_2$O"/TiCl$_4$/various methylindenes/MeCl$_2$/−72°C	Maréchal (1969b)
"H$_2$O"/TiCl$_4$/indene/4,7-dimethylbenzofuran/MeCl$_2$/−78°C	Zaffran and Maréchal (1970)
"H$_2$O"/TiCl$_4$/4,6-dimethylindene/4,7-dimethylindene/MeCl$_2$/−72°C	Maréchal et al. (1969b)
"H$_2$O"/TiCl$_4$/4,7-dimethylindene/4,5,6,7-tetramethylindene/MeCl$_2$/−72°C	Maréchal et al. (1969b)
"H$_2$O"/BF$_3$·OEt$_2$/cyclopentadiene/2-methylstyrene/−78°C	Imanishi et al. (1968)
"H$_2$O"/BF$_3$·OEt$_2$/cyclopentadiene/2-chloroethyl vinyl ether/MeCl$_2$, EtNO$_2$/−78°C	Kohjiya et al. (1972)
"H$_2$O"/BF$_3$·OEt$_2$/α-methylstyrene/2-chloroethyl vinyl ether/MeCl$_2$/−78°C	Masuda et al. (1970)
"H$_2$O"/BF$_3$·OEt$_2$/p-methoxystyrene/2-chloroethyl vinyl ether/MeCl$_2$/−78°C	Masuda et al. (1970)
"H$_2$O"/TiCl$_4$/benzofuran/4,7-dimethylbenzofuran/MeCl$_2$/−78°C[a]	Zaffran and Maréchal (1970)

[a] A yield depression is observed. Abbreviations: EtCl = ethyl chloride, MeCl$_2$ = methylene chloride, Tol = toluene, Bz = benzene.

cular weight depression of the copolymer compared to homopolymerizations of the respective monomers. The first analysis of this phenomenon is due to Higashimura (Imanishi et al., 1963b, 1965; Okamura et al., 1961, 1967) and further developments to Kennedy and Chou (1976a).

According to these investigators the molecular weight of homo- or copolymers is

$$DP = \frac{\text{rate of propagation}}{\text{rate of (transfer + terminaison)}} \tag{6.122}$$

In copolymerization,

$$DP = \frac{k_{p_{11}}[M_1^\oplus][M_1] + k_{p_{12}}[M_1^\oplus][M_2] + k_{p_{22}}[M_2^\oplus][M_2] + k_{p_{21}}[M_2^\oplus][M_1]}{k_{t_1}[M_1^\oplus] + k_{tr_{11}}[M_1^\oplus][M_1] + k_{tr_{12}}[M_1^\oplus][M_2] + {} \atop {} + k_{tr_{22}}[M_2^\oplus][M_2] + k_{tr_{21}}[M_2^\oplus][M_2] + k_{tr_{21}}[M_2^\oplus][M_1] + k_{t_2}[M_2^\oplus]}$$

$$\tag{6.123}$$

where k_p, k_{tr}, and k_t are rate constants for propagation, chain transfer to monomer, and termination, respectively, and the subscripts follow the usual convention. Since DP is in most carbocationic polymerization systems largely determined by chain transfer to monomer, the termination terms can be neglected. Thus

$$DP_{12} = \frac{r_1[M_1]^2 + 2[M_1][M_2] + r_2[M_2]^2}{(1/DP_1)r_1[M_1]^2 + A[M_1][M_2] + (1/DP_2)r_2[M_2]^2} \tag{6.124}$$

where DP_{12}, DP_1, and DP_2 are, respectively, DP of the copolymer and homopolymers 1 and 2; $A = k_{tr12}/k_{p12} + k_{tr21}/k_{p21}$ determines the extent of molecular weight depression of the copolymer relative to the homopolymers.

With the "H_2O"/$TiCl_4$/isobutylene-styrene/n-hexane–CH_2Cl_2 (1:1)/ $-78°C$ system $A = 6.9 \times 10^{-3}$ (Imanishi et al., 1965). In this example $A \gg 1/DP_1$ and/or $1/DP_2$ because the degrees of polymerization of both homopolymers are similar ($DP_{PIB} = 2000$ and $DP_{PSt} = 5500$). If $DP_1 \gg DP_2$, A approaches $1/DP_2$ but it can never approach $1/DP_1$. One of the terms in A must predominate whereas the other is insignificant and may be neglected. For the "H_2O"/$EtAlCl_2$/isobutylene–β-pinene/$EtCl$/$-50°$ to $-130°C$ system, $A = 6.7 \times 10^{-3}$ (Kennedy and Chou, 1976a).

According to the latter authors, on the introduction of a copolymerizable monomer into a carbocationic homopolymerization, the DP of the copolymer *must* decrease in reference to the DP's of the respective homopolymers provided both homopolymerizations produce high polymers and the copolymer contains an appreciable amount (e.g., 5%) of each monomer.

This assertion is derived by the following reasoning: in a copolymerization there must exist two propagating chain ends of different energies. According to Hammond's postulate for exothermic reactions (Hammond, 1955), the fastest event in a copolymerization is the one that involves the most reactive chain end $\sim M_1^{\oplus}$ and most stable monomer M_2, and produces either the most stable chain end $\sim M_2^{\oplus}$ (by cross-propagation) or most stable cation $M_2^{\oplus}$ plus macroolefin $\sim M_1^{=}$ (by cross-transfer):

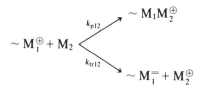

$$\sim M_1^{\oplus} + M_2 \Big\langle \begin{array}{c} \xrightarrow{k_{p12}} \sim M_1 M_2^{\oplus} \\ \xrightarrow{k_{tr12}} \sim M_1^{=} + M_2^{\oplus} \end{array}$$

Thus the rate constants of the fastest propagation and fastest chain transfer are k_{p12} and k_{tr12}, respectively. The molecular weight of the copolymer is determined by competition between these two reactions.

In carbocationic homopolymerizations and most probably in copolymerizations as well, $\Delta H_{tr}^{\ddagger} > \Delta H_p^{\ddagger}$, so that by lowering the energy content of the monomer from M_1 in a homopolymerization to M_2 in copolymerization, the activation energy difference of cross-transfer $\Delta E_{tr12}^{\ddagger}$ decreases more than that of cross-propagation $\Delta E_{p12}^{\ddagger}$, that is, $-\Delta H_{tr}^{\ddagger} > -H_p^{\ddagger}$. Thus according to Hammond's postulate the rate of cross-transfer *must* increase more than that of propagation, which *must* result in molecular weight depression in copolymer systems. In contrast, increasing the energy content of the monomer from M_2 in a homopolymerization to M_1 in copolymerization, the

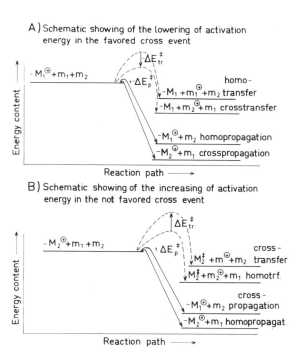

Figure 6.17 □ Energy diagrams.

activation energy of cross-transfer $\Delta H_{tr21}^{\ddagger}$ would lead to increased molecular weight. Obviously this latter scenario cannot hold because k_{tr12} dominates the denominator and k_{tr21} is negligible. Figure 6.17 illustrates these two possibilities.

This analysis leads to the following conclusions (Kennedy and Chou, 1976a). The molecular weight of a copolymer is always lower than that of the respective homopolymers obtained under identical conditions. The phenomenon of molecular weight depression of copolymers becomes increasingly pronounced at low temperatures because the term $\exp(-\Delta H^{\ddagger}RT)$ increases with decreasing temperatures. This phenomenon is relatively small when both copolymerizable monomers produce low molecular weight homopolymers under conditions identical to those used in their copolymerization, because in such systems the activation energy difference between propagation and transfer is small.

On the basis of a similar analysis with regard to the importance of unimolecular termination as the polymer yield-determining event, it was concluded (Kennedy and Chou, 1976a) that coinitiator efficiency (grams of polymer produced per mole of coinitiator) should also precipitously decrease with increasing amount of comonomer in the charge. Data obtained in the isobutylene–β-pinene system corroborated this postulate (Kennedy and Chou, 1976a).

☐ REFERENCES

Abkin, A. D., Sheinker, A. P., Yakovleva, M. K. and Mezhirova, L. P. (1961), *J. Polym. Sci.* **53**, 39.

Alfrey, Jr., T., and Goldfinger, (1944), *J. Chem. Phys.* **12**, 205.

Alfrey, T., and Price, C. C. (1947), *J. Polym. Sci.* **2**, 101.

Alfrey, Jr., T., and Wechsler, H. (1948), *J. Am. Chem. Soc.* **70**, 4266.

Alfrey, Jr., T., Arond, L. H., and Overberger, C. G. (1949), *J. Polym. Sci.* **4**, 539.

Alfrey, Jr., T. A., Rohrer, J. J., Mark, H. (1951), "Copolymerization," in *High polymers*, Intersciences, New York, Vol. 8, p. 12.

Amis, E. S. (1966), *Solvents effects on Reaction Rates and Mechanisms*, Academic Press, New York.

Andrews, L. Y., and Keefer, R. R. M. (1949), *J. Am. Chem. Soc.* **71**, 3644.

Andrews, L. Y. and Keefer, R. R. M. (1950), *J. Am. Chem. Soc.* **72**, 5034.

Anosov, V. A. and Korotkov, A. A. (1960), *Vysokomol. Soedin*, **2**, 354.

Anton, A., Zwegers, J., and Maréchal, E. (1970), *Bull. Soc. Chim. Fr.*, 1466.

Anton, A., and Maréchal, E. (1971a), *Bull. Soc. Chim. Fr.*, 2256.

Anton, A., and Maréchal, E. (1971b), *Bull. Soc. Chim. Fr.*, 3753.

Aoki, S., Harita, Y., Otsu, T., Imoto, M. (1965), *Bull. Chem. Soc. Jap.* **38**, 1928.

Asami, R., Hasegawa, K., Asai, N., Moribe, I., and Doi, A. (1976), *Polym. J.* **8**, 74.

Aso, C., and Ohara, O. (1967), *Makromol. Chem.* **109**, 161.

Aso, C., and Ohara, O. (1969), *Makromol. Chem.* **127**, 78.

Baird, N. C. (1970), *Theoret. Chim. Acta* (Berl.) **16**, 239.

Barre, F., and Maréchal, E. (1974), *C.R. Acad. Sci. C* **279**, 329.

Belliard, P., and Maréchal, E. (1972), *Bull. Soc. Chim. Fr.*, 4255.

Bergman, E. D., Fisher, E., and Pullman, B. (1951), *J. Chim. Phys.* **48**, 356.

Blin, Ph., Bunel, C., and Maréchal, E. (1978), *J. Chem. Res*, S 206, M 2619.

Bogomoenyi, V. Y., Dolgoplosk, B. A., and Chirikova, Z. P. (1964), *Dokl. Akad. Nauk SSSR* **159**, 1069.

Borg, P., and Maréchal, E. (1974), *C.R. Acad. Sci.* **278c**, 593.

Borg, P., and Maréchal, E. (1977), *J. Macromol. Sci. Chem.*, *J. Macromol. Sci. Chem.* **A11**, 897.

Brown, G. R., and Pepper, D. C. (1963), *J. Chem. Soc.* 5930.

Brown, G. R., and Pepper, D. C. (1965), *Polymer* **6**, 497.

Brackman, D. S., and Plesch, P. H. (1958), *J. Chem. Soc.*, 3563.

Branch, G. E. K. and Calvin, M. (1941), *Theory of Organic Chemistry*.

Bresler, L. S., Dolgoplosk, B. A., Kropacheiva, E. N., Nelson, N. V. N., and Nikitina, A. P. (1963), *Dokl. Akad. Nauk SSSR* **151**, 650.

Bunel, C., Cohen, S., Laguerre, J. P., and Maréchal, E. (1975), *Polym. J. (Jap.)* **7**, 320.

Bunel, C. (1981) J. Pol. Sci. Polym. Phys. and Phys. Chem. (in Press).

Caillaud, Ph., Huet, J. M., and Maréchal, E. (1970), *Bull. Soc. Chim.*, 1473.

Cesca, S., Priola, A., Bruzzone, M., Ferraris, G. and Giusti, P., (1975), *Makromol. Chem.* **176**, 2339.

Chan, R. K. S. and Meyer, V. E. (1968), in "The computer in polymer science," *J. Polym. Sci. C*, **25**, 11.

Cho, W. J., Bunel, C., and Maréchal, E. (1980), *J. Polym. Sci. Chem. Ed.* **18**, 1995.

Claus, J., and Maréchal, E. (1976), *Bull. Soc. Chim.* 1211.

Cohen, S., Belliard, P., and Maréchal, E. (1973), *Polymer* **14**, 352.

Cohen, S., and Maréchal, E. (1975), *J. Polym. Sci., Polym. Symp.* **52**, 83.

Cook, D. (1963), *Can. J. Chem.*, **41**, 522.

Coudane, J., and Maréchal, E. (to be published).

Cundall, R. B. (1963), *The Chemistry of Cationic Polymerization*, by P. H. Plesch, Ed., Pergamon Press, Chapter 15, p. 550.

Da Silva, J. C., and Maréchal, E. (1974), *Bull. Chem. Soc.*, 1272.

Denon, J. C. E. (1945), U.S. Patent 2,384,731.

Dimitrov, I. K., and Panayotov, I. M. (1971), *Comm. Dept. Chem. Bulg. Acad. Sci.* **4**, 137.

Dunphy, J. F., and Marvel, C. S. (1960), *J. Polym. Sci.*, **47**, 1.

Eizner, Yu. Ye., Skorokhodov, S. S., and Zubova, T. P. (1971), *Eur. Polym. J.* **7**, 869.

Eley, D. D., and Saunders, J. (1954), *J. Chem. Soc.*, 1677.

Fineman, M. and Ross, S. D. (1950), *J. Polym. Sci.* **5**, 259.

Florin, R. E. (1949), *J. Am. Chem. Soc.* **71**, 1867.

Florin, R. E. (1951a), *J. Am. Chem. Soc.* **73**, 4468.

Florin, R. E. (1951b), *Int. Cong. Pure Appl. Chem.*

Fontana, C. M., and Kidder, G. A. (1948), *J. Am. Chem. Soc.* **70**, 3745.

Foster, F. C. (1950), *J. Polym. Sci.* **5**, 369.

Franklin, J. L. (1968) in *Carbonium Ions*, G. A. Olah and P. Schleyer, Eds., Wiley-Interscience, New York, Vol. 1.

Fukui, K., Yonezawa, T., and Shingu, H. (1952), *J. Chem. Phys.* **20**, 722.

Fukui, K., Imamura, A., Yonezawa, T., and Nagata, C. (1954), *Bull. Chem. Soc. Jap.* **27**, 423.

Fukui, K., Yonezawa, T., and Nagata, C. (1957), *J. Chem. Phys.* **26**, 831.

Fukui, K., Nagata, C., Yonezawa, T., and Morikuma, K. (1961), *Bull. Chem. Soc. Jap.* **34**, 230.

Furukawa, J., Kobayashi, E., and Taniguchi, T. (1973), *Kyoto Daigaku Nippon Kagakuseni Kenkyusho Koenshu* **30**, 1.

Furukawa, J. (1974a), private communication.

Furukawa, J., Kobayashi, E., and Taniguchi, S., (1974b), *Bull. Inst. Chem. Res., Kyoto Univ.* **52**, 472.

Garreau, H., and Maréchal, E. (1973), *Eur. Polym. J.* **9**, 453.

Garreau, H., and Maréchal, E. (1973), *Eur. Polym. Sci., Poly. Symp.* **56**, 295.

Garreau, H., and Maréchal, E. (1977), *C.R. Acad. Sci.* **284c**, 107.

Hall, C. G. (1951), *Proc. Roy. Soc., Ser A* **205**, 541.

Ham, G. E. (1954a), *J. Polym. Sci.* **14**, 87.

Ham, G. E. (1954b), *J. Polym. Sci.* **14**, 484.

Ham, G. E. (1963), paper presented at 145th National Meeting American Chemical Society, New York, Sept. 1963; *Polym. Prepr.* **4**, 2, 224.

Ham, G. E. (1964), "Copolymerization," *High Polymers*, Wiley-Interscience, New York, Vol. 18.

Hammond, G. S. (1955), *J. Am. Chem. Soc.*, **77**, 334.

Harwood, H. J., and Ritchey, W. M. (1964), *J. Polym. Sci. Polym. Lett.* **2**, 601.

Harwood, H. J., Johnston, N. W., and Piotrowsky, H. (1968), *J. Polym. Sci. C* **25**, 23.

Hasegawa, K., and Asami, R. (1976), *Polym. J.* **8**, 276.

Hasegawa, K. I., and Asami, R. (1978), *J. Polym. Sci.* **16**, 1449.

Hatada, K., Nagata, K., and Yuki, H. (1970), *Bull. Chem. Soc. Jap.* **43**, 3195.

Heublein, G. (1971), *Symp. "Polymers—71" Varna, Bulg.* **1**, 24.

Heublein, G., and Romer, W. (1973), *Makromol. Chem.* **163**, 143.

Heublein, G., and Barth, O. (1974), *J. Prakt. Chem.* **316**, 649.

Heublein G. (1975a), *Zum. Ablauf Ionischer Polymerisationsreactionen*, Akademie-Verlag, Berlin.

Heublein, G., and Heublein, B. (1975b), *Faserforsch. Textiltech.* **26**, 107.

Heublein, G., Wondraezek, B., Toparkus, H., and Berndt, H. (1976a), *Faserforsch. Textiltech.* **27**, 57.

Heublein, G., and Schütz, E. (1976b), *Faserforsch. Textiltech.* **27**, 129.

Higashimura, T., and Okamura, S. (1960), *Chem. High Polym. Jap.* **17**, 635.

Higashimura, T., Tanaka, A., Miki, T., and Okamura, S. (1967), *J. Polym. Sci. A1* **5**, 1927.

Higashimura, T., Kusudo, S., Ohsumi, Y., Mizote, A. and Okamura, S. (1968a), *J. Polym. Sci.* **6**, 2511.

Higashimura, T., Kusudo, S., Ohsumi, Y., and Okamura, S. (1968b), *J. Polym. Sci. A1* **6**, 2523.

Higashimura, T., Okamura, S., Morishima, I., and Yonezawa, T. (1969a), *Polym. Lett.* **7**, 23.

Higashimura, T., Masuda, T., Okamura, S., Yonezawa, T. (1969b), *J. Polym. Sci. A1* **7**, 3129.

Higashimura, T., Kawamura, K., and Masuda, T. (1972), *J. Polym. Sci.* **10**, 85.

Higashimura, T., Kawamura, K., Masuda, T. (1973), *J. Polym. Sci. (Chem.)* **11**, 713.

Higashimura, T., and Kishiro, O. (1974), *J. Polym. Sci. Polym. Chem. Ed.* **12**, 967.

Higashimura, T., Kishiro, O., Matsuzaki, K., and Uryu, T. (1975), *J. Polym. Sci. Polym. Chem. Ed.* **13**, 1393.

Higashimura, T., Kishiro, O., and Takeda, T. (1976), *J. Polym. Chem. Ed.* **14**, 1089.

Higashimura, T., and Yamamoto, K. (1977), *J. Polym. Sci. Polym. Chem. Ed.* **15**, 301.

Hiroyasu, I., and Isao, Y. (1975), *Kobunshi Robunshi* **32**, 357.

Hoffman, R. (1963), *J. Chem. Phys.* **39**, 1397.

Hoffman, R. (1964), *J. Chem. Phys.* **40**, 2480.

Iino, M., and Tokura, N. (1964), *Bull. Chem. Soc. Jap.* **37**, 23.

Iino, M., and Tokura, N. (1965), *Bull. Chem. Soc. Jap.* **38**, 1094.

Imanishi, Y., Mizote, A., Higashimura, T., and Okamura, S. (1963a), *Polym. Sci. Jap.* **20**, 49, 58.

Imanishi, Y., Momiyama, Z., Higashimura, T., and Okamura, S. (1963b), *Chem. High Polym. Jap.* **20**, 369.

Imanishi, Y., Higashimura, T., and Okamura, S. (1965), *J. Polym. Sci.* A1 **3**, 2455.

Imanishi, Y., Yamane, T., Momiyama, Z., and Higashimura, T. (1966), *Chem. High Polym. Jap.* **23**, 152.

Imanishi, Y., Hara, K., Kohgiya, S. and Okamura, S. (1968), *J. Macromol. Sci. Chem.* **A2**, 1423.

Imanishi, Y., Imanura, H., and Higashimura, T. (1970), *Kobunshi Kagaku* **27**, 242.

Immergut, E. H., Kollman, G., and Malatesta, A. (1960), *Makromol. Chem.* **41**, 9.

Imoto, M., and Saotome, K. (1958), *J. Polym. Sci.* **31**, 208.

Imoto, M. (1969), "Historical development of the reactivity of vinyl monomers," in *Structure and Mechanism in Vinyl Polymerization*, T. Tsuruta and K. F. O'Driscoll, Eds., Dekker, New York, p. 1.

Ise, N. (1979), private communication.

Jones, J. I. (1951), *J. Appl. Chem.* **1**, 568.

Jones, G. D., Runyon, J. R., and Ong, J. (1961a), *Ind. Eng. Chem.* **53**, 297.

Jones, G. D., Runyon, J. R., and Org, J. (1961b), *J. Appl. Polym. Sci.* **5**, 452.

Kamath, P. M. and Haas, H. C. (1957), *J. Polym. Sci.* **24**, 143.

Kanoh, N., Gotoh, A., Higashimura, T., and Okamura, S. (1963), *Macromol. Chem.* **63**, 106.

Kanoh, N., Ibeda, K., Gotoh, A., Higashimura, T., and Okamura, S. (1965), *Makromol. Chem.* **86**, 200.

Karplus, M., and Pople, J. A. (1963), *J. Chem. Phys.* **38**, 2803.

Kawabata, N., Tsuruta, T., and Furukawa, J. (1962a), *Makromol. Chem.* **51**, 70.

Kawabata, N., Tsuruta, T., and Furukawa, J. (1962b), *Makromol. Chem.* **51**, 80.

Kawabata, N., Tsuruta, T., and Furukawa, J. (1963), *Bull. Chem. Soc. Jap.* **36**, 905.

Kelen, T., and Tüdős, F. (1974), *React. Kinet. Catal. Lett.* **1**, 487.

Kelen, T., and Tüdős, F. (1975), *J. Macromol. Sci. Chem.* **A9**, 1.

Kelen, T., Tüdős, F., Turcsanyi, B., and Kennedy, J. P. (1977), *J. Polym. Sci. Polym. Chem. Ed.* **15**, 3047.

Kennedy, J. P. (1964), "Cationic copolymerization," in *Copolymerization, High Polymers*, G. E. Ham. Ed., Wiley-Interscience, New York, Vol. 18, p. 283.

Kennedy, J. P., and Canter, N. H. J. (1967a), *J. Polym. Sci.* Al **5**, 2455.

Kennedy, J. P., and Canter, N. H. J. (1967b), *J. Polym. Sci.* Al **5**, 2712.

Kennedy, J. P., and Chou, T. (1975a), in *Cationic Polymerizations of Olefins*, J. P. Kennedy, Ed., Wiley, New York.

Kennedy, J. P., and Chou, T. (1975b), U.S. Patent 3,923,759.

Kennedy, J. P., Kelen, T., and Tüdős, F. (1975c), *J. Polym. Sci., Polym. Chem. Ed.* **13**, 2277.

Kennedy, J. P., and Chou, T. (1976a), *Adv. Polym. Sci.* **21**, 1.

Kennedy, J. P., and Chou, T. (1976b), *J. Macromol. Sci. Chem.* **A10**, 1357.

Kohjiya, S., Imanishi, Y., and Okamura, S. (1968), *J. Polym. Sci. Al* **6**, 809.

Kohjiya, S., Nakamura, K., and Yamashita, S. (1972), *Angew. Makromol. Chem.* **27**, 189.

Kollmar, H., and Smith, H. O. (1971), *Theor. Chim. Acta Berl.* **20**, 65.

Kucera, M., Hladky, E., and Majerova, K. (1967), *J. Polym. Sci. C* **16**, 257.

Laguerre, J. P., and Maréchal, E. (1974), *Ann Chim.* **9**, 163.

Landler, Y. (1950), *C.R. Acad. Sci.* **230**, 539.

Landler, Y. (1952), *J. Polym. Sci.* **8**, 63.

Laval, F. and Maréchal, E. (1977), *J. Polym. Sci., Polym. Chem. Ed.* **15**, 149.

Ledwith, A., and Wood, H. J. (1966), *J. Chem. Soc. B*, 753.

Lewis, F. M., Walling, C., Cummings, W., Briggs, E. R., and Mayo, F. R. (1948), *J. Am. Chem. Soc.* **70**, 1519.

Lipatova, T. E., Gantmakher, A. R., and Medvedev, S. S. (1956), *Zh. Phys. Chem.* **30**, 1752.

Lipatova, T. E., Gantmakher, A. R., and Medvedev, S. S. (1966), *Dokl. Akad. Nauk SSSR*, **100**, 925.

Lipscomb, N. T., and Matthews, W. K. (1971), *J. Polym. Sci. Al* **9**, 563.

Lyudvig, E. B., Gantmakher, A. R., and Medvedev, S. S. (1958), *Dokl. Akad. Nauk SSSR* **119**, 90.

Lyudvig, E. B., Gantmakher, A. R., and Medvedev, S. S. (1959a), *Symp. Macromol. Weisbaden*, Sect. III-A12.

Lyudvig, E. B., Gantmakher, A. R., and Medvedev, S. S. (1959b), *Vysokomol. Soedin.* **1**, 1333.

Maréchal, E., Basselier, J. J., and Sigwalt, P. (1964), *Bull. Soc. Chim. Fr.* 1740.

Maréchal, E., Bit, C. and Sigwalt, P. (1966), *Bull. Soc. Chim. Fr.*, 3487.

Maréchal, E., Richard, J. P., Menissez, J. P., and Zaffran, C. (1968a), *C.R. Acad. Sci. Paris, Ser. C* **266**, 1427.

Maréchal, E., Richard, J. P., Menissez, J. P., and Zaffran, C. (1968b), *C.R. Acad. Sci. Paris, Ser. C* **266**, 1635.

Maréchal, E. (1968c), *Bull. Soc. Chim. Fr.*, 1459.

Maréchal, E. and Evrard, P. (1969a), *Bull. Soc. Chim.*, 2039.

Maréchal, E. (1969b), *Int. Symp. Macromol. Chem. Budap.* **1**, 363.

Maréchal, E. (1969c), *C.R. Acad. Sci. Paris, Ser. C* **268**, 1121.

Maréchal, E. (1969d), *C.R. Acad. Sci. Paris, Ser. C* **269**, 752.

Maréchal, E., Zaffran, G., Zaffran, C., and Lepert, A. (1969e), *C.R. Acad. Sci. Paris, Ser. C* **268**, 1350.

Maréchal, E. (1970), *J. Polym. Sci. Al.* **8**, 2867.

Maréchal, E. (1973), *J. Macromol. Sci. Chem.* **A7**, 433.

Maréchal, E. (1976), *J. Polym. Sci., Polym. Symp.* **56**, 349.

Martin, R. H., Lampe, F. W., and Taft, R. W. (1965, 1966), *J. Am. Chem. Soc.* **87**, 2490; **88**, 1353.

Marvel, C. S., and Dunphy, J. F. (1960), *J. Org. Chem.* **25**, 2209.

Masuda, T., Higashimura, T., Okamura, S. (1970), *Polym. J.* **1**, 19.

Masuda, T., and Higashimura, T. (1971a), *Polym. J.* **2**, 29.

Masuda, T., and Higashimura, T. (1971b), *J. Polym. Sci. B* **9**, 783.

Masuda, T., and Higashimura, T. (1973a), *Macromolecules* **6**, 801.

Masuda, T., and Higashimura, T. (1973b), *Makromol. Chem.* **167**, 191.

Masuda, T. (1973c), *J. Polym. Sci.* **11**, 2713.

Masuda, T., Otsuki, M., and Higashimura, T. (1974a), *J. Polym. Sci., Polym. Chem. Ed.* **12**, 1385.

Masuda, T., Mori, T., and Higashimura, T. (1974b), *J. Polym. Sci. Polym. Chem. Ed.* **12**, 2065.

Mayen, M., and Maréchal, E. (1972), *Bull. Soc. Chim.*, 4662.

Mayo, F. R. and Lewis, F. M. (1944), *J. Am. Chem. Soc.* **66**, 1594.

Meier, R. L. (1950), *J. Chem. Soc.*, 3656.

Merz, E., Alfrey, T. A., and Goldfinger, G. (1946), *J. Polym. Sci.* **1**, 75.

Mizote, A., Higasminura, T., and Okamura, S. (1960), *Am. Chem. Soc. Div. Polym. Chem. Prepr.* **7**, 409.

Mizote, A., Tanaka, T., Higashimura, T., and Okamura, S. (1965), *J. Polym. Sci. Al* **3**, 2567.

Mizote, A., Tanaka, T., Higashimura, T., and Okamura, S. (1966a), *J. Polym. Sci. Al* **4**, 869.

Mizote, A., Tanaka, T., and Higashimura, (1966b), *Kobunshi Kagaku* **23**, 239.

Mizote, A., Higashimura, T., and Okamura, S. (1966c), *Am. Chem. Soc. Div. Polym. Chem. Prepr.* **7**, 409.

Mizote, A., Kusudo, S., Higashimura, T., and Okamura, S. (1967), *J. Polym. Sci. Al* **5**, 1727.

Mizote, A., Higashimura, T., and Okamura, S. (1968), *J. Polym. Sci. Al* **6**, 1825.

Mondal, M. A. S., and Young, R. N. (1971), *Eur. Polym. J.* **7**, 1575.

Montgomery, R. D., and Fry, C. E. (1968), *J. Polym. Sci. C* **25**, 59.

Nguyen Anh Hung and Maréchal, E. (1978), *Polymer* **19**, 1303.

O'Driscoll, K. F., and Kuntz, I. (1963), *J. Polym. Sci.* **61**, 19.

O'Driscoll, K. F. (1964), *J. Polym. Sci. A* **2**, 4201.

O'Driscoll, K. F., Yonezawa, T., and Higashimura, T. (1965a), *J. Polym. Sci.* **3**, 2215.

O'Driscoll, K. F. (1965b), *J. Polym. Sci.* **3**, 2223.

O'Driscoll, K. F., Yonezawa, T., and Higashimura, T. (1966a), *J. Macromol. Sci. Chem.* **1**, 17.

O'Driscoll, K. F., and Yonezawa, T. (1966b), *Rev. Macromol. Chem.* **1**, 1.

O'Driscoll, K. F. (1969), *Macromol. Sci. Chem. A* **3**, 307.

Okamura, S., Higashimura, T., and Takeda, K. (1961), *Kobunshi Kagaku* **18**, 389.

Okamura, S., Higashimura, T., Imanishi, Y., Yamamoto, R., and Kimura, K. (1967), *J. Polym. Sci. C* **16**, 2365.

Okuyama, T., Fueno, T., Nakatsuji, H., and Furukawa, Y. (1967), *J. Am. Chem. Soc.* **89**, 5826.

Okuyama, T., Fueno, T., Furukawa, J., and Uyeo, K. (1968), *J. Polym. Sci.* **6**, 993, 1001.

Okuyama, T., Fueno, T., and Furukawa, J. (1969a), *J. Polym. Sci. Al* **7**, 2433.

Okuyama, T., Fueno, T., and Furukawa, J. (1969b), *J. Polym. Sci. Al*, **7**, 3045.

Okuyama, T., Asami, N., and Fueno, T. (1970), *Bull. Chem. Soc. Jap.* **43**, 3553.

Olivier, M., and Maréchal, E., (1974a), *C.R. Acad. Sci.* **278c**, 757.

Olivier, M., and Maréchal, E. (1974b), *Bull. Soc. Chim. Fr.* 696, 699, 1605.

Otsuki, M., Masuda, T., and Higashimura, T. (1976), *J. Polym. Sci. Polym. Chem. Ed.* **14**, 1157.

Overberger, C. G., Arnold, L. H., and Taylor, J. J. (1951), *J. Am. Chem. Soc.* **73**, 5541.

Overberger, C. G., Arond, L. H., Tanner, D. H., Taylor, J. J., and Alfrey, Jr., T. (1952), *J. Am. Chem. Soc.* **74**, 4848.

Overberger, C. G., Ehrig, R. J., and Tanner, D. H. (1954), *J. Am. Chem. Soc.* **76**, 772.

Overberger, C. G., Tanner, D. H., and Pearce, E. M. (1958), *J. Am. Chem. Soc.* **80**, 4566.

Overberger, C. G., and Kamath, V. G. (1959), *J. Am. Chem. Soc.* **81**, 2910.

Overberger, C. G., and Kamath, V. G. (1963), *J. Am. Chem. Soc.* **85**, 446.

Owen, N. L., and Shepard, N. (1964), *Trans. Faraday Soc.* **60**, 634.

Panayotov, I. M., Dimitrov, I. K., and Bakerdjiev, I. E. (1969), *J. Polym. Sci. Al* **7**, 2421.

Panayotov, I. M., Velickova, R. S., Matev, N. (1974), *C.R. Acad. Bulg. Sci.* **27**, 1679.

Panayotov, I. M., and Heublein, G. (1977), *J. Macromol. Sci. Chem.* **A11**, 2065.

Pepper, D. C. (1974), *Makromol. Chem.* **175**, 1077.

Pople, J. A., Santry, D. P., and Segal, G. A. (1965a), *J. Chem. Phys.* **43**, S129.

Pople, J. A., and Segal, G. A. (1965b), *J. Chem. Phys.* **43**, S136.

Pople, J. A., and Segal, G. A. (1966), *J. Chem. Phys.* **44**, 3289.

Pople, J. A., and Gordon, M. (1967), *J. Am. Chem. Soc.* **89**, 4253.

Pople, J. A., and Beveridge, D. L. (1970), *Approximate Molecular Orbital Theory*, McGraw-Hill, New York.

Priola, A., Cesca, S., Ferraris, G., and Bruzzone (1975), *Makromol. Chem.* **176**, 1969.

Pullman, A., Pullman, B., Bergman, E. D., Bertier, G., Fisher, E., Hirshberg, Y., and Pontis, J. (1951), *J. Chim. Phys.* **48**, 359.

Quére, J. P., and Maréchal, E. (1971), *Bull. Soc. Chim.*, 2983.

Quére, J. P., and Maréchal, E. (1971b), *Bull. Soc. Chim.*, 2983.

Rehner, J., Zapp, R. L., and Sparks, W. J. (1953), *J. Polym. Sci.* **11**, 21.

Roothaan, C. C. J. (1951), *Rev. Mod. Phys.* **23**, 69.

Sakurada, I., Ise, N., Hayashi, Y., and Nakao, M. (1968), *Macromolecules* **1**, 265.

Saotome, K., and Imoto, M. (1958), *Kobunshi Kagaku* **15**, 368.

Sawamoto, M., Masuda, T., and Higashimura, T. (1976), *Makromol. Chem.* **177**, 2995.

Shostakoskii, M. F., Skvortsova, G. G., Zapunnaya, and Kozyrev, K. (1967), *Vysokomol. Soedin, A* **9**, 704; *Polym. Sci. USSR* **9**, 787.

Sigwalt, P. (1961), *J. Polym. Sci.* **52**, 15.

Sigwalt, P., and Maréchal, E. (1966), *Eur. Polym. J.* **2**, 15.

Smets, G., and De Haes, L. (1950), *Bull. Soc. Chim. Belge* **59**, 13.

Swain, C. G., and Langsdorf, W. P. (1951), *J. Am. Chem. Soc.* **73**, 2813.

Taft, R. W., Jr. (1952), *J. Am. Chem. Soc.* **74**, 5372.

Taft, R. W., Jr., Purlee, E. L., Riesz, P., and De Fazio, C. A. (1955), *J. Am. Chem. Soc.* **77**, 1584.

Tegge, B. R. (1953), U.S. Patent 2,643,993.

Thaler, W. A., and Buckley, J. D., Jr. (1976), *Rubber Chem. Technol.* **49**, 960.

Thomas, R. M., and Sparks, W. J. (1944), U.S. Patent, 2,356,128.

Tidwell, P. W., and Mortimer, G. A. (1965), *J. Polym. Sci. A* **3**, 369.

Tidwell, P. W., and Mortimer, G. A. (1970), *Rev. Macromol. Chem.* **5-2**, 135.

Tobolsky, A. V., and Boudreau, R. J. (1961), *J. Polym. Sci.* **51**, S 53.

Tokura, N., Matsuda, M., and Iino, M. (1963), *Bull. Chem. Soc. Jap.* **36**, 278.

Toncheva, V., Velichkova, R., and Panayotov, I. M. (1974), *Bull. Soc. Chim. Fr.*, 1033.

Tortai, J. P., and Maréchal, E. (1971), *Bull. Soc. Chim.*, 2673.

Tsuruta, T. (1969), in *Structure and Mechanism on Vinyl Polymerization*, T. Tsuruta and K. O'Driscoll, Eds., Marcel Dekker, New York, p. 27.

Tüdős, F., Kelen, T., Földes-Berezsnich, T., and Turcsányi, B. (1975), *React. Kinet. Cat. Lett.* **2**, 439.

Tüdős, F., Kelen, T., Földes-Berezsnich, T., and Turcsányi, B. (1976), *J. Macromol. Sci. Chem.* **A10**, 1513.

Visse, F. (1974a), Thesis, Rouen.

Visse, F., and Maréchal, E. (1974b), *Bull. Soc. Chim. Fr.* 387; *Polymer* **26**, 485.

Walling, Ch., and Briggs, E. R. (1945), *J. Am. Chem. Soc.* **67**, 1774.

Walling, C., Briggs, E. R., Wolfstirn, K. B., and Mayo, F. R. (1948), *J. Am. Chem. Soc.* **70**, 1537.

Yamamoto, K., and Higashimura, T. (1974), *J. Polym. Sci. Polym. Chem. Ed.* **12**, 613.

Yamamoto, K., and Higashimura, T. (1975), *Polymer* **16**, 815.

Yamamoto, K., and Higashimura, T. (1976), *J. Polym. Sci. Polym. Chem. Ed.*, **14**, 2621.

Yamashita, Y., Okada, M., and Hirota, M. (1969), *Angew. Makromol. Chem.* **9**, 136.

Yonezawa, T., Higashimura, T., Katagiri, K., Hayashi, K., Okamura, S., and Fukui, K. (1957), *J. Polym. Sci.* **26**, 311.

Yonezawa, T., Yamaguchi, K., and Kato, H. (1967), *Bull. Chem. Soc. Jap.* **40**, 536.

Yonezawa, T., Nakatsuji, H., and Kato, H. (1968), *J. Am. Chem. Soc.* **90**, 1239.

Zaffran, C., and Maréchal, E. (1970), *Bull. Soc. Chim.*, 3521.

Zwegers, J., and Maréchal, E. (1972), *Bull. Soc. Chim.*, 1157.

□ 7 □

Step-Growth
Polymerizations*

*Contributed by Robert W. Lenz, Chemical Engineering Department, University of Massachusetts, Amherst, Mass., and John E. Chandler, Chemical Research Division, American Cyanamid Co., Stamford, Conn.

7.1 □   INTRODUCTION

Hundreds of publications, some appearing in the literature more than a century ago, exist on carbocationic step-growth polymerizations of benzyl and related derivatives to prepare polybenzyls and analogous polymers:

$$
n \ H Ar \overset{R_1}{\underset{R_2}{C}} X \xrightarrow{\text{Lewis acid}} \left[Ar \overset{R_1}{\underset{R_2}{C}} \right]_n + HX \tag{7.1}
$$

where Ar is an aromatic nucleus (generally phenylene) and R_1 and R_2 are alkyl or proton substituent groups (Lenz, 1967). The initial intermediate is either a carbocation or an incipient carbocation, and the mechanism is generally that of an aromatic electrophilic substitution or Friedel–Crafts alkylation, as illustrated in equation 7.2 for the polymerization of benzyl chloride with a metal halide coinitiator. According to this mechanism the

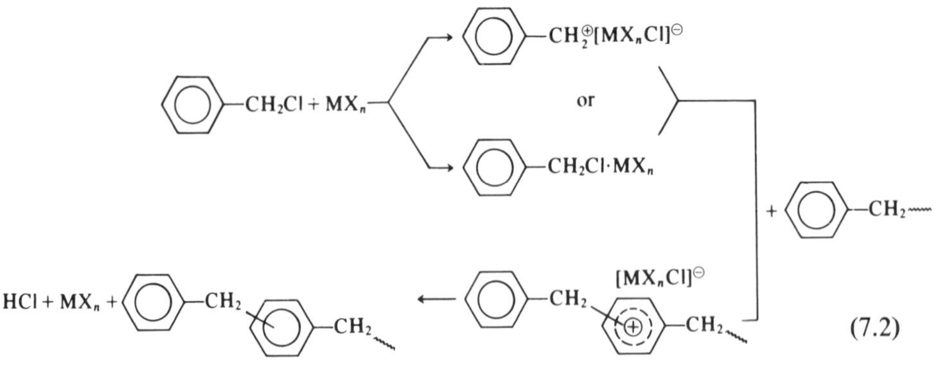

benzyl and the cyclohexadienyl carbocation may be formed (Olah, 1971). The latter is the usual intermediate proposed for aromatic electrophilic substitutions.

Other carbocationic step-growth polymerizations investigated include (1) the preparation of poly(arylene–alkylenes) by the closely related reaction of an alkylene dihalide with an aromatic compound as shown in equation 7.3 (Lenz, 1967):

$$
n \ ArH_2 + n X(CH_2)_y X \xrightarrow{\text{Lewis acid}} [Ar(CH_2)_y]_n + 2n \ HX \tag{7.3}
$$

(2) the preparation of polyarylenes, particularly polyphenylene, by an unusual oxidative coupling reaction, as shown in equation 7.4 (Lenz, 1967); and (3) the preparation of poly(arylene sulfones) by Friedel–Crafts sul-

fonation, as illustrated in equation 7.5 for the reaction of *m*-pheny-lenedisulfonyl chloride and diphenyl ether (Rose, 1974). Since the first two

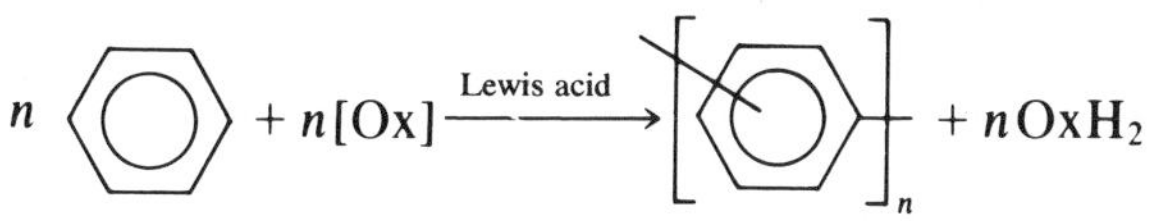

$$n \; \langle \bigcirc \rangle + n[Ox] \xrightarrow{\text{Lewis acid}} \left[\langle \bigcirc \rangle \right]_n + n\,OxH_2 \qquad (7.4)$$

reactions are of much less interest from the point of view of preparing useful linear polymers of reasonable molecular weight and the last reaction involves only the cyclohexadienyl carbocation intermediate, none of these reactions are covered in this review.

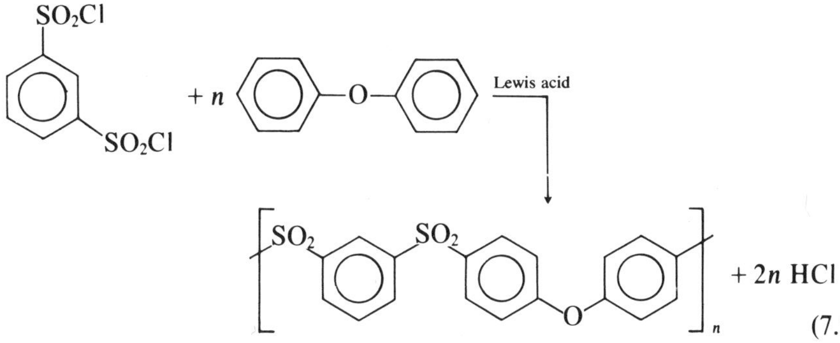

$$+ 2n \; HCl \qquad (7.5)$$

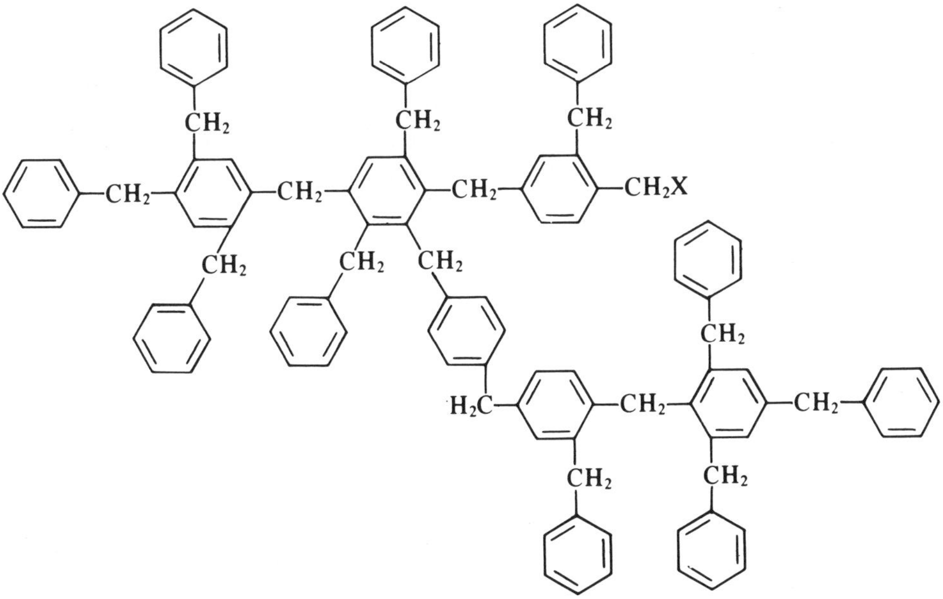

Generally, Friedel–Craft acids have been used to prepare polybenzyls from benzyl halides or related derivatives at temperatures of room temperature and above, and the products are almost always highly branched, amorphous polymers. A typical structural representation for these polymers is shown below (Lenz, 1967). To understand the factors that determine the formation of such a structure, one must first examine Friedel–Crafts alkylation of aromatic compounds, which is the general mechanism for formation of these polycondensation products.

7.2 ☐ REACTION MECHANISM

Friedel–Crafts alkylation of aromatic compounds is an electrophilic substitution on the aromatic ring, as described in equation 7.2. A generalized mechanism is given in equation 7.6. The intermediate π-complex, **III**, which is

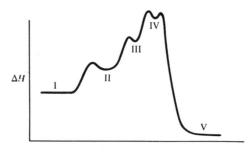

$$(7.6)$$

formed in reactions of substantial positional selectivity with relatively weak nucleophiles, as in the present case, is of low energy compared to the high energy δ-complex intermediate, **IV**. The reaction path diagram for these states can be represented as follows (Olah, 1971):

The alkylating agent can be an alkyl halide, an alcohol, or an olefin. The initiating systems are either protic acids or Friedel–Crafts acids. The reactivity of Friedel–Crafts acids is a function of the type of reactants, including both the alkylating agent and the substrate. For example, in the reaction of benzyl chloride with benzene and toluene, the order of coinitiator activity was found to be $AlCl_3 > SbF_5 > TiCl_4 > FeCl_3 > SbCl_5 > SnCl_4$ (Olah et al., 1972). The type of alkylating agent involved **II** would be mainly influenced by the structure of the alkyl halide reactant according to the stability order for carbenium ions (tertiary > secondary > primary and benzyl > alkyl); that is, increasing ion stability would favor free carbenium ion character in the reactive intermediate, **II**.

Substrate and Positional Selectivity

In the polymerization of benzyl chloride it would be desirable to prepare a crystalline, linear p-substituted high molecular weight polymer. This achievement would require high substrate selectivity (i.e., substitution on the end group and not on internal repeating units) and positional selectivity (i.e., substitution only at the para position of the terminal unit and not on the ortho or meta positions). Incomplete selectivity would yield either a branched or structurally irregular polymer with less desirable properties. Previous reports by Olah and co-workers on benzylation with $TiCl_4$ at 30°C indicated than an electron-donating alkyl group increases the rate of alkylation of the substrate: for example, the relative rates for benzylation of toluene and benzene were 6.3 : 1 (Olah et al., 1972). This ratio would be favorable for forming a linear polymer with predominantly disubstituted repeating units, but the substitution distribution for the reaction on toluene was 40.5% ortho, 4.3% meta, and only 55.2% para. Considering this distribution, it was quite surprising when Kennedy and Isaacson reported that the self-alkylation of benzyl chloride led to a crystalline polybenzyl. That is, in order for the polymer to be crystalline it must have contained a very high degree of para substitution (Kennedy and Isaacson, 1966). The polymerization was carried out at a much lower temperature, – 125°C, than the alkylation reactions, and this factor was believed by the authors to be responsible for the apparent linearity and crystallinity of the polybenzyl formed.

Steric and Substituent Effects

The substitution distribution does not depend solely on electronic effects of the type just described. Steric effects produced by ring substituents are also important, and the amount of ortho substitution, in particular, depends largely on the size of the substituents (Olah, 1964a). This effect was shown, for example, in the comparative isopropylation of toluene, cumene, and *tert*-butylbenzene. Condon alkylated these three substrates with propylene using an $AlCl_3$–nitromethane system at 40°C and found the ratio of the meta/para partial rate factors, prf, was nearly the same for each of the aromatic compounds (Condon, 1949). In contrast, the prf's for the ortho positions were 2.4, 0.37, 0.0, respectively, for the three compounds. Furthermore, the data listed by Olah (1964b) showed a decrease in reactivity of propylene and other alkenes toward different alkylbenzenes in the following order: toluene > ethylbenzene > isopropylbenzene > *sec*-butylbenzene > *tert*-butylbenzene. Volkov and Zavgorodnii (1959) also found that, with mixtures of benzene and alkylbenzenes, olefins were more reactive toward benzene. Francis (1948) attributed polyalkylation, where it occurred preferentially, to heterogeneity and not to perferential reactivity of the alkylated product; he reasoned that the catalyst layer preferentially extracted the preliminary reaction product. According to the literature dialkylbenzenes exhibit lower alkylation reactivity than monoalkylbenzenes; the following sequence of reactivities of toluene and the xylenes is representative (Olah, 1964b): toluene > *o*-xylene > *m*-xylene > *p*-xylene. This reactivity order would favor end group alkylation over internal unit alkylation in a polymerization, which would be necessary to the formation of linear polymers.

Somewhat conflicting information exists on the relative reactivities of trialkylbenzenes. The difficulties seem to stem from the occurrence of side reactions and rearrangements. Kirkland et al. (1958) reacted propylene with pseudocumene and obtained a complex mixture of products in 62% overall yield. Schlatter (1956), and Shimanko and Pokrovskaya (1959) alkylated 1, 2-dimethyl-4-*tert*-butylbenzene and obtained only 1-alkyl-2,3-dimethyl-5-*tert*-butylbenzene.

Olah (1964b) explained the substitution behavior of the above alkylbenzenes by suggesting that the steric requirements of the isopropyl and ethyl groups were modest in comparison to that of *tert*-alkyl groups. However, when mesitylene was isopropylated using $AlCl_3$, a rearranged product, 1,2,4-trimethyl-5-isopropylbenzene, was obtained in 85% yield (Pokrovskaya, 1959). Durene is also known to rearrange under strongly acidic conditions to isodurene (1,3,4,5-tetramethylbenzene), presumably by protonation to form a cyclohexadienyl cation followed by 1,2 methyl shift (Shine, 1967).

An alternative mechanism to explain the preparation of the 1,3,5-trialkyl product in the methylation of *m*-dialkylbenzenes would involve an alkylation–dealkylation sequence as proposed by Geissman (1968). According to this view first a 1,2,3,5-tetraalkyl product would be formed, which subsequently would lose the methyl group at the 2-position.

These results would suggest that the preparation of a linear (that is, exclusively disubstituted) polymer would be favored by steric hindrance toward further substitution on the internal, disubstituted repeating units compared to substitution on the monosubstituted end group. Steric hindrance would be maximized if the repeating units were *p*-substituted and greatly so if the benzyl carbon atom were also substituted. This expectation was found to hold to an unusually high degree by Kuo and Lenz (1977), who studied the substitution distribution in the reaction of benzyl

Table 7.1 □ **Model Compound Reactions to Predict Substitution Distributions in the Polymerization of Benzyl Chloride and α-Methylbenzyl Chloride** (equation 7.8)

Equation 7.8		Reaction Conditions			Substitution Distribution (%)		
R	R'	Coinit.	Solvent	Temp. (°C)	Para	Ortho	Meta
H	H	$TiCl_4$	CH_3NO_2	25	50	29	16
		$AlCl_3$	C_2H_5Cl	−78	47	43	10
CH_3	H	$TiCl_4$	CH_3NO_2	25	91	5	4
		$AlCl_3$	C_2H_5Cl	−78	97	0.5	2.5
CH_3CH_3		$TiCl_4$	CH_3NO_2	25	94	1	5
		$AlCl_3$	C_2H_5Cl	−78	99	0.5	0.5

chloride and α-methylbenzyl chloride with diphenyl methane and 1,1-diphenylalkanes (equation 7.8). This reaction was a model for the polymerization of benzyl chloride and α-methylbenzyl chloride, and the results, collected in Table 7.1 show that a methyl group in the α-position has a

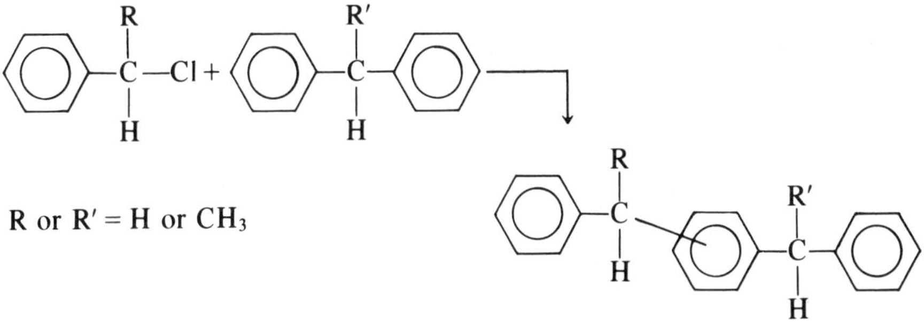

R or R' = H or CH_3

large effect on the substitution distribution at low temperatures. The steric effect was essentially quantitative or positional specific when both reactants had α-methyl substituents; this result accounts *in part* for the formation of crystalline polymer in the low-temperature polymerization of α-methylbenzyl chloride.

7.3 □ POLYBENZYLS

Friedel and Crafts (1885) first prepared insoluble polymers in 1885 from benzyl chloride (equation 7.9). Ingold and Ingold (1928) polymerized benzyl

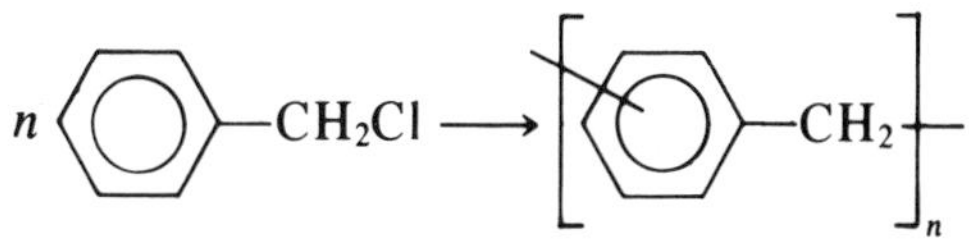

$$(7.9)$$

fluoride using catalytic amounts of concentrated sulfuric acid and found the glassy product to have an overall composition of $(C_7H_6)_n$. Henne and Leicester (1938) also reported the polymerization of benzyl fluoride and proposed that a polymer with a pendant phenyl group was formed:

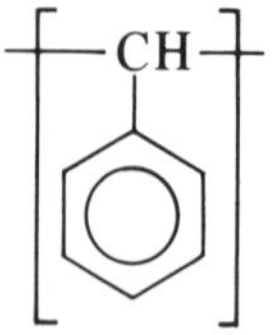

Jacobson (1932) investigated the nature of the polymer prepared from benzyl chloride using aluminum chloride, ferric chloride, and stannic chloride. Of these three coinitiators only stannic chloride gave a completely soluble polymer, and Jacobson proposed a structure having a p-phenylenemethylene repeating unit for this product (equation 7.10). His explanation for the insolubility of the products obtained by using the other two coinitiators was the formation of a three-dimensional network. Benzyl alcohol was also polymerized by protic acids in this work, and it was believed that the structure of the product was the same as that obtained from benzyl halides. Dermer and Hopper (1941) studied 40 different inorganic chlorides in the polymerization of benzyl chloride, but they did not determine the structure of the thermoplastic resins prepared. Lebedev et al. (1949) prepared low molecular weight polymers (1900 to 2100) using aluminum chloride, ferric chloride, and zinc chloride with benzyl chloride at room temperature. Pyrolysis of these polymers indicated dihydroanthracene end groups, and the following intramolecular alkylation was proposed to explain this result:

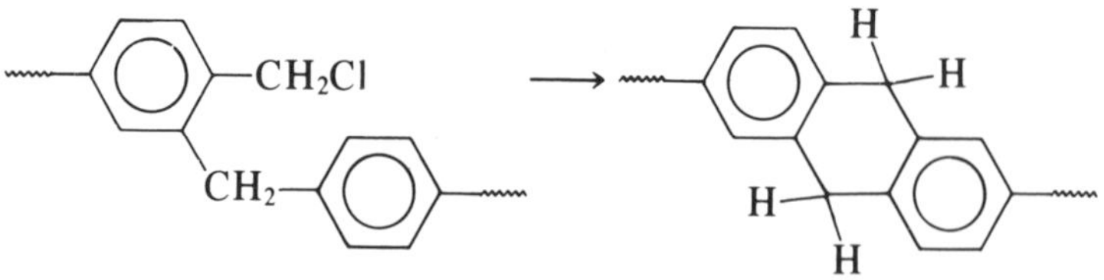

$$(7.10)$$

IR spectroscopy, X-ray diffractions, oxidation, and pyrolysis studies by Haas et al. (1955) confirmed that only amorphous, highly branched polymers were formed in these polymerizations. These authors also prepared a solid using durylmethyl chloride with iron oxide at 100°C. This product was apparently crystalline and melted over the range 260 to 280°C. It was believed to be linear but with only four repeating units. More recently, Ballard et al. (1970) prepared poly-2,3,5,6-tetramethylbenzyl with a molecular weight of approximately 2400 using stannic chloride coinitiator, and this polymer softened or melted above 350°C. Moore (1964) polymerized 2,5-dimethylbenzyl chloride (equation 7.11) with aluminum

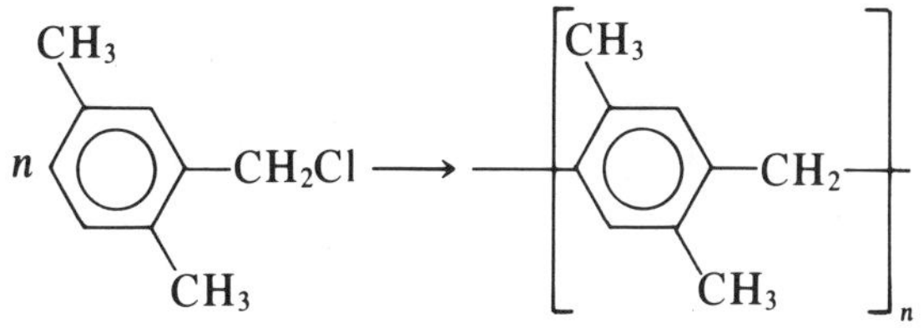

$$n \; \text{(structure)} \; \text{—CH}_2\text{Cl} \longrightarrow \text{—(structure)—CH}_2\text{—} \Big]_n \qquad (7.11)$$

chloride in nitrobenzene at room temperature. X-Ray diffraction revealed that this product was also crystalline. Pappas and Valkanas (1974) and Pappas et al. (1976) carried out similar studies and obtained polymers with melting points up to 240°C but with molecular weights less than 2000.

Mararigna and Montaudo (1964) prepared amorphous polybenzyl from benzyl chloride with a molecular weight of 6400 using stannic chloride at 200°C in an autoclave. As mentioned above, Kennedy and Isaacson (1966–1968) were the first to report the preparation of crystalline polybenzyl at low temperatures by the use of Friedel–Crafts acids or modified Ziegler–Natta catalysts. The crystalline polybenzyl prepared from benzyl chloride with aluminum chloride in ethyl chloride at −130°C melted at 142°C. NMR and IR analyses reportedly confirmed the linearity of the polymer, but no details were given.

Crystalline poly(α-methylbenzyl) was prepared by the same workers from 1-chloroethylbenzene (equation 7.12) using the same low-temperature

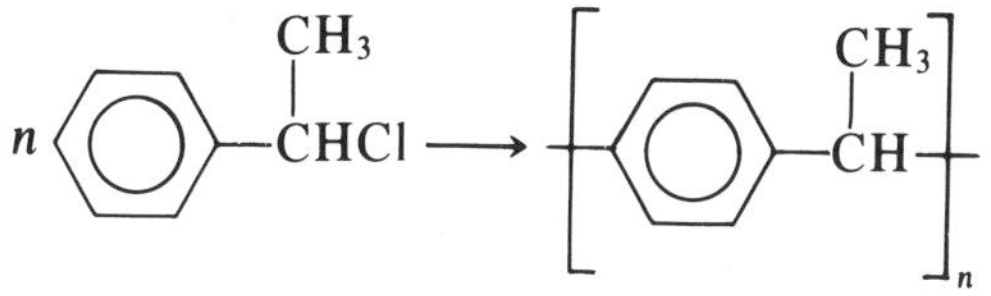

$$n \; \text{(structure)} \; \text{—CHCl} \longrightarrow \text{—(structure)—CH—} \Big]_n \qquad (7.12)$$

conditions, but the yields and molecular weights were low, 10 to 15% and 2400, respectively. They also prepared crystalline poly(2,5-dimethylbenzyl) at −78°C. However, when Montaudo et al. (1970) repeated Kennedy's work with benzyl chloride, they failed to obtain crystalline polymer under the same conditions.

Kuo and Lenz (1977) reinvestigated the low-temperature polymerization of benzyl chloride, α-methylbenzyl chloride, and 2,5-dimethylbenzyl chloride (equations 7.10 to 7.13). For benzyl chloride, the observations of Montaudo and co-workers were confirmed that the polybenzyl obtained was generally noncrystalline under the conditions reported by Kennedy and Isaacson. Indeed, model compound studies indicated that under these conditions (AlCl₃, −125°C ethyl chloride), a product with only 67% para content would be obtained (Kuo and Lenz, 1976). Hence the polymer must be irregular in substitution distribution and quite likely branched also.

In contrast, the polymers from both α-methylbenzyl chloride and 2,5-dimethylbenzyl chloride were both quite crystalline. In fact, poly(2,5-

dimethylbenzyl) was so highly crystalline that it was difficult to keep it in solution during preparation, and as a result, only relatively low molecular weight products were formed (Kuo and Lenz, 1977). However, even though the highest molecular weight obtained for this polymer by Kuo and Lenz was less than 5000, crystalline melting points higher than 300°C were observed. In order to reduce the high degree of crystallinity and increase the solubility of this polymer, Chandler (1978) studied the copolymerization of 2,5-dimethylbenzyl chloride with 2-methyl-5-isopropylbenzyl chloride.

The polymerization of α-methylbenzyl chloride below -100°C has been studied extensively (Kuo and Lenz, 1976; Chandler, 1978; Chandler et al., 1979). The polymerization, when carried out in this temperature range with AlCl$_3$, was found to be unpredictable and gave irreproducible yields, molecular weights, and crystalline properties. In many cases, polymers with bimodal molecular weight distributions were found and the melting points varied from 130 to 250°C. This variation was apparently due to variations in product molecular weights and stereoregularities (Chandler et al., 1979).

X-Ray diffraction investigations carried out on oriented films of crystalline poly(α-methylbenzyl) indicated that the polymer was isotactic (Bruno et al., 1975):

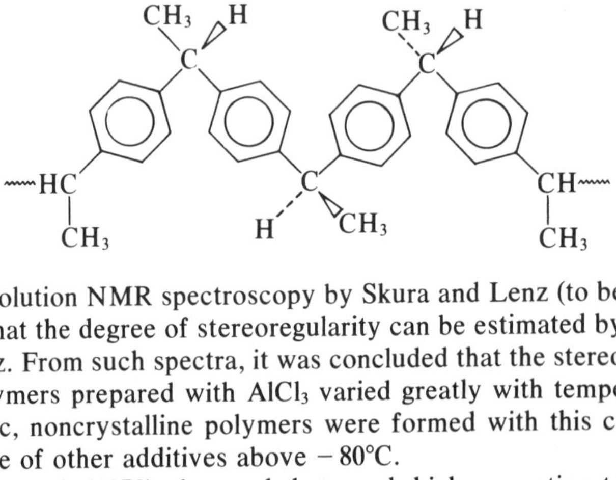

High-resolution NMR spectroscopy by Skura and Lenz (to be published) indicates that the degree of stereoregularity can be estimated by ^{1}H spectra at 270 MHz. From such spectra, it was concluded that the stereoregularities of the polymers prepared with AlCl$_3$ varied greatly with temperature, and only atactic, noncrystalline polymers were formed with this coinitiator in the absence of other additives above -80°C.

Chandler et al. (1979) observed that much higher reaction temperatures could be used, however, for the preparation of crystalline (α-methylbenzyl) if the AlCl$_3$ coinitiator was complexed with nitro compounds in ethyl chloride. Temperatures as high as -65°C gave highly stereoregular polymers in quantitative yields under these conditions.

Skura and Lenz (to be published) found that the polymerization of α-chloroethylbenzene was very fast even at -90°C and maximum molecular weight was achieved within 5 min. Longer reaction times resulted in a decrease in the number average molecular weight and a narrowing of the molecular weight distribution. The latter was presumably due to transbenzylation at long reaction times, which are known to occur in Friedel–Crafts reactions of benzyl chloride with $AlCl_3$ (Olah, 1972).

Skura and Lenz (to be published) polymerized α-ethylbenzyl chloride under the same conditions. According to ^{1}H-NMR poly(α-ethylbenzyl) was also highly stereoregular, but the melting point, 95°C, was much lower than that obtained for poly(α-methylbenzyl).

Fritz and Rees (1972) attempted to polymerize cumyl chloride, but they obtained only very low molecular weight products, which they attributed to the formation of indane structures by the reaction of a normal and an unsaturated end group, as shown in equation 7.13. They were able,

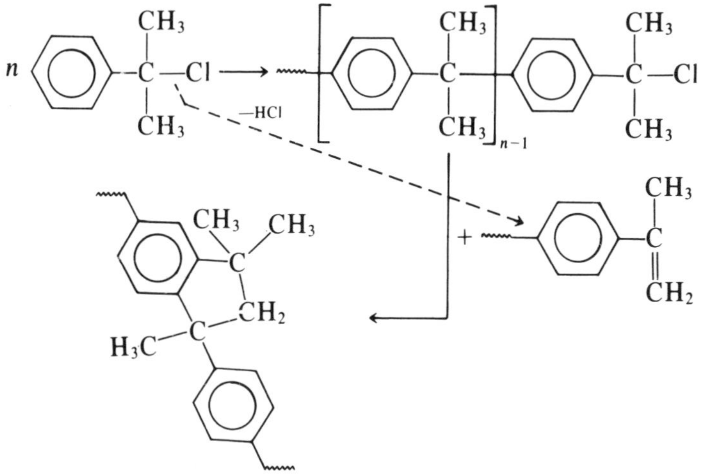

$$(7.13)$$

however, to obtain high molecular weight polymers from the reaction of either $m-$ or p-bis(2-chloroisopropyl) benzene with aromatic compounds such as naphthalene, diphenyl ether, diphenylmethane, and related monomers, as shown in equation 7.14.

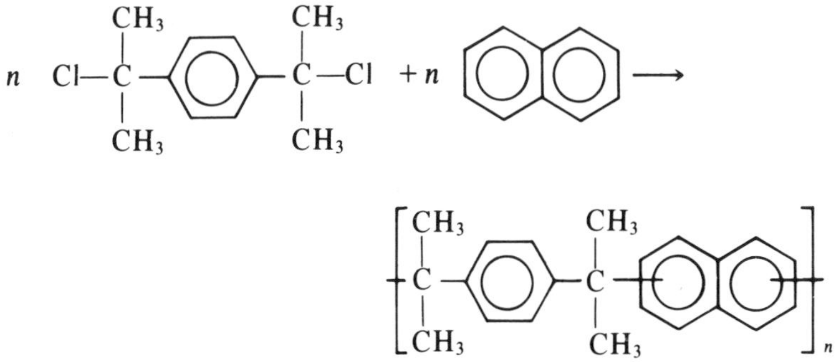

$$(7.14)$$

Special three-component, modified Friedel–Crafts acid systems were used below -20°C. The polymers had high glass temperatures and useful mechanical properties, but they were not reported to be crystalline, possibly because of variations in substitution distribution along the chain.

In conclusion, it appears that after approximately 100 years of research on the preparation of polybenzyls and related polymers, sufficient control has now been achieved in this condensation polymerization reaction to prepare high molecular weight polymers with desirable properties. These polymers could soon find applications as either amorphous or crystalline engineering thermoplastics, with either high glass transition temperatures or melting points.

□ REFERENCES

Ballard, D. G. H., Hollyhead, W. B., and Jones, R. (1970), *Eur. Polym. J.* **6**, 1619.

Bruno, G., Montaudo, G., Hien, N. V., Marchessault, R. H., Sundararajian, P. R., Chandler, J. E., and Lenz, R. W. (1975) *J. Polym. Sci. Polym. Lett.* **13**, 559.

Chandler, J. E. (1978), Ph.D. Dissertation, University of Massachusetts, Amherst, Mass.

Chandler, J. E., Johnson, B. H., and Lenz, R. W., (1979), *Polym. Prepr.* **20**, 858.

Condon, F. E., (1949) *J. Am. Chem. Soc.* **71**, 3544.

Dermer, O. C., and Hopper, E. (1941), *J. Am. Chem. Soc.* **63**, 3525.

Francis, A. W. (1948), *Chem. Rev.* **43**, 257.

Friedel, C., and Crafts, J. M. (1885) *Bull. Soc. Chim. Fr.* **43**, 53.

Fritz, A., and Rees, R. W. (1972), *J. Polym. Sci. A-1* **10**, 2365.

Geissman, T. A. (1968), *Principles of Organic Chemistry*, 3rd ed., Freeman, San Francisco, p. 581.

Haas, H. C., Livingston, D. I., and Saunders, M. (1955), *J. Polym. Sci.* **7**, 503.

Henne, A. L., and Leicester, H. M. (1938) *J. Am. Chem. Soc.* **60**, 864.

Ingold, C. K., and Ingold, E. H. (1928), *J. Chem. Soc.*, 2249.

Jacobson, R. A. (1932), *J. Am. Chem. Soc.* **54**, 1513.

Kennedy, J. P., and Isaacson, R. B. (1966), *J. Macromol. Chem.* **1**, 541.

Kennedy, J. P., and Isaacson, R. B. (1967), U.S. Patent 3,346,514.

Kennedy, J. P., and Isaacson, R. B. (1968), U.S. Patent 3,418,259.

Kirkland, E. V., Funderburk, O. P., and Wadsworth, F. T. (1958), *J. Org. Chem.* **23**, 163.

Kuo, J., and Lenz, R. W. (1976), *J. Polym. Sci., Polym. Chem. Ed.* **14**, 2749.

Kuo, J., and Lenz, R. W. (1977), *J. Polym. Sci. Polym. Chem. Ed.* **15**, 119.

Lebedev, N. N., Korshak, V. V., and Tsipershtein, M. A. (1949), *J. Gen. Chem. USSR Engl. Transl.* **19**, 647.

Lenz, R. W. (1967), *Organic Chemistry of Synthetic High Polymers*, Wiley-Interscience, New York, pp. 227ff.

Mararigna, P., and Montaudo, G. (1964), *Ann. Chim.* **54**, 217.

Montaudo, G., Bottino, F., Caccamese, S., and Finocciaro, P. (1970), *J. Polym. Sci. A-1*, **8**, 2453.

Moore, J. E. (1964), *Polym. Prepr.* **5**, 203.

Olah, G. A. (1964a), *Friedel-Crafts and Related Reactions*, Vol. II, Wiley-Interscience, p. 8.

Olah, G. A. (1964b), in G. A. Olah (1964a), p. 19.

Olah, G. A. (1971), *Acc. Chem. Res.* **4**, 240.

Olah, G. A., Kobayashi, S., and Tashiro, M. (1972), *J. Am. Chem. Soc.* **94**, 7448.

Pappas, N. A., and Valkanas, G. N. (1974), *J. Polym. Sci. Polym. Chem. Ed.* **12**, 2567.

Pappas, N. A., Valkanas, G. N., and Diamanti-Kontsida, E. T. (1976), *J. Polym. Sci. Polym. Chem. Ed.* **14**, 1241.

Pokrovskaya, E. S. (1959), *Tr. Inst. Nefti, Akad. Nauk SSSR* **13**, 5; *Chem. Abstr.*, (1959) **55**, 9849a.

Rose, J. B. *Polymer* (1974), **15**, 456.

Schlatter, M. J. (1956) "Joint symposium on synthetic fuels and chemicals," *Am. Chem. Soc., Div. Petrol. Chem., Prepr. Symp.* **1**, 77.

Shimanko, N. A., and Pokrovskaya, W. S. (1959), *Dokl. Akad. Nauk SSSR.* **129**, 1313.

Shine, H. J. (1967), *Aromatic Rearrangements*, Elsevier, New York, p. 10.

Skura, J. F., and Lenz, R. W. (to be published).

Volkov, R. N., and Zavgorodnii, S. V. (1959), *Zh. Obshch. Khim.* **29**, 3672–3677.

□ 8 □

Sequential (Block and Graft) Copolymers

8.1 ☐ INTRODUCTION

The aim of this chapter is to survey and organize available information on the synthesis of block and graft copolymers by carbocationic techniques. Cationic syntheses in which oxonium, immonium, and sulfonium ions are involved fall outside the scope of this chapter.

Remarkable advances have been made recently in the field of synthesis of sequential copolymers by carbocationic processes. Progress was possible because of the increased understanding of the mechanism of elementary steps in carbocationic polymerizations, particularly in the understanding of the chemistry of ion generation and initiation. Further, the recognition of chain transferless carbocationic polymerizations (see Section 4.3) led to the preparation of unadulterated sequential copolymers, that is, block and graft copolymers free of homopolymer contaminants.

Impetus for research in these areas arose because of the spectacular developments in the commercial exploitation of a great variety of sequential copolymers, notably the glassy–rubbery–glassy triblocks prepared by anionic living polymerization techniques. Although anionic techniques can be used for the precise tailoring of polymer sequences, molecular weights, and molecular weight distributions, the number of monomers which lend themselves for the necessary high degree of mechanistic control is very small, that is, styrene and a few conjugated dienes. In contrast, carbocationic polymerizations are effective for the polymerization of a large variety and number of monomers; however, the precision of molecular tailoring with this technique has not yet reached the degree of sophistication attainable by anionic methods and required for the controlled microarchitecture of sequential copolymers.

8.2 ☐ A NOTE ON TERMINOLOGY

The architecture of sequential copolymers may be very complex and the concise naming of these structures may be extremely cumbersome. We have adopted Ceresa's nomenclature (1962) to describe sequential copolymers.

Additional terminology that has proved quite useful gives a clue as to the synthesis details of sequences. The words "from" and "onto" indicate the *direction* of synthesis. Thus "grafting from" describes a process in which an active site generated along a polymer backbone starts to propagate a monomer in the system and thus leads to branches attached to the backbone. Similarly, in "blocking from" the active site is at a terminus of a polymer and polymerization starting at this site leads to a growing block copolymer. "Grafting onto" occurs when a growing polymer chain attacks another polymer which carries a reactive functional group and thus branches are produced. Similarly, in "blocking onto" the propagating polymer chain reacts with another polymer which has a reactive function at a terminus. Schematically these processes may be illustrated as follows:

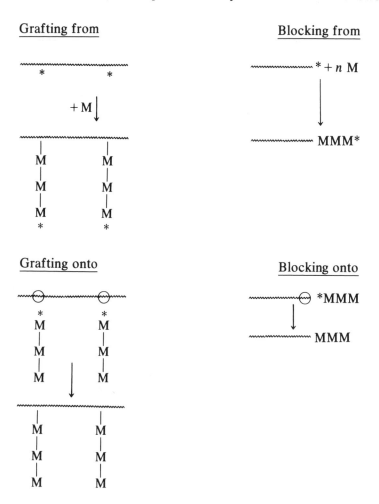

Grafting from

Blocking from

Grafting onto

Blocking onto

where the asterisk indicates an active site (carbocation), M is monomer, and ⊛ is a site capable of reacting with the growing species, that is, unsaturation, or phenyl substituent.

Graft copolymers can also be obtained by copolymerization of macromers with conventional vinyl monomers (see Section 8.4).

In addition to these routes, two or more polymer sequences can also be linked by a variety of linking or coupling techniques. In these methods preformed blocks carrying active functionalities are linked either directly or by a small coupling agent to sequential copolymers.

From the point of view of synthesis the preparation of block or graft copolymers occurs by fundamentally similar processes and the techniques and concepts used are identical. The distinguishing feature between block and graft copolymer synthesis is the locus of the active site necessary for the linking of the different sequences: in block copolymer synthesis this site is at the terminus (or termini) and in graft copolymers the sites are along the chain that is to become the backbone or trunk polymer.

8.3 □ BLOCK COPOLYMERS

It is more difficult to create a single active site at a terminus of a chain than many sites along the chain; consequently the number of options open for the synthesis of block copolymers are far fewer than those for grafts. This may explain why reports on the synthesis of block copolymers by cationic techniques started to appear only after the principle of controlled initiation by the use of alkylaluminum coinitiators became known. The discovery of controlled initiation of carbocationic polymerizations by combination of active organic halide/alkylaluminum compounds (Kennedy and Baldwin, 1968) opened the road to the synthesis of block copolymers by this technique.

Synthesis of Block Copolymers

The first block copolymer synthesized by a carbocationic technique was a diblock of polyisobutylene and polystyrene prepared by a blocking-from route (Peyrot, 1972; Jolivet and Peyrot, 1973). The process involved three steps:

1 The synthesis of a polyisobutylene with a benzyl head group:

$$C_6H_5CH_2Cl + n\,CH_2{=}C(CH_3)_2$$

$$\xrightarrow{Et_2AlCl} C_6H_5CH_2{-}CH_2{-}C(CH_3)_2{-}CH_2{-}\overset{\oplus}{C}(CH_3)_2 + Et_2AlCl_2^{\ominus}$$

2 Chloromethylation of the benzyl head group:

$$C_6H_5CH_2{-}PIB \xrightarrow[SnCl_4]{CH_3OCH_2Cl} ClCH_2{-}C_6H_4CH_2{-}PIB$$

3 Polymerization of styrene from the head group using Et_2AlCl coinitiator:

$$PIB{-}CH_2C_6H_4{-}CH_2Cl + St \xrightarrow[-25°C]{Et_2AlCl} PIB{-}CH_2C_6H_4{-}CH_2{-}PSt$$

If the linkage of the two sequences by an internal $-CH_2C_6H_4CH_2-$ group is neglected, this structure may be regarded as poly(isobutylene-b-styrene).

The purpose of Peyrot and Jolivet evidently was the demonstration of a workable synthesis of a new structure, not its purification or characterization. Although only an extremely limited amount of information is available (only three experiments giving some relative selective solvent extraction data), the chemical steps employed should lead to the architecture claimed. The mechanism of benzyl halide/alkylaluminum-initiated carbocationic polymerizations has been the subject of recent fundamental investigations which suggest that under optimal conditions the synthesis of essentially pure blocks by this technique may not be impossible.

The technique developed by Kennedy and Melby (1974a, 1975a) for the synthesis of polystyrene and polyisobutylene diblocks is similar to the above method inasmuch as it uses alkylaluminum coinitiators; however, in other aspects it is fundamentally different. These authors exploited for the preparation of diblocks the principle of "selective sequential cationic initiation."

Kinetic experiments showed that the polymerization of isobutylene by t-BuCl/Me$_3$Al is several orders of magnitude faster than that by t-BuBr/Me$_3$Al (Kennedy et al., 1973). Thus the possibility arose of initiating the polymerization of isobutylene by a di-tertiary bromo–chloro halide under selective conditions, that is, conditions under which only the tertiary chlorine would be mobilized and the tertiary bromine would survive. It was theorized and subsequently demonstrated that the polymerization of a second monomer could be induced by the tertiary bromine. An initiator was synthesized (2-bromo-6-chloro-2,6-dimethylheptane):

$$\text{Cl-}\underset{\underset{\text{CH}_3}{|}}{\overset{\overset{\text{CH}_3}{|}}{\text{C}}}\text{-CH}_2\text{-CH}_2\text{-CH}_2\text{-}\underset{\underset{\text{CH}_3}{|}}{\overset{\overset{\text{CH}_3}{|}}{\text{C}}}\text{-Br}$$

which combines a tertiary chlorine and bromine in one molecule, and the tertiary halogens are sufficiently apart to influence each other by induction. In the course of preliminary experiments conditions have been worked out under which the chlorine and bromine could be selectively substituted by alkylaluminum compounds, thus simulating selective polymerizations. Guided by a variety of model studies sequential polymerizations of styrene and isobutylene were carried out (Kennedy and Melby, 1974a, 1975a). The synthesis sequence included the following steps:

1 Polymerization of styrene (St):

$$\text{Cl-}\underset{\underset{\text{CH}_3}{|}}{\overset{\overset{\text{CH}_3}{|}}{\text{C}}}\text{-(CH}_2\text{)}_3\text{-}\underset{\underset{\text{CH}_3}{|}}{\overset{\overset{\text{CH}_3}{|}}{\text{C}}}\text{-Br} + \text{St} \xrightarrow[\text{EtCl},-80°\text{C}]{\text{Et}_3\text{Al}} \text{polystyrene-}\underset{\underset{\text{CH}_3}{|}}{\overset{\overset{\text{CH}_3}{|}}{\text{C}}}\text{-(CH}_2\text{)}_3\text{-}\underset{\underset{\text{CH}_3}{|}}{\overset{\overset{\text{CH}_3}{|}}{\text{C}}}\text{-Br}$$

2 Removing unreacted initiator and coinitiator; adding isobutylene IB, and inducing polymerization by Et_2AlCl:

$$
\text{polystyrene} - \overset{\overset{\displaystyle CH_3}{|}}{\underset{\underset{\displaystyle CH_3}{|}}{C}} - (CH_2)_3 - \overset{\overset{\displaystyle CH_3}{|}}{\underset{\underset{\displaystyle CH_3}{|}}{C}} - Br + IB \xrightarrow[\substack{CH_2Cl_2/C_6H_{14} \\ \text{from } -45 \text{ to } -65°C}]{Et_2 AlCl}
$$

$$
\text{polystyrene} - \overset{\overset{\displaystyle CH_3}{|}}{\underset{\underset{\displaystyle CH_3}{|}}{C}} - (CH_2)_3 - \overset{\overset{\displaystyle CH_3}{|}}{\underset{\underset{\displaystyle CH_3}{|}}{C}} - \text{polyisobutylene}
$$

This synthesis became possible as a consequence of two key findings: the large (orders of magnitude) rate difference of halide substitution by trimethylaluminum between *tert*-butyl chloride and bromide (Kennedy et al., 1973) referred to above, and the polymerization of styrene in the absence of chain transfer by the use of certain alkylaluminum compounds (Kennedy and Rengachary, 1974b; Kennedy and Smith, 1974c). The latter information was essential for the "clean" homopolymer-free synthesis of polystyrene possessing a terminal bromine, PSt–Br. Since more vigorous conditions are necessary to achieve initiation of isobutylene from the tertiary bromine site than from the chlorine site, a stronger Lewis acid, Et_2AlCl, and somewhat higher temperatures, from -45 to $-60°C$, were employed to effect the second step. Since the possibility for chain transfer in isobutylene polymerization initiated by the $PSt–Br/Et_2Al$ system exists, the possibility for homo-polyisobutylene formation also arises. Consequently a selective solvent extraction procedure was developed to separate the pure poly(styrene-*b*-isobutylene) from the crude product.

Selective sequential cationic initiation has also been employed for the synthesis of bigraft copolymers (see below).

A poly(styrene-*b*-isobutylene) prepared by using a polystyrene sequence of $\bar{M}_n = 20,000$ and containing 79/21 polystyrene/polyisobutylene by NMR gave an overall $\bar{M}_n$ of 35,000. The block copolymer formed cloudy solutions in *n*-pentane (a solvent only for PIB) and in methyl ethyl ketone (a solvent only for PSt). Apparently in these solvents, the soluble sequence forces the insoluble one attached to it in solution. In cyclohexane at room temperature slightly hazy solutions were obtained which became clear above 35°C, the theta temperature of PSt. In good solvents for both sequences, for example, toluene, benzene, or CCl_4, visually clear solutions were obtained.

Films cast from benzene solutions were homogeneous and partly transparent. Films cast from cyclohexane were striped, presumably owing to phase separation (cyclohexane is a poor solvent for PSt below 35°C). The copolymer exhibited two T_g's at 96 and $-74°C$, characteristic of PSt and PIB, respectively.

The intrinsic viscosity $[\eta]$ of the block copolymer as a function of temperature using toluene in the range from 15 to 55°C is shown in Figure 8.1. Such a discontinuous temperature profile is characteristic of sequential copolymers (Dondos, 1966). Materials for which $[\eta]$ increases with increasing temperatures may be of interest as motor oil additives.

A different technique for the preparation of block copolymers became possible recently by understanding the mechanism of termination in BCl$_3$-coinitiated olefin polymerizations (Kennedy et al., 1976b, 1978). It has been demonstrated that H$_2$O/BCl$_3$ efficiently initiates the polymerization of isobutylene and styrene (Kennedy et al., 1976b, 1977d) and that termination occurs by chlorination of the growing cation (Kennedy et al., 1977e):

$$\underset{\underset{CH_3}{|}}{\overset{\overset{CH_3}{|}}{\text{wwCH}_2\!-\!\text{C}}}{}^{\oplus}\!\text{BCl}_4^{\ominus}(\text{or BCl}_3\text{OH}^{\ominus}) \rightarrow \underset{\underset{CH_3}{|}}{\overset{\overset{CH_3}{|}}{\text{wwCH}_2\!-\!\text{C}}}\!-\!\text{Cl} + \text{BCl}_3(\text{or BCl}_2\text{OH})$$

Conditions have been found under which chain transfer is absent and chain breaking is exclusively by termination by chlorination. Although direct analytical proof for a terminal –Cl atom in a high polymer is exceedingly difficult to obtain because of its very low concentration, model experiments, ^{1}H-NMR spectroscopy with low molecular weight models, and kinetic and chemical evidence (Kennedy et al., 1977b) substantiate this postulate. Indeed, direct confirmation for terminal tertiary chlorine terminus was obtained by synthesizing poly(isobutylene-*b*-styrene) by the following process:

1 Synthesis of the postulated polyisobutylene with the tertiary chlorine terminus:

$$\text{H}_2\text{O/BCl}_3 + \text{CH}_2\!\!=\!\!\text{C(CH}_3)_2 \xrightarrow[-78°C]{CH_2Cl_2} \text{H} \text{ww PIB ww CH}_2\!-\!\underset{\diagdown CH_3}{\overset{\diagup CH_3}{\text{C}}}\!\!-\!\text{Cl} + \text{BCl}_2\text{OH}$$

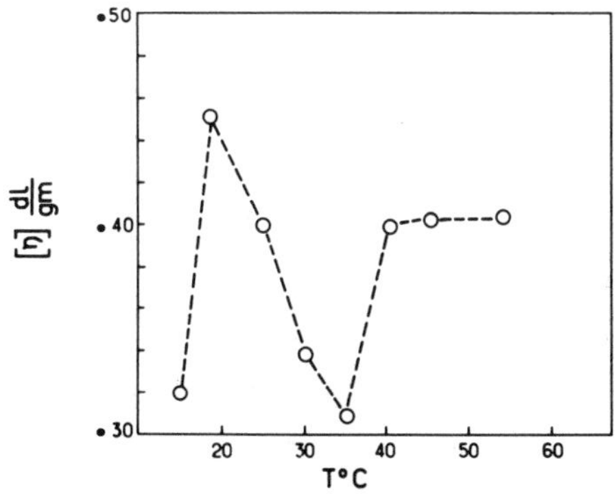

Figure 8.1 □ Effect of temperature on the intrinsic viscosity of poly(styrene-*b*-isobutylene) in toluene solvent.

2 Purification of this prepolymer, addition of styrene, and blocking from the *tert*-chloro terminus by introducing Et$_2$AlCl:

$$\text{H} \sim\!\!\sim \text{PIB} \sim\!\!\sim \text{CH}_2\!-\!\underset{\underset{\text{CH}_3}{|}}{\overset{\overset{\text{CH}_3}{|}}{\text{C}}}\!-\!\text{Cl} + \text{St} \xrightarrow[\text{CH}_2\text{Cl}_2\ -40°C]{\text{Et}_2\text{AlCl}} \text{H} \sim\!\!\sim \text{PIB} \sim\!\!\sim \text{CH}_2\!-\!\underset{\underset{\text{CH}_3}{|}}{\overset{\overset{\text{CH}_3}{|}}{\text{C}}}\!\sim\!\!\sim \text{PSt} \sim\!\!\sim$$

A similar series of experiments with styrene in the first step and isobutylene in the second was also carried out (Kennedy et al., 1978) and yielded poly(styrene-*b*-isobutylene).

The pure block copolymers were obtained by selective solvent extraction and characterized as to composition, molecular weight, and molecular weight distribution. The effect of temperature on intrinsic viscosity (toluene solvent, from 20 to 60°C) of a poly(styrene-*b*-isobutylene) sample showed similar discontinuities to that illustrated in Figure 8.1.

These block copolymers could be obtained only by the intermediacy of prepolymers carrying active (tertiary, benzylic) chlorine termini. In this sense the synthesis of these diblocks is analytical proof for the presence of active chlorine termini in polyolefins obtained by the H$_2$O/BCl$_3$ initiator system.

Recently these developments have been extended to the synthesis of diblock copolymers carrying a useful function (e.g., unsaturation) at the head group (Kennedy et al., 1979a, 1980a). It was discovered that BCl$_3$ is an efficient coinitiator in conjunction with a variety of cationogenic substances, that is, substituted allylic and α-substituted benzylic chlorides. For example, the combination of 1-chloro-3-methyl-2-butene/BCl$_3$ can be used to initiate the polymerization of isobutylene which in turn, along the lines presented above, can initiate the sequential polymerization of styrene yielding an α-functionalized diblock:

$$\underset{\underset{\text{CH}_3}{|}}{\overset{\overset{\text{CH}_3}{|}}{\text{C}}}\!=\!\text{CH}\!-\!\underset{\underset{\text{Cl}}{|}}{\text{CH}_2} + i\text{-}C_4H_8 + BCl_3 \rightarrow \underset{\underset{\text{CH}_3}{|}}{\overset{\overset{\text{CH}_3}{|}}{\text{C}}}\!=\!\text{CH}\!-\!\text{CH}_2 \sim\!\!\sim \text{PIB} \sim\!\!\sim \text{CH}_2\!-\!\underset{\underset{\text{CH}_3}{|}}{\overset{\overset{\text{CH}_3}{|}}{\text{C}}}\!-\!\text{Cl}$$

$$\underset{\underset{\text{CH}_3}{|}}{\overset{\overset{\text{CH}_3}{|}}{\text{C}}}\!=\!\text{CH}\!-\!\text{CH}_2 \sim\!\!\sim \text{PIB} \sim\!\!\sim \text{CH}_2\!-\!\underset{\underset{\text{CH}_3}{|}}{\overset{\overset{\text{CH}_3}{|}}{\text{C}}}\!\sim\!\!\sim \text{PSt} \sim\!\!\sim$$

The head unsaturation may be used in further, specific derivatizations and could lead to a host of interesting materials. For example, hydrochlorination of the head group would give rise to a tertiary chlorine atom which could be used to initiate the subsequent blocking of another polystyrene sequence by the use of an alkylaluminum coinitiator (see above). This would be an avenue toward the synthesis of poly(styrene-b-isobutylene-b-styrene), a new very promising triblock copolymer in view of the commercially successful styrene–diene–styrene thermoplastic elastomer triblocks.

An efficient synthesis of poly(α-methylstyrene-b-isobutylene-b-α-methylstyrene) has recently been achieved (Kennedy and Smith, 1980c). The key to this synthesis was the preparation by the inifer technique of telechelic polyisobutylenes containing $-CH_2C(CH_3)_2Cl$ end groups on both termini (see Section 4.3).

Another possible route toward polyisobutylene containing triblocks would be by combination of anionic–cationic techniques. The first step would be the synthesis of a "living" polystyrene which could be linked to an appropriately substituted benzyl dihalide, for example, $X(CH_3)_2C-C_6H_4-C(CH_3)_2X$:

$$\text{wwCH}_2-\underset{\underset{C_6H_5}{|}}{\text{CHLi}} + X(CH_3)_2C-C_6H_4-C(CH_3)_3X$$

$$\rightarrow \text{wwCH}_2-\underset{\underset{C_6H_5}{|}}{\text{CH}}-(CH_3)_2C-C_6H_4-\underset{\underset{C_6H_5}{|}}{\text{C}(CH_3)_2} X + LiX$$

Reactions similar to this one have been studied by Burgess et al. (Burgess et al., 1977) for the sequential block copolymerization of poly(styrene-b-tetrahydrofuran), the discussion of which falls outside the scope of this chapter. The above product could be used in conjunction with BCl_3 to initiate transferless polymerization of isobutylene which terminates by chlorination and produces a treminally functionalized diblock:

$$PSt-(CH_3)_2C-C_6H_4-C(CH_3)_2X + n\text{-}CH_2=\underset{\underset{CH_3}{|}}{\overset{\overset{CH_3}{|}}{C}} \xrightarrow{BCl_3}$$

$$PSt-(CH_3)_2C-C_6H_4-C(CH_3)_2\text{ww}PIB\text{ww}\,CH_2-\underset{\underset{CH_3}{|}}{\overset{\overset{CH_3}{|}}{C}}-Cl$$

which in turn could initiate the cationic polymerization of styrene by the use of an alkylaluminum compound coinitiator.

Higashimura et al. (1979) prepared poly(p-methoxystyrene-b-isobutyl vinyl ether) in a two-step sequential monomer addition process. According to these authors the $I_2/pMeOSt/CCl_4$ system produces long-lived carbocations:

$$I_2 \rightleftharpoons I^{\oplus} \cdots I_3^{\ominus} \xrightarrow{\;p\text{-MeOSt}\;} ICH_2\overset{\oplus}{-}\overset{}{C}H \cdots I_3^{\ominus}$$

benzene ring

OCH₃

which are able to induce the block copolymerization of a subsequently added monomer, for example, IBuVE. The best blocking efficiency was obtained in nonpolar medium; however, even under optimal conditions blocking efficiency was 39%, indicating the presence of an unacceptable extent of chain transfer for blocking by sequential addition. It is to be hoped that his extremely promising lead will be continued, improved, and extended.

Finally, evidence has been presented that demonstrates the existence of diblock copolymers of ethyl or isobutyl vinyl ether and N-vinylcarbazole (Rooney et al., 1976). The synthesis was based on results of fundamental studies which indicated that the polymerization of ethyl vinyl ether initiated by $(C_6H_5)_3CCl \cdot SbCl_5$ is nonterminating (Chung et al., 1975; Ledwith et al., 1975); that is, after ~90% monomer conversion a fresh aliquot of monomer added to the system polymerized with a rate similar to the first (Chung et al., 1975). The first batch of vinyl ether monomer polymerized extremely rapidly and it was very difficult to obtain less than complete conversion in a pure system.

In the blocking experiment, the polymerization of ethyl or isobutyl vinyl ethers was induced by $(C_6H_5)_3CCl \cdot SbCl_5$ and, subsequently, blocking from was effected by introducing N-vinylcarbazole. The authors attempted to find conditions under which chain transfer is minimized; however, this search was only partially successful and the final product was probably badly contaminated by homopolymer sequences. Relatively pure block copolymer fractions have been obtained by selective solvent extraction and some of these fractions exhibited unusual, for block copolymers characteristic, intrinsic viscosity versus temperature trends (Rooney et al., 1976).

Correct assessment of relative polymerizabilities of styrene and aziridines (and thietanes), combined with the knowledge that polystyrenes prepared by $HClO_4$ in CH_2Cl_2 at $-78°C$ contain polystyrene-perchlorate (PSt–OClO₃) end groups, was exploited by Bossaer et al. (1977) for the synthesis of interesting polystyrene–polyaziridine block copolymers, for example, 1-*tert*-butyl- or 1-(2-phenylethyl)-2-methylaziridine. The authors noted that perchlorate esters of PSt are unable to induce the polymerization of styrene, but, are able to initiate the polymerization of stronger nucleophiles, for example, aziridines and thietanes. Syntheses were carried out in two steps: (1) preparation of "dormant" PSt–OClO₃ and (2) initiation of aziridine (or thietane) polymerization by blocking from PSt–OClO₃. The following set of equations helps to visualize the mechanism:

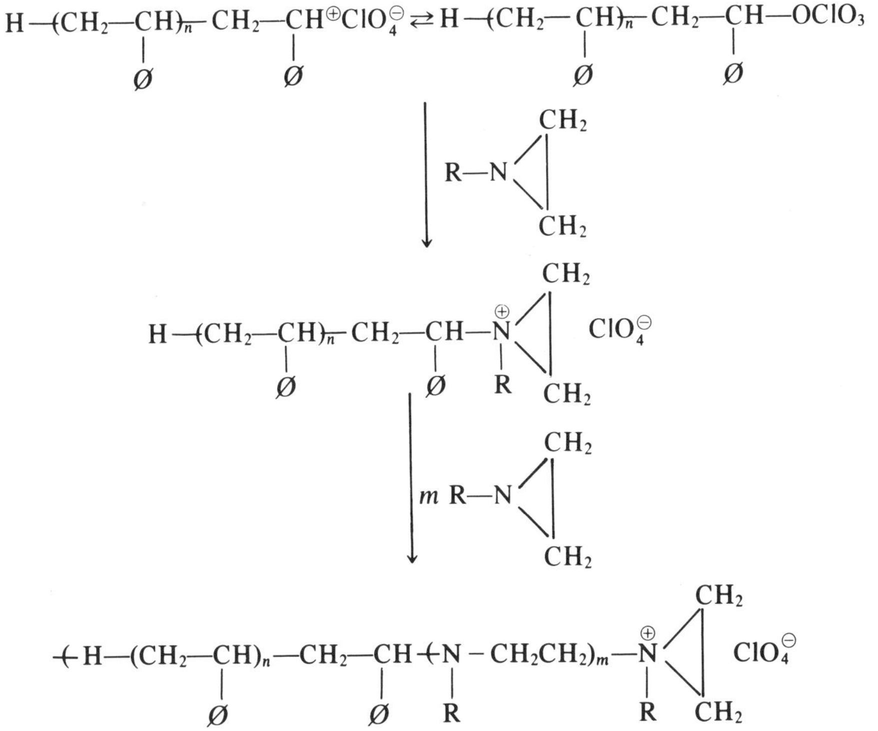

$$H-(CH_2-CH)_n-CH_2-CH^{\oplus}ClO_4^{\ominus} \rightleftarrows H-(CH_2-CH)_n-CH_2-CH-OClO_3$$

where R = *tert*-butyl. Convincing solubility and GPC data substantiate the claim that true uncontaminated block copolymers have been obtained (Bossaer et al., 1977).

A Summary of Block Copolymers

Although the field of block copolymer synthesis by carbocationic techniques is barely six years old, it shows remarkable growth. This area was opened by researchers recognizing the usefulness of controlled initiation of carbocationic polymerizations by alkylaluminum coinitiators. Further advances have been made by applying increased understanding of termination of BCl₃-coinitiated polymerization. A combination of these developments assures the synthesis of a variety of new block copolymers with potentially useful properties.

Progress in this area will be determined solely by the rate of growth of understanding mechanistic details of carbocationic polymerizations in general and those of initiation and termination steps in particular. Although the chances of finding truly "living" carbocationic systems leading to

Table 8.1 □ Block Copolymers by Carbocationic Techniques

Name	Synthesis Steps Used	Remarks	Reference
Poly(isobutylene-b-styrene)	1. $C_6H_5CH_2Cl/Et_2AlCl/i\text{-}Bu^=$ 2. Chloromethylation of head group 3. $St + Et_2AlCl$	Connecting group $-CH_2-C_6H_4-CH_2-$; Qualitative study	Peyrot (1972) Jolivet and Peyrot (1973)
Poly(styrene-b-isobutylene)	1. $Cl(CH_3)_2C-(CH_2)_3-C(CH_3)_2Br/Et_3Al/St$ 2. $PSt\text{-}Br + i\text{-}Bu^= + Et_2AlCl$	Connecting group $-(CH_3)_2C-(CH_2)_3-C(CH_3)_2-$; selective sequential initiation	Kennedy and Melby (1974a, 1975a)
Poly(isobutylene-b-styrene)	1. $H_2O/BCl_3/i\text{-}Bu^=$ 2. PIB-tert-Cl/St/Et_2AlCl		Kennedy et al. (1978)
Poly(styrene-b-isobutylene)	Same except addition sequence of $i\text{-}Bu^=$ and St reversed		Kennedy et al. (1978)
Poly(isobutylene-b-styrene) with unsaturated head group	Same except using 1-chloro-3-methyl-2-butene initiator instead of H_2O	Head group $(CH_3)_2C=CH-CH_2-$	Kennedy et al. (1979a, 1980d)
Poly(isobutylene-b-α-methyl-styrene)	Same except using α-methyl-styrene in second step		Kennedy et al. (1979a)
Poly(ethyl vinyl ether-b-N-vinylcarbazole)	1. $(C_6H_5)_3CCl \cdot SbCl_5/EVE$ 2. NVC	Low yield, qualitative study	Rooney et al. (1976)
Poly(isobutyl vinyl ether-b-N-vinylcarbazole)	Same except using isobutyl vinyl ether	Same	Rooney et al. (1976)

Polymer	Procedure	Comments	Reference
Poly(p-methoxystyrene-b-iso-butyl vinyl ether)	1. p-MeOSt/I$_2$/CCl$_4$/−15°C 2. Addition of IBuVE	Although "living" pzn. claimed, low blocking efficiency ($<39\%$) indicates transfer	Higashimura et al. (1979)
Poly(styrene-b-1-tert-butyl-aziridine) Poly(styrene-b-1-(2-phenyl-ethyl)-2-methyl-aziridine) Poly(styrene-b-3,3-dimethyl-thietane)	1. Synthesis of "dormant" PSt–OClO$_3$ 2. Addition of heterocyclic cpd		Bossaer et al. (1977)
Poly(α-methylstyrene-b-iso-butylene-b-α-methyl-styrene)	1. Telechelic PIB dichloride synthesis by inifer technique 2. Blocking α-MeSt from tert-chloro termini		Kennedy and Smith (1980c)

sequential blocking such as achievable by certain anionic systems are rather dim, the search for cationic systems leading to controlled initiation and termination is continuing since it became quite apparent that such control may lead to many valuable multiblocks.

Several routes to interesting rubbery/glassy diblock copolymers have already been worked out, and these materials exhibit some unusual characteristics. Table 8.1 is a compilation of block copolymers prepared by carbocationic techniques to date.

Potential applications of high molecular weight diblocks may be as blending or emulsifying agents for the mixing of homopolymers. Applications for low molecular weight diblocks may be in lube or motor oil additives, viscosity improvers, thickening agents. Further in the future, one can clearly perceive avenues toward the preparation of di- or triblocks of rubbery/glassy, crystalline/rubbery, or glassy/crystalline polymer sequences with even more promising combinations of physical–mechanical properties.

8.4 □ GRAFT COPOLYMERS

Generalities

Cationic graft copolymerizations is an exciting, growing field of synthetic polymer chemistry, which provides a variety of avenues toward a large number of unique materials with unusual combinations of properties (Kennedy, 1977a). This technique leads to many graft copolymers that cannot be produced by any other method(s), for example, the graft consisting of a poly(vinyl chloride) backbone and butyl rubber branches (PVC-g-IIR), or that of a polybutadiene backbone and polyisobutylene branches (PBd-g-PIB).

Bigrafts, that is, grafts which on one common backbone carry two different branches, are a unique group of materials that have been prepared to date only by carbocationic techniques. Some of these products, for example, a graft comprising a random ethylene–propylene copolymer backbone attached to which are a polystyrene sequence and a polyisobutylene sequence (EPDM-g-PSt-g-PIB), exhibit unexpected solubility, viscosity, and stress–strain properties.

In many instances grafts that can be obtained by other techniques, for example, poly(vinyl chloride-g-styrene) and poly[(ethylene-co-propylene)-g-styrene], can be produced faster, more conveniently, and with fewer side reactions or by-products via cationic techniques than by other methods.

Cationic grafting may lead to pure or essentially pure grafts, obviating laborious homopolymer extraction. Graft copolymerizations, particularly free radical grafting, usually yield low grafting efficiencies and produce, in addition to the graft product, homopolymer(s) or other undesirable by-products.

Although carbocationic graft copolymerization is a relatively young discipline, a large amount of information is already available. Progress became particularly rapid in this field after the discovery in 1965 (Kennedy and Baldwin, 1968; Kennedy, 1971) that certain alkylaluminum compounds in conjunction with certain halogen-containing high polymers are efficient coinitiators for the polymerization of olefins to high molecular weight products at high rates.

A sizable body of original data concerning synthesis, characterization, and physical properties of cationically prepared graft copolymers have recently been organized and published (Kennedy, 1977a). This source also contains a comprehensive literature list and discussion of early developments, that is, experiments carried out before the principle of controlled carbocationic initiation by alkylaluminums (Kennedy and Baldwin, 1968) became known. Prior to this discovery carbocationic grafting methods yielded large amounts of homopolymers together with only insignificant quantities of grafts. Readers seeking information relative to early grafting-from or grafting-onto syntheses should consult Kennedy 1977a, Chapter 1. As shown in this analysis efficient graft copolymer syntheses to date can be achieved only by grafting-from techniques. Very recently a special case of an efficient grafting-onto method was developed for the synthesis of poly(butadiene-g-styrene) (see below).

Synthesis Principles and Graft Characteristics

Early attempts of graft syntheses have been analyzed and criticized in detail (Kennedy, 1977a). With regard to grafting-from hydrogenated backbones, the use of conventional strong Friedel–Crafts acids, for example, $AlCl_3$, BF_3, and $TiCl_4$, led to severe backbone degradation and large amounts of homopolymer formation because these coinitiators readily induce polymerization of cationically active monomers in the presence of protogenic impurities such as moisture. After sporadic work in the 1950s, this field languished until the discovery (Kennedy and Baldwin, 1968) that these disadvantages can be avoided by the use of certain organoaluminum compounds. Thus it has been demonstrated that the milder Friedel–Crafts acids di- and triorganoaluminum compounds do not tend to degrade halogenated backbones and they coinitiate the polymerization of cationic monomers only upon the purposeful addition of suitable cationogens, for example, tertiary, allylic, and benzylic halides. In contrast to the aforementioned strong Friedel–Crafts acids, these alanes do not coinitiate polymerization of even highly reactive monomers in the presence of adventitious protogens, and they can be stirred with reactive olefins in "open" systems (i.e., systems not particularly protected from moisture) without polymerization ensuing. However, on addition of a suitable cation source to quiescent monomer/alane coinitiator systems, immediate and sometimes even explosive polymerization ensues and high polymer is formed. If the cationogen is a macromolecule that contains active halogen, for example, allylic, tertiary, or benzylic chlorine, initiation occurs at the polymer and graft copolymerization commences:

$$P\text{--}X + M + R_2AlX \rightarrow [P^{\oplus}R_2AlX_2^{\ominus} + M] \rightarrow PM^{\oplus}R_2AlX^{\ominus}$$
$$PM^{\oplus} + nM \rightarrow P\text{--}(M)_n\text{--}M^{\oplus}$$

where PX is active halogen-containing polymer and M is cationically active monomer.

The first detailed study on carbocationic graft copolymerizations employing alkylaluminum coinitiators concerned the synthesis of a thermoplastic elastomer poly[(ethylene-co-propylene)-g-styrene], that is, a rubbery EPM backbone carrying glassy branches (Kennedy and Smith, 1974c). A systematic examination of the effect of reaction parameters such as temperature, solvent polarity, and nature of alkylaluminum coinitiator on grafting efficiency was carried out. Conditions have been developed under which grafting efficiencies greater than 85% can be achieved.

The backbone was prepared by lightly chlorinating a commercial EPM sample (1 to 5% chlorine), dissolving it in an inert polar solvent or solvent mixture, introducing the monomer, and finally inducing grafting by adding the alkylaluminum (Et$_2$AlCl, Et$_3$Al, Me$_3$Al) coinitiator. Schematically,

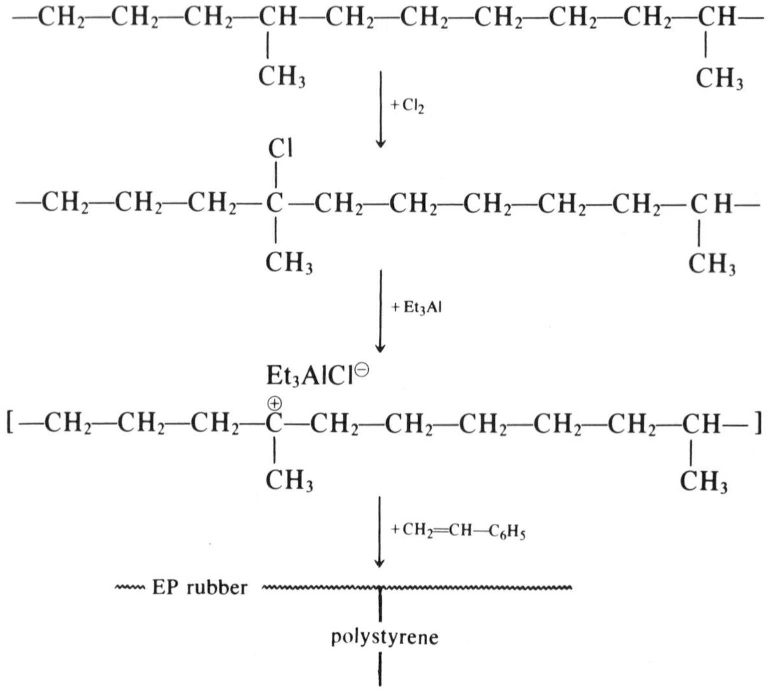

Using essentially the same technique a large variety of new grafts have been prepared by a number of investigators. Table 8.2 is a compilation of reasonably well-characterized graft and bigraft copolymers prepared by carbocationic techniques since 1968. Although insufficient space is available to treat comprehensively all the systems investigated, a brief discussion of generalizable observations and problems is in order.

Grafting initiation requires the simultaneous presence of a macromolecular halide initiator, a suitable coinitiator that is inert toward trace quantities of cationogenic impurities (e.g. H_2O), and monomer. The macromolecular initiator must contain active halides, for example, allylic, benzylic, or tertiary chlorines. The active halides may be introduced into the polymer by lightly halogenating suitable chains as in the case of EPM or by chloromethylating pendant rings such as in polystyrene; they may arise during polymerization, as is the case with polychloroprene or poly(vinyl chloride); or they may be introduced by copolymerizing a suitable comonomer, for example, p-chloromethylstyrene. For example, conventional free radical polymerization of vinyl chloride or chloroprene produces a sufficient number of active chlorines (e.g., allylic or tertiary in the case of vinyl chloride, tertiary allylic with chloroprene) in commercially available macromolecules for efficient grafting. These carbenium ion precursors in the presence of suitable, mainly aluminum-containing, Lewis acids give rise to macromolecular carbocations which in turn induce grafting.

The three components, that is, macromolecular initiator, coinitiator, and monomer, must be carefully matched for optimal performance. For example, styrene can be readily grafted from lightly chlorinated EPM with Et_2AlCl, Me_3Al, or Et_3Al. Fastest rates and highest yields are obtained with Et_2AlCl; however, highest grafting efficiencies are produced by Et_3Al. In this regard the usefulness of Me_3Al falls between the other two alkylaluminums (Kennedy and Smith, 1974c). The grafting of styrene from lightly chlorinated butyl rubber (chlorobutyl rubber) is much faster and gives higher yields and grafting efficiencies than from bromobutyl rubber (Kennedy and Charles, 1977b). Also, the grafting of styrene from chlorobutyl rubber readily occurs in hydrocarbon solvent alone; however, under essentially the same conditions, grafting cannot be induced from lightly chlorinated EPM (Kennedy and Smith, 1974c). The original literature and Kennedy, 1977a should be consulted for a discussion of these observations.

The extent of grafting is quantitatively expressed by grafting efficiency, GE ($100 W_g/(W_g + W_h)$, where W is the weight and subscripts g and h indicate graft or homopolymer); considerable effort has been spent studying the effect of reaction variables on this parameter. The effect of conversion on GE is revealing. In several systems under certain conditions GE is close to 100% and independent of conversion (Kennedy and Smith, 1974c; Kennedy and Charles, 1977b; Oziomek and Kennedy, 1977; Ambrose and Newell, 1979), whereas in other instances GE decreases with

Table 8.2 □ Graft Copolymers Prepared by Carbocationic Techniques

Graft Copolymers	Reference
Elastomeric backbones	
1. Elastomeric branches	
Poly(butadiene-g-isobutylene)[a]	Kennedy (1976a)
Poly(chloroprene-g-isobutylene)	Kennedy and Metzler (1977f)
Poly[chloroprene-g-(isobutylene-co-isoprene)]	Kennedy and Baldwin (1968)
Poly(chloroprene-g-isobutyl vinyl ether)	Kennedy and Plamthottam (1979b)
Poly[(styrene-co-butadiene)-g-isobutylene][a]	Oziomek and Kennedy (1977)
Poly[(isobutylene-co-isoprene)-g-chloroprene][a]	Kennedy and Baldwin (1968)
2. Glassy branches	
Poly[(ethylene-co-propylene)-g-styrene][a]	Kennedy and Smith (1974c)
Poly[(ethylene-co-propylene)-g-indene][a,b]	Sigwalt et al. (1976)
Poly[(isobutylene-co-isoprene)-g-styrene][a,b]	Kennedy and Charles (1977b)
Poly[(isobutylene-co-isoprene)-g-α-methylstyrene][a]	Kennedy and Baldwin (1968), Kennedy (1976a)
Poly[(isobutylene-co-isoprene)-g-p-chlorostyrene][a]	Kennedy and Baldwin (1968)
Poly[(isobutylene-co-isoprene)-g-indene][a]	Kennedy (1976a), Sigwalt et al. (1976)
Poly[(isobutylene-co-isoprene)-g-(indene-co-α-methylstyrene)][a]	Sigwalt et al. (1976)
Poly[(isobutylene-co-isoprene)-g-acenaphthylene][a]	Kennedy (1976a)
Poly[(isobutylene-co-p-chloromethylstyrene)-g-indene]	Pary et al. (1978)
Poly[isobutylene-co-p-chloromethylstyrene)-g-α-methylstyrene]	Pary et al. (1978)
Poly(chloroprene-g-styrene)	Kennedy and Baldwin (1968)
Poly(butadiene-g-α-methylstyrene)[a]	Ambrose and Newell (1979)
Chlorosulfonated polyethylene-g-polystyrene	Kennedy and Baldwin (1968)
Poly[(p-chloromethylstyrene-co-butadiene)-g-styrene]	Janovic and Saric (1976)

3. Two branches (bigrafts)
   A. A glassy and an elastomeric branch
      Poly[(ethylene-co-propylene-co-1,4-hexadiene)-g-styrene-g-isobutylene][a,b]   Kennedy and Vidal (1975b, c), Vidal and Kennedy (1976)
      Poly[(ethylene-co-propylene-co-1,4-hexadiene)-g-α-methylstyrene-g-   Vidal and Kennedy (1976)
        isobutylene][a,b]
   B. Two glassy branches
      Poly[(ethylene-co-propylene-co-1,4-hexadiene)-g-styrene-g-α-methylstyrene][a,b] Kennedy and Vidal (1975b, c); Vidal and Kennedy (1976)

Glassy backbones

1. Elastomeric branches
   Poly(vinyl chloride-g-isobutylene)   Kennedy and Davidson (1977c)
   Poly[vinyl chloride-g-(isobutylene-co-isoprene)]   Kennedy and Davidson (1976b)
   Chloromethylated polystyrene-g-polyisobutylene   Jolivet and Peyrot (1973); Franta et al. (1977);
   Chlorinated polyvinyl chloride-g-polybutadiene   Heublein et al. (1977)
   Chlorinated polycyclopentadiene-g-polybutadiene   Heublein and Freitag (1978)
2. Glassy branches
   Poly(vinyl chloride-g-styrene)   Kennedy and Nakao (1977g)
   Poly(vinyl chloride-g-2-methyl-2-oxazoline)   Trivedi and Schulz (1980)

---

[a] Lightly chlorinated backbone used.
[b] Lightly brominated backbone used.

increasing conversions. Independence of GE of conversion is most likely due to rapid termination and relatively slow chain transfer. In consonance with this proposition very high GE's were usually obtained by using Et_3A or Me_3Al, that is, with coinitiators exhibiting fastest hydridating or methylating capability, or at low temperatures where transfer is slowest. It has also been noted that under certain conditions GE decreases with increasing conversions (Kennedy and Charles, 1977b; Kennedy and Davidson, 1977c; Kennedy and Metzler, 1977f). This trend is understandable considering that graft initiation can occur only at the macromolecular carbenium ion and GE *must* be 100% at infinitesimal conversion. Exceptions to this rule indicate peculiarities in the kinetics, which may be revealing for a deeper understanding of the system (Sigwalt et al., 1976; Kennedy and Davidson, 1977c).

Recent research has shown that grafting efficiencies close to 100% can be achieved in the poly(chloroprene-g-α-methylstyrene) and chlorobutyl rubber-g-poly(α-methylstyrene) systems by the use of proton traps (Kennedy and Guhaniyogi, 1980b).

The polarity of the medium strongly affects grafting details and it may even become decisive for grafting. For example, the grafting of styrene from lightly chlorinated EPM with Et_2AlCl does not occur in n-hexane at $-50°C$; however, grafting occurs readily using a medium containing methyl chloride in addition to hydrocarbon solvent. Indeed, a linear relation seems to prevail between polar solvent, for example, CH_3Cl, content and grafting efficiency (Kennedy and Smith, 1974c; Kennedy and Charles, 1977b). In general, grafting is faster and grafting efficiencies are higher using polar solvent than nonpolar.

The effect of temperature on grafting was often the subject of experimental inquiry. In general, lowering the temperature increases grafting efficiency and reduces grafting rate. These effects are understandable because lowering the temperature reduces the relative rates of chain transfer and initiation.

As would be expected, increasing monomer concentration increases grafting rate and branch concentration. For example, increasing the styrene concentration in the range from 0.5 to 1.5 M increases the rate almost eightfold and polystyrene branch concentration from 16 to 31% (Kennedy and Smith, 1974c).

Carbocations are notoriously reactive species and may undergo a variety of ill-controlled reactions. Detailed investigations concern the elucidation of unexpected phenomena during cationic grafting and interpreting them in terms of carbocationic reactivity (Sigwalt et al., 1976; Kennedy, 1977a). For example, in experiments in which aromatic compounds participate, for example, styrene, indene, and polystyrene, the possibility of ring alkylation exists, which may result in undesirable crosslinking, particularly at high monomer conversions (Kennedy, 1977a; Kennedy and Charles, 1977b).

Recently it has been found (Kennedy, 1976a) that insoluble grafts of chlorobutyl rubber-g-polystyrene obtained under a variety of experimental conditions can be completely and permanently solubilized by brief hot milling. Table 8.3 shows some representative data. After milling, toluene-insoluble samples became completely soluble and intrinsic viscosities could be determined. Further, hot milling markedly improves tensile strength, modulus, and elongation. Conceivably, severely entangled graft copolymers that refuse to dissolve, even on prolonged shaking in toluene at room temperature, disentangle and become soluble under high shear and temperature.

Attention is drawn to the relatively low (0.6 to 0.7) intrinsic viscosity of these grafts. It is very difficult to account for these modest values since the intrinsic viscosity of the "naked" chlorobutyl backbone was 1.18 dl/g (diisobutylene "dimer", 20°C) and 40 to 60% polystyrene was grafted from it. That degradation probably did not occur is indicated by the remarkably high tensile, modulus, and elongation values and by the increase in these values during milling. Similarly, peculiar intrinsic viscosity trends have also been observed with the PVC-g-PSt system (Kennedy and Nakao, 1977g).

It is worth mentioning that clear molded sheets of chlorobutyl-g-PSt can be obtained by using silicone-coated Mylar films. Teflon-coated aluminum foils with satin or glossy finish gave inferior clarity and gloss (Kennedy, 1976a).

The possibility of grafting by ring alkylation has been studied (Kennedy, 1976a) by mixing equal weights of polystyrene and chlorobutyl rubber in a common solvent (methylcyclohexane, and n-hexane–ethyl chloride mixtures) and adding alkylaluminum (Et$_2$AlCl and Et$_3$Al) in the range −50 to +22°C. This alkylation method may be regarded as a coupling technique. Conditions can be found under which polystyrene is readily alkylated by chlorobutyl; however, the reaction is invariably accompanied by severe molecular weight degradation of the rubber, which cannot be prevented by altering conditions, for example, decreasing the temperature or changing solvent polarity or concentrations.

The schematic energy profile of grafting from, coupling (alkylation), and backbone degradation occurring in the chlorobutyl/styrene/Et$_2$AlCl system shown in Figure 8.2 helps to visualize the energetics of these competing reactions.

The ground state shows the most likely active chlorine site in chlorobutyl. The next higher energy stage is the macrocation obtained after halogen loss to the Lewis acid. In the absence of polystyrene or styrene, chlorobutyl would rapidly degrade to a variety of lower molecular weight fragments (Kennedy and Phillips, 1970). On balance, the cleavage of the macrocation represents a gain in entropy and leads to improved solvation and to the formation of branched olefins with the simultaneous loss of a C–C bond.

Table 8.3 □ The Effect of Milling on Poly[(isobutylene-co-isoprene)-g-styrene][a]

Milling 20 min at 120°C	Analysis			Physical Characteristics				
	GE (%)	Relative Composition IIR/PSt(w/w)	$[\eta]^{20}$ dimer (dl/g)	Appearance of Molded Sheets	Tensile (kg/cm^2)	Modulus (300%) (kg/cm^2)	Elong. (%)	Mooney 8 at 125°C
Before	73	59/41	Insoluble	Clear, transparent	107	94	420	
After			0.69	Clear, transparent	133	115	425	32
Before	76	60/40	Insoluble	Clear, transparent	98	84	450	
After			0.61	Clear, transparent	120	95	480	
Before	77	43/57	Insoluble	Hazy	100	—	60	
After			0.66	Hazy	137	—	240	40.5

[a] Synthesis: $Et_2AlCl/EtCl$-n-C_6H_{14} (45:65) − 15°C; [St] = 0.58 M; St conversion = 60–90%.

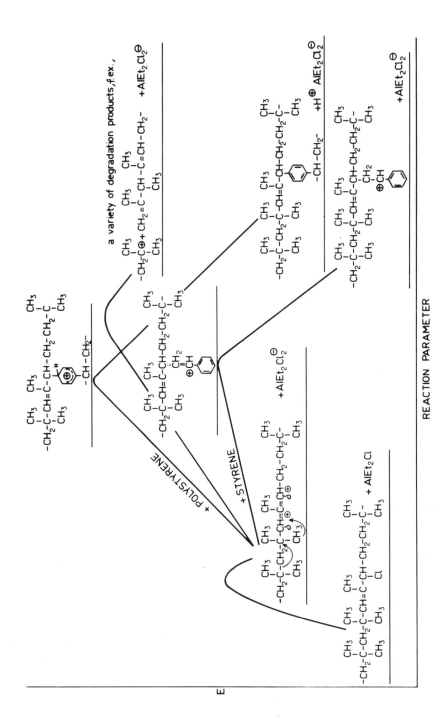

Figure 8.2 □ Comparative energy profiles of alkylation, degradation, and grafting reactions involving chlorobutyl.

In the presence of polystyrene the possibility for alkylation exists; however, this route is not favorable. Energy is required to compensate for the loss of aromaticity in the transition state; also the C–H bond lost is stronger than the C–C bond gained. Steric compression in the transition state must also significantly retard the rate of alkylation relative to macromolecular cleavage.

Grafting styrene from the macrocation is energetically most favorable: a new C–C bond is gained and except for some decrease in entropy very little if any energy is sacrificed. The transition state is far less compressed than that involved in alkylation.

Direct competition experiments showed that grafting styrene from chlorobutyl is greatly preferable to alkylation of polystyrene. In these experiments chlorobutyl rubber was stirred with mixtures of styrene and polystyrene and coinitiator (Et_2AlCl) was added. Product analysis showed preferential grafting, that is, respectable grafting efficiencies and thermoplastic elastomer properties (stress–strain). In contrast, if styrene monomer was omitted from the charge, alkylation accompanied by severe chlorobutyl degradation was obtained and although coupling did occur the products had no strength and showed the presence of a low molecular weight component (degraded rubber).

The products obtained by alkylation are ill-defined mixtures, probably containing a large amount of "block" copolymers which arise during cationic chain cleavage:

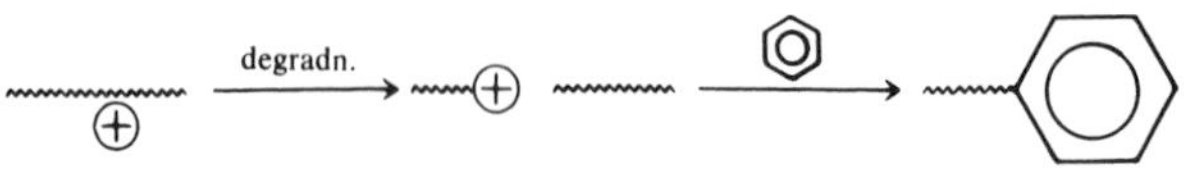

Bigraft Copolymers

The principle of selective sequential initiation used to prepare some block copolymers (Section 8.3) has been extended to the synthesis of graft copolymers (Kennedy and Vidal, 1975b, c; Vidal and Kennedy, 1976). It was theorized that the orders of magnitude difference (Kennedy and Melby, 1975a) of methylation rates of *tert*-butyl chloride and bromide by Me_3Al can be exploited for the synthesis of bigraft copolymers, that is, grafts built with two different branch polymer sequences attached to a common third backbone sequence. The plan was to chlorobrominate a suitable polymer and thus to obtain active (tertiary or allylic) chlorine and bromine sites, and to use this polymer in conjunction with suitable alkylaluminum coinitiators to initiate selectively first the polymerization of monomer A from the chlorine sites and, subsequently, the second polymerization of monomer B from the bromine sites. Schematically,

Step 1:

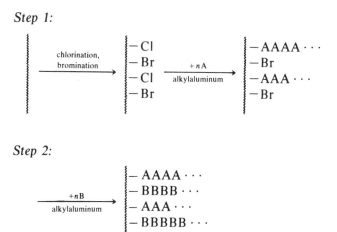

Step 2:

A random copolymer of ethylene–propylene–1,4-hexadiene (47.5–49.5–3.0) (EPDM, Nordel®) was selected to function as backbone. It was first chlorinated and subsequently brominated by t-BuOCl and t-BuOBr, which produced the needed allylic halogens. Using this backbone a variety of detailed studies have been carried out (Kennedy and Vidal, 1975b). Usually styrene was the first monomer that was grafted (Et$_2$AlCl, n-heptane/ethyl chloride solvent, $-30°C$) and after purification and characterization this active bromine-containing monograft was subjected to the second grafting step using α-methylstyrene or isobutylene (Et$_2$AlCl, n-heptane/ethyl chloride, $-20°C$). The correct names of the products obtained are poly(ethylene-co-propylene-co-1,4-hexadiene)-g-styrene-g-α-methyl-styrene] and poly[(ethylene-co-propylene-co-1,4-hexadiene)-g-styrene-g-styrene-g-isobutylene]. The most demanding operation was the purification of the bigrafts to obtain pure materials for characterization and physical property studies.

A bigraft containing PαMeSt and PIB branches on an EPDM backbone has also been described (Vidal and Kennedy, 1976).

A variety of bigraft compositions were prepared and their properties, for example, stress–strain behavior, investigated. Interestingly, bigrafts showed only two glass temperatures (T_g). For example, an EPDM-g-PSt-g-PαMeSt containing 63% EPDM, 19% PSt, and 18% PαMeSt showed a T_g for the EPDM phase at $\sim -53°C$ and only one high temperature T_g at $\sim 139°C$. An EPDM-g-PSt-g-PIB (47–32–21%) exhibited T_g's at -35 and 100°C (Kennedy and Vidal, 1975c; Vidal and Kennedy, 1976). Evidently, the PSt and PαMeSt phases coalesce in the first bigraft, and the EPDM and PIB phases approach the behavior of a microhomogeneous phase in the second. Interesting blending options can be envisioned by the use of these three-component two-phase materials.

Surface Grafting

The principle of graft synthesis by the grafting-from technique using alkylaluminum coinitiators has been extended to heterogeneous systems. It was theorized and subsequently demonstrated that various solids carrying suitable cation precursors on or close to the surface could be used to initiate carbocationic grafting or olefins. The first indication that hetero-grafting was indeed feasible involved the use of PVC slurries and isobutylene in aliphatic hydrocarbon media (Thame et al., 1972). The lead was extended by Vidal et al. (1977), who showed that PVC slurries and films in n-heptane–1,2-dichloroethane mixtures readily initiate surface grafting of isobutylene and styrene in the presence of Et_2AlCl at $-20°C$. Scanning electron microscopy revealed sandwiches of two outer layers (PIB) surrounding an inner core (PVC). The outer skins could be physically peeled off the core and characterized separately. Recently Vidal et al. (1980) demonstrated grafting isobutylene from chemically modified silica surfaces. Thus surface –OH groups of well-characterized silica have been treated with $Cl–Si(CH_3)_2CH_2CH_2C_6H_4CH_2Cl$ to provide

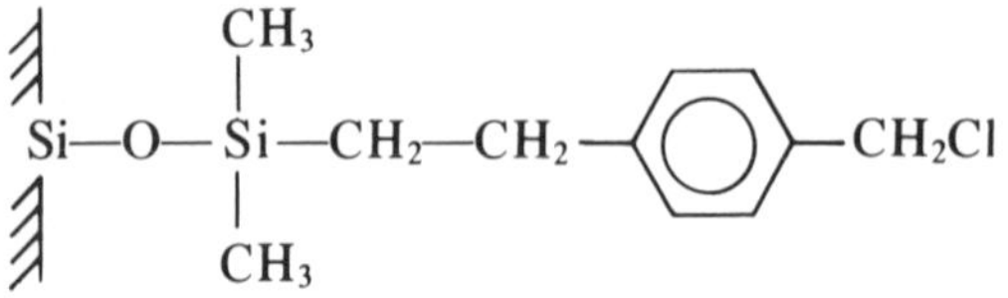

and the solids have been subsequently employed to initiate the polymerization of isobutylene in conjunction with Et_2AlCl in various slurries.

Grafting of olefins from plasma-chlorinated polypropylene fibers has been demonstrated (Simionescu et al., 1979). Crystalline polypropylene fibers were purified by solvent extractions to remove additives introduced during textile fiber manufacture and chlorinated by sending the fiber through a cold-plasma chamber at various speeds. The chlorinated fiber emerging from the glow-discharge plasma apparatus was carefully washed and subsequently used to initiate grafting of styrene, 2-vinylnaphthalene, 4-vinylbiphenyl, and N-vinylcarbazole in n-heptane or n-heptane–1,2-dichloroethane slurries at $-16°C$. Characterization studies including scanning electron microscopy indicated surface grafting.

An Efficient Grafting Onto: The Synthesis of Poly(Butadiene-g-Styrene)

Advantages and disadvantages of grafting from and grafting onto have been analyzed (Kennedy, 1977a) and it was concluded that grafting onto is a relatively inefficient synthesis method that must result in large amounts of homopolymer contamination. In order to avoid homopolymer formation, grafting onto demands the attack of a growing carbocation on a functional group, for example, unsaturation, on a backbone prior to chain transfer to monomer *and* the immediate cessation of propagation on link up with the backbone polymer. In addition to these demands, which are extremely difficult to meet, thermodynamically unfavorable polymer–polymer intermingling is required to ensure the addition of the attacking growing chain. Thus it is not very surprising that grafting-onto experiments invariably yielded poor grafting efficiencies.

Recently Kennedy and Delvaux (1981) surprisingly found efficient gel-free grafting onto in the t-BuCl/Et$_2$AlCl/styrene/PBd/ethyl chloride–n-heptane/-35 to $45°C$ system. Research leading to this development started by grafting styrene from chlorinated polybutadiene (Cl–PBd) in the presence of Et$_2$AlCl; however, gelation invariably occurred at $\sim 10\%$ conversion. Evidently the polybutadienyl cation $-CH_2-CH=CH-\overset{\oplus}{C}H-$ formed by loss of an allylic chlorine from Cl–PBd is able to approach pendant or in-chain double bonds on polybutadiene chains and bring about gelation. Schematically,

Cl–PBd + Et$_2$AlCl →

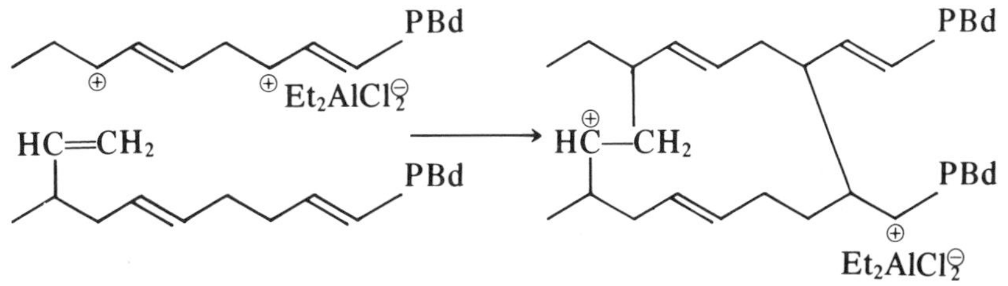

In contrast, in grafting-onto experiments the polymerization of styrene initiated by t-BuCl–Et$_2$AlCl combinations in the presence of polybutadiene under various solvent/temperature conditions occurs in the absence of gelation. Grafting efficiencies up to $\sim 50\%$ can be obtained readily. According to model experiments grafting involves the attack of growing polystyrene cations on 1,4 or 1,2 unsaturations in the PBd chain, followed by rapid termination (mainly by hydridation) of the aliphatic carbocation; for example,

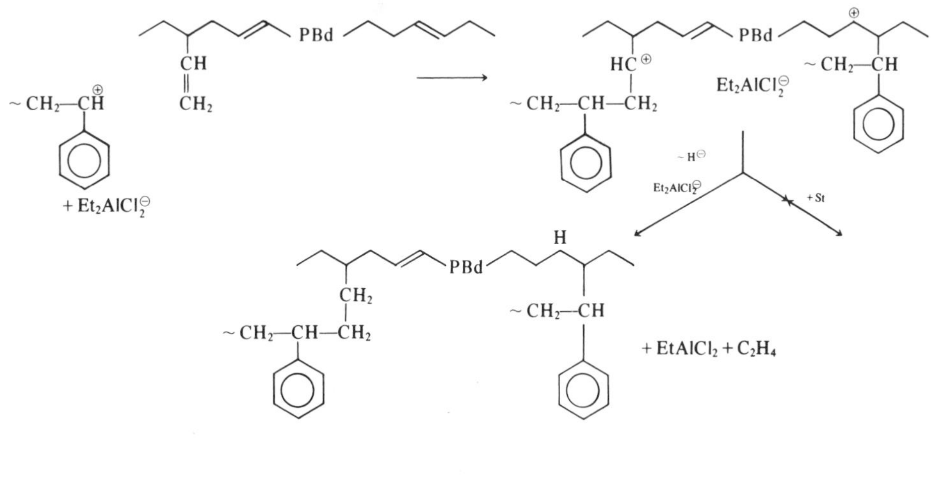

Evidently, steric hindrance created around the newly formed cation by the three adjacent polymer sequences (two due to the polybutadiene backbone and one to the PSt branch) prohibits propagation or crosslinking, that is, the approach of another monomer or polymer segment. The most favorable event subsequent to grafting onto is termination by hydridation (Kennedy and Delvaux, 1981).

Graft Blocks

The discovery that BCl_3-coinitiated isobutylene polymerization terminates by chlorination of the propagating tertiary carbenium ion and yields $-CH_2C(CH_3)_2Cl$ end groups led not only to intersecting possibilities in block copolymer synthesis (Section 8.3) but also to the synthesis of more complex sequential copolymers, for example, the assembling of graft blocks. In this regard it is also important that BCl_3 is much less sensitive to initiation with adventitious protogenic impurities than, say, $AlCl_3$ or BF_3; indeed, depending on conditions (temperature, medium, polarity) BCl_3 demands the purposeful addition of efficient cationogens for initiation.

Thus recently the first synthesis of a graft block was accomplished (Kennedy and Plamthottam, 1979d) by exploiting the possibilities offered by coinitiation with BCl_3. The first step in this synthesis was the grafting of isobutylene from polychloroprene by BCl_3. The graft, the PIB branches of which contained tertiary chlorine termini, was subsequently mixed with αMeSt in a suitable solvent, and blocking from was effected by the introduction of Et_2AlCl coinitiator. The tertiary chlorine terminus in conjunction with Et_2AlCl efficiently initiates the polymerization of αMeSt:

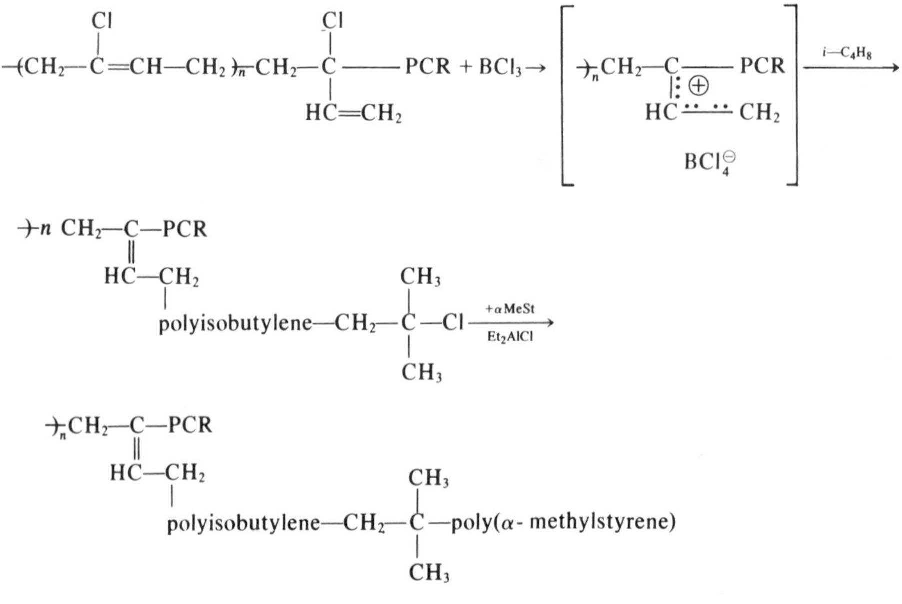

where PCR indicates polychloroprene rubber, whose major repeating unit is shown only once for the sake of brevity in the first formula (the amount of allylic structures in PCR is ~ 1.5%). Grafting is visualized to occur at the sterically less encumbered primary allylic site.

Grafts by Macromers

In addition to the more conventional graft copolymer synthesis methods, grafts can also be produced by the use of macromers (macromolecular monomers). The first graft synthesis, that is, the synthesis of poly(butyl acrylate-g-isobutylene, to involve carbocationic techniques has recently been accomplished (Kennedy and Frisch, 1980a).

The first step toward the synthesis of this graft copolymer was the preparation of a polyisobutylene macromer, that is, a polyisobutylene containing a styryl head group, by the following route:

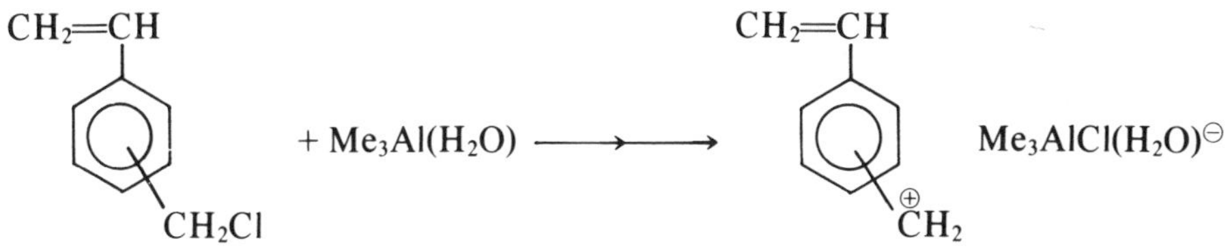

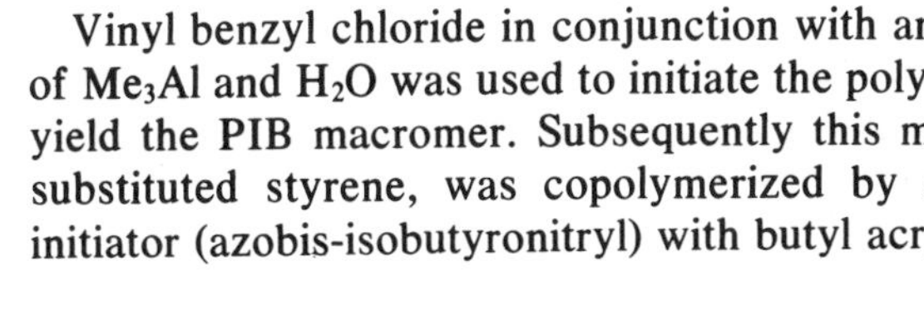

$$CH_2{=}CH{-}C_6H_4{-}CH_2Cl + Me_3Al(H_2O) \longrightarrow CH_2{=}CH{-}C_6H_4{-}\overset{\oplus}{C}H_2 \quad Me_3AlCl(H_2O)^{\ominus}$$

$$CH_2{=}CH{-}C_6H_4{-}\overset{\oplus}{C}H_2 \xrightarrow{+St} CH_2{=}CH{-}C_6H_4{-}CH_2\ polystyrene$$

Vinyl benzyl chloride in conjunction with an initiating system consisting of Me_3Al and H_2O was used to initiate the polymerization of styrene and to yield the PIB macromer. Subsequently this macromer, which in fact is a substituted styrene, was copolymerized by a conventional free radical initiator (azobis-isobutyronitryl) with butyl acrylate:

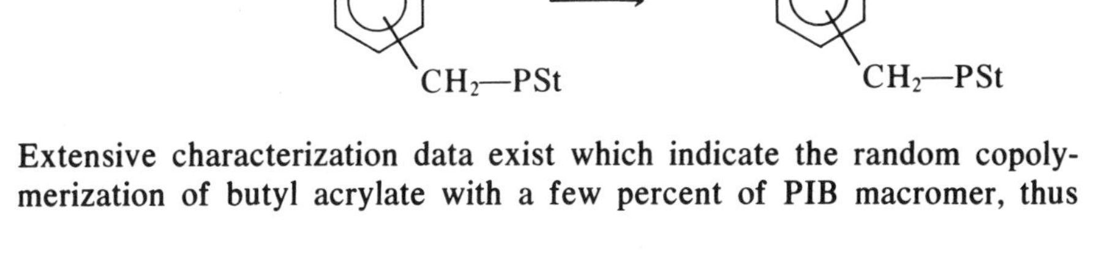

$$CH_2{=}CH{-}COOBu + CH_2{=}CH{-}C_6H_4{-}CH_2{-}PSt \xrightarrow{R^{\cdot}} {\sim}CH_2{-}CH({-}COOBu){-}CH_2{-}CH({-}C_6H_4{-}CH_2{-}PSt){-}$$

Extensive characterization data exist which indicate the random copolymerization of butyl acrylate with a few percent of PIB macromer, thus yielding a new graft copolymer whose structure (except the styryl units in the trunk) is essentially that of a poly(butyl acrylate-g-isobutylene).

Conclusions

Although sporadic efforts for the synthesis of graft copolymers by carbocationic methods were made some 20 years ago (for a review of early, now obsolete, investigations consult Kennedy, 1977a), interest in this field has recently suddenly increased mainly because of two developments: (1) the recent spectacular commercial success of sequential copolymers, notably those containing polystyrene and polydiene blocks, and (2) recently developed cationic syntheses together with an increased insight into the mechanism of carbocationic polymerization mechanisms. Although an analysis of commercial implications is beyond the scope of this book, a few brief remarks about the second reason, that is, new synthesis methods and new mechanistic information, are in order.

Sustained recent research into cationic graft copolymerizations could not have come about without the discovery of controlled carbocationic initiation (Kennedy and Baldwin, 1968). Prior to the finding that carbocationic initiation of olefins can be effected cleanly and rapidly by two-component initiator systems (e.g., RCl/Et_2AlCl), experiments directed toward the cationic synthesis of grafts, or any sequential copolymers for that matter, were doomed to failure. These efforts remained unsuccessful because conventional cationic coinitiators, for example, $AlCl_3$ and BF_3, produced more homopolymers than desired grafts. The fundamental difficulty was uncontrolled initiation by impurities, mainly by ubiquitous water. For example, Kockelbergh and Smets (1958), who attempted to initiate the polymerization of isobutylene by bromomethylated polystyrene plus $AlBr_3$, must have produced much more homopolymer by initiation with water than graft by the benzyl cation:

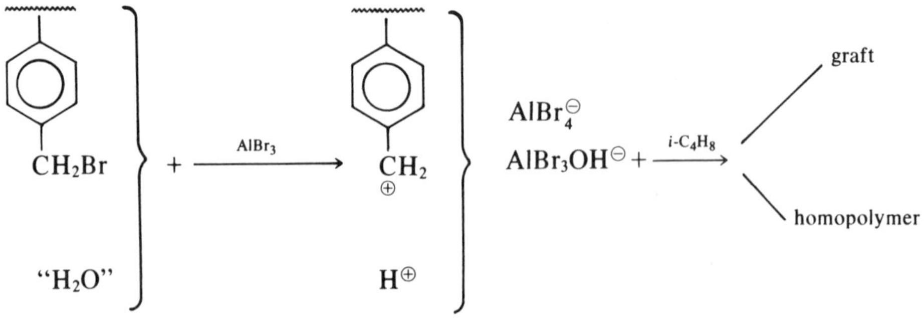

Grafting initiation became controllable and efficient after it was shown that certain Lewis acids, for example, Et_3Al and Et_2AlCl, do not coinitiate the polymerization of olefins in the presence of trace amounts of water and that initiation with these agents requires a cation source (cationogen) purposefully added in quantities sufficient to effect initiation. Grafting efficiencies increased dramatically by the use of these coinitiators and in several instances complete monomer utilization (100% GE) could be obtained.

It is quite apparent that against the backdrop of today's preparative capabilities and projections, among the three basic methods for the synthesis of graft copolymers, that is, grafting from, grafting onto, and coupling of sequences, only grafting from is of promise for efficient syntheses; the other two methods have been and probably will remain only of limited significance. The fundamental difficulty with grafting onto is that in order to obtain the desired graft copolymer two different polymer sequences must be mixed; however, thermodynamics in most cases forbids homogeneous polymer blends or solutions. Polymer sequences of reasonably high molecular weights mix only in exceptional cases either in solution or in the solid phase. Coupling of sequences is even more difficult since in these instances, in addition to the problem of miscibility, the question of stoichiometry between the functional groups on the sequences to be coupled also arises. Grafting from is relatively free from these difficulties: branch-forming monomer molecules usually can readily penetrate the backbone polymer and after graft initiation, and in the absence of specific polymer–monomer or polymer–polymer interactions, the kinetics of grafting approaches that of a homopolymerization. Thus it appears to be almost impossible to reach 100% grafting efficiency by either grafting onto or coupling methods; however, well-documented cases exist in which complete grafting by grafting-from occurs.

Although controlled initiation led to substantial improvement in grafting efficiencies by eliminating adventitious homopolymer formation via initiation, homopolymer still could arise by chain transfer. This problem was alleviated by the finding that with certain coinitiators (Et_3Al, Me_3Al, BCl_3), termination was faster than chain transfer. Experimentally in these systems homopolymerization was essentially completely eliminated; however, yields remained low, owing to fast termination. For example, the grafting of polystyrene from chlorinated poly(ethylene-co-propylene) with Et_3Al gave nearly 100% GE at −35°C or below. Evidently termination by hydridation in these instances is faster than chain transfer. Blocking and grafting in the presence of proton traps has yielded much improved GE's.

As a consequence of these developments, progress in the field of carbocationic grafting was remarkable over the past decade and at present the synthetic chemist has many options to tailor-build a large variety of graft copolymers. Table 8.2 is a compilation of grafts synthesized since 1968, since the discovery of controlled cationic initiation.

Interesting extensions of these grafting possibilities are bigrafts and graft blocks (Section 8.4).

In view of the large number of relatively inexpensive monomers, backbones, and coinitiators employable by cationic techniques, continued interest in this field seems to be assured. Future research should focus on the adaptation of these grafting methods to slurry or solventless systems, which are more cost-effective than syntheses in solution. Some preliminary experiments along these lines indicate that grafting from PVC or silica slurries (Section 8.4) is possible. In principle it should be feasible to develop grafting methods using solids on the mill, extruder, or internal mixer, leading to substantial cost reductions.

□ REFERENCES

Ambrose, R. J., and Newell, J. J. (1979), *J. Polym. Sci., Polym. Chem. Ed.* **17**, 2129.

Bossaer, P. K., Goethals, E. J., Hackett, P. J., and Pepper, D. C. (1977), *Eur. Polym. J.* **13**, 489.

Burgess, F. J., Cunliffe, A. V., MacCallum, J. R., and Richards, D. H. (1977), *Polymer* **18**, 726.

Ceresa, R. J. (1962), *Block and Graft Copolymers*, Butterworths, London.

Chung, Y. J., Rooney, J. M., Squire, D. R., and Stannett, V. T. (1975), *Polymer* **16**, 527.

Dondos A. (1966), *Makromol. Chem.* **99**, 275 (1966).

Franta, E., Rempp, P., and Atshar-Taromi, F. (1977) *Makromol. Chem.* **178**, 2139.

Heublein, G., Freitag, W., and Thiel, H.-J. (1977), *Faserforsch. Textiltech.* **28**, 505.

Heublein, G., and Freitag, W. (1978), *J. Pract. Chem.* **320**, 725.

Higashimura, T., Mitsuhashi, M., and Sawamoto, M. (1979), *Macromolecules* **12**, 178.

Janovic, Z., and Saric, K. (1976), *Croat. Chem. Acta* **48**, 59.

Jolivet, Y., and Peyrot, J. (1973) *Int. Symp. Cationic Polym. Rouen, Fr., Sept., Abstr.* C18.

Kennedy, J. P., and Baldwin, F. P. (1968), Belg. Patent 701,850 (Jan. 26, 1968); U.S. Patent 3,904,708 to ESSO Research and Engineering Co. (Sept. 9, 1975).

Kennedy, J. P., and Phillips, R. P. (1970), *J. Macromol. Sci. Chem.*, **A4**, 1759.

Kennedy, J. P. (1971), paper presented at *23rd IUPAC Meet., Boston, Macromol. Prepr.* **1**, 105.

Kennedy, J. P., Desai, N. V., and Sivaram, S. (1973), *J. Am. Chem. Soc.* **95**, 6386.

Kennedy, J. P., and Melby, E. G. (1974a), *Polym. Prepr.* **15**, 180.

Kennedy, J. P., and Rengachary, S. (1974b), *Adv. Polym. Sci.* **14**, 1.

Kennedy, J. P., and Smith, R. R. (1974c), *Recent Advances in Polymer Blends, Grafts and Blocks*, L. H. Sperling, Ed., Plenum Press, New York, pp. 303–359.

Kennedy, J. P., and Melby, E. G. (1975a), *J. Polym. Sci. Chem. Ed.* **13**, 29.

Kennedy, J. P., and Vidal, A. (1975b), *J. Polym. Sci. Chem. Ed.*, **13**, 1765.

Kennedy, J. P., and Vidal, A. (1975c), *J. Polym. Sci. Chem. Ed.* **13**, 2269.

Kennedy, J. P. (1976a), unpublished results, Akron, Ohio.

Kennedy, J. P., and Davidson, D. L. (1976b), *J. Polym. Sci. Chem. Ed.*, **14**, 153.

Kennedy, J. P., Feinberg, S. C., and Huang, S. Y. (1976c), *Polym. Prepr.* **17**, 194.

Kennedy, J. P., Ed. (1977a), *Cationic Graft Copolymerization, J. Appl. Polym. Sci., Appl. Polym. Symp.* **30**.

Kennedy, J. P., and Charles, J. J. (1977b), *J. Appl. Polym. Sci. Appl. Polym. Symp.* **30**, 119.

Kennedy, J. P., and Davidson, D. L. (1977c), *J. Appl. Polym. Sci. Appl. Polym. Symp.* **30**, 13.

Kennedy, J. P., Huang, S. Y., and Feinberg, S. C. (1977d), *J. Polym. Sci. Chem. Ed.* **15**, 2801.

Kennedy, J. P., Huang, S. Y., and Feinberg, S. C. (1977e), *J. Polym. Sci. Chem. Ed.* **15**, 2869.

Kennedy, J. P., and Metzler, D. K. (1977f), *J. Appl. Polym. Sci. Appl. Polym. Symp.* **30**, 141.

Kennedy, J. P., and Nakao, M. (1977g) *J. Appl. Polym. Sci. Appl. Polym. Symp.*, **30**, 73.

Kennedy, J. P., Feinberg, S. C., and Huang, S. Y. (1978), *J. Polym. Sci. Chem. Ed.*, **16**, 243.

Kennedy, J. P., Huang, S. Y., and Smith, R. A. (1979a), *Polym. Bull.* **1**, 379.

Kennedy, J. P., and Plamthottam, S. S. (1979b), *J. Macromol. Sci. Chem.*, **A14**, 729.

Kennedy, J. P., and Plamthottam, S. S. (1979c), *Polym. Prepr.* **20**, 98.

Kennedy, J. P., and Frisch, K. C., Jr. (1980a) *Preprints, Macro Florence, IUPAC*, **vol. 2**, 162.

Kennedy, J. P., and Guhaniyogi, S. (1980b), unpublished data.

Kennedy, J. P., and Smith, R. A. (1980c), *J. Polym. Sci. Polym. Chem. Ed.* **18**, 1523.

Kennedy, J. P., Huang, S. Y., and Smith, R. A. (1980d), *J. Macromol. Sci. Chem.* **A14**, 1085.

Kennedy, J. P., and Delvaux, M.-J. (1981), *Adv. Polym. Sci.*, **38**, 141.

Kockelbergh, G., and Smets, G. (1958), *J. Polym. Sci.* **33**, 227.

Ledwith, A., Lockett, E., and Sherrington, D. (1975), *Polymer* **16**, 31.

Oziomek, J., and Kennedy, J. P. (1977), *J. Appl. Polym. Sci. Appl. Polym. Symp.* **30**, 91.

Pary, B., Tardi, M., Polton, A., and Sigwalt, P. (1978), *New Developments in Ionic Polymerization, Symp. Strasbourg, Feb. 27–March 2, Abstr.*, 52.

Peyrot, J. (1972), Germ. Patent 2,161,859 to Comp. Francaise de Raffinage (June 29, 1972).

Rooney, J. M., Squire, D. R., and Stannett, V. W. (1976), *J. Polym. Sci. Chem. Ed.* **14**, 1877.

Sigwalt, P., Polton, A., and Miskovic, M. (1976), *J. Polym. Sci. Polym. Symp.* **56**, 13.

Simionescu, C. I., Deneš, F., Percec, V., Totlin, M., and Kennedy, J. P. (1979), *Prepr. Makro Mainz* **1**, 448.

Thame, N. G., Lundberg, R. D., and Kennedy, J. P. (1972), *J. Polym. Sci. A1* **10**, 2507.

Trivedi, P. D., and Schulz, N. (1980), *Polym. Bull.* **3**, 37.

Vidal, A., and Kennedy, J. P. (1976), Polym. *Lett.* **14**, 489.

Vidal, A., Donnet, J. B., and Kennedy, J. P. (1977), *Polym. Lett.* **15**, 585.

Vidal, A., Guyot, A., and Kennedy, J. P. (1980), *Polym. Bull.* **2**, 315.

□9□
Macromolecular Engineering by Carbocationic Polymerization

This chapter gives a brief glance at the history of carbocationic poly-merization, a more detailed analysis of recent progress, and finally some cautious predictions about the immediate future of this science. The conclusion is reached that the era of macromolecular engineering by cationic polymerization techniques has dawned and that research and development in the near future will increasingly exploit the understanding of the detailed mechanism of carbocationic polymerizations for the pre-paration of new materials that have a combination of desirable (and sometimes unique) properties or advantageous processing characteristics.

9.1 □ A GLANCE AT THE PAST

In contrast to the history of political and social developments, which often undergoes remarkable gyrations and reinterpretations, the history of science is fortunately much less fragile. Thus much of the history of carbocationic polymerizations that was assembled and analyzed some five years ago (Kennedy, 1975a) does not need repetition. Very briefly, the first stirrings of carbocationic polymerizations occurred at the end of the eighteenth century, and except for occasional sporadic reports by classical workers such as Cannizzaro, Berthelot, Vislicenius, Friedel, and Crafts, whose papers could now be construed to be polymer papers, the field lay dormant till the dawn of polymer science and Staudinger's research in the period 1920–1930. This "prehistoric" era ended in 1934 when Whitmore in a landmark paper interpreted the reaction between strong acids and olefins in terms of carbenium ions and outlined in less than 1000 words the modern mechanism concept of carbocationic polymerizations (Whitmore, 1934).

The second major event that shaped the history of this field was the research done in the 1930's by R. M. Thomas and his associates at the Standard Oil Development Company, which culminated in the discovery (Thomas and Sparks, 1944) and commercialization of butyl rubber. From a perspective of some 40 years the two pillars of carbocationic poly-merizations are Whitmore and Thomas; the former provided the scientific–organic chemical basis and the latter first demonstrated large-scale tech-nological–commercial feasibility of this field.

This era of the pioneers was followed by the era of maturity and consolidation during the 1940s. This decade was characterized by a variety of fundamental contributions, most notably the discovery of "cocatalysis" and the role of protogenic impurities by Polanyi, Plesch, and Evans et al., and stereocontrol by Schildknecht et al. (Kennedy, 1975a). By the mid-1950s the field becomes a coherent, distinct discipline inspiring and occupying a variety of researchers all around the globe.

Until very recently, however, both scientists and technologists of cationic polymerizations had to be satisfied with finding conditions for the synthesis of suitably high molecular weight materials and generating a measure of understanding of conversion (yield) and molecular weight control. Limited insight into the mechanism of elementary reactions prevented a more sophisticated approach so that macromolecular engineering, even modest molecular weight distribution control, was out of the question. Researchers in both academe and industry "took what they got" and there was very little detailed information available to improve this situation. The barrel became dangerously close to empty by the end of the 1960s, at which time practically all the readily available monomers had been polymerized by all kinds of (Brønsted and Lewis) acids and their combinations, and the major polymer structures had been determined. It did not take much to realize that syntheses of new, usually expensive, cationically polymerizable monomers or the laborious amassing of kinetic trivia, particularly in the absence of detailed chemical understanding, were futile exercises except for sporadic academic inquiry. The future of carbocationic polymerizations was (is) clearly not in these directions.

This historical account of carbocationic polymerizations (Kennedy, 1975a) ended with an examination of representative researches in progress during 1972–1974; however, it did not progress to an analysis of the present with an eye on the future. Reviewing progress during the recent past brought matters much more sharply into focus and the shape of future developments only dimly visible in 1974 became quite distinct by 1979–1980.

A quiet revolution is in progress in carbocationic polymerizations, and the era of macromolecular engineering by cationic techniques is upon us. The aim of macromolecular engineering is the synthesis of useful polymer structures by generating and subsequently exploiting mechanistic understanding of elementary steps, that is, initiation, propagation, chain transfer, and termination of polymerizations. Macromolecular engineering is rather old in free radical initiated polymerizations and is currently in full bloom in anionic polymerizations (witness the syntheses of sequential copolymers by organolithium techniques) because the mechanisms of these polymerizations are relatively simple and consequently readily controllable and exploitable. In contrast, the mechanism of cationic, particularly carbocationic, polymerizations remained until very recently poorly understood, uncontrollable, and unexploitable. During the last 5 to 10 years this situation has dramatically improved and the quantity and quality of mechanistic information relative to the elementary steps in cationic polymerization became sufficient for exploitation by the macromolecular engineer.

The tip of the iceberg of macromolecular engineering by carbocations can already be perceived. Operational elements of this emerging science, such as controlled initiation, selective initiation, inifers, quasi-living systems, and controlled termination, that is, elements that have already been exploited in the preparation of new structures and composites, are highlighted below.

9.2 ☐ ELEMENTS OF CATIONIC MACROMOLECULAR ENGINEERING

Controlled Initiation

From the point of view of macromolecular engineering, the major handicap of working with conventional Friedel–Crafts acid coinitiators, for example, $AlCl_3$, $AlBr_3$, BF_3, and $TiCl_4$, is that these materials rapidly induce α-olefin polymerizations by protonation in the presence of adventitious traces of protogenic impurities, usually H_2O (uncontrolled initiation) (see also Section 4.1) so that the polymers contain unreactive "sterile" CH_3 head groups. The discovery (Kennedy and Baldwin, 1965) of inexpensive coinitiators such as Et_2AlCl and Et_3Al, that is, coinitiators that do not induce polymerizations in the presence of traces of moisture but require purposeful addition of more than trace amounts of cationogens, led to a revival of this stodgy science and to the synthesis of polymers with "engineered" head groups by "controlled initiation." For example, macromolecules with unsaturated head groups can be readily obtained by using allyl halide initiators, say, $CH_3CH=CHCH_2Cl$, in conjunction with Et_2AlCl coinitiator, that is, with the $CH_3CH=CH\overset{\oplus}{C}H_2/Et_2AlCl_2^{\ominus}$ initiating ion pair (see also Section 4.1). Similarly, phenyl head groups can be obtained by the use of appropriately substituted benzyl halide initiators (Kennedy, 1968; Reibel et al., 1979). A feasible way of introducing cyclopentadiene head groups into polymers is to initiate polymerization with dicycopentadienyl chloride in the presence of Et_2AlCl and subsequently cracking the protective dimer (Kennedy et al., 1980a):

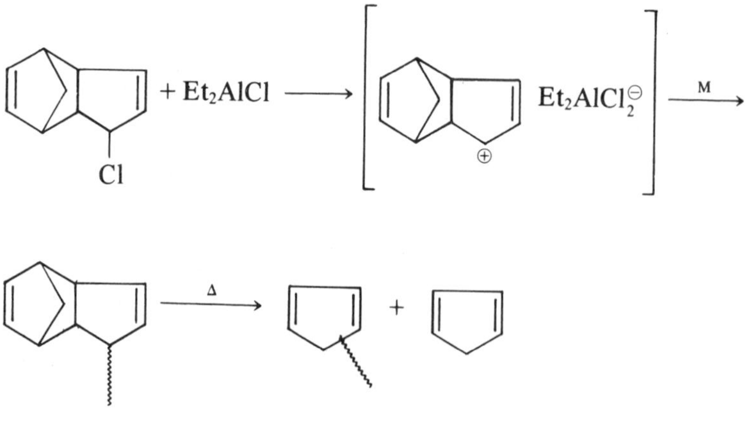

Polymers with cyclopentadienyl head groups are valuable intermediates for the preparation of many new materials.

Recently controlled initiation was used for the preparation of polyisobutylenes, PIB, and poly(α-methylstyrenes), PαMeSt, with $-Si(CH_3)_{3-n}Cl_n$

and $-Si(CH_3)_{3-n}H_n$ head groups HSi–PIB and HSi–PαMeSt (Kennedy and Chang, 1980b). The first step toward this objective was the synthesis of a series of new initiators combining two functions in one molecule: a Si–Cl or Si–H group and a benzylic chlorine, for example, $Cl(CH_3)_2Si–CH_2CH_2C_6H_4CH_2Cl$ or $H_3Si–CH_2CH_2C_6H_4CH_2Cl$. Next, conditions were defined under which cationic polymerizations could be initiated by the $-C_6H_4CH_2Cl$ moiety (Me$_3$Al coinitiator, CH$_2$Cl$_2$ diluent, low temperatures) without destroying the Si–Cl or Si–H functions. Model experiments proved valuable during these investigations and showed the way toward conditions under which quantitative survival of Si–Cl and Si–H head groups during carbocationic polymerization could be achieved. Quantitative NMR and IR spectroscopy indicated that one SiH group per PIB and PαMeSt chain can be obtained by the use of, for example, the $H(CH_3)_2SiCH_2CH_2C_6H_4CH_2Cl/Me_3Al/i - C_4H_8$ or $CH_2 = C(CH_3)C_6H_5$ system under a variety of conditions. Polymers such as HSi–PIB and HSi–PαMeSt combine the unique synthetic possibilities offered by carbocationic polymerizations and the chemistries of Si–Cl and Si–H bonds. For example, hydrosilylation of polybutadiene by HSi–PαMeSt would be a promising route for the preparation of desirable new thermoplastic elastomer graft copolymers.

The principle of controlled initiation has been extended to the synthesis of novel graft copolymers by the use of macromolecular initiators (see also Section 8.4). Just as small alkyl or tertiary halides are efficient cationic initiators in conjunction with certain alanes or BCl$_3$ in open systems, macromolecular halides containing active, that is, allylic, benzylic, or tertiary, halogens may also initiate polymerization under identical conditions. For example, poly(vinyl chloride) PVC or polychloroprene CR are excellent initiators for the polymerization of isobutylene or styrene in conjunction with Et$_2$AlCl or Et$_3$Al and yield PVC-g-PSt, PVC-g-PIB, etc. (Kennedy and Davidson, 1977b; Kennedy, 1977a). Graft copolymers prepared by controlled initiation prepared to date are compiled in Table 8.2.

The principle of "sequential cationic initiation" was employed for the synthesis of the first well-characterized hydrocarbon diblock copolymer prepared by a carbocationic technique (see also Section 4.1). Fundamental research showed that the rate of initiation of olefin (e.g., isobutylene) polymerization with t-BuCl/Et$_2$AlCl was many orders of magnitude faster than with t-BuBr/Et$_2$AlCl (Kennedy et al., 1973). A molecule was built that contained a tertiary chlorine and bromine $Cl(CH_3)_2CCH_2CH_2CH_2C(CH_3)_2Br$, and conditions were developed under which polymerization of a monomer started exclusively from the chlorine site leaving the bromine site unchanged (Kennedy and Melby, 1975b). Subsequently, by slightly changing conditions, the bromine site was activated to initiate the polymerization of a second monomer. Poly(isobutylene-b-styrene) has been prepared in this manner

(Kennedy and Melby, 1975b). Schematically,

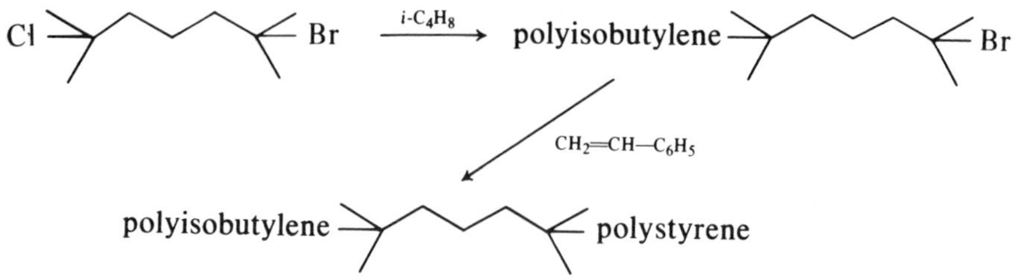

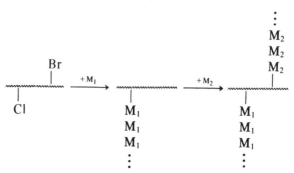

Selective cationic initiation was also employed for the synthesis of new bigraft copolymers (see also Section 8.4). Thus an ethylene–propylene copolymer rubber (EPM) was sequentially chlorinated and brominated so that it contained a few active chlorines and bromines along the backbone. This chlorobrominated backbone was used selectively to initiate the polymerization of first one monomer M_1, and then another, M_2. By this technique combinations of polystyrene, poly(α-methylstyrene), and polyisobutylene branches have been attached to ethylene–propylene copolymer backbones (Kennedy and Vidal, 1975c):

$$
\begin{array}{ccccc}
& & & & \vdots \\
& & & & M_2 \\
& & & & M_2 \\
& Br & & & M_2 \\
& | & \xrightarrow{+M_1} & & | \\
\text{~~~~~~~} & | & & \text{~~~~~~~} \xrightarrow{+M_2} \text{~~~~~~~} & | \\
& Cl & & M_1 & M_1 \\
& & & M_1 & M_1 \\
& & & M_1 & M_1 \\
& & & \vdots & \vdots \\
\end{array}
$$

The physical–mechanical properties of some of these materials were quite unexpected (Vidal and Kennedy, 1976). For example, the bigraft poly[(ethylene-co-propylene)-g-isobutylene-g-styrene] exhibited only two T_g's. Evidently the rubbery domains coalesced; however, the mixed rubbery domain segregated of the glassy domain.

Propagation

The macromolecular engineer is concerned only indirectly with propagation (Section 4.2). The paramount aim is for propagation to proceed uninterrupted from controlled initiation until controlled termination. The goal is to control initiation, that is, create a useful head group, and then to maintain uneventful propagation until controlled chain breaking (chain transfer or termination) gives rise to a useful end group.

The macromolecular engineer is usually interested in lengthening the time of uninterrupted propagation in order to prepare high molecular weight products. However, this last demand is quite flexible since many highly desirable materials are of medium or low molecular weights, for example, telechelic liquid prepolymers. α,ω-bifunctional prepolymers have recently become preparable by carbocationic techniques (Section 9.2). Since carbocationic propagation is a rather exothermic low activation energy demanding process (Section 4.2), it proceeds rapidly under mildest conditions; the experimental challenge is not to maintain propagation but to delay and control chain breaking.

Control of Chain Transfer

Chain transfer, more precisely chain transfer to the monomer (Section 4.3), or monomer transfer for short, is the bane of the macromolecular engineer. The fundamental problem is that monomer transfer invariably leads to an unreactive "sterile" CH_3- head group:

$$\sim \overset{|}{\underset{|}{C}}-\overset{|}{\underset{|}{C}}^{\oplus} \rightarrow \sim \overset{|}{C}=\overset{|}{\underset{|}{C}} + H^{\oplus}$$

$$H^{\oplus} + CH_2{=}\overset{|}{\underset{|}{C}} \rightarrow CH_3{-}\overset{|}{\underset{|}{C}}^{\oplus}$$

Another undesirable feature of monomer transfer is the formation of asymmetric molecules, that is, chains with a CH_3- head group and an unsaturated or other (indanyl, with styrene derivatives) dissimilar end group. Unfortunately, because of the inherently unstable nature of growing carbenium ions, chain transfer to monomer is "built-in" in carbocationic polymerizations.

The most significant achievement of macromolecular engineering is the effective control of chain transfer to monomer. This can occur by four techniques: by the inifer concept, by proton traps, by quasi-living polymerization, and by controlled termination. The following sections highlight these techniques.

The Inifer Method

Inifers are bifunctional initiators and transfer agents that fulfill two functions simultaneously: they are bifunctional initiators that effect controlled initiation *and* they are bifunctional transfer agents that produce desirable end groups (Section 4.3).

Inifers can be employed for the synthesis of biterminally functional "telechelic" polymers (Kennedy and Smith, 1979h, 1980e). Scheme 9.1

conveys the mechanism of the inifer concept and outlines the route to an α, ω-difunctional polymer (XRX = inifer).

Scheme 9.1 ☐ Bifunctional Initiator–Chain Transfer Agent

Ion Generation

$$XRX + MeX_n \rightleftharpoons XR^{\oplus} + MeX_{n+1}^{\ominus}$$

Cation and propagation

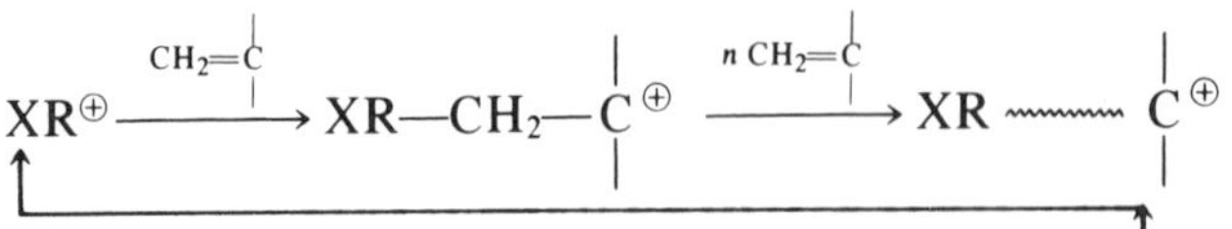

Chain transfer with inifer

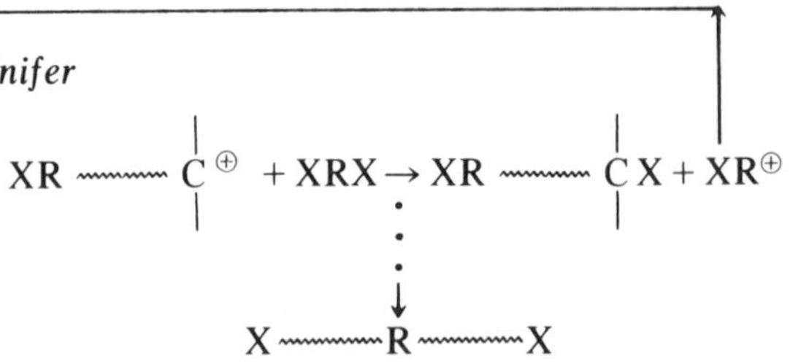

Requirement: $R_{tr,i} > R_{tr,M}$

This principle has been exploited for the synthesis of α,ω-ditertiary chlorine containing polyisobutylenes (Kennedy and Smith, 1979h):

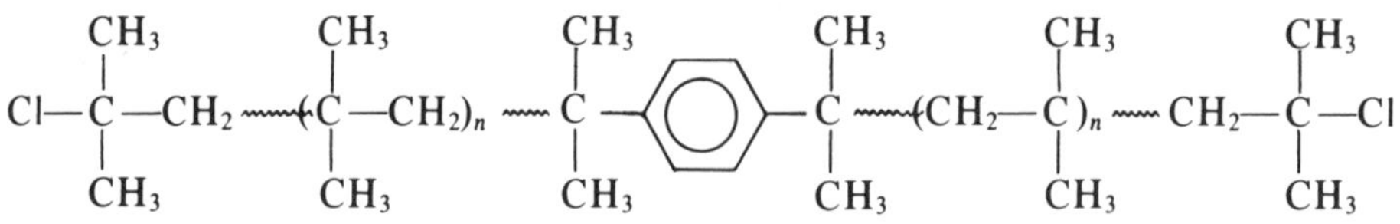

Cl—PIB—Cl

The structure of this telechelic molecule has been thoroughly determined (Section 4.3)

The synthesis of this telechelic polymer is based on a combination of the following operational elements: (1) rapid initiation of isobutylene polymerization by the cumyl chloride ($C_6H_5(CH_3)_2Cl$)/BCl_3 system, (2) highly efficient chain transfer activity by dicumyl chloride $Cl(CH_3)_2CC_6H_4C(CH_3)_2Cl$ in isobutylene polymerization, and (3) termination by chlorination of isobutylene polymerization coinitiated by BCl_3 and formation of –$CH_2C(CH_3)_2Cl$ endgroup.

It was anticipated and subsequently demonstrated that the $Cl(CH_3)_2CC_6H_4C(CH_3)_2Cl/BCl_3/i\text{-}C_4H_8$ "inifer system" produces the above telechelic polymer. The detailed mechanism involved in the synthesis is discussed in Section 4.3.

According to a thorough analysis of the mechanism in Schemes 9.1 and 4.3, the molecular weight dispersity $\bar{M}_w/\bar{M}_n$ of Cl–PIB–Cl prepared by the inifer technique must be 1.5. Experiments corroborated that this is indeed the case (Fehérvári et al., 1979, 1980).

Suitably high molecular weight ($\bar{M}_n = 20$ to 50×10^3) Cl–PIB–Cl's have been used for the preparation of PαMeSt-b-PIB-b-PαMeSt thermoplastic elastomer triblock copolymer. The synthesis was accomplished by initiating the polymerization of αMeSt by employing the Cl–PIB–Cl as a double-headed initiator in conjunction with Et_2AlCl coinitiator (Kennedy and Smith, 1979h, 1980f):

$$Cl\!-\!PIB\!-\!Cl + \text{excess } Et_2AlCl \rightarrow {}^{\oplus}PIB^{\oplus} + 2Et_2AlCl_2^{\ominus}$$

$$\Big/ \alpha \text{ MeSt}$$

PαMeSt-b-PIB-b-PαMeSt

Conditions have been worked out accurately to control the molecular weight of Cl–PIB–Cl from ~500 to ~50,000. The low molecular weight telechelic products proved to be of particular interest for the preparation of a large variety of prepolymers. Thus methods have been developed for the synthesis of polyisobutylene α,ω-dienes by regioselective dehydrohalogenation by sterically hindered bases (Kennedy et al., 1979c). Subsequently, the liquid α,ω-diolefin was converted to the α,ω-primary diol (hydroboration followed by alkaline oxidation) and according to sensitive analytical methods the functionality of the telechelic diol remained 2.00 (Iván et al., 1980): Scheme 9.2 shows these transformations. The α,ω-telechelic diolefin II gave the corresponding α,ω-diepoxide upon oxidation with peracid and α,ω-silanes upon hydrosilylation with $HSi(CH_3)_2Cl$, followed by reduction with $LiAlH_4$ (Kennedy and Chang, 1980b).

Scheme 9.2 □ **Synthesis of α,ω-Di(hydroxyl)polyisobutylene**

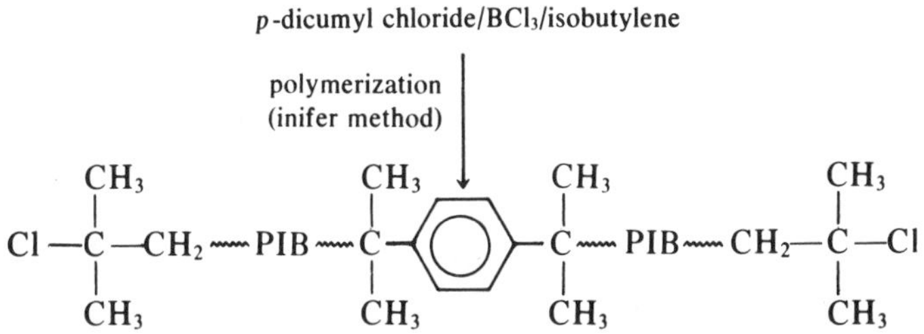

p-dicumyl chloride/BCl₃/isobutylene

polymerization
(inifer method)

I

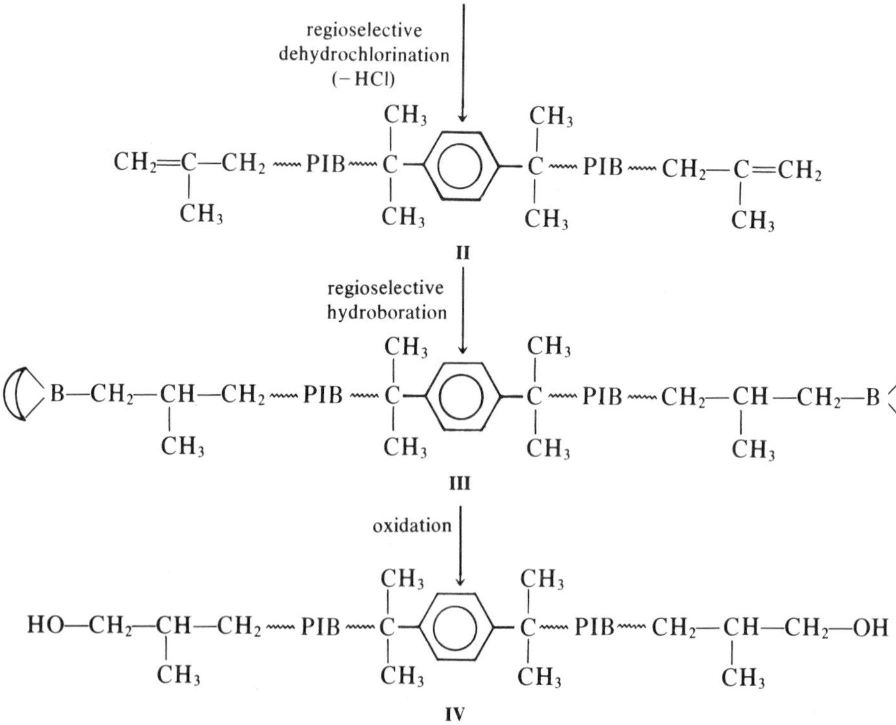

regioselective
dehydrochlorination
(– HCl)

$$CH_2{=}\underset{\underset{CH_3}{|}}{C}{-}CH_2{\,\sim\!\!\sim\,}PIB{\,\sim\!\!\sim\,}\underset{\underset{CH_3}{|}}{\overset{\overset{CH_3}{|}}{C}}{-}\bigcirc{-}\underset{\underset{CH_3}{|}}{\overset{\overset{CH_3}{|}}{C}}{\,\sim\!\!\sim\,}PIB{\,\sim\!\!\sim\,}CH_2{-}\underset{\underset{CH_3}{|}}{C}{=}CH_2$$

II

regioselective
hydroboration

$$B{-}CH_2{-}\underset{\underset{CH_3}{|}}{CH}{-}CH_2{\,\sim\!\!\sim\,}PIB{\,\sim\!\!\sim\,}\underset{\underset{CH_3}{|}}{\overset{\overset{CH_3}{|}}{C}}{-}\bigcirc{-}\underset{\underset{CH_3}{|}}{\overset{\overset{CH_3}{|}}{C}}{\,\sim\!\!\sim\,}PIB{\,\sim\!\!\sim\,}CH_2{-}\underset{\underset{CH_3}{|}}{CH}{-}CH_2{-}B$$

III

oxidation

$$HO{-}CH_2{-}\underset{\underset{CH_3}{|}}{CH}{-}CH_2{\,\sim\!\!\sim\,}PIB{\,\sim\!\!\sim\,}\underset{\underset{CH_3}{|}}{\overset{\overset{CH_3}{|}}{C}}{-}\bigcirc{-}\underset{\underset{CH_3}{|}}{\overset{\overset{CH_3}{|}}{C}}{\,\sim\!\!\sim\,}PIB{\,\sim\!\!\sim\,}CH_2{-}\underset{\underset{CH_3}{|}}{CH}{-}CH_2{-}OH$$

IV

Polyfunctional inifers provide new avenues toward unique tribranched macromolecular stars:

$$^XR^X + A \rightleftharpoons {^X_XR}^\oplus + XA^\ominus$$

$$^X_XR^\oplus + nM \rightarrow {^X_XR}{\,\sim\!\!\sim\,}M^\oplus + {^X_XR}^X \underset{\Longrightarrow}{\Longrightarrow} \quad \begin{matrix} X{\,\sim\!\!\sim\,}R{\,\sim\!\!\sim\,}X \\ | \\ X \end{matrix}$$

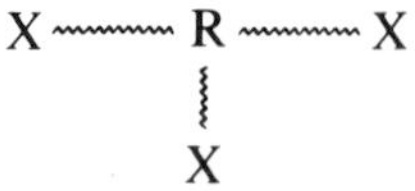

The terminally functional tristar polymers can be used as intermediates (initiators) for the synthesis of triblock stars:

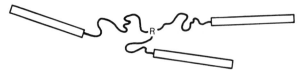

The synthesis of tristar isobutylene-α-methylstyrene block copolymers is well under way (Kennedy et al., 1981).

Proton Traps

Chain transfer may be controlled by the use of proton traps. These materials exhibit extraordinary specificity toward protons and are able to intercept protons arising during chain transfer to monomer. For example, α-methylstyrene polymerizations coinitiated by BCl_3, $SnCl_4$, or $TiCl_4$, for example, in the presence of 2,6-di-*tert*-butyl pyridine (DTBP) result in reduced yields but much higher molecular weights and narrower molecular weights distributions than control runs carried out in the absence of this highly hindered amine (Kennedy and Chou, 1979e) (Section 4.3). Indeed, by the use of DTBP it was possible to obtain highest $\bar{M}_n$'s (depending on conditions up to 10^6) and lowest $\bar{M}_w/\bar{M}_n$ (1.2 to 1.6) values (Kennedy and Chou, 1980c).

The action of proton traps may be rationalized by a competition between monomers and proton trap for the proton arising during chain transfer:

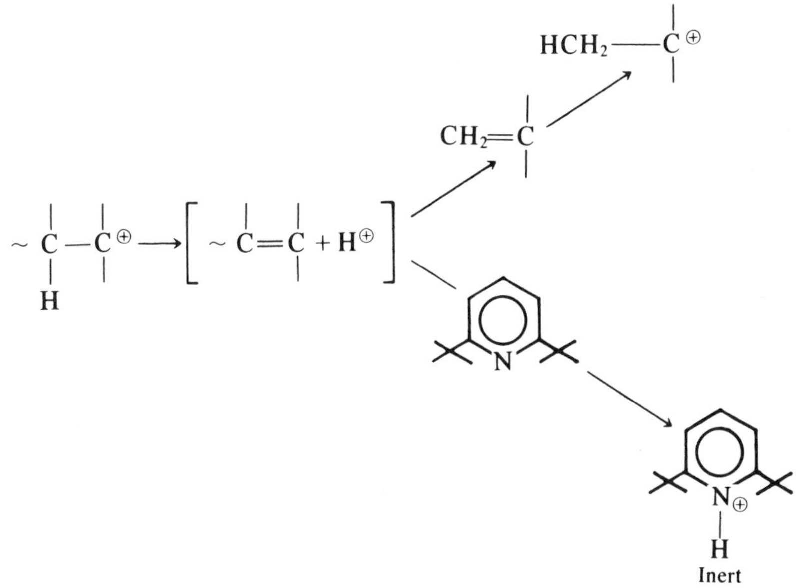

In the presence of DTBP the proton is trapped at a rate approaching diffusion control. Although the process is kinetically chain transfer, that is, the rate-determining event is proton expulsion, the kinetic chain cannot continue so that proton trapping is in fact a termination reaction or "terminative proton entrapment" (Kennedy and Chou, 1979e, 1980c). Proton traps have recently been used to increase grafting efficiencies close to 100% (see 8.4 and 4.3).

Quasi-Living Polymerization

Living anionic and oxonium ion polymerizations are extremely versatile synthetic options in the hands of the macromolecular engineer for the preparation of well-defined tailor-made polymers. By definition, living polymerizations exist if the rates of termination and transfer are zero (R_t and $R_{tr} = 0$), and the role of initiation is much faster than that of propagation ($R_i \gg R_p$). Combination of these criteria leads to the well-known molecular weight equation $\overline{DP}_n = [M]/[I]_0$ where $[M]$ and $[I]_0$ are monomer and initiator concentrations, respectively. Diagnostic proofs of living polymerization systems are linear $\overline{DP}_n$ versus conversion plots starting at the origin, the formation of block copolymers by sequential monomer addition, and narrow molecular weight distributions ($\bar{M}_w/\bar{M}_n \sim 1.0$).

In spite of efforts by several groups of investigators, the existence of living carbocationic polymerizations has not yet been demonstrated. The main problem is the presence of chain transfer (and frequently termination) in carbocationic polymerization systems. Recent insight into the mechanism of chain-breaking processes led to the proposition that conditions close to that of living could be achieved in the presence of reversible chain transfer and termination steps (Faust et al., 1980). It was postulated that reversible chain transfer and termination could be obtained provided $[M]$ would be lower than that of the reactivatable polymer chain end (terminus with unsaturation) throughout the experiment. In the presence of excess monomer, transfer to monomer could not be prevented since equilibria 9.1 and 9.2 that control monomolecular and bimolecular chain transfer to monomer reactions would be strongly displaced to the right:

$$\sim \overset{\overset{\displaystyle H}{\displaystyle |}}{\underset{|}{C}} - \overset{|}{\underset{|}{C}}{}^{\oplus} MeX_{n+1}^{\ominus} \rightleftharpoons \sim \overset{|}{C} = \overset{|}{\underset{|}{C}} + H^{\oplus}MeX_{n+1}^{\ominus} \tag{9.1}$$

(followed by rapid protonation of monomer), and

$$\sim \overset{\overset{\displaystyle H}{\displaystyle |}}{\underset{|}{C}} - \overset{|}{C}{}^{\oplus} MeX_{n+1}^{\ominus} + C = C \rightleftharpoons \sim \overset{|}{C} = \overset{|}{\underset{|}{C}} + H - \overset{|}{\underset{|}{C}} - \overset{|}{C}{}^{\oplus}MeX_{n+1}^{\ominus} \tag{9.2}$$

However, in the presence of vanishingly small amounts of monomer these reactions would indeed become reversible.

The requirement of reversible termination could be met in systems in which termination (i.e., collapse of the propagating cation/counteranion pair) is in equilibrium with ionization:

$$\sim \overset{\overset{\displaystyle H}{\displaystyle |}}{\underset{\underset{\displaystyle H}{\displaystyle |}}{C}} - \overset{|}{\underset{|}{C}}{}^{\oplus} MeX_{n+1}^{\ominus} \rightleftharpoons \sim \overset{|}{\underset{|}{C}} - \overset{|}{\underset{|}{C}} - X + MeX_n$$

If monomer is added at a sufficiently slow and constant rate to carbocationic systems in which equilibria 9.1 to 9.3 operate so as to maintain the concentration of monomer M below that of olefin-ended chains $M_n^=$, that is, so that $[M]/[M_n^=] \ll 1$, the incoming monomer is consumed exclusively by propagation (addition to the carbocation) and $\bar{M}_n$ increases at the same rate of monomer addition. Under these so-called "quasi-living" conditions, $\bar{M}_n$ versus monomer addition plots would be linear with positive slopes reminiscent of $\bar{M}_n$ versus time or conversion plots obtained in truly living anionic or oxonium ion systems.

In living polymerizations all the propagating species are in fact active, "living," all the time. In contrast, in quasi-living carbocationic polymerizations chain transfer and termination may be present but these processes are reversible and the only monomer-consuming processes are initiation and propagation. The chain ends are not required to be active all the time; they may be temporarily dormant, but they must remain reactivatable by protonation and be capable of resuming propagation. Since termination and transfer are reversible and monomer consumption is by propagation, to the observer it would appear that $R_t = 0$ and $R_{tr} = 0$. The following scheme helps to visualize requirements for quasi-living carbocationic polymerization:

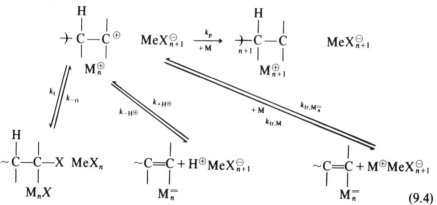

$$(9.4)$$

where $MeX_{n+1}^{\ominus}$ is the counterion and MeX_n is the coinitiator and the rate constants are defined in the scheme above.

The essential requirements of quasi-living carbocationic polymerizations are that kinetic termination, proton loss, and bimolecular chain transfer to monomer be reversible and that the only significant monomer-consuming event be addition to the active chain end, that is, propagation.

A comprehensive kinetic scheme has been developed (Faust et al., 1980) that describes quasi-living polymerization in quantitative terms. The molecular weight equation obtained for ideal quasi-living conditions $\bar{M}_n = M_{total}/[I]_0$, where $M_{total} =$ cumulative amount of monomer added, is very similar to that describing truly living anionic systems, hence the term quasi-living.

Figure 9.1 shows the $\bar{M}_n$ versus monomer input plot obtained in an experiment in which α-methylstyrene was slowly and continuously added to $C_6H_5(CH_3)_2Cl/BCl_3$ initiating system in $CH_2Cl_2 +$ methylcyclohexane solvent at $-50°C$. The presence of quasi-living conditions is indicated by the linearity and zero origin of the plot.

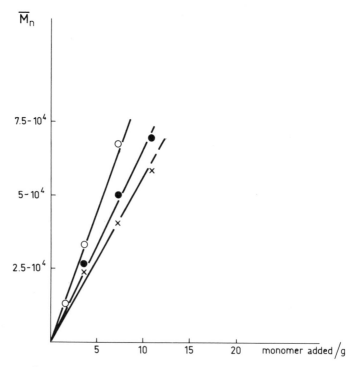

Figure 9.1 ☐ $\bar{M}_n$ versus monomer input plot of $C_6H_5C(CH_3)_2Cl/BCl_3/\alpha MeSt$ polymerization system at different initiator concentrations: × and ○ $[I]_0 = 1.88 \times 10^{-4}$; ○ $= 9.4 \times 10^{-5}$.

The synthetic value of quasi-living carbocationic systems is considerable because it could lead to control of molecular weight and molecular weight distribution and could open new avenues toward the preparation of block copolymers.

Controlled Termination

Controlled termination exists when the rate of termination is faster than that of chain transfer; in this manner chain transfer can be avoided (Kennedy et al., 1978).

The key to progress in this area was the mechanistic understanding of termination by ion pair collapse, that is, the neutralization of the growing carbenium ion by transferring to it a fragment of the counteranion $\sim C^\oplus + AX^\ominus \rightarrow \sim CX + A$. Increased insight into the details of mechanisms of termination led to the synthesis of varieties of new terminally functional polymers (see also Section 4.4). Systems have been found in which useful functions, such as chlorine (Kennedy et al., 1978), phenyl (Kennedy and Chung, 1980d), cyclopentadienyl (Kennedy and Castner, 1979a, b), and vinyl (Mandal and Kennedy, 1978), were transferred to the growing carbocation *before chain transfer to monomer occurred*. For example, the collapse of the growing polyisobutylene ion pair $\sim \overset{\oplus}{C}(CH_3)_2BCl_4^{\ominus}$ resulted in a terminal tertiary chlorine $\sim C(CH_3)Cl$ plus BCl_3 faster than chain transfer to monomer (Kennedy et al., 1978).

Termination by chlorination yielded a macromolecule with a terminal tertiary chlorine which was then used to initiate the polymerization of a second monomer, styrene. The end product was a linear block copolymer of isobutylene and styrene (Kennedy et al., 1978) (Section 8.3):

$$H_2O/BCl_3/i\text{-}C_4H_8 \rightarrow PIB\text{\tiny vvv} CH_2\underset{\underset{CH_3}{|}}{\overset{\overset{CH_3}{|}}{C}}{}^{\oplus}BCl_4^{\ominus} \rightarrow PIB\text{\tiny vvv}CH_2\underset{\underset{CH_3}{|}}{\overset{\overset{CH_3}{|}}{C}}\!-\!Cl + BCl_3$$

$$PIB\text{\tiny vvv}CH_2\underset{\underset{CH_3}{|}}{\overset{\overset{CH_3}{|}}{C}}\!-\!Cl + styrene + Et_2AlCl \rightarrow PIB\text{\tiny vvv}CH_2\underset{\underset{CH_3}{|}}{\overset{\overset{CH_3}{|}}{C}}\!-\! \boxed{PSt}$$

Termination by ion pair collapse with alkylaluminum-coinitiated poly-merizations turned out to be particularly fruitful. For example, the t-BuCl/dimethylcyclopentadienylaluminum (Me$_2$CPDAl) system was found to initiate isobutylene polymerization under controlled conditions and ter-minate by cyclopentadienylation (Kennedy and Castner, 1979a, b). Thus the products contained t-Bu head groups and CPD end groups. Similarly, vinyl (Mandal and Kennedy, 1978) and phenyl (Kennedy and Chung, 1980d) termini have been introduced by t-BuCl/(CH$_2$=CH)$_3$Al and t-BuCl/(C$_6$H$_5$)$_3$Al combinations, respectively.

The relative rapidity of the termination, occurring before chain transfer, is particularly important for the synthesis of "clean" block and graft copolymers, that is, sequential polymers unadulterated by homopolymers (see also Section 8.1). Consider the case of a grafting reaction. Since initiation is controlled and grafting *must* commence at the halogenated backbone, initiation will not be a source of homopolymer formation. In contrast, chain transfer, if it occurs, will lead to homopolymer. Since termination is faster than chain transfer, homopolymer cannot form and grafting efficiencies close to 100% are obtained. Many instances have been described which conform with this scenario (Kennedy, 1977a; Sigwalt et al., 1976).

The similarity between the mechanisms of alkylative termination and derivatization of chlorinated polymers has been exploited for the pre-paration of modified polymers. For example, phenylation of PVC (Ken-nedy and Ichikawa, 1974), vinylation of chlorinated ethylene–propylene copolymer rubber (Kennedy and Mandal, 1977c), cyclopentadienylation of

PVC (Iván et al., 1979), chlorinated EPM, and chlorobutyl rubber (Kennedy and Castner, 1979a, b) have been described and the materials characterized. For example, cyclopentadienylation of chlorobutyl by Me$_2$CPDAl produced remoldable rubbery networks (Kennedy and Castner, 1979b). Thermal reformability of networks is due to the reversible nature of Diels–Alder condensation of pendant GPD groups:

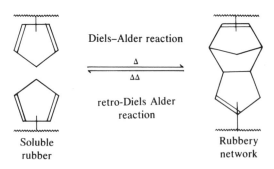

9.3 ☐ COMBINATION OF ELEMENTS AND SUMMARY

Although remarkable advances have already been made toward the synthesis of new desirable structures by controlling one elementary step, the synthesis of graft copolymers by controlled initiation is a case in point (see Section 8.4), the tremendous preparative potential of cationic macromolecular engineering can be far better exploited by simultaneously mobilizing more than one controlled elementary step. Several examples have already been touched upon in the preceding sections, that is, the synthesis of biterminally chlorinated polyisobutylenes by use of inifer, or the synthesis of a polyisobutylene containing a (CH$_3$)$_3$C head group and cyclopentadiene end group by controlled initiation and controlled termination by the t-BuCl/Me$_2$CPDAl initiating system. Additional examples that have been investigated follow.

Combining principles of controlled initiation and controlled termination and making use of the Cl$_2$/BCl$_3$/i-C$_4$H$_8$ system (Kennedy and Chen, 1979d) yielded

$$\text{Cl—CH}_2\text{—}\underset{\underset{\text{CH}_3}{|}}{\overset{\overset{\text{CH}_3}{|}}{\text{C}}}\text{~~~ PIB~~~CH}_2\text{—}\underset{\underset{\text{CH}_3}{|}}{\overset{\overset{\text{CH}_3}{|}}{\text{C}}}\text{—Cl}$$

This α,ω-bifunctional polymer provided the cationogen in conjunction with Et$_2$AlCl for controlled initiation of styrene polymerization and yielded the block copolymer (Kennedy and Chen, 1979d)

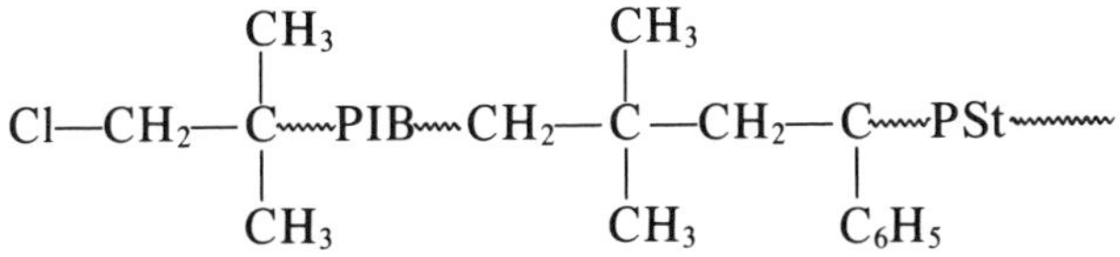

$$\text{Cl—CH}_2\text{—}\overset{\overset{\displaystyle CH_3}{|}}{\underset{\underset{\displaystyle CH_3}{|}}{C}}\text{∿PIB∿CH}_2\text{—}\overset{\overset{\displaystyle CH_3}{|}}{\underset{\underset{\displaystyle CH_3}{|}}{C}}\text{—CH}_2\text{—}\overset{}{\underset{\underset{\displaystyle C_6H_5}{|}}{C}}\text{∿PSt∿}$$

Combination of controlled initiation with controlled termination in the $(CH_3)_2C{=}CH{-}CH_2Cl/(CH_2CH)_3Al/i\text{-}C_4H_8$ system gave rise to (Mandal and Kennedy, 1978)

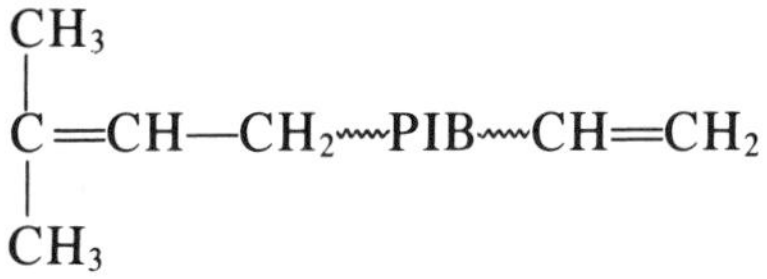

$$\overset{\overset{\displaystyle CH_3}{|}}{\underset{\underset{\displaystyle CH_3}{|}}{C}}{=}CH{-}CH_2\text{∿PIB∿}CH{=}CH_2$$

Similarly, combination of controlled initiation with controlled termination in the $(CH_3)_2C{=}CH{-}CH_2Cl/BCl_3/i\text{-}C_4H_8$ system produced (Kennedy et al., 1979c)

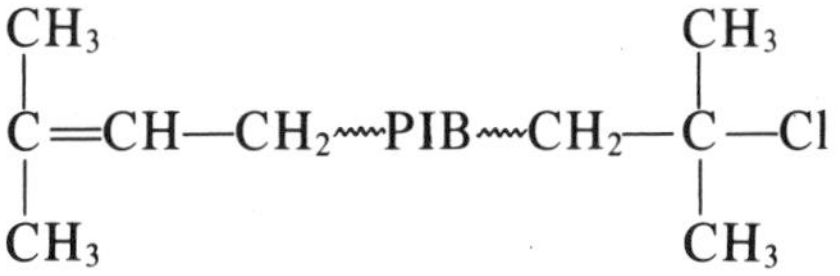

$$\overset{\overset{\displaystyle CH_3}{|}}{\underset{\underset{\displaystyle CH_3}{|}}{C}}{=}CH{-}CH_2\text{∿PIB∿}CH_2{-}\overset{\overset{\displaystyle CH_3}{|}}{\underset{\underset{\displaystyle CH_3}{|}}{C}}{-}Cl$$

which subsequently under controlled initiation conditions in the presence of Et_2AlCl initiated α-methylstyrene polymerization and yielded the head-functionalized diblock (Kennedy et al., 1979c):

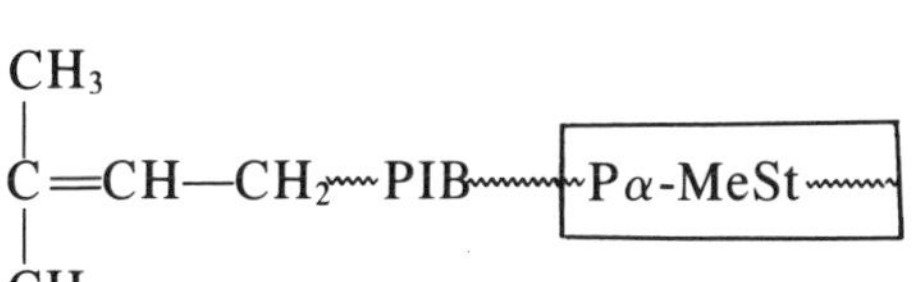

$$\overset{\overset{\displaystyle CH_3}{|}}{\underset{\underset{\displaystyle CH_3}{|}}{C}}{=}CH{-}CH_2\text{∿PIB∿}\boxed{{\sim}P\alpha\text{-MeSt∿}}$$

Still, combination of controlled initiation with termination using PVC initiator, BCl_3 coinitiator, and isobutylene monomer yielded grafts containing chlorine-ended branches ready for further functionalization (Gupta and Kennedy, 1979).

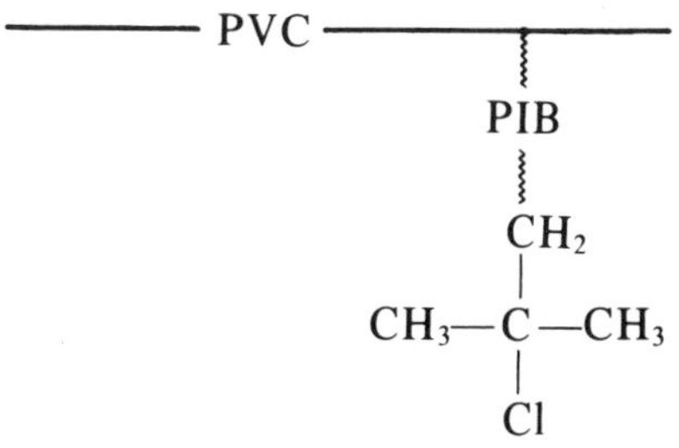

$$\begin{array}{c}
\text{————— PVC —————} \\
| \\
\vdots \\
\text{PIB} \\
\vdots \\
CH_2 \\
| \\
CH_3{-}\underset{\underset{\displaystyle Cl}{|}}{C}{-}CH_3
\end{array}$$

Table 9.1 □ Accomplishments of Cationic Macromolecular Engineering

Head Group	Controlled Initiation (Head Group Control) — Example of Initiating System Introducing Head Group	Controlled Chain Transfer (Suppressing Chain Transfer to Monomer) — Product	Means to Obtain Product	Controlled Termination (End Group Control in the Absence of Chain Transfer) — End Group or Pendant Group in Modified Polymer	Example of Counteranion Introducing End Group or Pendant Group
H– $(CH_3)_3C$–	HCl/Et$_2$AlCl t-BuCl/Me$_3$Al	Terminally bifunctional polymers	Inifers	–H –CH$_3$	Et$_3$AlCl$^{\ominus}$ Me$_3$AlCl$^{\ominus}$
$C=C-C$	Allyl chlorides Et$_2$AlCl or BCl$_3$	Narrowed MWD PαMeSt	Proton traps	–CH=CH$_2$	(CH$_2$CH)$_3$AlCl$^{\ominus}$
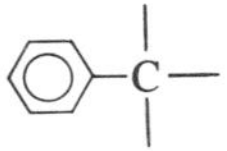 (benzyl / norbornene structures)	Benzyl halides Et$_2$AlCl or BCl$_3$			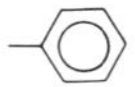 (phenyl ring) (cyclopentene)	(C$_6$H$_5$)$_3$AlCl$^{\ominus}$
(structures)	Cl/Et$_2$AlCl				Me$_2$CPDAlCl$^{\ominus}$
Cl– BI– P–	Cl$_2$/Me$_3$Al or BCl$_3$ Br$_2$/BCl$_3$ or Et$_2$AlCl P-Cl/Et$_2$AlCl or BCl$_3$ (P = active chlorine-containing polymer)	Linearly increasing $\bar{M}_n$ with conversion	Quasi-living polymerization technique	–Cl	BCl$_4^{\ominus}$
H–Si(CH$_3$)$_2$–	HSi(CH$_3$)$_2$–(CH$_2$)$_2$– C$_6$H$_4$CH$_2$Cl/Me$_3$Al				

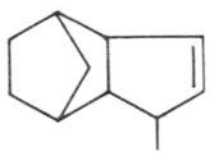

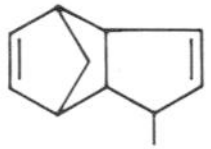

Table 9.2 □ Cationic Macromolecular Engineering Combination of Operational Elements

---

Initiation
  Head-functionalized polymers (H)〜〜〜

  Grafts ▌〜〜〜〜〜

Termination
  End-functionalized polymers 〜〜〜(E)

  Modified polymers 〜〜(E)〜〜
                        (E)

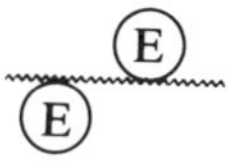

Initiation + termination
  Asymmetric telechelic polymers (H)〜〜〜(E)

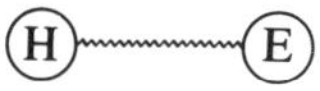

  End-functionalized grafts ▌〜〜〜(E)

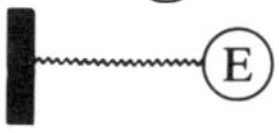

Sequential initiation + termination
  Diblocks 〜〜 R —▭ ▌〜〜

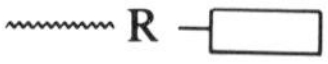

  Bigrafts ▭ ▌

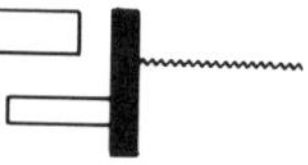

Initiation + chain transfer (inifer)
  Symmetric telechelic polymers (E)〜〜 R 〜〜(E)
  Asymmetric telechelic polymers (H)– R 〜〜〜(E)

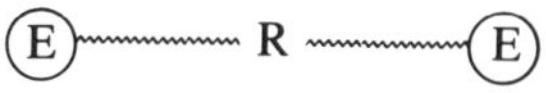

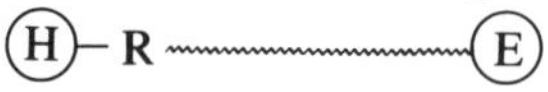

Initiation + chain transfer (proton sponge)
  MWD control; $\bar{M}_w/\bar{M}_n = 1.2–2.0$

Initiation + termination + initiation
  Diblocks 〜〜▭

  Head-functionalized blocks (H)〜〜▭

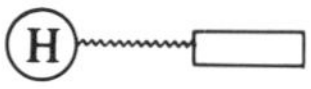

  Graft blocks ▌〜〜▭

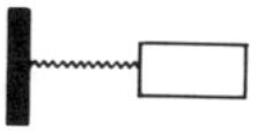

Initiation + termination + initiation + termination
  End-functionalized blocks 〜〜▭—(E)

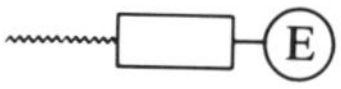

  Head-and-end-functionalized blocks (H)〜〜▭—(E)

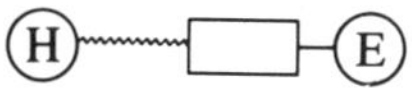

Initiation + chain transfer (inifer) + initiation
  Triblocks ▭—〜〜—▭

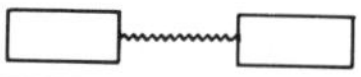

The first well-characterized graft block copolymer was prepared by combination of a controlled initiation, termination, and initiation sequence. The first step of the synthesis involved the grafting of polyisobutylene from polychloroprene rubber CR by BCl_3 and yielded polyisobutylene branches carrying active chlorine termini. The synthesis was completed by initiating the polymerization of α-methylstyrene at the graft branch termini (Kennedy and Plamthottam, 1979g):

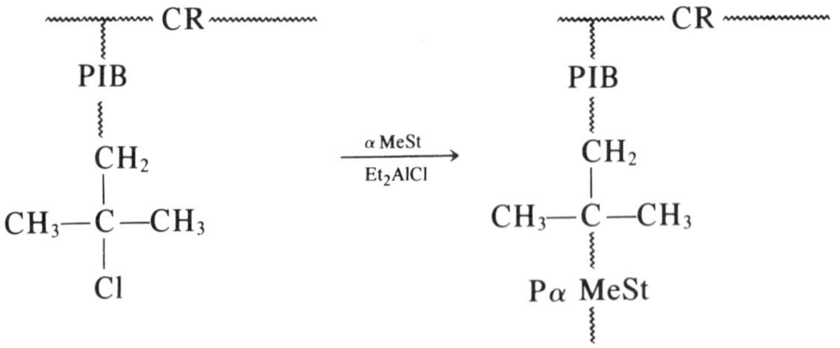

These few initial examples illustrate the versatility and scope of cationic macromolecular engineering. Tables 9.1 and 9.2 show controlled elementary events that have already been utilized and the functions and polymers they produce.

Obviously, permutation of these possibilities leads to an endless variety of new structures and topologies, and the only limiting factor in the construction of new materials is the imagination of the molecular engineer. Clearly the time is ripe for cooperation between the materials scientist and the carbocationic polymer chemist to exploit jointly new synthetic possibilities for the engineering of new, unique composites and structures.

□   REFERENCES

Faust, R., Fehérvári, Á., and Kennedy, J. P. (1980), to be published.

Fehérvári, Á., Kennedy, J. P., and Tüdős, F. (1979), *Polym. Prepr.* **20**, 320.

Fehérvári, Á., Kennedy, J. P., and Tüdős, F. (1981), *J. Macromol. Sci., Chem.* **A15**, 215.

Gupta, S. N., and Kennedy, J. P., *Polym. Bull.* (1979) **1**, 253.

Iván, B., Kennedy, J. P., Kelen, T., and Tüdős, F. (1979), *Polym. Bull.* **2**, 415.

Iván, B., Kennedy, J. P., and Chang, V. S. C. (1980), *J. Polym. Chem. Ed.*, **18**, 3177.

Kennedy, J. P., and Baldwin, F. P. (1965), Belg. Patent 663,320 (April 30, 1965).

Kennedy, J. P. (1968), in *Polymer Chemistry of Synthetic Elastomers.* J. P. Kennedy, and E. G. M. Tornquist, Eds., Wiley-Interscience, New York, p. 291.

Kennedy, J. P., Desai, N. V., and Sivaram, S. (1973), *J. Am. Chem. Soc.* **95**, 6386.

Kennedy, J. P., and Ichikawa, M. (1974), *Polym. Eng. Sci.* **14**, 322.

Kennedy, J. P. (1975a), *Cationic Polymerization of Olefins: A Critical Inventory*, Wiley-Interscience, New York, 1975, p. 7.

Kennedy, J. P. and Melby, E. G., (1975b), *J. Polym. Sci. Chem. Ed.*, **13**, 29.

Kennedy, J. P., and Vidal, A. (1975c), *J. Polym. Sci., Chem. Ed.* **13**, 1765.

Kennedy, J. P., Ed. (1977a), *Cationic Graft Copolymerization Polym. Sci., Appl. Polym. Symp.* **30**, 1.

Kennedy, J. P., and Davidson, D. (1977b), *J. Appl. Polym. Sci. Appl. Polym. Symp.* **30**, 13.

Kennedy, J. P., and Mandal, B. M. (1977c), *Polym. Lett.* **15**, 595.

Kennedy, J. P., Huang, S. Y., and Feinberg, S. C. (1978), *J. Polym. Sci. Chem. Ed.* **16**, 243.

Kennedy, J. P., and Castner, K. F. (1979a), *J. Polym. Sci. Chem. Ed.*, **17**, 2055.

Kennedy, J. P., and Castner, K. F. (1979b), U.S. Patent 4138,441 (Feb. 6, 1979).

Kennedy, J. P., Chang, V. S. C., Smith, R. A., and Iván, B. (1979c), *Polym. Bull.* **1**, 575.

Kennedy, J. P., and Chen, F. J. Y. (1979d), *Polym. Prepr.* **20**, 310.

Kennedy, J. P., and Chou, R. T. (1979e), *Polym. Prepr.* **20**, 306.

Kennedy, J. P., Huang, S. Y., and Smith, R. A. (1979f), *Polym. Bull.* **1**, 371.

Kennedy, J. P., and Plamthottam, S. (1979g), *Polym. Bull.* **20**, 98.

Kennedy, J. P., and Smith, R. A. (1979h), *Polym. Prepr.* **20**, 316.

Kennedy, J. P., Carlson, G., and Riebel, K. (1980a), unpublished data.

Kennedy, J. P., and Chang, V. S. C. (1980b), *Polym. Prepr.* **21**, 146.

Kennedy, J. P., and Chou, R. T. (1980c), *Polym. Prepr.* **21**, 148.

Kennedy, J. P., and Chung, D. Y. L. (1980d), *Polym. Prepr.* **21**, 150.

Kennedy, J. P., and Smith, R. A. (1980e), *J. Polym. Sci. Chem. Ed.* **18**, 1539.

Kennedy, J. P., Ross, L. R., Lackey, J. E., and Nuyken, O. (1981), *Polym. Bull.* **4**, 67.

Mandal, B. M., and Kennedy, J. P. (1978), *J. Polym. Sci. Chem. Ed.* **16**, 833.

Reibel, L., Kennedy, J. P., and Chung, D. Y. (1979), *J. Polym. Sci. Chem. Ed.* **17**, 2757.

Sigwalt, P., Polton, A., and Miscovic, M. (1976), *J. Polym. Sci. Polym. Symp.* **56**, 13.

Thomas, R. M., and Sparks, R. W. (1944), U.S. Patent 2,356,128 to Standard Oil Development Co.

Vidal, A., and Kennedy, J. P. (1976), *Polym. Lett.* **14**, 489.

Whitmore, F. C. (1934), *Ind. Eng. Chem.* **26**, 94.

□10□
Industrial Processes and Technological Aspects

10.1 ☐ INTRODUCTION

Industrial processes that involve carbocationic polymerizations are of the greatest economic–technological significance. Industrial consequences of carbocationic polymerizations are very important to the applications engineer or compounder because literally many thousands of formulations include cationic polymers; however, this broad field is much less visible to the scientist trained to deal with well-defined, usually homogeneous structures and controlled reactions. These processes embrace not only polymerizations to high molecular weight products but also oligomerizations of aliphatic and cyclic olefins, diolefins, acetylenes and nonhydrocarbons, and "polymer gasolines." Discussion of preparation methods of gasolines and light oils, though decidedly carbocationic processes, falls outside the scope of this chapter; these processes aimed at the production of isobutylene dimers and trimers for fuels, fuel additives, and lube oils are better discussed in specialized treatises on petrochemistry. Somewhat aged but still useful reviews of this subject are those by Schaad (1955) and Oblad et al. (1958).

This chapter focuses on carbocationic polymerizations yielding low (from $\bar{M}_n \sim 500$), moderately high (from $\bar{M}_n \sim 1000$), to high molecular weight products (up to $M_n \sim 10^5$ to 10^6) finding uses in thousands of applications, that is, additives, viscosity index improvers, primers, coatings, sealants, caulking agents, tackifiers, modifiers, electrical insulators, gaskets, molded compounds, elastomers, and resins.

A thorough review of the scientific literature failed to uncover meaningful up-to-date information on industrial processes employing carbocationic polymerizations. The literature searcher very soon notices the paucity of systematic scientific data and comes to the conclusion that producing companies jealously guard their own empirically acquired information. Another eye-opener is the observation that cationic polymerization industries are still practicing archaic processes and producing the same old staples developed many years, sometimes even decades, ago. Evidently the marketability of these venerable products is due to their favorable price–performance characteristics; even the recent violent price increases of basic petrochemicals had only little influence on the favorable prognostication of the future of these materials. The data shown in Table 10.1 recently compiled by Debreczeni (1979) contain some revealing information in this regard. It appears that need–pull or demand–push has not yet triggered systematic product innovation and industrial research in this field.

Rapid process/product mortality on the marketplace also renders difficult the surveying of the field of industrial processes or of commercially available products made by carbocationic polymerizations. Products often are introduced without much fanfare and later are quietly dropped as a consequence of marketing strategies. Product grades and modifications change rapidly owing to shifting market demand and/or feedstocks. Thus

Table 10.1 □ **Supply and Demand for Isobutylene and Butenes 1973–1990**[a].

Million Pounds	1973	1978	1983	1985	1990
U.S. Production					
Normal butylenes	18,510	19.260	20.790	20,530	18,250
Isobutylene	9,965	10,890	12,070	12,100	10,390
U.S. Demand					
Normal butylenes	2,644	2,133	1,527	1,379	1,263
Isobutylene[b]	1,002	1,109	2,205	2,397	3,650
Butyl and halobutyl rubbers	370	350	378	400	440
Polybutenes	400	462	560	610	760
Polyisobutylene	22	27	43	52	80

[a] After Debreczeni (1979).
[b] In addition to those listed, other uses for isobutylenes are in di- and tri-isobutylenes, heptenes, methyl *tert*-butyl ether, and specialty chemicals, for example.

instead of attempting to assemble the latest information on available product lines (up-to-date product information is available in specialized journals), this chapter focuses on time-proven products and mainly concerns principles of manufacturing methods, structure of commercially available products, characteristics, and uses.

After a survey of commercial products/processes by carbocationic polymerization techniques one is struck by the realization that in spite of our deep understanding of many aspects of cationic mechanisms and the large number of relatively inexpensive cationically active monomers and initiators (cf. Chapter 4), only few systems have reached the commercial arena.

10.2 □ ISOBUTYLENE-BASED CARBOCATIONIC POLYMERIZATIONS

Isobutylene is one of the principal petrochemical raw materials used for the manufacture of chemicals and polymers. The major use of isobutylene is in the manufacture of butyl rubber and derivatives, polyisobutylenes, polybutenes, and di- and triisobutylene. Indeed, the first major petrochemical operation was the manufacture of polyisobutylene (Schriesheim and Kirshenbaum, 1978). As stated in the introduction, di- and triisobutylenes, whose major applications are as lube oil additives, surfactants, and plasticizers, for example, are not considered further. In contrast, low, medium, or high molecular weight polymers, copolymers, and derivatives of polyisobutylene are important and versatile materials made by carbocationic polymerizations. About 80% of the isobutylene consumed in the United States goes into polymers. Of this, about 50% goes into butyl rubber, about 25% into various low molecular weight polyisobutylenes and polybutenes made of C_4 refinery streams, and about 5% into high molecular weight polyisobutylenes (see Table 10.1).

Depending on their molecular weights polyisobutylenes range from liquids, viscous oily products, slow flowing semisolids, to tough, nonflowing elastomers.

The supply of isobutylene seems to be assured in the near future and will far exceed forecast chemical demand (Debreczeni, 1979), particularly from olefin-producing steam crackers. It is interesting to contemplate that the two new uses of isobutylene, that is, making methyl methacrylate and methyl-*tert*-butyl ether, neither of which has yet reached the commercial stage, may account for half of 1990 chemical consumption of isobutylene (Debreczeni, 1979).

Isobutylene is in a sense an "ideal" monomer. It is an extraordinarily reactive cationic monomer under all manners of acidic conditions, and it is one of the few monomers whose polymerization can be readily controlled to give from the lowest oligomers, through medium molecular weights, up to the highest polymers, that is, up to molecular weights in the millions (Kennedy, 1975). Disturbing side reactions, for example, hydride shifts and isomerizations, are essentially absent due to the absence of tertiary hydrogens in the chain or monomer. Thus from the chemical (but not from the kinetic) vantage point the polymerization of isobutylene is a rather uncomplicated process.

With regard to the early history of isobutylene-based carbocationic polymerizations the reader is referred to various older and quite recent detailed reviews (Güterbock, 1959; Plesch, 1963; Kennedy and Kirshenbaum, 1971; Kennedy, 1975).

Low Molecular Weight Polyisobutylenes

Polyisobutylenes with molecular weights of less than about 10,000 are usually regarded to be of relatively low molecular weight materials. Unfortunately, the trade nomenclature of these products is confusing and misleading; the general term *polybutenes* is commonly used for relatively low molecular weight isobutylene polymers of viscous oily consistency. These products represent a rather heterogeneous mixture of various structures for example, homopolymers and copolymers, terminally unsaturated, internally unsaturated, and saturated structures, linear, and possibly branched molecules. They are commonly made from C_4 refinery streams containing predominantly isobutylene together with other olefins, that is, 1-butene and *trans*- and *cis*-2-butene. The term polybutenes would imply that these products are derived from butenes (1-butene and 2-butenes) excluding isobutylene, which is clearly not the case. A better term would be polybutylenes; however, this term is commonly used in the trade to describe isotactic poly(1-butene) made by Ziegler–Natta catalysts. In contrast, the term *polyisobutylenes* signifies products prepared by polymerization of purified (~99 wt%) isobutylene streams. Although unjustified and confusing to the uninitated this terminology is now generally used by the trade. Since our mission here is to survey the industrial scene, we reluctantly accept this custom.

Polybutenes

Manufacture

Polybutenes are manufactured from C_4 olefin refinery streams of catalytic cracking or steam cracking of petroleum. These C_4 fractions (B–B, or butane–butylene fractions), after butadiene has been removed, generally contain a few percent propane–propylene, about 10% 1-butenes, 10 to 15% *cis*- and *trans*-2-butene, 15 to 30% (sometimes as high as 60%) isobutylene and about the same amount of *n*-butane, about 30% isobutane, and the balance molecules higher than C_5. Streams obtained as a by-product of ethylene manufacture are usually used. Steam crackers should increase their share of butylenes supply in the future. For polybutene manufacture isobutylene is not removed from the C_4 stream; however, the removal of other olefins is essential for polyisobutylene production.

The major current producers of low molecular weight polybutenes are Cosden Oil and Chemical Co., Amoco Chemicals Corp., a subsidiary of the Standard Oil Co. (Indiana), Oronite Division of the California Chemical Co. in the United States; Badische Anilin- und Soda Fabrik (BASF) A.-G. in Germany, BP Chemicals (U.K.) Ltd in Great Britain, and Furukawa Kagaku in Japan.

Producers in this field are extremely cautious about disclosing substantive chemical or engineering information. However, examination of details available in trade and patent literature, as well as structural characteristics of the products, provides the serious student significant insight into these operations.

It appears that the basic polymerization process in use today is a rather crude $AlCl_3$ slurry-initiated continuous polymerization of refinery B–B feeds in the temperature range from ~$-60°C$ to ambient. Usually the C_4 feed is dried, desulfurized, and cooled, and together with a stream of initiator ($AlCl_3$) injected into a reactor. The polymerization is exothermic so that the reactor has to be cooled. The effluent is neutralized (this process is not trivial; it is quite difficult to decompose and remove sometimes quite large amounts—$< 1\%$—of $AlCl_3$ from the effluent), purified (fractionation and filtration), packaged, and stored.

The earliest patent disclosure that mentions the polymerization of olefins including isobutylene was applied for in 1923 and issued in 1930 (Ricard, 1930). This inventor used a suspension of $AlCl_3$ in petroleum ether or gasoline and mentions the formation of a viscous liquid which is regarded as the active ingredient. No additional "promoters" (H_2O, t-BuCl, $HCCl_3$, HCl) have been used. Since most contemporary polybutene producers are still using either $AlCl_3$ slurries or complexes, this disclosure appears to be seminal for the whole field.

A few details of Cosden Oil's polybutene process have been described (Mark and Orr, 1956; Labine, 1960). This process uses a dry desulfurized B–B refinery stream from catalytic cracking. The feed is passed through bauxite driers into a surge drum, cooled to a preselected temperature from -70 to $-20°F$, and sent to the reactor. Higher molecular weight products are obtained at lower temperatures. The initiating system is solid $AlCl_3$ slurried in an inert hydrocarbon diluent and is "activated" by gaseous HCl. The reactor effluent is introduced to a settler to remove $AlCl_3$ residues and is recycled. The product is washed with water to remove $AlCl_3$ and HCl, dried with bauxite, and purified through glass filters before storage. The molecular weights of Cosden's polybutenes range from ~ 660 to ~ 1520. A simple flow chart of the process is shown in Figure 10.1.

According to a relevant patent (Jackson, 1960) the Cosden process uses unusually high concentration, 10 to 20 wt %, of $AlCl_3$ in the slurry activated by HCl or other agents that can produce HCl *in situ* (H_2O, $CHCl_3$). The effluent is recycled at least four to eight times of feed. As an example, a recycle ratio of 8:1 is used to prepare ~ 500 molecular weight product and the recycle ratio increases to 10:1 to produce ~ 1450 molecular weight material. Evidently longer residence times are needed to get higher molecular weights; however, the cost of the material must also increase.

Only by studying the patent literature may one glean details of the Oronite (California Chemical Co.) and Indopol [Standard Oil Company (Indiana)] processes. By evaluating Oronite's patents one may surmise that their best product in terms of overall product characteristics and lowest cost may be obtained by following Lavine and Folsom's (1949) patent, according to which the polymerization is due to $AlCl_3$ dissolved in hot liquid butane. The specifications emphasize the homogeneous nature of the process and the absence of solid $AlCl_3$ in the reactor. The butane used to dissolve $AlCl_3$ must be free of olefins to avoid tar formation. The use of a promoter is not mentioned in the patent. However, some protogenic materials (moisture) most probably are present in the system; otherwise $AlCl_3$ would be completely insoluble even in hot n-butane. $AlCl_3$ concentration in the reactor is in excess of 0.1% of the charge (0.14 to 0.18%).

An early patent relevant to the Indopol process (Evering et al., 1946) describes the use of a liquid aromatic-free hydrocarbon/$AlCl_3$ complex partially spent by isomerization or alkylation processes for the polymerization of C_4 fractions. The C_4 feed is continuously passed upward through the liquefied $AlCl_3$ complex without stirring. Internal cooling coils provide temperature regulation. The liquid $AlCl_3$ complex is obtained of light saturated hydrocarbons and HCl. Fresh complex is produced by

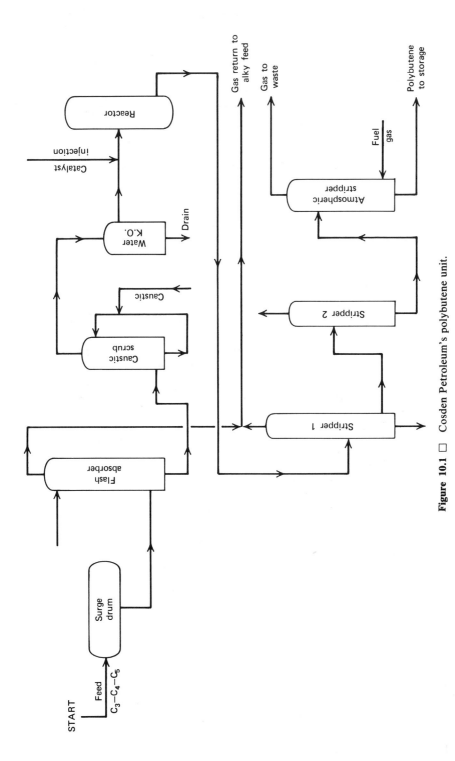

Figure 10.1 □ Cosden Petroleum's polybutene unit.

471

adding solid $AlCl_3$ to the spent catalyst and the fresh $AlCl_3$ is rapidly taken up by the liquid. This operation is reminiscent of the preparation of "Friedel–Crafts sludges" of alkylation–isomerization processes. Evidently 1- and 2-butenes are incorporated into the polybutene and their amount increases with conversion. Extensions/improvements of this early patent include a stirred reactor (Russum, 1954), a new method for makeup catalyst addition (Fragen, 1954), and recycling of the complex catalyst dispersed in light liquid hydrocarbon rather than a homogeneous phase (Yahnke and Healy, 1954).

Table 10.2 summarizes some aspects of manufacturing characteristics of major polybutene processes. Evidently these processes are fundamentally similar and differ only in empirical details. All processes are continuous-flow operations using feeds largely free of sulfur and nitrogen compounds and cursorily dried by chemical drying agents, for example, bauxite, or azeotropic distillation. The feed is often cooled prior to its continuous introduction into the reactor. The reactors are agitated, cooled vessels; two or three reactors may be joined in series. They usually operate in the range of -10 to $+80°C$ and up to 2 MPa (20 kg/cm²). Cooling is provided either by direct heat exchange or internally by refluxing the charge. $AlCl_3$ appears to be the coinitiator in all instances and the initiator, protogenic material(s), either is provided by ubiquitous impurities or is introduced purposefully in the form of an "activator," for example, HCl, H_2O, or $HCCl_3$. The activator is also fed continuously into the reactor either in solution or as a hydrocarbon slurry. The unspent $AlCl_3$ or residue is removed from the effluent by settling and/or reaction with ammonia, amines, methanol, water, or aqueous soldium hydroxide. The stream finally passes through clay or

Table 10.2 □ **Characteristics of Polybutene Manufacturing Processes**

Company	Feedstock	Temperature and Control (°F)	Initiating System	Continuous Introduction
Cosden	C_4–C_5	-45 to $+60$; external cooling of action, slight pressure	$AlCl_3$, 10–20 wt % HCl activator	$AlCl_3$ slurry in dry light polymer
Oronite	C_4	-30 to 80; feed cooled, adiabatic solvent evapn.	Claimed to be soluble $AlCl_3$ in n-butane	$AlCl_3$ solution in hot butane
Indopol	C_4	0–40; cooling by internal coils, ~100 psi	liquid $HCl/AlCl_3/$ hydrocarbon complex	Solid $AlCl_3$ added to recycled complex

bauxite columns to remove AlCl₃ residues and coloring matter. In the distillation section the *n*-butane, isobutane, and unconverted butenes are flashed-off in one column and the low molecular weight polybutenes are taken overhead in other columns. Peculiarly, very little research seems to be going on in such obvious areas as increasing initiator efficiency or simplifying initiator removal by the use of suitable coinitiator solvents, coagents, or supports, for example.

Molecular Weight Control of Polybutenes

Polymer viscosity and molecular weight are regulated by controlling the polymerization temperature and AlCl₃ concentration. Molecular weights are also influenced by feed composition, but this variable cannot be changed. Since isobutylene is the most reactive feed component, practically all of it is converted to polymer. The conversion of butenes in the feed is usually low. Decreasing the polymerization temperature results in higher selectivity of isobutylene polymerization, that is, less butene copolymerization. The butenes tend to terminate the polyisobutylene propagating chain, and hence reduce molecular weights. The molecular weight distributions of several commercial polybutylene grades are shown in Figure 10.2. The bimodal molecular weight distribution (H-25 grade Amoco Polybutylene) is due to characteristic processing details: in the first stage of the polymerization (e.g., in one reactor), the isobutylene in the feed is converted to a fairly high molecular weight polymer whereas very little of the butenes copolymerize. Following the disappearance of isobutylene from

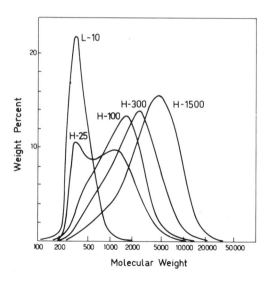

Figure 10.2 □ The molecular weight distributions of polybutenes (indopols) manufactured by Amoco Chemicals Corporation. (Symbols designate commercial grades.)

the feed, in a separate stage or reactor, the butenes are copolymerized, which yields a much lower molecular weight copolymer and leads to a bimodal molecular weight distribution. The low molecular weight fractions could be removed by vacuum or steam distillation.

Structure, Properties, and Uses

The composition and structure of copolymers provide important insight as to the polymerization mechanism by which they were prepared. Peculiarly, there is an almost total lack of quantitative information on copolymerization of isobutylene with butenes. Examination of poison coefficient/transfer agent studies (Kennedy and Squires, 1967; also see Section 4.4) suggests that 1-butene is a pure but rather mild poison whereas the 2-butenes (on the basis of 2-octene data) would be expected to function as both poisons and transfer agents. This leads to the proposition that as far as polybutene manufacture is concerned, 1-butene will reduce the yield but will little affect average molecular weight. Yield reduction could be overcome by increased $AlCl_3$ input and other conventional measures. It is expected that polybutene termini will have characteristic groups originating from 1-butene. 2-Butenes will reduce yield somewhat (poisoning) and molecular weight (transfer activity) severely. Considering the thermodynamic stability difference between *cis*- and *trans*-2-butene, the former is expected to be a more reactive poison and transfer agent, that is, to give lower yields and molecular weights, than the latter. Changing the temperature of polymerization should not affect these fundamental effects.

One may regard the polybutene process as an inefficient isobutylene polymerization carried out in the presence of poisons and transfer agents. The butenes are mainly consumed by alkylation by the growing isobutylene cation or by hydride donation followed by isobutylene cationation, that is, initiation.

Since the C_4 streams used to prepare polybutenes contain besides isobutylene appreciable amounts of 1-butene and 2-butenes, these polymers are in fact copolymers of isobutylene and small amounts of butenes. The overall composition of polymers is governed by the reactivity ratios of the component monomers (cf. Chapter 6). And although reactivity ratios are strongly affected by polymerization conditions, that is, counteranion, temperature, and medium polarity, the reactivity ratio of isobutylene is much higher than those of the comonomers in the feed, a fact that cannot be affected by changing conditions.

The structure of polybutenes has only very recently become the subject of scientific investigation. Puskas et al. (1976) used NMR and IR spectroscopy and hydrogenation techniques to gain insight into the nature of unsaturation in some commercial materials. Similarly to Kennedy (1975), these authors also criticized Imura et al.'s (1964) earlier results. According to Puskas et al. (1976) with regard to unsaturation analyses commercial polybutenes and polyisobutylenes obtained by AlCl₃ coinitiator are essentially identical and both contain tetrasubstituted olefin. Although a variety of suggestions could be made to account for this interesting conclusion, speculation is premature in the absence of confirmatory evidence.

Since polybutene production usually occurs at or close to 100% olefin conversion, product composition and molecular weight change with conversion in a predictable manner and largely contribute to product heterogeneity. Unfortunately the considerable scientific understanding of relevant reactivities and kinetics has not yet been brought to bear on these problems.

Polybutenes are colorless, odorless viscous liquids soluble in hydrocarbons, chlorinated hydrocarbons, and ethers. The low molecular weight grades flow freely at room temperature, whereas the higher molecular weight grades (>1000) require heating for pumping. The latter are excellent lubricants and electrical insulators. Polybutenes are "nondrying," that is, they do not crosslink when exposed to air, and they are thermally stable up to 280 to 300°C. The lowest molecular weight polybutenes ($\bar{M}_n \sim 300$) may become oxidized to a small extent when exposed to air at room temperature; however, the more viscous grades are impermeable to gases so that they do not get oxidized under similar conditions. As a precaution polybutenes are usually blanketed with an inert gas during storage.

The most important use of polybutenes is in the manufacture of motor oil additives. Although some polybutenes may be blended directly into lubricating oils they usually serve as chemical intermediates for additive synthesis. The unsaturation provides the functionality for postpolymerization syntheses, for example, ene reaction with maleic anhydride. Other major uses are in formulations of sealants, caulks, coatings, adhesives, and laminating agents; in high-voltage electrical cables as impregnating oils and pipe oils; as lubricants for sheet metal cutting; and as compressor oil.

Polybutenes are nontoxic and nonirritating and thus find applications in paper treatment and lubricants that come in contact with food products. They have clearance from the Federal Drug and Food Administration for such indirect food-contact use.

Polisobutylenes

In contrast to polybutenes produced by the use of C₄ refinery streams which are ill-defined copolymers of isobutylene plus butenes, polyisobutylenes are pure homopolymers of isobutylene. The literature, particularly patent literature, is extremely rich in disclosures of isobutylene polymerization processes and extensions and improvements of basic patents. Scientific aspects of isobutylene polymerizations are discussed elsewhere in this book.

A variety of polyisobutylene grades are commercially available. The main difference between these products is their molecular weight, which determines their ultimate use. Isobutylene is one of the very few monomers that can be readily polymerized to practically any molecular weight level from liquid oligomers to tough elastomers, and all these materials exhibit characteristic useful properties rendering them valuable for a variety of end uses. As discussed elsewhere in this book (cf. Chapters 4 and 5) the most important, although not exclusive, parameter that controls molecular weight is the polymerization temperature.

Manufacture

There are two fundamentally different ways to produce low molecular weight polyisobutylenes: by direct polymerization at relatively high temperatures and by degradation of higher molecular weight products. The latter process is obviously less cost-efficient. Two producers of these materials are Exxon Chemical Co. in the United States (Vistanex LM and Paratone N) and BASF in Germany (Oppanol B1 and B3).

The BASF process for the manufacture of Oppanol B3 has been outlined in Güterbock's (1959) book. A liquid mixture of isobutylene and isobutane (internal coolant) is brought into contact with a stream of BF_3 dissolved in CH_3OH (Otto et al., 1939). The polymerization occurs at the boiling point of isobutane at $\sim -10°C$. The boiling solvent provides efficient internal cooling in view of the highly exothermic polymerization and the importance of temperature regulation for molecular weight control. Catalyst residues are removed in a separator. Subsequently the effluent is degassed, washed, rectified (i.e., low boiling components are removed by vacuum), bleached, filtered, and dried. According to its specification chart $\bar{M}_n = 820$ for Oppanol B3.

Exxon's process for the manufacture of low molecular weight polyisobutylenes consists of cooling an isobutylene stream to the desired temperature, contacting the stream with a slurry of $AlCl_3$ in n-hexane in a well agitated reactor(s), and quenching the polymerization with alcohol or alkali (Goering and Mistretta, 1956). The molecular weight is regulated by careful temperature control. A simplified flow chart is shown in Figure 10.3. The fresh isobutylene makeup feed is mixed with recycled hexane and isobutylene, dried and cooled, and injected continuously with the $AlCl_3$ slurry into the reactor. The effluent is treated with aqueous alkali to quench the reaction and remove catalyst residues. The stream is heated to flash-off isobutylene for recycling.

For Vistanex LM production the solution is filtered, the n-hexane is flashed-off and recycled, and the product is filled into cans. For Paratone N production the polyisobutylene in n-hexane solution is blended with a neutral oil, the residual monomer, light ends, and n-hexane are removed, and the product is filtered before packaging in cans.

The difference between Vistanex LM (LM = low molecular weight) and

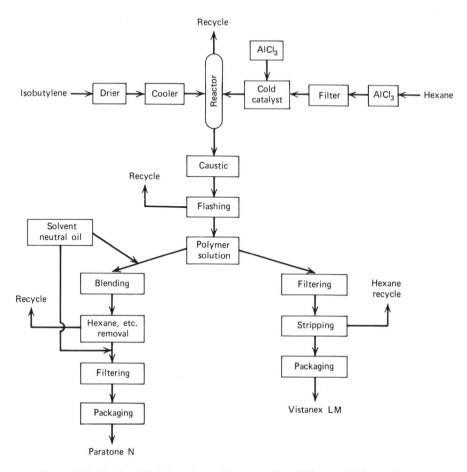

Figure 10.3 □   Simplified flow sheet of Paratone N and Vistanex LM manufacture.

Paratone N is in their molecular weight: Vistanex LM grades are colorless viscous liquids whose $\bar{M}_v$'s range from 50,000 to 75,000. These materials are used in adhesives, sealants, waxes, electric insulating compounds, and blends with asphalt. Paratone N, a trade name indicating polyisobutylene in a light oil solution, has a $\bar{M}_v$ of ~120,000. It is used as viscosity index improver in lubricating oils and automotive transmission fluids.

Another option for the preparation of low molecular weight poly-isobutylenes is high temperature–high shear degradation of higher molecular weight materials (Güterbock, 1959). Although a wide variety of molecular weights can be obtained by controlled degradation, that is, by controlling the shear gradient, residence time in the extruder, and temperature, this process must be economically less attractive than direct synthesis from monomer.

Structure

The overall in-chain structure of low molecular weight polyisobutylenes has been firmly established [linear, head-to-tail $-CH_2C(CH_3)_2-$ enchainment]; however, there is room for better definition of the termini. It has been established that polyisobutylenes contain various types of terminal unsaturations and that these functions determine some very important chemical properties, that is, oxidative stability and the nature and scope of derivatives. Thus the exact knowledge of end group structures is of great importance to the application chemist or engineer. Commonly it is anticipated that the head group of polyisobutylenes obtained under controlled conditions (i.e., reasonably low conversions, low temperatures such as from -20 to $-78°C$) is $(CH_3)_3\sim$ and the end structures are $\sim CH_2C(CH_3)=CH_2$ or $\sim CH=C(CH_3)_2$. The saturated head group is due to initiation or chain transfer by proton, $CH_2=C(CH_3)_2 + H^{\oplus} \rightarrow (CH_3)_3C^{\oplus}$, whereas the end groups are due to proton elimination, $\sim CH_2\overset{\oplus}{C}(CH_3)_2 \rightarrow$ $\sim CH_2C(CH_3)=CH_2$ or $\sim CH=C(CH_3)_2$ (Kennedy, 1975). In this sense polyisobutylenes may be regarded to be monoolefins. Closer examination, however, reveals that commercial polyisobutylenes contain less, sometimes substantially less, than one double bond per molecule. Although a number of possibilities could be invoked, details of this problem have not yet been discussed.

A large amount of information relative to the detailed structure of polyisobutylene termini has been assembled and examined (Kennedy, 1975). Of particular interest in this context is a recent publication by Puskas et al. (1976), which concerns ^1H-NMR and IR examination of terminal unsaturations in low molecular weight polyisobutylenes prepared at $\sim 100\%$ conversion at relatively high temperatures (5 to 25°C). According to these workers the nature of coinitiator used (BF_3 or $AlCl_3$) affects end group structure. For example, the major end group obtained in the presence of $AlCl_3$ was proposed to be $\sim C(CH_3)=CHCH_3$ (internal vinylidene group), the formation of which is difficult to explain. Under the particular polymerization conditions used, that is, complete conversion and relatively high temperatures, unanticipated isomerizations and alkylations may occur, which may help to explain these observations.

Properties and Uses

The properties and applications of polyisobutylenes and polybutenes are quite similar; however, polyisobutylene is more expensive than polybutenes because it is manufactured by the use of pure isobutylene feeds. Polyisobutylenes may be viewed as premium polybutenes and are used where increased product homogeneity is of value. As dictated by the cost performance ratio, the market for low molecular weight polyisobutylenes is much smaller than for the less expensive polybutenes; for example, in 1978 close to 20 times more polybutenes than polyisobutylenes were produced (cf. Table 10.1).

The lowest molecular weights polyisobutylenes in the market, Oppanol B1 and B3 (clear viscous liquids, with $\bar{M}_n \sim 300$ and 820, respectively), are used as viscosity index improvers, electrical insulating oils, and adhesives. The somewhat higher molecular weight products Oppanol B10 and B15 and Vistanex LM grades ($\bar{M}_n$ range from 8000 to 13,000) are clear, permanently tacky semisolids and are used in blends with waxes, caulking compounds, viscosity index improvers, adhesives, and sealants, for example.

Medium and High Molecular Weight Polyisobutylenes

Manufacture

Two processes are in use today for the manufacture of medium and high molecular weight polyisobutylenes: BASF's belt process ("Bandverfahren") in West Germany and Exxon's cold slurry process in the United States and France (Socabu). The trade names of the products are Oppanol (BASF) and Vistanex MM (Exxon).

The main difference between the BASF and Exxon processes is the coinitiator used and the manner of cooling: the BASF process uses BF_3 and 'internal" cooling, whereas that of Exxon employs $AlCl_3$ and cooling by heat exchangers. The heart of the BASF process (Otto et al., 1940) is a closed-loop moving belt guided over rolls at both ends like a conveyor belt. Liquid isobutylene and ethylene about 1:1 in volume are sprayed on the cold stainless steel trough moving in a long tube and BF_3 diluted with liquid ethylene is introduced close to the monomer port (see Figure 10.4). The polymerization takes place in boiling ethylene at $-104°C$. The polymerization is almost instantaneous and the heat of polymerization is

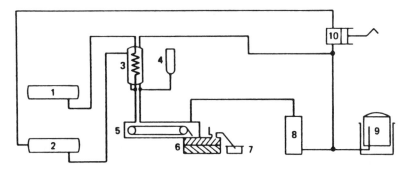

Figure 10.4 □ Process for the manufacture of Oppanol B.1, liquid isobutylene; 2, liquid ethylene; 3, cooler–mixer; 4, BF_3–ethylene reservoir; 5, continuous belt; 6, compacting rolls; 7, product; 8, CaO to purify ethylene recycle; 9, gasometer; 10, compressor.

efficiently removed by flashing-off the internal coolant ethylene. The flash-off ethylene is purified and recycled. The thick layer of polymer that is formed travels on the belt, is scraped off, and falls on a steam-heated compacting double screw extruded to remove entrapped gases, that is, ethylene, isobutylene, and BF_3.

In contrast, the Exxon cold slurry process for the manufacture of high molecular weight Vistanex grades employs a stirred overflow reactor cooled to the desired temperature by external and internal heat exchangers (Thomas and Sparks, 1944). The temperature is regulated by controlling the evaporation of the coolant in the compressors and may be as low as $-100°C$. A cooled liquid isobutylene/methyl chloride mixture is fed at the bottom of the reactor in the vicinity of the rapidly moving impeller and close to a second entry port feeding the initiator, a stream of $AlCl_3$ in methyl chloride. The polymerization is virtually instantaneous, and since polyisobutylene is insoluble in methyl chloride at low temperatures the polymer precipitates in the form of a fine slurry. The finely dispersed slurry particles ensure efficient heat removal. The slurry stream overflows into a flash tank where it is brought into contact with steam and hot water, which decomposes $AlCl_3$ and flashes-off solvent and unreacted monomer. The latter are rectified and recycled, the water and residual volatiles in the polymer are removed in a heated extruder, and the product is ready for finishing.

Structure, Properties, and Uses

In spite of unconfirmed sporadic reports (mainly in the Russian literature) to the contrary, it is generally accepted that the structure of medium or high molecular weight polyisobutylene is the conventional head-to-tail enchainment $-CH_2C(CH_3)_2-$. With regard to possible head and end groups, the reader should consult the preceding sections. The lower polymerization temperatures employed for the manufacture of high molecular weight polyisobutylene than of low molecular weight materials would lead one to expect that the structures of the former are more homogeneous and uniform than those of the latter.

The properties of polyisobutylene have been thoroughly investigated from the scientific and technological point of view; a discussion of this subject is beyond the scope of this chapter.

Briefly, polyisobutylenes of $M_n < 10,000$ are clear, permanently tacky, slow-flowing liquids or solids. Highest molecular weight grades are tough elastomers. They resemble unvulcanized natural rubber in tensile strength, elasticity, and electrical properties. As essentially saturated hydrocarbons, they are resistant to heat and chemical attack. They are excellently soluble in aliphatic and cycloaliphatic solvents, ethers, $HCCl_3$, CCl_4, and CS_2, and somewhat less so in aromatic solvents.

Polyisobutylenes exhibit excellent barrier properties and are often blended with other polymers to reduce their permeability. The higher its molecular weight, the lower its permeability. These polymers have excellent dielectric properties and high resistance to moisture absorption or penetration. Thus they are well suited as electrical insulators.

Polyisobutylenes are used in a wide range of adhesives and caulking and sealing compounds, and in blends with waxes, polyethylene, polypropylene, and other polymers to improve their flexibility particularly at low temperatures. They are used as tackifiers in greases, in motor oils to improve viscosity index, in chewing gum bases, as cable and insulating oils to reduce moisture transmission, as impregnant for leather to improve wear, in tank and ditch linings, in membranes on roofs, and as moisture barriers in construction. Blended with mineral-filled asphalt compounds, polyisobutylenes improve their physical properties and resistance to weathering.

Isobutylene Copolymers, Terpolymers, and Derivatives

Butyl Rubber

The most important isobutylene derivative, indeed the most important product made commercially by carbocationic techniques, is butyl rubber, a copolymer of isobutylene with small amounts of isoprene and its derivatives, particularly halobutyl rubbers.

The discovery and early history of these products have been recounted by one of the discoverers (Thomas, 1969) and Kennedy (1975). Many details of butyl rubber manufacture, characterization properties, and uses have been published (Thomas et al., 1940; Thomas and Peterson, 1954; Buckley, 1959; Kennedy, 1968; Kennedy and Kirshenbaum, 1971; Stucker and Higgins, 1977).

Manufacture

Butyl rubber is a random copolymer of isobutylene and ~1.5 mole % isoprene; the latter serves to introduce sufficient unsaturation for vulcanization. The copolymerization is extremely rapid in the presence of AlCl$_3$. To reduce the rate of reaction, and to obtain the product in a conveniently handleable discrete particle form, the charge is diluted with methyl chloride.

In practice a precooled feed of ~27 vol % mixed monomers and ~73 v/v CH$_3$Cl is introduced at the bottom of a large vertical stainless steel reaction vessel of some 14,000 to 15,000 l capacity. Butyl plants usually have several of these reactors, some in use and some being cleaned. The reactors have cooling jackets and internal cylindrical concentric cooling elements of stainless steel containing the liquid ethylene coolant. The reactors are insulated with a thick layer of insulating material. The monomer feed is continuously pumped in and up the center of the cooling element to flow around and between it and the outer cooling surfaces. The charge is stirred by a set of propelled blades activated by a strong motor. The prechilled (~ −96°C) cooinitiator, 0.2 to 0.3 wt % AlCl₃ in CH₃Cl solution, is fed through several (three) separate nozzles. Reaction is virtually instantaneous and highly exothermic upon mixing the AlCl₃ stream with the monomer feed. The copolymer is formed in the form of a fine slurry (hence the term "cold slurry" process). The process is centrally controlled from a control house. Each reactor is connected to about ~ 12 recording instruments. The course of the polymerization is followed by monotoring the pressure developed by the ethylene vapors boiled off from the cooling elements. If the pressure of boiling ethylene is increasing or getting too high the feeding rate of the AlCl₃ stream is reduced or the AlCl₃ stream is diluted. The rate of reaction is controlled by controlling the rate of AlCl₃ input or AlCl₃ concentration.

Temperature control is of utmost importance: the feeds are prechilled to ~ −85°C prior to entry into the reactor. The charge is kept between −96 and −101°C and the ethylene in the cooling elements a few degrees below this temperature.

The slurry exits at the top of the reactor via a short tube of ~ 3 in. in diameter into a thin pipe of ~ 15 in. diameter and 5 ft long which passes into the top of a flash tank (a vertical cylinder of ~ 130,000 1 capacity). The effluent containing the rubber particles, isobutylene, and methyl chloride is met by steam and hot water in the flash tank, where the unreacted monomer and solvent flash-off and the rubber precipitates in water at ~ 65°C at the bottom of the flash drum. Zinc stearate processing aid (~ 2% on rubber) and a stabilizer (phenyl-β-naphthlamine) are introduced in the flash tank. The volatiles are condensed to remove moisture, compressed, refrigerated, and distilled for recovery of isobutylene and methyl chloride. Isoprene is consumed by the copolymerization process and the small amount of unconverted diene is discarded. The butyl rubber emerges from the flash tank in the form of white flocks in hot water. It is filtered, dried, finished, and packaged. A simplified flow sheet of a butyl plant is shown in Figure 10.5.

Butyl rubber is manufactured by Exxon Chemical Co. and Columbian Carbon in the United States, by Esso Petroleum Ltd. in the United Kingdom, by Polymer Corp. in Canada and Belgium, by Socabu in France, and by Japan Butyl Co. in Japan. Butyl is also manufactured in the Soviet Union and plans to build butyl plants in Italy and Brazil are under consideration.

Structure, Properties, and Uses

Commercial butyl rubbers are random head-to-tail copolymers of isobutylene (~ 98.5 mole %) and isoprene (~ 1.5 mole %). The isoprene units are

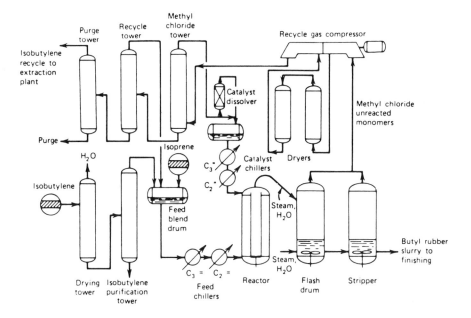

Figure 10.5 □   Plant flow sheet for butyl production.

mainly in trans-1,4 enchainment:

$$-CH_2\overset{\overset{\displaystyle CH_3}{|}}{\underset{\underset{\displaystyle CH_3}{|}}{C}}-CH_2-\overset{\overset{\displaystyle CH_3}{|}}{C}=\overset{\underset{\displaystyle H}{|}}{C}-CH_2-$$

The structure of butyl rubber was the subject of numerous investigations and has been reviewed (Kennedy and Kirshenbaum, 1971).

Butyl rubber is a general-purpose elastomer. It can be crosslinked with sulfur and various other curing systems. The low degree of unsaturation imparts high resistance to thermal degradation, ozone attack, chemicals, and aging. It exhibits the lowest resilience among commercially available elastomers. Owing to this high damping property, butyl is extensively used in engine and body mounts and in truck and trailer springs.

Butyl rubber has excellent barrier properties and is extensively used in inner tubes, and (in the form of its halogenated form) as inner liners in tubeless tires. Owing to its chemical inertness it is used in chemical-tank linings, protective clothing, gloves, window channels, hoses, gaskets, and electrical insulation (Buckley, 1959).

Liquid Butyl: Manufacture, Properties and Uses

Today liquid butyl rubber is made by depolymerization, most likely high shear/temperature extrusion, of high molecular weight butyl rubber and sold under the trade name Kalene 800 by Hardman Inc. The product is a somewhat gray flowable viscous liquid with $\bar{M}_n = 10,000$ to $20,000$, $\bar{M}_v \sim 30,000$; mole unsaturation ~ 3.5 mol %. Except for its low viscosity and somewhat higher unsaturation level, overall properties are similar to those of the high molecular weight product. This material finds uses as architectural sealant, adhesive binder, and electrical encapsulating and potting agent, among others. It is of some historical interest that the Enjay Chemical Company produced and marketed for a few years (from 1969 till about 1974) liquid butyls (and chlorobutyls) made by direct polymerizations containing 4.2% isoprene units. This material was made by a catalyst system different from $AlCl_3$ and higher temperatures than used in other butyls (Anonymous, 1969).

Isobutylene–Isoprene–Divinylbenzene Terpolymers

Whereas conventional butyl rubbers are soluble, isobutylene–isoprene–divinylbenzene terpolymers are partially insoluble because they are cross-linked. This is a desirable property for certain sealant and adhesive applications where partial crosslinking prevents sag and cold flow (Walker et al., 1974).

These terpolymers are prepared similarly to conventional butyl rubbers by terpolymerization with divinylbenzene in the charge. This bifunctional monomer leads to partial crosslinking during the synthesis.

The terpolymers can be cured by peroxides. Conventional butyl rubbers (or polyisobutylene) degrade under the influence of peroxides. Evidently, available $CH_2=CH-C_6H_4-$ groups in the terpolymer provide reactive cure sites for peroxides.

Unlike regular butyl rubber, the uncured terpolymer does not cold-flow. The terpolymer can be blended with regular butyl and the blends exhibit increased green strength and resistance to cold flow. Ozone resistance of the terpolymer is superior to regular butyl at comparable unsaturation level. The peroxide-cured terpolymer can be used in blends with other polymers, for example, nitrile rubber, ethylene–propylene copolymers, and polyethylene. In addition to the inherent advantageous properties of conventional butyl rubber peroxide-cured terpolymers exhibit low compression set and excellent heat resistance. These materials are used in electrical insulation, moldings, caulks, gaskets, automotive and architectural sealant tapes, and pressure-sensitive adhesives.

The moderately crosslinked terpolymers are produced and sold by Polysar Ltd. and are available as 20 and 50% crosslinked grades. These terpolymers compete with conventional butyls moderately crosslinked in a mixer or on a mill produced by Columbian Carbon of Cities Service Co.

Halogenated Butyl Rubbers

Chloro- and bromobutyls are undoubtedly the commercially most important butyl rubber derivatives. These rubbers are made by direct halogenation of butyl rubber solutions in n-hexane at ambient temperature (Baldwin et al., 1961). By careful control of chlorine concentration in the reactor severe molecular weight breakdown can be avoided and allylic chlorination can be achieved. Predominantly allylic substitution and insignificant addition to the double bond occur, so that the original amount of unsaturation in the butyl rubber is not much decreased, if at all, by the chlorination process. The desired chlorine per double bond ratio is 1.0.

Chlorination in the reactor is visualized to proceed by a competitive radical/ionic mechanism in which the dominant component is by far the ionic one. Among the structures that may form, the two most important are

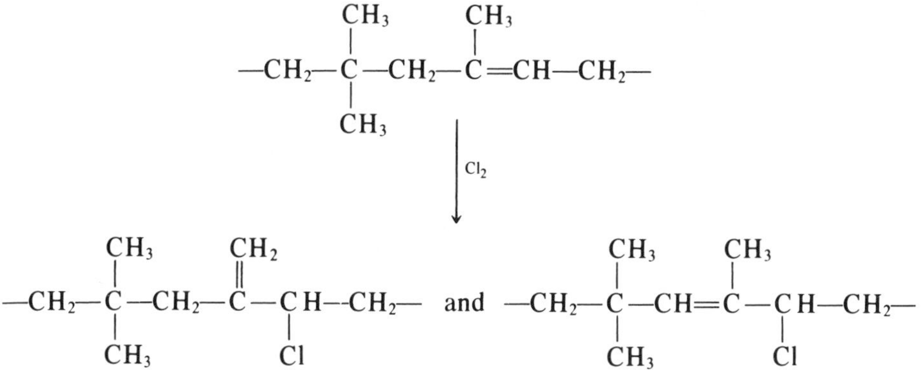

$$-CH_2-\underset{\underset{CH_3}{|}}{\overset{\overset{CH_3}{|}}{C}}-CH_2-\overset{\overset{CH_3}{|}}{C}=CH-CH_2-$$

$$\downarrow Cl_2$$

$$-CH_2-\underset{\underset{CH_3}{|}}{\overset{\overset{CH_3}{|}}{C}}-CH_2-\underset{\underset{Cl}{|}}{\overset{\overset{CH_2}{||}}{C}}-CH-CH_2- \quad \text{and} \quad -CH_2-\underset{\underset{CH_3}{|}}{\overset{\overset{CH_3}{|}}{C}}-CH=\overset{\overset{CH_3}{|}}{C}-\underset{\underset{Cl}{|}}{C}H-CH_2-$$

Although the chlorine content in chlorobutyl is only $\sim 1.2\%$, the effect of even such a small amount of chlorine on the chemical and physical properties of the product is remarkable. First, chlorine imparts cure versatility. Owing to the presence of reactive allylic chlorines *and* unsaturation, chlorobutyl rubber can be crosslinked with a variety of ZnO or sulfur cures or other methods (Anonymous, 1961). Second, the rate of cure of chlorobutyl is faster than that of the parent butyl rubber owing to the presence of reactive allylic chlorines. Third, chlorobutyls have less tendency for cure reversion than butyl rubber. Fourth, chlorobutyl is more compatible with conventional rubbers than the nonhalogenated product and can be convulcanized with natural rubber, SBR, neoprene, nitrile rubber, and conventional butyl rubber.

Chlorobutyl is used extensively where barrier properties, resistance to heat, oxygen, oxygenated solvents, acids, and bases, versatility and speed of curing, and covulcanizability with other elastomers are needed. One of the most important applications of chlorobutyl is in inner liners of tubeless tires, but other uses, for example, tire curing bags, tire sidewalls, hoses, conveyor belts, molded articles, and gaskets are also of considerable commercial significance.

Brominated butyl rubber (or bromobutyl) is in many respects similar to the chloro derivative; however, it is claimed to be more cure reactive; that is, it offers a faster cure. Further, it has superior adhesion, flex fatigue, and heat resistance, although the latter three properties are much more compound dependent (Law, 1974).

Chlorobutyl rubber is produced and marketed by Exxon in the United States and by Polysar in Canada; bromobutyl rubber is manufactured and sold by Polysar and B. F. Goodrich (Hycar 2202).

Miscellaneous Polyisobutylene Derivatives

From time to time various polyisobutylene derivatives have appeared on the market or reached the market development stage only to be quietly interred later owing to dwindling demand or lack of demand. This section concerns a brief obituary of recently deceased shorter- or longer-lived polyisobutylene derivatives focusing on the chemical possibilities they offered.

Carboxy-Terminated Polyisobutylene

Carboxy-terminated polyisobutylene (CTPIB) (Baldwin et al., 1969) has been prepared by degradative ozonization in the presence of pyridine of high molecular weight isobutylene-*co*-piperylene (2 to 4 mol %):

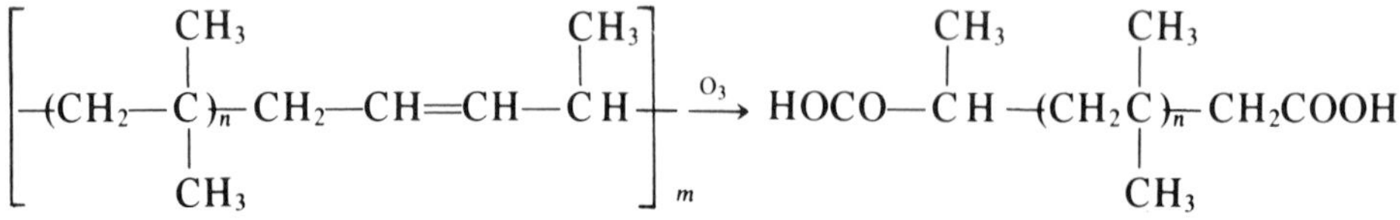

$$\left[-(CH_2-\underset{\underset{CH_3}{|}}{\overset{\overset{CH_3}{|}}{C}})_n-CH_2-CH=CH-\underset{}{\overset{\overset{CH_3}{|}}{C}}H-\right]_m \xrightarrow{O_3} HOCO-\overset{\overset{CH_3}{|}}{C}H-(CH_2\underset{\underset{CH_3}{|}}{\overset{\overset{CH_3}{|}}{C}})_n-CH_2COOH$$

In the absence of pyridine large amounts of ozonides and peroxides form, which are difficult to convert to useful acid termini. CTPIB's were viscous liquids with a molecular weight range from 1800 to 3500 and an average functionality of ~ 1.8 COOH groups per chain. This predominantly diacid could be converted to networks by various reactions such as with epoxides or aziridine.

Hydroxy-Terminated Polyisobutylene

Hydroxy-terminated polyisobutylenes were prepared from the product of ozonolytic degradation (see above) by direct reduction with lithium aluminum hydride in ethyl ether, or, more practically, by reaction with propylene oxide (Zapp et al., 1970). In the latter case secondary hydroxy ester functionality was obtained:

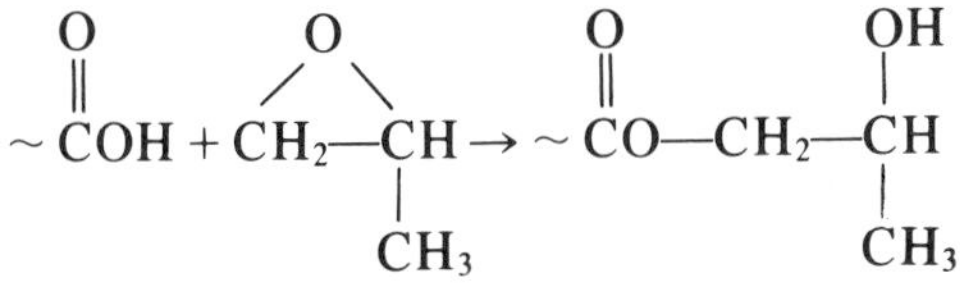

$$\sim \overset{\overset{\displaystyle O}{\|}}{C}OH + \overset{\overset{\displaystyle O}{/\backslash}}{CH_2-CH} \rightarrow \sim \overset{\overset{\displaystyle O}{\|}}{C}O-CH_2-\overset{\overset{\displaystyle OH}{|}}{CH}$$
$$\qquad\qquad\qquad \overset{|}{CH_3} \qquad\qquad\qquad\qquad \overset{|}{CH_3}$$

These materials were reacted with isocyanates for the preparation of networks.

Conjugated Diene Butyl

Conjugated diene butyl CDB was obtained by the controlled dehydrochlorination of chlorinated butyl rubber (Baldwin and Gardner, 1977):

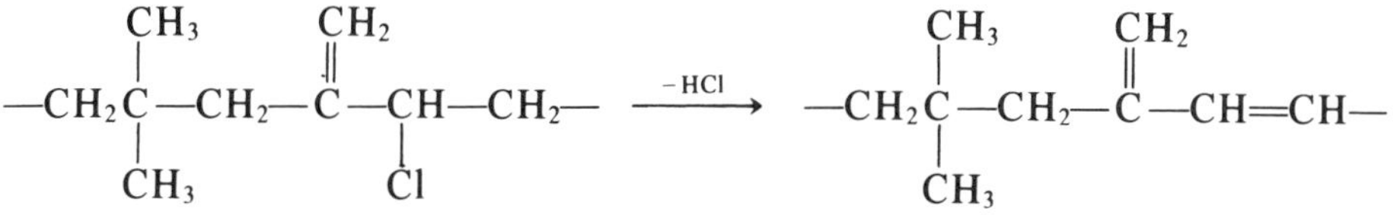

The versatile conjugated double bond system in the chain was used to "graft cure," for example, with methacrylic acid or styrene, which led to transparent rubbers exhibiting only one T_g at $-59°C$.

S-Polymer

S-Polymers were high molecular weight copolymers of isobutylene and styrene produced by low temperature copolymerization and marketed by the Enjay Company (Newberg et al., 1948). These materials were transparent plastic sheets intended for food packaging, particularly fresh and dried fruits, on account of their excellent barrier properties. In addition these products were evaluated for coatings, particularly paper coating and impregnating, adhesives, plasticizers, and sealants.

Isobutylene–Cyclopentadiene Copolymers

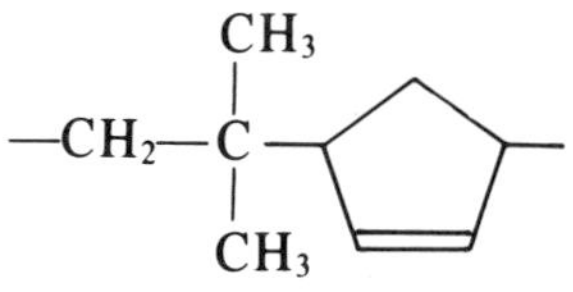

High molecular weight rubbery isobutylene–cyclopentadiene copolymers containing up to ~40% cyclopentadiene have been produced (Thaler and Buckley, 1976) and made available in market development quantities by Exxon. These random copolymers with the overall structure exhibited excellent ozone resistance, fast curing character with conventional recipes, and good compatibility with conventional high functionality rubbers, such as SBR, PBD, and natural rubber.

Butyl Latex

For a while butyl rubber has also been available in the form of aqueous latex (Enjay, 1968). The latex contained ~2.5% anionic emulsifier and some CH_2O preservative. This product was used in paper coatings, textile treatment, adhesives, roof coatings, and so forth.

10.3 □ HYDROCARBON RESINS

Hydrocarbon resins embrace a wide variety of low molecular weight ($M_n \leqslant 2000$) products used in numerous applications, for example, hot melt coatings, adhesives, tackifying agents, printing inks, and additives in rubbers. Hydrocarbon "resins" is somewhat of a misnomer since they include viscous liquids and soft solids as well as hard, brittle thermoplastics. The discussion of these materials is divided into two sections, petroleum resins of petrochemical origin and terpene resins of botanochemical source S.

Petroleum Resins: Feeds, Manufacture, Varieties

Petroleum resins are low molecular weight products obtained by polymerization of olefins or mixtures of olefins produced by cracking of petroleum distillates. Petroleum resin production started in ~1950 and was triggered by the rapid acceptance and expansion of coumarone–indene resins (see below). Petroleum resins have largely displaced the older coumarone–indene resins and they are now employed wherever the latter were used, that is, adhesives, coatings, inks, and rubber additives.

Depending on the nature of the starting materials, three large classes of petroleum resins may be distinguished: *aliphatic resins* produced from C_4 to C_6 fractions containing various olefins, isoprene, piperylene, and saturated paraffins; *aromatic resins* obtained from C_8 to C_{10} feeds containing styrene, α-methylstyrene, indene, and vinyltoluene together with dicyclopentadiene and alkyl-substituted aromatics; and *dicyclopentadiene resins* obtained from streams containing largely dicyclopentadiene.

Early developments of hydrocarbon resins including petroleum resins, cyclopentadiene resins, and terpene resins, for example, have been summarized (Powers, 1966; Findlay, 1968) and an updated review is under preparation (Vredenburgh et al., 1980).

Petroleum resins are conventionally produced by contacting particular feeds with $AlCl_3$, less frequently with BF_3 or other Friedel–Crafts acids, sometimes in the presence of various "promoters" such as HCl and *t*-BuCl, at various temperatures (from −20 to +100°C) depending on the available feedstock and desired end product. Polymerizations are quenched by various bases, for example, water, alkalies, lime, and ammonia, and volatiles are removed by distillation.

The products range from viscous liquids to brittle thermoplastic resins with softening points up to 180 to 190°C. Some of the most important characteristics of contemporary resins are softening point, color, and compatibility with wax and ethylene–vinyl acetate copolymer. High softening points, highest color, and a high degree of miscibility ("wax cloud point") are desirable properties.

Among petroleum resin producers in the United States are Exxon Chemical Co., Goodyear, Hercules, Neville Chemical, Northwest Industries Co. (Velsicol), and Reichhold Chemicals.

With regard to aliphatic resins, it is difficult but not impossible to engineer their ultimate properties. The key factor is always feed composition, which can vary within very wide limits and depends on the source of the crude oil and cracking conditions. For example, the nature and amount of olefins, the olefin/diolefin ratio, and presence or absence of piperylene or isoprene in the feed determine ultimate resin properties. Resin characteristics, for example, softening point, tackification power, and rheological properties, can be modified to a certain degree by introducing aromatic monomers or dicyclopentadiene. Recently increasing attention is being paid to molecular weight distribution control by a variety of means.

Aromatic resins (boiling range of monomers 140 to 200°C) contain significant quantities of polyindenes, similar to coumarone–indene resins. Coumarone–indene resins or so-called coal tar resins were one of the earliest synthetic resins. They were produced commercially around 1920 chiefly for floor covers, tiles, and sheets. These resins are made from a variety of fractions obtained in coking by the use of Friedel–Crafts acids, usually AlCl$_3$. Older processes used H$_2$SO$_4$; however, AlCl$_3$ or BF$_3$ give higher softening and lighter materials. Acid residues are removed by washing with bases. In the United States Hercules and Neville Chemicals manufacture these resins.

Dicyclopentadiene resins are usually manufactured by thermal polymerization, that is, in the absence of initiator, of various dicyclopentadiene-containing feeds that may contain styrene or indene and may yield quite high softening polymers (up to 175°C). Cyclopentadiene feeds are also responsive toward acid-initiated oligomerizations; however, these processes yield products of somewhat lower softening points, for example, 30 to 130°C, which are used as plasticizers.

In addition to these low-cost petroleum resins of ill-defined structures produced of feeds containing a variety of active and inactive ingredients, markets also exist for higher quality resins manufactured from pure feeds containing essentially one monomer or two copolymerizable monomers, for example, styrene and α-methylstyrene.

Indeed, some of these materials are not "resins" but sufficiently well-defined low molecular weight polymers or oligomers. One of the outstanding characteristics of these products is their lack of color. Examples of these products are polystyrene (Piccolastic® of Hercules Inc.) and poly(α-methylstyrene) (Amoco Resin 18). The latter is a colorless material that, depending on its molecular weight ($\bar{M}_n = 685$, 790 and 960), softens at 99, 119, and 144°C. It is used in high gloss overprint varnishes and as a processing aid and additive.

Properties and Uses

Petroleum resins are seldom used alone as such; rather they are employed as additives in conjunction with other materials, or as improvers of specific properties. The most important characteristic of petroleum resins is their low cost: their price was ~30 ¢/lb in 1979.

Among the polymer properties that determine end uses of petroleum resins and that can be controlled to a certain extent by polymer–chemical techniques are softening point, molecular weight and its distribution, melt viscosity, solubility, and color. Requirements of the users in all these categories are rather modest and could be readily attained by the polymer property designer or compounder, were it not for the low cost demand of the manufacturer. Thus the focus of property engineering recently has been shifting from the compounder to the chemist–engineer whose assignment is to gain increased insight into the manufacturing processes and thus to produce higher-quality materials by direct synthesis. Modern instrumentation, in particular gel permeation chromatography, and advances in the understanding of reaction mechanism are providing much needed help in this area.

Petroleum resins are used literally in thousands of applications such as in adhesives, coatings, inks, varnishes, paints, paper, plastics, rubber, and textiles. They come in all kinds of forms, for example, powder, flake, crushed nuggets, bead, solid block, molten, dispersions, and solutions. Perhaps the single most important use of petroleum resins is in the low-cost hot melt adhesive market. Aliphatic resins are widely employed as tackifying agents in pressure-sensitive adhesives with various conventional or modern thermoplastic rubbers. Petroleum resins are used to stiffen paper in packaging and in floor coverings, that is, a base polymer such as polyethylene or polypropylene or ethylene–ethyl acrylate copolymer blended with petroleum resins for tack, improved adhesion, and heat and UV stability.

Other uses are in sealants, caulks, and mastics, again in combination with various rubbers. Petroleum resins are also used extensively in protective coatings, paints, and varnishes to improve or impart a variety of desirable properties such as penetrating characteristics, water resistance, leafing power, and gloss. Further, these materials are employed in tire manufacture and in hoses, shoe soles, molded rubber goods, and wire insulation to improve flex crack resistance, elongation, and tensile properties. Another large application area is as binders in all kinds of printing inks. In addition, numerous, sometimes exotic, specific applications have been described (Vredenburgh et al., 1980).

Polyterpene Resins*

Terpene resins are low molecular weight hydrocarbon resins, the monomers of which are produced by plants. These resins, used by the adhesive, sealant, wax coating, and investment casting industries, fall into three major categories: pressure-sensitive adhesives, hot melt adhesives and coatings, and elastomeric sealants. Pressure-sensitive adhesives include solvent, emulsion, and hot melt pressure-sensitive adhesives and rubber cements. Hot melt adhesives include hot melt adhesives, coatings, and investment waxes; elastomeric sealants include sealants, chaulks, and can end cements. Each of these categories requires specific types of tackifying resins.

Polyterpene resins are produced commercially by solution polymerization of terpene monomers, for example, β-pinene, α-pinene, and dipentene. In a typical continuous polymerization terpene, xylene and powdered $AlCl_3$ are constantly metered into a primary stirred reactor. About 100 parts of monomer and solvent are employed with 2 parts of $AlCl_3$. Since a large amount of heat is generated during the polymerization, feed rates and cooling are adjusted to maintain the temperature in the 40 to 50°C range.

The polymerizing mixture is continuously pumped into a secondary reactor where the remaining 10% or so of the reaction is completed. The $AlCl_3$ is inactivated by a turbulent water stream, following which the organic phase is separated. After further washing to remove residual inorganic remnants, the xylene solution enters the rundown stage. The resin can be recovered by either batch or continuous processing. In batch rundown, xylene is distilled from the resin solution by gradually raising the temperature to about 230°C with application of vacuum during the final stage. The base resin softening point (ring and ball method, ASTM E28-58T) is generally about 115°C and can be adjusted upward to 135°C by subjecting the molten resin to steam sparging. This technique removes part or all of the lower molecular weight dimers and trimers.

*Contributed by E. R. Ruckel, Arizona Chemical Company, Stamford Conn.

β-Pinene Resins

Initiation and isomerization–propagation in β-pinene polymerization may be visualized as follows:

Initiation

$$\text{``H}_2\text{O''} + \text{AlCl}_3 \longrightarrow \text{``H}^{\oplus}[\text{AlCl}_3\text{OH}]^{\ominus}\text{''} \equiv \text{H}^{\oplus}\text{G}^{\ominus}$$

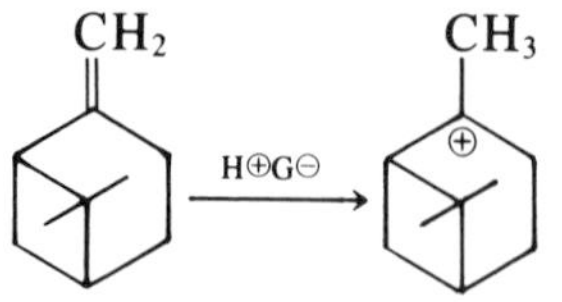

Propagation

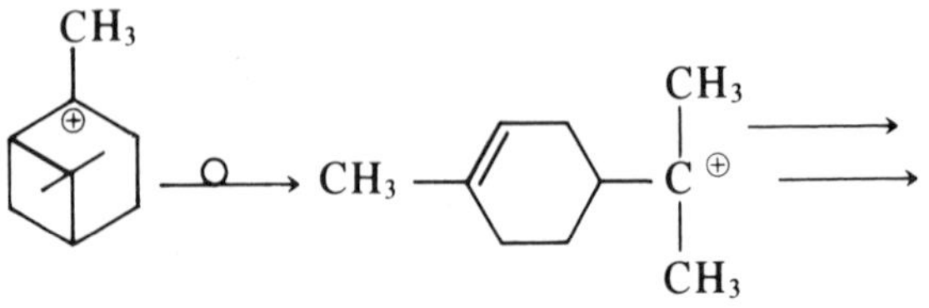

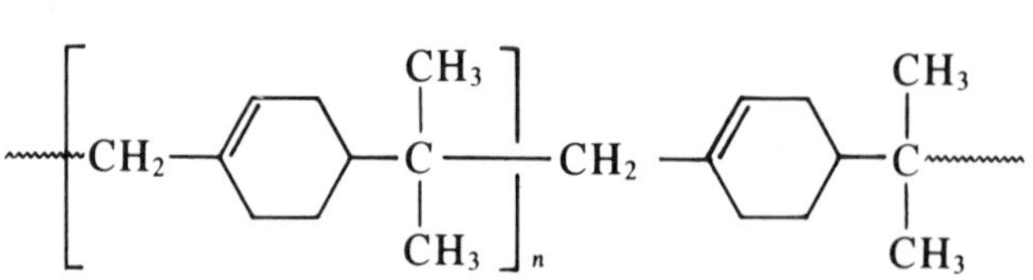

Repeat unit

Commercial β-pinene feeds contain ~75% β-pinene, with the remainder dipentene, α-pinene, and minor levels of other terpenes. The main repeat unit is a ring having 1,4 disubstitution which results in an extended chain having a large hydrodynamic volume in solution. This polymer is equivalent to a perfectly alternating copolymer of isobutylene and cyclohexene. Peracid oxidation indicates approximately one olefin per mer unit.

Chain transfer is operative. A possibility is rearrangement to the sterically hindered nonpropagating camphene cation by proton expulsion. This process would give rise to camphenic end groups (Takata et al., 1966):

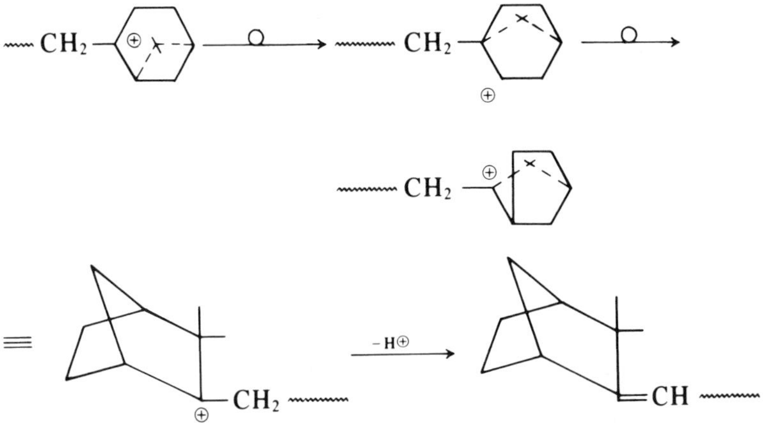

Camphenic end group

This rearrangement is analogous to the acid-catalyzed Wagner–Meerwein isomerization of α- or β-pinene proposed by Williams and Whittaker (1971):

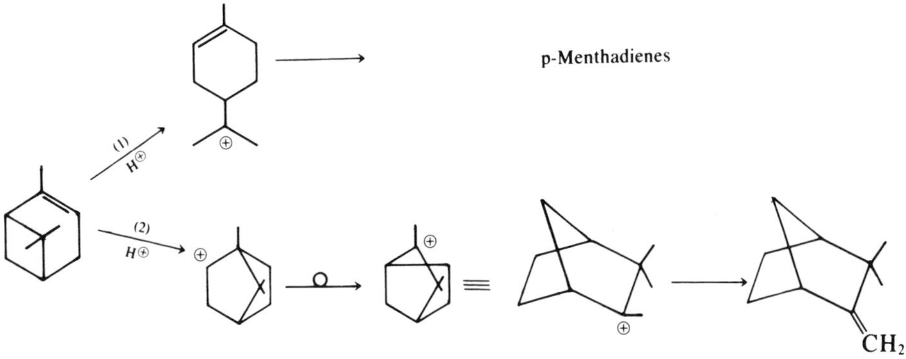

p-Menthadienes

Chain transfer to aromatic compounds may also occur:

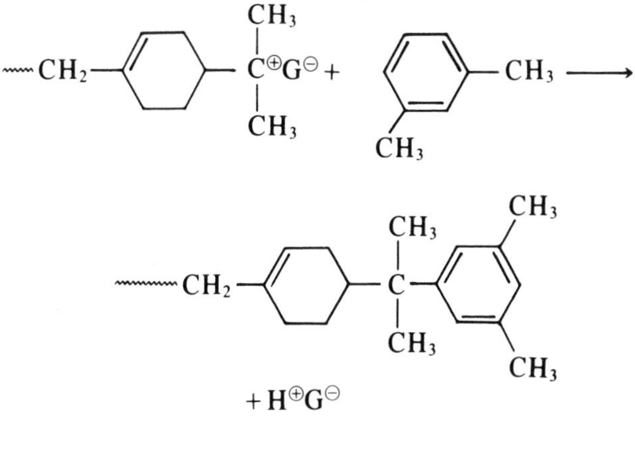

$+ H^{\oplus}G^{\ominus}$

Roberts and Day (1950) have also observed that the molecular weight obtained in the polymerization of α- or β-pinene and dipentene is low, that is, 1000 to 2000.

Dipentene Resins

Initiation is similar to that described for β-pinene, and propagation through the terminal methylene group would be predicted:

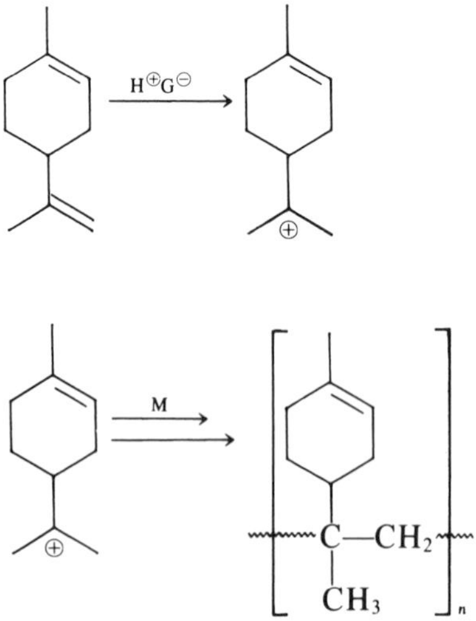

Analysis of olefin content by NMR, ozonolysis, and perbenzoic acid oxidation, however, indicates that approximately one-half of the mer units are unsaturated. The endocyclic double bond is consumed in some manner during the polymerization. To substantiate this theory, the structurally similar model compound 8,9-p-menthene was polymerized under identical conditions. Only dimer was obtained. The double bond in the ring thus permits polymerization of dipentene.

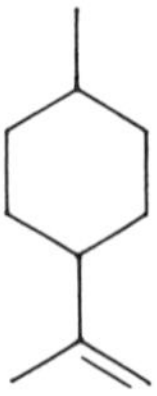

8,9-p-Menthene

Alternatively, the polymerization of dipentene may proceed by initiation at the trisubstituted olefin followed by cyclopolymerization to yield a structural unit similar to that proposed by Butler et al. (1965) for the polymerization of the related 1-methylene-4-vinylcyclohexane:

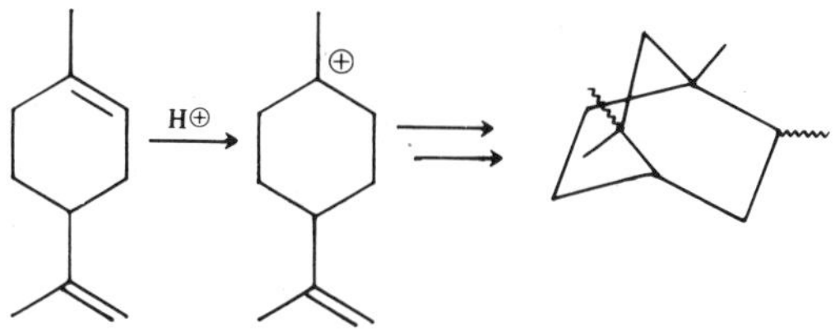

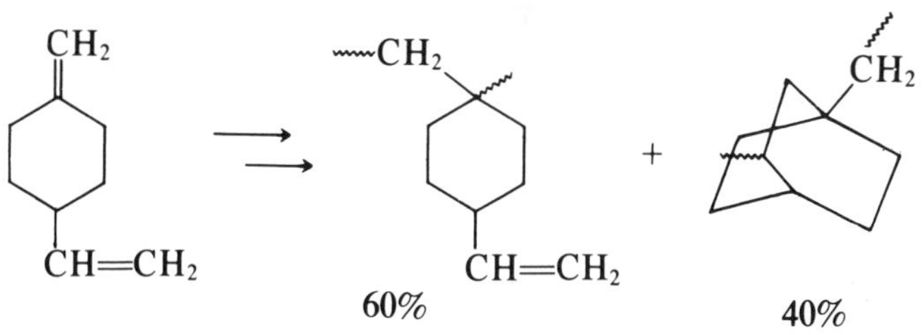

60%   40%

More probably, protonation yields pendant isopropyl carbenium ion which may attack the double bond of the penultimate mer unit and thus form a ring; subsequent polymerization proceeds from the penultimate mer unit:

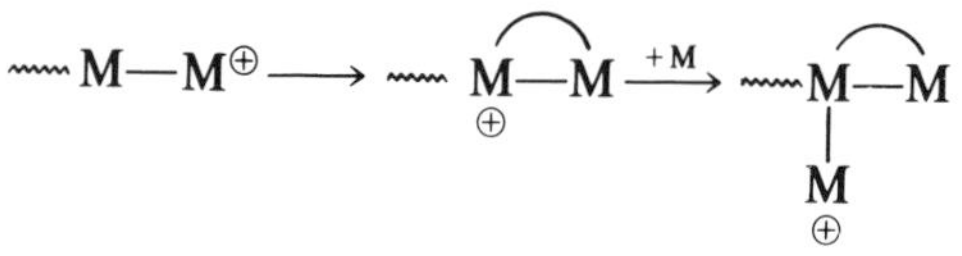

A structural representation based on this postulation is as follows (the dotted bonds are formed during polymerization and the arabic numerals indicate their sequence of formation):

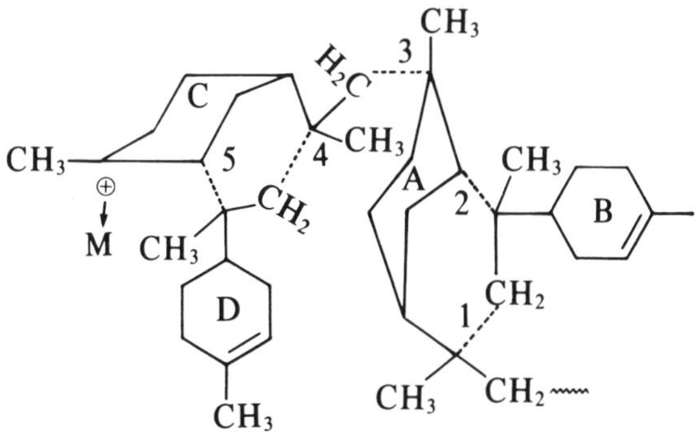

Either of these mechanisms satisfactorily explains the presence of only one double bond per every two to three mer units (Rucket et al. 1975).

α-Pinene Resins

This monomer is the most difficult of the common terpenes to polymerize because it does not possess an exocyclic methylene group. Although α-pinene forms the same initial carbenium ion as β-pinene, the propagation step is difficult for steric reasons. The presence of an adjuvant (Ruckel and Wang, 1971; Wojcik and Ruckel, 1971) is required to eliminate the formation of large amounts of dimer which otherwise would form during the polymerization.

The peracid oxidation of α-pinene resin shows that approximately two-thirds of the mer units contain an olefin, indicating that in the remaining one-third, the four-membered ring probably expands and results in a saturated mer unit possessing the [2:2:1] bicyclic system. Accordingly, the two proposed mer structures in an α-pinene resin are as follows:

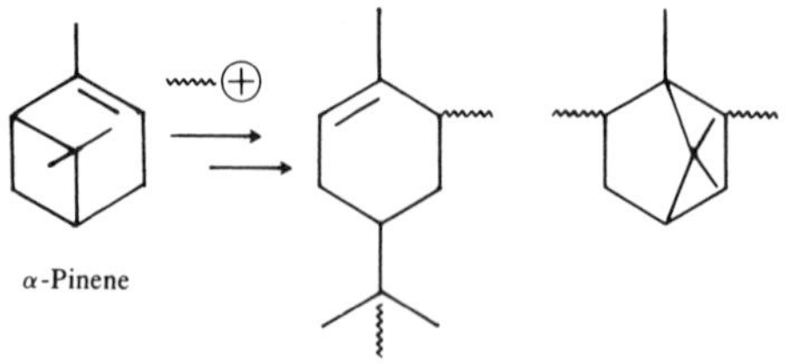

α-Pinene

Resin Characteristics

The number average molecular weights of representative commercial terpene resins, determined by vapor pressure osmometry, are presented in Table 10.3. Feed streams for the dipentene and α-pinene resins were 95 to 99% pure. The feed stream for the β-pinene was already described.

For a specified softening point, a dipentene resin has a lower molecular weight than a β-pinene resin, indicating that the dipentene polymer structure is more rigid and more compact than that of a β-resin. The density of dipentene resins is in fact higher than that of β-resins, 0.998 to 0.974. Although

Table 10.3 ☐ M_n of Typical Terpene Resins

R & B (°C)	β-Resin	Dipentene Resin	α-Resin
85	815	570	725
100	870	675	775
115	1030	720	815
125	1110	760	830
135	1230	810	870
M_w/M_n of 115	1.9	1.4	1.4

dipentene resins have a higher softening point/mer unit, they have a smaller hydrodynamic volume and hence form solutions of lower viscosity.

The density of α-resins, at 0.976, is very close to that of β-resins; however, the molecular weight and hydrodynamic volume are closer to those of dipentene resins than β-resins. For this reason, the bulk properties of α-resins resemble those of dipentene resins with the exception of thermal stability. The thermal stability of α-resins is poorer because of the partial steric interaction due to the 1,3 substitution on the cyclohexene ring and absence of any double-stranded placements.

The structures of β-pinene and dipentene resins are quite different and impart different properties to blends. The chain of a β-resin is more extended and flexible than that of a dipentene resin. Thus, dipentene resins should exhibit a lower viscosity than β-resins at equal degree of polymerization. The viscosity of dipentene resins also has a greater dependence on temperature; it is reduced to a greater extent than a comparable β-resin. Dipentene resins are also more compatible with ethylene–vinyl acetate copolymers. The cloud point, obtained in a compatibility test using a 10:20:20 blend of resin/wax/ethylene–vinyl acetate copolymer, is about 90°C and can be compared with about 175°C for a β-resin. Thus these resins are preferred to β-pinene resins for hot melt adhesives. Since formulations employing dipentene resins reach compatible cloud-free liquid state at a lower temperature, less oxidation is likely to occur. Dipentene resins have also been found to be more color stable than β-pinene resins, which probably reflects the presence of fewer oxidatively vulnerable olefinic sites. Dipentene resins also tend to be thermally stable, presumably because of their multiple strand structure.

The cloud point of α-pinene resins is about 115°C, again reflecting the closer similarity and molecular weight distribution to dispentene-based resins.

Production

The U.S. production of tackifying resins for domestic merchant sale in 1979 was over 200 million pounds. Of this about two-thirds consisted of petroleum-based hydrocarbon resins, the balance shared about equally by terpene resins and rosin esters.

The consumption of terpene resins during 1979 by the U.S. adhesive industry is estimated at 30 to 40 million pounds. Of this ~42% is β-pinene-derived polyterpene resins, 35% dipentene-derived polyterpene resins, 15% terpene–phenolic resins, and 8% α-pinene-based polyterpene resins. A large export market also exists.

There are presently seven producers of terpene resins. Four of these suppliers, Arizona, Crosby, Hercules, and Reichhold, are basic in polymer-grade α-pinene, β-pinene, and dipentene terpene monomers. Durez, Natrochem, and Schenectady must obtain their feedstocks from the merchant market.

Applications

There are four major classes of terpene resins. Each class contributes different adhesive properties; therefore, specific terpene resins are preferred for specific adhesive systems. Weyman (1965) described the influence of dipentene and β-pinene on the adhesive characteristics of formulations based on natural and SBR rubber.

Terpene resins are used by adhesive manufacturers to impart tack to both solvent-based and hot melt adhesive systems, to provide high gloss, good moisture/vapor transmission resistance, and good flexibility for wax coating, and to impart toughness to investment casting waxes. Terpene resins are excellent light color tackifiers of natural rubber, polyisoprene, and ethylene–vinyl acetate copolymers and are used in both solvent-based and hot melt adhesive systems.

β-Pinene Resins

The estimated consumption of β-pinene resins in 1979 was about 20 to 25 million pounds. β-Pinene resins are used mainly in the manufacture of solvent-based adhesive tapes, labels, and can end sealants. These resins have excellent adhesive properties, for example, tack, shear, and peel strengths, good color, color retention, and excellent oxidation and ultraviolet light resistance. β-Pinene resins are primarily used with elastomers such as natural rubber and polyisoprene.

Dipentene Resins

The estimated 1979 domestic consumption of dipentene-based polyterpene resin was 15 to 20 million pounds. Dipentene resins are used mainly in the manufacture of hot melt adhesives and coatings. These resins have excellent color stability, oxidation resistance, odor, and hot track properties. Dipentene polyterpene resins are used to tackify elastomers used in the hot melt adhesive and coating industry, for example, ethylene–vinyl acetate copolymers and styrene–butadiene–styrene block copolymers.

α-Pinene Resins

The estimated 1979 consumption of α-pinene resins was 2 to 4 million pounds. α-Pinene resins are excellent tackifiers of, for example, SBR and styrene–isoprene–styrene block copolymers. These resins have excellent color and color stability and are used in both hot melt and solvent-based adhesive systems.

Terpene–Phenolic Resins

These modified terpene resins having increased polarity are prepared by the boron trifluoride-catalyzed condensation of a terpene with phenol or cresol. The estimated 1979 consumption of terpene–phenolic resins was 5 million pounds. These resins are excellent tackifiers of, for example, acrylic resins, SBR, and natural and neoprene rubber. They have light color and good color stability. Many hot melt adhesive and coating manufacturers were found to be using terpene–phenolic resins. Terpene–phenolic resins are superior to polyterpene resins when better elongation and toughness properties in hot melt adhesives are required.

10.4 □ POLYBUTADIENE OILS

A yellowish, low molecular weight ($\bar{M}_n = 1000$, $\bar{M}_w = 50,000$ to $100,000$, $\bar{M}_w/\bar{M}_n > 50$) polybutadiene oil is marketed by DuPont (Budium®) as a beer and beverage can interior body base coat (primer) for three-piece cans. This product is made by a two-step polymerization process (Ikeda, 1957). In the first step butadiene is polymerized to a liquid at 0 to 5°C with a mixture of $BF_3/Et_2O/H_2O$ (1:1:0.9) in a naphtha solution. Subsequently, in the second step, anhydrous BF_3 is introduced until the desired viscosity is reached. The reaction is quenched by methanol, which is later removed by distillation. The product contains linear ($\sim 80\%$) and cyclized ($\sim 20\%$) structures and the linear component consists of trans-1,4 ($\sim 80\%$) and 1,2 ($\sim 20\%$) enchainments. The polymer is spray-coated on aluminum or steel rolled stocks by the can manufacturer. Since the beverage can market is rapidly being taken over by new technology using two-piece cans, the days of this product seem to be numbered.

10.5 □ VINYL ETHER-BASED INDUSTRIAL POLYMERIZATION PROCESSES AND PRODUCTS

Vinyl ethers can be homopolymerized only by cationic techniques; thus the poly(vinyl ether) industry is largely a cationic one. Copolymerization of vinyl ethers with acrylates, maleic anhydride, vinyl acetate, or vinyl chloride, for example, can proceed by free radical techniques; however, the discussion of these processes falls outside the scope of this section. Brief surveys concerning alkyl vinyl ethers and their polymerization behavior, history, and general significance have appeared from time to time; and reader is referred to these for overview and background (Schildknecht, 1970; Field and Lorenz, 1970; Bikales, 1971; Russell, 1977).

Technological information relative to the industrial polymerization processes of alkyl vinyl ethers is practically nonexistent. The student of this field has to assemble laboriously meaningful details from the patent and trade literature.

The seminal disclosure of vinyl ether polymerizations is a 1936 German patent (Müller-Cunradi and Pieroh, 1936) assigned to the now defunct I.G. Farbenindustrie describing the polymerization of highly purified isobutyl vinyl ethers by gaseous BF_3. It is of some interest that this patent, more than 40 years old, must be still the basis of BASF's vinyl ether polymerization process, presumably making use of their "Bandverfahren" also employed for their polyisobutylene process (see Section 10.2). Similarly to the diisobutylene process, the polymerization of lower alkyl vinyl ether, for example, isobutyl vinyl ether, also ensues instantaneously upon contact with gaseous BF_3 and the high heat of reaction is efficiently removed by "internal" cooling, that is, by allowing the liquid propane diluent to flash-off, as described in the aforementioned patent.

The monomer must be carefully purified prior to polymerization, and traces of protic impurities that severely depress molecular weight and process efficiency must be removed. A more recent German patent (Mohr et al., 1978) is directed toward achieving this objective by the use of special driers prepared by impregnating Al_2O_3 (~ 75 wt %) with NaOH or KOH ($\sim 25\%$) and activating the absorbent by heating to 250 to 300°C in a stream of inert gas. The alkyl vinyl ether polymers prepared by the use of these drying installations are claimed to be clear and colorless, an objective difficult to achieve in large-scale operations.

The GAF Corporation is the only current source of vinyl ethers monomers and polymers (under the trade name Grantrez) in the United States. The monomers are made by the high-pressure acetylene route; however, there is a dearth of information as to polymerization process details. In West Germany, BASF is manufacturing and selling vinyl ethers and their polymers (trade name Lutonal).

Poly(methyl vinyl ether) (Gantrez-M, Lutonal-M) is a colorless, tacky, amorphous semisolid. Importantly, it is soluble in both cold water and aromatic hydrocarbons but insoluble in aliphatic solvents. The water solubility of poly(methyl vinyl ether) is due to hydrogen bonding between the ether linkage and H_2O. On heating, the hydrogen bridges are destroyed and solubility decreases. For example, Lutonal M40 becomes water insoluble at 28°C and above this temperature it precipitates from solution. This critical temperature is known as the "cloud point"; it is affected by solvents, additives, and their concentrations. The polymer is compatible (forms clear homogeneous films) with various other commercial polymers, for example, methyl cellulose, polyvinylpyrrolidone, epoxy resins, and phenol–formaldehyde resins. Trade literature from GAF and BASF provides copious information on these details.

Poly(ethyl vinyl ether) (Lutonal-A) and poly(isobutyl vinyl ether) (Lutonal-I, -IC) are highly viscous oils and, depending on their molecular weight, tacky solids or rubbery materials, respectively. These materials are soluble in ethers, esters, ketones, and hydrocarbon solvents; however, they are insoluble in water. Poly(ethyl vinyl ether) is soluble in alcohols, but poly(isobutyl vinyl ether) only in the higher homologues.

All these materials are used as light-stable modifiers of polymers, non-migrating plasticizers, adhesives, or coatings. Poly(methyl vinyl ether) improves wetting on difficult-to-bond surfaces. It is an excellent tackifier in pressure-sensitive tapes, it improves adhesion to metal and glass, and it shows strong adhesion specificity for both high and low surface free-energy substrates. It can be used, for example, as a general-purpose adhesive, hot melt adhesive, heat seal, in glass laminating, surgical tapes, and surface coatings for a variety of materials including glass or metal. Poly(ethyl vinyl ether) and poly(isobutyl vinyl ether) exhibit similar properties; they are used to enhance and improve flexibility of adhesive films. A typical application is as adhesives for rugs, linoleum, and floor tiles. They are blended with bitumen and waxes to enhance the flexibility and low-temperature characteristics of these materials. Poly(isobutyl vinyl ether) blends with beeswax and resins are used as glues in insect traps.

□ REFERENCES

Anonymous (1961), *Enjay Butyl HT*, Synthetic Rubber Division, Enjay Chemical Co., New York.

Anonymous (1969), *Chem. Eng. News*, April 21, 15.

Baldwin, F. P., Buckley, D. J., Kuntz, I., and Robison, S. B. (1961), *Rubber Plast, Age* **42**, 500.

Baldwin, F. P., Burton, G. W., Griesbaum, K., and Hanington, G. (1969), *Adv. Chem. Ser.* **91**, 448.

Baldwin, F. P., and Gardner, I. J., (1977), *Chemistry and Properties of Crosslinked Polymers*, Academic Press, New York, p. 273.

Bikales, N. B. (1971), *Encyclopedia of Polymer Science and Technology*, Vol. 14, Wiley-Interscience, New York, p. 511.

Buckley, D. J. (1959), *Rubber Chem. Technol.* **32**, 1475.

Butler, G. B., Miles, M. L., and Bray, W. S., (1965), *J. Polym. Sci. Part A*, **3**, 723.

Debreczeni, E. C. (1979), study quoted, *Chem. Eng. News*, April 16, 12.

Enjay Chemical Company (1968), *Enjay Butyl Latex 80-21*, Bulletin 012.

Evering, B. L., d'Ouville, E. L. and Carmody, D. R. (1946), U.S. Patent 2,407,873 (Sept. 17, 1946).

Field, N. D., and Lorenz, D. H. (1970), in *Vinyl and Diene Monomers*, E. C. Leonard, Ed., Wiley-Interscience, Vol. 1, p. 365.

Findlay, J. (1968), *Encyclopedia of Polymer Science and Technology*, Wiley-Interscience, New York, **9**, 853.

Fragen, N. (1954), U.S. Patent 2,677,001 (April 27, 1954).

Goering, H. G., and Mistretta, V. F. (1956), U.S. Patent 2,739,143.

Güterbock, H. (1959), *Polyisobutylene*, Springer, Berlin.

Ikeda, C. K. (1957), U.S. Patent 2,777,890 to DuPont (Jan. 15, 1957).

Imura, K., Endo, R., and Takeda, M., (1964), *Bull. Chem. Soc. Jap.*, **37**, 874.

Jackson, W. K. (1960), U.S. Patent 2,957,930 (Oct. 25, 1960).

Kennedy, J. P., and Squires, R. G. (1967) *J. Macromol. Sci. Chem.*, **A1**, 995.

Kennedy, J. P. (1968), in *Polymer Chemistry of Synthetic Elastomers*, J. P. Kennedy and E. G. M. Tornqvist, Eds., Wiley-Interscience, New York, p. 291.

Kennedy, J. P., and Kirshenbaum, I. (1971), in *Vinyl and Diene Monomers*, E. C. Leonard, Ed., New York, Part 2, Chap. 3, p. 691.

Kennedy, J. P. (1975), in *Cationic Polymerization of Olefins: A Critical Inventory*, Wiley-Interscience, New York.

Labine, R. A. (1960), *Chem. Eng.* **67**, 98.

Lavine, I. F., and Folsom, L. T. (1949), U.S. Patent 2,484,384 (Oct. 11, 1949).

Law, C. (1974), *Rubber World*, August, 48.

Mark, D., and Orr, A. R., (1956) *Pet. Refiner* **35**, 185.

Mohr, H., Schmitt, W., Güterbock, H., and Pfannmüller, H. (1978), German Patent to BASF.

Müller-Cunradi, M., and Pieroh, K. (1936) German Patent 634,295 to I. G. Farbenindustrie.

Newberg, R. G., Briggs, J. R., and Fairclough, W. A., (1948) *Mod. Packag.*, November, 339.

Oblad, A. G., Mills, G. A., and Heinemann, H. (1958), in *Catalysis* Vol. VI, P. H. Emmet, Ed., Reinhold, New York, p. 341.

Otto, M., Güterbock, H., and Melan, W. (1939) German Patent 868,293.

Otto, M., Güterbock, H., and Hellemans, A., (1940) German Patent 697,482.

Plesch, P. H. (1963), in *The Chemistry of Cationic Polymerization*, P. H. Plesch, Ed., Macmillan, New York, chap. 4, p. 139.

Powers, P. O., (1966), *Kirk-Othmer Encyclopedia of Chemical Technology*, A. Standen, Ed., Wiley-Interscience, New York, 2nd ed., **11**, p. 242.

Puskas, I., Banas, E. M., and Nerheim, A. G. (1976), *J. Polym. Sci. Symp.* **56**, 191.

Ricard, E. (1930), U.S. Patent 1,745,028 (Jan. 28, 1930) (application July 24, 1923).

Roberts, W. J., and Day, A. R. (1950), *J. Am. Chem. Soc.* **72**, 1226.

Ruckel, E. R., and Wang, L. S. (1971), U.S. Patent 4,011,385.

Ruckel, E. R., Arlt, H. G., Jr. and Wojcik, R. T. (1975), *Adhesion Science and Technology*, Vol. 9A, L. H. Lee, Ed.

Russell, K. E. (1977), *High Polym.* **29**, 306.

Russum, L. W. (1954), U.S. Patent 2,677,000 (April 27, 1954).

Schaad, R. E. (1955), in *The Chemistry of Petroleum Hydrocarbons*, B. T. Brooks, S. S. Kurtz, Jr., C. E. Boord, and L. Schmerling, Eds., Reinhold, New York, p. 221.

Schildknecht, C. E. (1970) *Kirk-Othmer Encyclopedia of Chemical Technology* **21**, 412.

Schriesheim, A., and Kirshenbaum, I. (1978), *Chemtech*, May, 310.

Smyers, W. H. (1942), U.S. Patent 2,274,749.

Stucker, N. E., and Higgins, J. J. (1977), in *Handbook of Adhesives*, 2nd ed., pp. 255–272.

Takata, A., Otsu, T., and Imoto, M. (1966), *Kogyo Kagaku Zasshi, J. Chem. Soc., Jap. Ind. Chem. Sect.* **69**, 715.

Thaler, W. A., and Buckley, D. J., Sr., (1976), *Rubber Chem. Tech.* **49**, 960.

Thomas, R. M., Lightbown, I. E., Sparks, W. J., Frolich, P. K., and Murphree, E. V. (1940), *Ind. Eng. Chem.* **32**, 1283.

Thomas, R. M., and Sparks, W. J. (1944), U.S. Patent 2,356,128.

Thomas, R. M., and Peterson, W. H. (1954), *Rubber World*, May–June, 16.

Thomas, R. M. (1969), Goodyear Medal Address, Los Angeles Calif., May 1, 1969, *Rubber Chem. Technol.* **42** G87.

Vredenburgh, W. A., Penn, J. Y., and Holohan, J. F., Jr. (1980), in *Kirk–Othmer Encyclopedia of Chemical Technology*, Vol. 12, Wiley-Interscience. Sons, Inc., in preparation.

Walker, J., Wilson, G. J., and Kumbhani, K. J. (1974) *J. Inst. Rubber Ind.*, April, 1.

Weyman, H. P. (1965), *Naval Stores Rev.*, February, 6.

Williams, C. M., and Whittaker, J. (1971), *J. Chem. Soc.*, 688.

Wojcik, R. T., and Ruckel, E. R. (1971), U.S. Patent 4,016,346.

Yahnke, R. L., and Healy, J. W. (1954), U.S. Patent 2,677,002 (April 27, 1954).

Zapp, R. L., Serniuk, G. E., and Minckler, L. S. (1970), *Rubber Chem. Technol.* **43**, 1154.

Index